REVIEWS in MINERALOGY   Volume 21

Geochemistry and Mineralogy of
RARE EARTH ELEMENTS

B.R. LIPIN & G.A. MCKAY, EDITORS

The authors and editors:

William V. Boynton
Dept. of Planetary Sciences
University of Arizona
Tuscon, Arizona 85721

Douglas G. Brookins
Dept. of Geology and Institute
of Meteoritics
University of New Mexico
Albuquerque, New Mexico 87131

Donald M. Burt
Dept. of Geology
Arizona State University
Tempe, Arizona 85287

Fredrick A. Frey
Dept. of Earth, Atmospheric
and Planetary Sciences
Massachusetts Institute of Technology
Cambridge, Massachusetts 02139

Richard I. Grauch
U. S. Geological Survey
Box 25046 Denver Federal Center
Mail Stop 973
Denver, Colorado 80225

Gilbert N. Hanson
Dept. of Earth and Space Sciences
State University of New York
Stony Brook, New York 11794

Larry A. Haskin
Dept. of Earth and Planetary Sciences
Washington University
St. Louis, Missouri 63130

Anthony N. Mariano
48 Page Brook Road
Carlisle, Massachusetts 01741

Gordon A. McKay, Co-editor
NASA/Johnson Space Center
Code SN4
Houston, Texas 77058

Scott M. McLennan
Dept. of Earth and Space Sciences
State University of New York
Stony Brook, New York 11794

P. Jonathan Patchett
Dept. of Geosciences
University of Arizona
Tucson, Arizona 85721
- - - - - - - -

Bruce R. Lipin, Co-editor
U. S. Geological Survey
National Center, MS 954
Reston, Virginia 22092

Paul H. Ribbe, Series Editor
Dept. of Geological Sciences
Virginia Polytechnic Institute
and State University
Blacksburg, VA 24061

Published by
THE MINERALOGICAL SOCIETY OF AMERICA
Washington, D.C.

REVIEWS IN MINERALOGY

(Formerly: SHORT COURSE NOTES)

ISSN 0275-0279

Volume 21: *GEOCHEMISTRY AND MINERALOGY*

OF RARE EARTH ELEMENTS

ISBN 0-939950-25-1

ADDITIONAL COPIES of this volume as well as those listed below may be obtained at moderate cost from the MINERALOGICAL SOCIETY OF AMERICA, 1625 I Street, N.W., Suite 414, Washington, D.C. 20006 U.S.A.

GEOCHEMISTRY AND MINERALOGY OF
RARE EARTH ELEMENTS

FOREWORD and ACKNOWLEDGMENTS

The authors of this volume presented a short course on the rare earth elements to about 80 participants in San Francisco, California, December 1-3, 1989, just prior to the fall meeting of the American Geophysical Union. This was the nineteenth in a series of short courses that the Mineralogical Society of America (MSA) began sponsoring in 1974. It was the brain-child of Robert Hazen and David Stewart, and the latter was the agent who recruited Bruce Lipin (and later, Gordon McKay) to undertake this monumental work. This book joins a long list of *Reviews in Mineralogy* (see opposite page for details) that MSA has made available to the scientific community at reasonable cost.

The editors and authors thank Mrs. Marianne Stern and Mrs. Margie Sentelle at Virginia Tech for their tireless efforts (under considerable pressure) in preparation of camera-ready copy for this volume. We also thank Mr. Todd N. Solberg for helping with word-processing problems on the computer.

Paul H. Ribbe
Series Editor
Blacksburg, Virginia
October 31, 1989

TABLE OF CONTENTS

Page

Chapter 1   **W. V. Boynton**

COSMOCHEMISTRY OF THE RARE EARTH ELEMENTS:
CONDENSATION AND EVAPORATION PROCESSES

Chapter 2   **P. J. Patchett**

RADIOGENIC ISOTOPE GEOCHEMISTRY
OF RARE EARTH ELEMENTS

Chapter 3   **G. A. McKay**

PARTITIONING OF RARE EARTH ELEMENTS BETWEEN MAJOR
SILICATE MINERALS AND BASALTIC MELTS

Chapter 4   G. N. Hanson

AN APPROACH TO TRACE ELEMENT MODELING
USING A SIMPLE IGNEOUS SYSTEM AS AN EXAMPLE

Chapter 5   W. F. McDonough & F. A. Frey

RARE EARTH ELEMENTS IN UPPER MANTLE ROCKS

vi

Chapter 8   D. G. Brookins
AQUEOUS GEOCHEMISTRY OF RARE EARTH ELEMENTS

Chapter 9   L. A. Haskin
RARE EARTH ELEMENTS IN LUNAR MATERIALS

vii

COMPOSITIONAL AND PHASE RELATIONS AMONG
RARE EARTH ELEMENT MINERALS

ECONOMIC GEOLOGY OF RARE EARTH MINERALS

Appendix   A. N. Mariano

CATHODOLUMINESENCE EMISSION SPECTRA OF
RARE EARTH ELEMENT ACTIVATORS IN MINERALS

COSMOCHEMISTRY OF THE RARE EARTH ELEMENTS:
CONDENSATION AND EVAPORATION PROCESSES

INTRODUCTION

The geochemist studies rocks in order to determine both their origin and what they can tell us about processes that form larger bodies. The cosmochemist also studies rocks for the same purpose, except that these rocks are from bodies in the solar system other than Earth. Most of these rocks come to us from space in the form of meteorites.

The job of understanding the information contained in meteorites is in some ways simpler than that for terrestrial rocks, but in many ways it is more complicated. It can be simpler since sometimes there have been relatively few processes that have acted on the rock. Examples of this case occur for some of the igneous meteorites such as the basaltic achondrites that appear to have had just a single stage of igneous fractionation (Consolmagno and Drake, 1977). In most cases, however, the lack of field relationships, the limited amount of sample available, and the lack of a known astrophysical context make interpretation of the data difficult.

Rather than serve as a review of all REE data in meteorites, this chapter is intended to illustrate the principles of cosmochemistry. This work will concentrate on a small fraction of the data from meteorites that can illustrate some applications of the REE different from those familiar to trace-element geochemists. Discussion of the general significance of the rare-earth elements (REE) in all meteorites has been provided in a review by Boynton (1983).

Meteorites

Meteorites are divided into three broad categories: stones, irons, and stoney irons. The stones are further divided into chondrites and achondrites. Most of the achondrites are igneous rocks, and the principles used to apply the REE in their study are not significantly different from that of other igneous rocks discussed in this book. The chondrites, on the other hand, are unlike any terrestrial rocks. In this chapter, I shall attempt to show how we can use the REE in these rocks to look back in time to try to understand the processes that formed our solar system.

The chondrites contain many different components (Fig. 1) that record a variety of processes that occurred as the solar system was forming (see Wasson, 1974, and Dodd, 1981, for more details). Most chondrites contain abundant, small (~1 mm.), spheres of once molten silicate droplets called chondrules. Some of the different classes of chondrites also contain inclusions of various types. Some of these inclusions were clearly molten; others clearly were not. Some of the more interesting inclusions for REE abundances are rich in Ca, Al and other refractory

2

Figure 1. The Allende meteorite contains a variety of objects that record events that took place when our solar system formed 4.6 billion years ago. The small spherical objects are chondrules; the large objects are mostly Ca,Al-rich inclusions. The latter record evidence of high-temperature fractionations of the REE between the solar nebular gas and grains. The slab is about 15 cm across.

elements. These Ca, Al-rich inclusions (CAI) appear to have formed at high temperatures very early in the formation of the solar system (see MacPherson et al., 1988, for a review). The most interesting REE data from chondrites have come from studies of various chondrules and inclusions of these meteorites; with a few exceptions, whole rock REE abundances of chondrites have long ceased to be important except for the determination of average solar system abundances (see below).

The chondrites are divided into various classes and types that have experienced different degrees of alteration after they formed. Some have experienced a significant amount of aqueous alteration (Zolensky and McSween, 1988), while others have experienced substantial thermal metamorphism (McSween et al., 1988). One of the biggest problems facing the cosmochemist is to be able to read through any processing that may have occurred in the meteorite parent body to find the clues to understanding processes that occurred before the meteorite bodies--and the planets--formed. As we shall see, the REE offer one of the best ways of distinguishing the primary cosmochemical processes from the later, secondary, processes that occurred on the parent body.

Astrophysical context for interpretation of cosmochemical data

Although the framework for interpretation of the processes that formed the chondrites is not so clear as it is for many terrestrial systems, we do, nevertheless, have such a framework that has served well to help interpret the data. As we shall see, parts of this framework are an over-simplification of what must have occurred, but it is important to understand this cosmochemical dogma before we go further.

Solar nebula. The solar system is thought to have formed from a cloud of interstellar gas and dust that collapsed under its own gravity to form a slowly rotating disk with the protosun at its center. This disk concept of the solar nebula was originally proposed by Kant (1755), but has been expanded upon by many investigators (see Wood and Morfill, 1988, for a review). The disk became heated during the collapse, but the extent of temperature rise is far from clear. The dust vaporized to varying degrees, depending on the ambient nebular temperature, which is a function of both time and distance from the sun. As will be seen, understanding the nature of heating events in the early solar system is one area in which REE have had a major impact.

Solar abundances. The matter that made up the solar system was well mixed. Even though the interstellar dust is made by a variety of nucleosynthetic processes, each making elements with vastly different isotopic ratios, to first order, i.e. to within $\pm$ 1%, all objects in the solar system are composed of elements with identical isotopic ratios, suggesting that the starting material out of which the solar system formed was homogeneous. There is an important exception to this homogeneity in the isotopic composition of oxygen (Clayton, 1981; Thiemens, 1988), but it does not negate the important conclusion that the cloud out of which the solar system formed must have had a relatively uniform composition, and that variations that we see must be due to chemical and physical changes that occurred in the process of forming the solar system. The sun, which contains most of the mass of the solar system, is assumed to have the composition of this starting material. The chondrites have the remarkable property that, in spite of the many processes that made the components of the chondrites and the alteration that occurred to the chondrites after they were formed, they contain a nearly unfractionated sample of the condensible elements in the sun. There are small differences in composition among different groups of chondrites, but based on several lines of evidence, we believe CI chondrites are the most like solar abundances (see Anders and Grevesse, 1989, for a review of solar abundances). The differences in composition are for broad groups of elements that have basic properties in common such as volatility and affinity for the metallic phase (Larimer and Wasson, 1988a,b; Palme et al., 1988a,b). For normalization of terrestrial REE data, it makes little difference whether CI chondrite or average chondrite values are used, but for most cosmochemical applications, CI chondrite values, which are lower than average chondrites by a factor of 1.267 (Boynton, 1983), are preferred since they represent the composition of the starting material for processes that we shall consider.

COSMOCHEMICAL PROPERTIES OF THE REE

Other chapters of this book deal with the familiar geochemical properties of the REE, but for cosmochemical applications of the REE, it is important to consider other properties in addition. The REE properly include Sc, Y, and the lanthanide elements from La to Lu, but we shall follow the common geochemical practice and exclude Sc which behaves unlike the other REE.

The lanthanides have very similar geochemical properties because in the 3+ ionic form in which they are found in nature, they differ only in the number of electrons in the 4f shell. This shell has little or no involvement in forming chemical bonds, so the elements have very similar chemical properties. The major difference in the elements of interest to geochemists is their size, which monotonically decreases with increasing atomic number due to the increased nuclear charge that is only partially shielded by the 4f electrons. The size differences account for the ability of the REE to substitute in different minerals with an affinity that is a smooth function of atomic number. The only exceptions are for Eu and Ce, which can have a significant stability as Eu^{2+} and Ce^{4+}, and thus have a difference in charge as well as size to distinguish them from the other REE.

In many of the cosmochemical processes that are of interest, we need to consider the volatilities of the REE in addition to their size and ionic charge. Their volatilities are important since we will be considering processes in which the REE are originally in a gaseous state in the solar nebula and then condense to form solid grains, or the reverse process in which the REE are vaporized from grains. Boynton (1975) showed that the volatilities of the REE do not show the modest variations that are a smooth function of size typical of igneous fractionation. As will be seen, the variations in volatility are not a smooth function of size and are far from modest, as they range over a factor of 10^8 in systems of interest! In this chapter, we shall learn how to calculate their volatilities, why they are so different, and how these data can provide important information about the processes that acted while our solar system was forming.

REE condensation reactions

The REE volatilities will be calculated, with some modifications, following the procedure of Boynton (1975). As will be shown later, in the solar nebula most of the REE exist in the gas as the monoxides. The condensation reactions can be written as:

$$MO_{(g)} + \tfrac{1}{2}H_2O_{(g)} \rightarrow MO_{1.5(s)} + \tfrac{1}{2}H_{2(g)} , \tag{1}$$

where M represents any of the REE. The equilibrium constant for this reaction is:

$$K_1 = \frac{a_{MO_{1.5}} \cdot (f_{H_2})^{1/2}}{f_{MO} \cdot (f_{H_2O})^{1/2}} = \exp\left[\frac{-\Delta G_1^0}{RT}\right] , \tag{2}$$

where a is the thermodynamic activity of the solid REE sesquioxide, and f is the fugacity of the gaseous species. Because the pressures in the solar nebula are very low, generally 10^{-6} to 10^{-3} atm., the ideal gas law applies, and one can thus substitute partial pressures for fugacities of the gaseous species. Assuming that the partial pressures are known and that sufficient thermodynamic data are available to calculate ΔG_1^0, one can calculate the temperature where the activity of the REE oxide equals unity and will thus become a stable condensate as a pure phase. This temperature, of course, is just a lower limit to the temperature at which a particular REE will condense, since the REE will condense at higher temperatures and lower activities by forming solid solutions with other suitable host phases.

Activity coefficients

In order to calculate the concentration of the REE in a host phase, we need to know the activity coefficient, γ, which relates mole fraction, X, to activity by

$$X_{MO_{1.5}} = a_{MO_{1.5}}/\gamma_{MO_{1.5}} . \tag{3}$$

Activity coefficients are parameters that tell how well an element or molecule is accommodated in the host phase compared to its standard state, in this case the pure $MO_{1.5}$ sesquioxide. Activity coefficients are greater than unity for host phases for which the element is not well suited, and they are less than unity for elements that are more stable in the host phase than in their pure phase. An example of activity coefficients presumably much greater than unity is for REE in olivine, where there are no large sites available; an example where they are less than unity occurs for P dissolved in metallic Fe, which is favored since bonds can form to make Fe_3P, a stable phosphide.

If we knew both what host phases were present in the solar nebula during condensation and the REE activity coefficients in these phases, we could calculate the concentration of the REE in each host phase at any temperature and our job would be over. Unfortunately, though we feel we have some understanding of what phases may be present, we do not know the REE activity coefficients in these phases.

There are several approaches that one can follow to address this problem. Boynton (1975) noted that knowledge of the absolute concentrations of the REE are not so important as long as one can calculate relative concentrations or REE ratios. He suggested that ratios of REE activity coefficients could be estimated from the known mineral-preference effects of the host phases for different REE. Davis and Grossman (1979), on the other hand, preferred to assume that the REE were condensing in ideal solid solution in the host phase. Ideal solid solution assumes that the activity coefficients are unity, i.e. that the host minerals show no relative preference for the REE on the basis of their size and that in absolute compatibility, the host phase is indistinguishable from the pure REE sesquioxide. The latter approach, although it ignores the known mineral-preference effects, has the advantage that the absolute temperature of condensation for a given set of REE abundances can be calculated if the pressure in the nebula is known. Unfortunately, if the assumption of ideality is incorrect, these temperatures will be wrong! The REE belong to a group referred to as incompatible elements by geochemists because they generally are not readily accepted by most host phases, so the assumption of ideality is almost certainly incorrect.

Let us now consider the magnitude of the differences by which the activity coefficients could differ from unity. Although it is difficult to measure absolute activity coefficients, in some cases, it is possible to determine ratios of activity coefficients. Johnston (1965) measured the ratio of Ce^{4+}/Ce^{3+} in silicate glasses at known oxygen fugacity. Boynton (1978) used these data to show that the ratio of activity coefficients of Ce^{4+}/Ce^{3+} was 1430 at 1200°C. Because this ratio is much greater than unity, the effect of activity coefficients in this system is clearly not negligible. Drake and Boynton (1988) used a similar approach and showed that the ratio of activity coefficients of Eu^{3+}/Eu^{2+} at 1470°C in a silicate liquid was even greater with a value of 2400. This liquid was in equilibrium with hibonite, $CaAl_{12}O_{19}$, an important REE host phase in CAI. They noted that it was clear that ideality could not apply to both Eu^{2+} and Eu^{3+}, but made the much less restrictive assumption that the activity coefficients of the trivalent REE in the liquid are equal, i.e. that the liquid showed no preference for REE on the basis of their size. Although even this assumption cannot be defended rigorously, it is more likely to be correct for the liquid than a similar assumption made for minerals, which have sites of well-defined size. Because of the inverse relationship between activity coefficients and concentration (Equation 3), and because activities of an element are equal in all phases at equilibrium, it follows that the activity coefficients of the REE in the hibonite are proportional to the reciprocal of the REE hibonite/liquid distribution coefficients.

In order to estimate the absolute value of the activity coefficients of the REE in the hibonite, Drake and Boynton (1988) had to make one additional assumption--that the activity coefficient of Eu^{+2} is unity in the liquid. They justified this assumption on the basis of the +2 charge being compatible with Ca^{2+}, but noted that this assumption could still be incorrect by a factor of 10 or more. What they showed with these assumptions was that the activity coefficients for REE in hibonite ranged from 330 for La to 24,000 for Yb.

The point of this discussion is that knowledge of the absolute value of the REE activity coefficients is uncertain at best. Assumptions that the REE form an ideal solution, i.e. that their activity coefficients are unity, may be incorrect by more than a factor of 10^4! We will take the approach that mineral/liquid distribution coefficients can be used to estimate *relative* activity coefficients by assuming the liquid has no preference based on size, but recognize that even this approach may introduce some errors. As will be seen, however, errors in the slope of the activity coefficient vs. size relationship as much as a factor of 10 are of little concern. Even though there have been many heated discussions at scientific meetings over the past ten years on the subject of the proper treatment of activity coefficients, it may be fair to say that the ideal-solution approach is waning, as Davis and MacPherson (1988) have now also begun to use hibonite mineral-preference effects to estimate REE volatilities.

Partial pressures

In order to use Equation 2, we need to be able to determine the partial pressure of the gaseous compounds. Before we consider the REE, we will consider the major species, H_2 and H_2O. The solar abundances of several important elements are given in Table 1. As can be seen, H and He are by far the most abundant, followed by C, N, O, and Ne, followed by everything else. Helium, N, and Ne, are not important for this discussion and will be ignored. Because C forms such a strong bond with O, at high temperatures nearly all of the C is in the form of CO. Because there is an excess of O over C, the balance of O is available for other compounds. Some of this excess combines with other elements like Si and Al, but because they are so much less abundant than O, most of the O combines to form H_2O. Ignoring the minor oxygen tied up with metals and remembering that it takes two H atoms to make H_2, we can calculate the ratio of pressure of H_2/H_2O as:

$$\frac{p_{H_2}}{p_{H_2O}} = \frac{\frac{1}{2} A_H}{A_O - A_C} , \tag{4}$$

where A refers to the solar abundance of the element. Because H_2O is the dominant source of available oxygen, Equation 1 was written as it was, with H_2O and H_2 rather than using O_2 as the source of oxygen. As long as oxygen is more abundant than carbon and metallic elements that react with it, the H_2/H_2O ratio buffers the O_2 pressure.

Table 1. Solar system abundances of the elements[1]

Element	Abundance[2]
H	2.79×10^{10}
He	2.71×10^{9}
C	1.01×10^{7}
N	3.13×10^{6}
O	2.38×10^{7}
Ne	3.44×10^{6}
Mg	1.074×10^{6}
Si	1.00×10^{6}
Fe	9.00×10^{5}
La	0.446
Lu	0.0367

[1]From Anders and Grevesse (1989).
[2]Relative to Si = 10^6 atoms.

Table 2. REE parameters related to volatility[1]

	$^2F_{MO}$	D_M	$D_{hib/melt}$	γ_M	D_M/γ_M
La	1.00	1.0	7.2	1.00	1.00
Ce	0.29	0.47	5.9	1.22	0.39
Pr	1.00	2.6	4.9	1.48	1.75
Nd	1.00	0.75	4.0	1.80	0.42
Sm	0.80	1.3	2.7	2.7	0.49
Eu	0.34	0.0014	2.0	3.60	0.00039
Gd	1.00	65	1.5	4.8	13
Tb	1.00	200	0.95	7.6	26
Dy	0.95	350	0.61	11.9	29
Ho	0.98	380	0.39	19	20
Er	0.97	4300	0.25	29	137
Tm	0.51	640	0.16	46	14
Yb	$<1.5 \times 10^{-6}$	0.54	0.10	72	0.0075
Lu	1.00	32000	0.06	113	280

[1]Columns 1 and 2 are from Boynton (1975) except Ce adjusted for $CeO_{2(g)}$; column 3 is from Drake and Boynton (1988). Column 1 is the fraction of gaseous REE present as monoxide; the other form is the monatonic vapor in all cases except Ce, where it is the dioxide. Column 2 is the solid/gas distribution coefficients relative to La = 1. Column 4 is the estimated activity coefficients relative to La = 1.

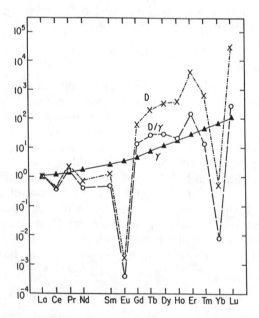

Figure 2. The relative solid/gas distribution coefficients, D, are dependent on the REE volatility and give the ratio of the REE in the solid to that in the gas, relative to La, for a hypothetical mineral that has no size preference for the REE. The relative activity coefficients, γ, are possible values expected of hibonite, $CaAl_{12}O_{19}$, a host phase for REE during high-temperature condensation from the solar nebula. The values of D/γ will be used in this work to describe the gas/solid partitioning of the REE. Expressing the volatilities as a distribution coefficient allows simple calculations of nebular REE patterns using equations derived for igneous crystallization or melting. Because the REE span such a wide range in volatility, in a way that is not a smooth function of size, it is possible to distinguish nebular fractionations from later igneous or metamorphic fractionations that may have occurred in a planetary body.

The partial pressures of the gaseous REE are a little more difficult to calculate because they can exist in the gas in more than one form. Although the monoxide is generally the most stable specie, for some elements the atomic vapor or dioxide is preferred. The amount in each form can be determined by equations such as:

$$M_{(g)} + H_2O_{(g)} \rightarrow MO_{(g)} + H_{2(g)} \; . \tag{5}$$

From the equilibrium constant for this reaction,

$$K_5 = \frac{p_{MO} \cdot p_{H2}}{p_M \cdot p_{H_2O}} \; , \tag{6}$$

one can calculate the ratio of the monatomic REE to the monoxide. For Ce, a similar reaction can be written that relates the monoxide to the gaseous dioxide. If we let F_{MO} equal the fraction of gaseous REE present as monoxide, we can write an expression for its partial pressure, analogous to Equation 4 for H_2O, as

$$\frac{p_{MO}}{p_{H_2}} = \frac{F_{MO} \cdot A_M}{\tfrac{1}{2} A_H} \; . \tag{7}$$

Solid/gas distribution coefficients

We have now discussed all of the aspects necessary to calculate volatilities of the REE. As discussed above, because of the problems with activity coefficients, we can only calculate relative volatilities with any confidence. For this reason, Boynton (1975) introduced the concept of a relative solid/gas distribution coefficient. If we combine Equations 2, 3 and 7, we can express the concentration of the REE in the solid as

$$X_{MO_{1.5}} = \frac{F_{MO} \cdot A_M \cdot p_{H_2}}{\tfrac{1}{2} A_H} \cdot \frac{K_1 \cdot (p_{H_2O})^{1/2}}{\gamma_{MO_{1.5}} \cdot (p_{H_2})^{1/2}} \; . \tag{8}$$

Because we are interested in ratios of REE, we can write equations similar to Equation (8) for two REE and take the ratio. The result for Sm and La is:

$$\frac{X_{SmO_{1.5}}}{X_{LaO_{1.5}}} = \frac{\gamma_{La} \cdot K_{Sm} \cdot F_{SmO}}{\gamma_{Sm} \cdot K_{La} \cdot F_{LaO}} \cdot \frac{A_{Sm}}{A_{La}} \; . \tag{9}$$

The quantity on the left is the ratio of REE in the solid; the right most term is the ratio of REE in the gas--if no significant amount of REE condensation has occurred. We then define a relative solid/gas distribution coefficient for Sm as:

$$D_{Sm} = \frac{K_{Sm} \cdot F_{SmO}}{K_{La} \cdot F_{LaO}} \; . \tag{10}$$

We can now write an equation that relates the ratio of REE in the solid to that in the gas as:

$$\frac{Sm_{(s)}}{La_{(s)}} = \frac{\gamma_{La}}{\gamma_{Sm}} \, D_{Sm} \, \frac{Sm_{(g)}}{La_{(g)}} \, , \tag{11}$$

assuming bulk equilibrium between the solid and gas.

Equation (11) is convenient because we have separated the two very different effects on the REE volatility. The middle term on the right, the relative solid/gas distribution coefficient, is independent of the host phase in which the REE is condensing. It is a function only of the thermodynamic data and the H_2/H_2O ratio, which determines the oxygen fugacity and thus the fraction of gaseous REE present as the monoxide. Note that D_M is dependent on temperature but not on nebular pressure. The first term on the right, the ratio of activity coefficients, is dependent only on the characteristics of the host mineral and is thus constrained to be a smooth function of REE size.

Equation (11) is also convenient since we can use all the equations derived for igneous fractionation (see chapter by Hanson, which describes the geochemical equivalent processes). Thus fractional crystallization is analogous to fractional condensation and partial melting is analogous to partial vaporization. The only difference, of course, is that we can only calculate ratios of REE.

Table 2 and Figure 2 present the calculated values of the distribution coefficients and activity coefficients at 1650K; Table 2 also provides values for the fraction of gaseous REE present as the monoxide. Uncertainties in the thermodynamic data produce a factor of two to three uncertainty in the D values. This temperature was chosen since Grossman (1972) showed it was the temperature at which perovskite, $CaTiO_3$, which is a good host for REE, condenses out of the solar nebula at a total pressure of 10^{-3} atm. At lower temperatures, the general shape of the pattern does not change; the fractionations just become greater. The activity coefficients are the reciprocal of interpolated value of the hibonite/melt distribution coefficients determined by Drake and Boynton (1988), normalized to La = 1. Although these data were taken at 1740K, the expected differences at 1650K are not large compared to the uncertainties in the assumptions in deriving activity coefficients from them.

As can be seen, there are two important features of the D values; they are not a smooth function of size, and the extent of the fractionations can be extreme. Because the REE volatilities are not a smooth function of size, nebular fractionation based on REE volatility can be readily distinguished from later planetary fractionations. In addition, because even the general fractionation between light and heavy REE is so great, it is hard to imagine any mineral preference effect completely negating the volatility difference between La and Lu.

Why are the REE volatilities so different?

The reason that the cosmochemical properties of the REE are so different from their geochemical properties depends on their state in the gas phase. As noted earlier, the REE exist in the gas phase as the monoxides, dioxides, or monatomic species. In these cases, the formal charge on the atom is +2, +4, and 0, respectively. In the condensed form, however, the REE exist as the sesquioxides with a +3 charge. Thus, in going from the gaseous to solid state, the REE change their electronic configuration; whereas in geochemical fractionations, the REE have the same electronic configuration independent of whether they are in a liquid or a solid. It is the differences in the stability of the electronic states in the gas phase that gives rise to the very different REE volatilities.

It would be natural upon first looking at Figure 2 to assume that the big difference in volatilities of Eu and Yb compared to their neighbors was due to their stability in the +2 valence state. Although this is the explanation for the Eu anomalies seen in igneous fractionations, it does not apply here. In fact, the calculation for Eu assumes all of the Eu is condensing as Eu^{3+}, but, under the highly reducing condition of the solar nebula ($fO_2 = 10^{-16.4}$), it is clear that much Eu^{2+} will also be present in the solid. Since Eu^{2+} will condense *in addition to* the amount of Eu^{3+} present, its presence will increase the total amount of Eu condensed and therefore actually decrease its volatility making Eu less anomalous compared to its neighbors. These two elements, Eu and Yb, are particularly volatile, not because of any stability in the 2+ form in the solid, but because in the gaseous state only they have an enhanced stability due to the 4f electronic shell being half-filled and completely-filled, respectively.

CALCULATED REE PATTERNS

Early condensates

We now have sufficient information to calculate possible REE patterns established by cosmochemical gas-solid fractionations. The first case we shall consider is one in which the REE are initially present completely in the gas phase at high temperatures. As the gas cools, the REE condense into an initial condensate with a REE pattern described by Equation (11). We shall assume the activity coefficient relationship of Table 2 and Figure 2, but one must remember that other minerals with different activity coefficients may form in addition to, or instead of, hibonite. We shall consider concentration in units of fraction in the solid, thus Equation (11) becomes:

$$\frac{Sm_{(s)}}{La_{(s)}} = \frac{\gamma_{La}}{\gamma_{Sm}} D_{Sm} \frac{1-Sm_{(s)}}{1-La_{(s)}}, \tag{12}$$

We shall then choose a fraction of La in the solid and use Equation (12) to calculate the fraction of each of the other REE in the solid.

The results for three different values of La$_{(s)}$ are given in Figure 3. For small amounts condensed, the REE pattern looks like a plot of the values of D/γ, but as more of the REE condense out of the gas, the fractionation between the more refractory elements decreases, as they become totally condensed. Obviously, if the condensation were to proceed to the point where the REE were essentially totally condensed, the fractionation between the REE in the solid would disappear.

It should be noted, of course, that because bulk equilibrium between solid and gas was assumed, one would get identical results if, instead of the gas cooling to form a solid, a solid were heated to form a gas. This result follows from the fact that an equilibrium configuration is independent of the path by which equilibrium is obtained.

Removing REE in the gas

It is important to look at the REE composition of the gas in equilibrium with the early condensate. This gas can be considered the cosmochemical equivalent of a basaltic liquid in equilibrium with its source. Figure 4 contains plots of REE remaining in the gas in equilibrium with the solids of Figure 3. There the fractionation of the refractory elements is more obvious, as the more volatile elements are still nearly totally in the gas.

COMPARISON WITH METEORITIC DATA

There are many examples of REE patterns in Ca,Al-rich inclusions (CAI) that appear to be established by gas/solid fractionations. Mason and Martin (1977) classified some of these into groups, and Boynton (1983) discussed some of their implications. In this work, we will concentrate just on the most important of these REE patterns.

Ultra-refractory component

Figure 5 shows the REE determined in a small CAI from the Murchison meteorite (Boynton et al., 1980). It shows a pattern similar to that expected in early nebular condensates. The heavy, refractory, REE are the most enriched, followed by the light, more volatile REE, followed by the two most volatile REE, Eu and Yb, which are the most depleted. Also shown are REE abundances calculated using the D/γ values of Table 2. The values are calculated by first choosing the value for La$_{(s)}$ that provides the observed La/Lu ratio, then normalizing the entire plot to the CI normalized abundances of La and Lu. As can be seen, the agreement between theory and observation is quite good. The calculated values for Eu and Yb are much lower than observed for reasons that are understood, but we will come back to this point later. If the activity coefficient relationship were adjusted within reasonable limits, the agreement would improve, but considering the factor of 2 or 3 uncertainty in the D values and a desire to keep this treatment simple, we will not try to improve the fit.

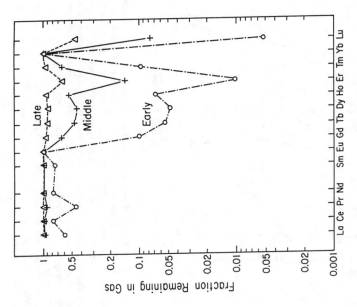

Figure 3. The fraction of each REE condensed in grains is calculated at three arbitrary stages in the condensation sequence. The lowest curve, Figure 2. The lowest curve represents the early stage of condensation. In the highest curve, the most refractory REE are nearly totally condensed and their ratios in the grains are nearly solar (chondritic). Note that these REE patterns are indistinguishable from those expected in grains that are a residue of a partial vaporization.

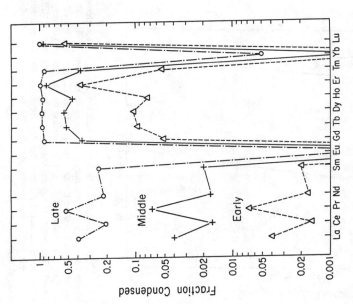

Figure 4. The fraction of each REE remaining in the gas are also calculated at three arbitrary stages in the condensation sequence. The lowest curve represents the late stage of condensation. Here Eu, Yb, and the light REE are present in nearly chondritic ratios in the early stages of condensation.

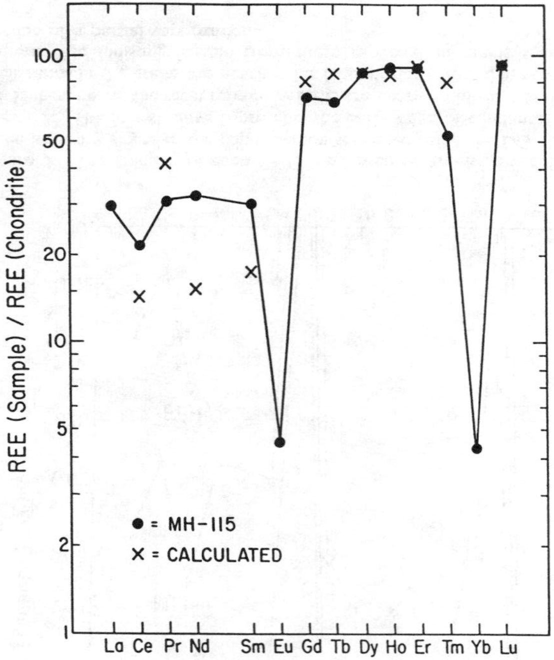

14

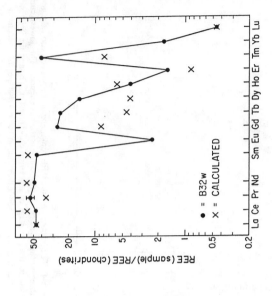

Figure 6. REE abundances in a Ca,Al-rich inclusion from the Allende meteorite show a greater enrichment of less refractory REE and several abundance anomalies. These strange abundances, particularly the Tm anomaly, cannot be explained as a result of an igneous fractionation process, but they are similar to those expected to be remaining in the nebular gas after a partial condensation (Fig. 4). This inclusion, therefore, appears to have sampled the fractionated nebular gas in equilibrium with an ultra-refractory condensate, similar to that in Figure 5, but Eu and Yb, the two most volatile REE, are also depleted because they were left behind when the grains that make this inclusion were isolated from the gas. Data are from Conard (1976).

Figure 5. REE abundances in a refractory inclusion from the Murchison meteorite show a greater enrichment in the more refractory REE. The values calculated based on the D/γ values of Figure 2 agree well with the data. Calculated values of Eu and Yb are too low to plot on this graph (see text). The relatively uniform enrichment of the most refractory REE near 100 times chondrites suggests that they are nearly totally condensed, and that this inclusion represents the first 1% of the total condensible material in the solar system. Data are from Boynton et al. (1980).

Because of the agreement between theory and observation, and because we know of no other way to form such a fractionated pattern, it seems clear that MH-115 had its REE pattern established as a result of a gas-solid fractionation process. We cannot say whether it was formed as a condensate or as a residue from partial vaporization, but it is clearly a nebular, not a planetary, product. For simplicity, we shall refer to MH-115 as a condensate. Its mineralogy is also consistent with its formation in the nebula at high temperatures. It was described by Boynton et al. (1980) as being composed of mostly spinel and hibonite with minor perovskite. It is perhaps worth noting, if it gives one more confidence in these calculations, that this ultra-refractory component was actually postulated to exist (Boynton, 1975) in connection with the formation of the Group II inclusions discussed below, five years before it was discovered.

This object provides strong evidence for high temperatures in the solar nebula. We do not know the maximum temperature precisely, of course, but one can show that it clearly must be on the order 1400K or more. These data may be thought to address the question of the ambient temperature in the solar nebula referred to in the introduction, but we cannot rule out a very brief event, perhaps as short as a few seconds, being responsible for the fractionation.

The observation that the most refractory REE, which are not strongly fractionated from each other, have CI-chondrite normalized abundances near 100 is significant. Remembering that CI chondrites are our best estimate of the composition of the condensible elements in the solar nebula, these data suggest that at least the most refractory REE are nearly totally condensed but that 99 percent of the total condensible mass, mostly O, Mg, Si, and Fe, remain in the gas. Thus, if the temperatures would have cooled further, there would be no more Lu available to condense, and its abundance in the solid would decrease as it was diluted by major elements that continue to condense.

Group II inclusions

Another interesting REE pattern found in meteorites is shown in Figure 6. It is from a CAI called B-32W (Conard, 1976) found in the Allende meteorite. It shows a very strong enrichment in the light REE relative to the heavy with a normalized La/Lu over 100. In addition, it shows a negative Eu anomaly and a strong positive Tm anomaly. This type of REE pattern was first found by Tanaka and Masuda (1973) and was later shown by Martin and Mason (1974) to be characteristic of a group of CAI found in Allende that they called Group II inclusions, though the Tm anomaly was unknown until B-32W was found. The anomaly has since been found in all Group II inclusions in which Tm is analyzed. This Tm anomaly is unknown in any natural terrestrial or lunar sample and clearly requires an extraordinary origin.

The origin of the Group II REE pattern was explained by Boynton (1975). It looks similar to the REE pattern expected of the more volatile REE remaining in the gas in equilibrium with an ultra-refractory condensate (Fig. 4). Exceptions occur for the two most volatile REE, Eu and Yb, which are depleted relative to La in the Group II pattern, even though they are the two most enriched REE in the gas. The pattern can be explained if the Group II grains condensed out of the gas left behind after an ultra-refractory condensate formed, but these Group II grains were then also isolated from the gas before the two most volatile REE, Eu and Yb, had fully condensed. The Group II REE pattern is thus a slice out of the middle of the REE condensation sequence, deficient in both the most refractory and the most volatile REE.

Figure 6 also shows calculated values based on the relative solid/gas distribution coefficients of Table 2. The values are calculated as they were for the ultra-refractory component by determining the fraction of La condensed, La$_{(s)}$, to match the observed La/Lu ratio. This time, however, it is the ratio in the gas that needs to match the observed Group II La/Lu ratio. Because Eu and Yb are depleted due to a second gas/solid fractionation, their values are not calculated. This second fractionation will also take a small amount of the light REE, but this effect is also ignored in this example. The calculated values agree well with the observations if we remember that there is a factor of two or three uncertainty in the distribution coefficients. The enrichment in light REE relative to the heavy ones and the Tm anomaly are explained by the REE volatilities. For simplicity, the same activity coefficient relationship was used for this calculation as for the calculation of the ultra-refractory condensate in Figure 5. Using the same activity coefficients is not required, however, as there may have been other minerals, such as perovskite and melilite, present in the ultra-refractory component that formed in the region of the solar nebula from which this particular Group II inclusion formed. See Boynton (1983, Fig. 3.3) for an example of a much better fit to these same data based on a more modest change of REE activity coefficient with ionic radius expected for perovskite or melilite. Davis and Grossman (1979) also devoted a lot of attention to fitting the observed data with calculated values and presented figures for 24 different Group II CAI.

The CI-normalized concentrations of Eu and Yb are generally found to be very similar in CAI. This observation is not expected based on the REE volatilities since Eu is so much more volatile than Yb. It was mentioned earlier that allowing for condensation of Eu^{+2} will decrease the volatility of Eu, but it would be remarkable if it were decreased to exactly match that of Yb. Davis and Grossman (1979) suggested that Eu and Yb would be brought in together in a late condensing component that contained no REE other than Eu and Yb. Boynton (1983) suggested that if the grains began to be separated from the nebula gas as condensation was progressing, rather than instantaneously as the model calculations assume, then the last condensing REE could come from a smaller reservoir with a lower gas/grain ratio and then condense completely. In either case, it is clear that the Group II inclusions record a complex series of gas/solid fractionation events.

FUN inclusions

Another small group of CAI have been found in the Allende meteorite that are called FUN inclusions because they have isotopic ratios that are strongly *F*ractionated relative to terrestrial ratios and also contain a component of *U*nknown *N*uclear origin (Wasserburg et al., 1977). Explaining the origin of these inclusions, particularly trying to account for their isotopic composition, has been very difficult (see Lee, 1988, for a review) and is certainly beyond the scope of this work. Two of these inclusions, HAL (Tanaka et al., 1979) and Cl (Conard, 1976) have large depletions in Ce abundances, by factors 15 and 3, respectively. These large Ce anomalies cannot be explained by the REE volatilities as given in Table 2, but they can be understood if the oxygen fugacity is increased (Boynton, 1978a).

REE condensation as a function of oxygen fugacity. The anomalies have been explained by considering the effect of oxygen fugacity on REE volatility (Boynton, 1978a). As mentioned earlier, oxygen fugacity effects the REE volatility because it effects F_{MO}, the fraction of the gaseous REE present as the monoxide. Another way to understand the effect of oxygen is to consider the condensation reaction of three different REE written in a somewhat different form than Equation (1):

$$LaO_{(g)} + 1/4\ O_{2(g)} \rightarrow LaO_{1.5(s)}, \tag{13}$$

$$CeO_{2(g)} \rightarrow CeO_{1.5(s)} + 1/4\ O_{2(g)}, \tag{14}$$

$$Yb_{(g)} + 3/4\ O_{2(g)} \rightarrow YbO_{1.5(s)}; \tag{15}$$

Equation (15) is appropriate for both Eu and Yb, as both of these elements have the monatomic specie as the dominant gaseous form. It can be seen that increasing the oxygen fugacity (or partial pressure) drives Reaction 13 to the right making La more refractory, but it drives Reaction 14 to the left making Ce more volatile. An increase in oxygen also makes Yb more refractory, even more so than it does for La since oxygen appears with a larger coefficient in Reaction 15. Thus an increase in oxygen fugacity makes Ce more volatile and Yb and Eu more refractory than they were with a solar oxygen fugacity. The other REE, that have mostly the monoxide in the gas phase, do not change in volatility *relative to La*. The effect of an increase in oxygen on the volatility of Eu relative to La is small because a relatively small increase in oxygen increases the $EuO_{(g)}/Eu_{(g)}$ ratio to the point where $EuO_{(g)}$ dominates and the volatility of Eu thereafter does not change relative to La.

The change in volatility with more reducing conditions can also be considered by inspection of Table 2. Relative to La, Ce gets more refractory, but only by a small factor before it becomes mostly monoxide in the gas, and Sm, Eu, Tm, and Yb get more volatile. Although we have a need to postulate conditions far more oxidizing than solar to explain the FUN inclusions, there are no meteorite data that require conditions far more reducing than solar[1], although modest deviations from solar oxygen fugacity cannot be excluded.

Astrophysical scenarios to alter oxygen fugacity. The enhanced volatility of Ce in an oxidizing environment provides an explanation of the Ce anomalies observed in the two FUN inclusions, but we need to consider means by which such an environment could be established.

Highly oxidizing conditions can be established in outflows from stars that have been through enough nuclear processing to convert their hydrogen to oxygen. These outflows can occur as the star is evolving or they can occur violently as the star explodes as in a supernova. Highly reducing environments, on the other hand, can be produced around stars that have converted their hydrogen only as far as carbon. Grains made around other stars, of course, would need to find their way into our solar system for us to see them in meteorites. Because of the large amount of nuclear processing, these grains would have an isotopic composition unlike normal solar system material. Although evidence for a tiny fraction of some of this type material has been found in CAI, particularly the FUN CAI (Wasserburg et al., 1977), no grains have been found that are dominated by such material. This observation makes it difficult to relate the supernova to the cause of the oxidizing environment present when the FUN REE patterns were established.

Large changes in the oxygen fugacity can also be generated by physical/chemical processes in the nebula. Highly reducing conditions can be established if there is a small change in the solar C/O ratio from 0.4 (Table 1) to 1.0 (Larimer, 1975). Under these conditions, graphite is a stable phase and the oxygen fugacity is controlled by a graphite/carbon monoxide buffer to a value a factor of 10^{15} lower than the solar value. Achieving a more modest reducing environment is difficult and requires a precise control of the C/O ratio near 0.9. Although it is likely that some classes of meteorites, like the enstatite chondrites, may have formed in such an environment (Larimer and Bartholomay, 1979), the author is aware of no viable mechanisms to achieve these conditions.

Generating more oxidizing environments, on the other hand, appears to be easily accomplished. Wood (1967) proposed a mechanism in connection with the amount of Fe^{2+} in meteoritic olivines and pyroxenes, whereby grains are separated from the hydrogen-rich nebular gas, and are then reheated to partially vaporize and generate an oxygen-rich gas. This mechanism has recently been extensively discussed by Kring (1988) concerning the varying redox conditions in the solar nebula that were recorded by the constituents of chondrules.

[1]Davis and Grossman (1979), apparently not recognizing that Sm would also get more volatile, postulated reducing conditions to account for the observed depletion of Yb over Sm, since their ideal solution model had no other means to account for the data.

Rims on CAI

Wark and Lovering (1977) described a series of about 5 layers, totalling only 50 μm in thickness, that form rims around virtually all CAI. The origin of these rims has been very difficult to understand, but the REE have shed some light on the problem.

Boynton and Wark (1984, 1985, 1987) and Wark and Boynton (1987) physically removed several rims from CAI and analyzed both them and the underlying CAI for REE abundances. What they found in each CAI was that the REE were enriched in the rim relative to the underlying material by constant factors, ranging from 2 to 5, except for Eu and sometimes Yb which were less enriched. This observation was true regardless of the type of REE pattern found in the CAI. Figures 7 and 8 show data from two of the inclusions. HAAG-1 (Fig. 7) has a Group II REE pattern that is distinct from all other known Group II patterns in that Lu is more abundant than the less volatile REE Dy and Ho. As can be seen, the rim has virtually the identical pattern except for the uniform enrichment of all the REE but Eu. Inclusion 818b (Fig. 8) is a common type of inclusion, with a melilite mantle and a pyroxene core, that formed by crystallization as a molten droplet (Stolper and Paque, 1986). The mantle shows a light REE enrichment and positive Eu anomaly that is determined by the igneous partitioning of REE into melilite. The pyroxene-rich core has the complementary enrichment of heavy REE. The rim has the identical igneously derived pattern except for the constant enrichment factor for all but Eu and Yb.

These data are important because they provide very strong evidence that the rim is derived from the underlying material. As shown in this chapter, it is not possible to condense REE out of the gas with a smoothly fractionated REE pattern. In addition, if the rim were unrelated to the interior, it is hard to imagine what process would be clever enough to put a Group II rim or a Group II inclusion and an igneous pattern on an igneous inclusion. Boynton and Wark proposed that the rims formed from the interior as a result of an intense, rapid heating event that melted just the surface of the CAI and vaporized an appreciable fraction of this melt. This vaporization explains the uniform enrichment of the REE and the exceptions for Eu and Yb, which were also partially vaporized.

WHAT HAVE WE LEARNED FROM THE REE?

The ability of the REE to unambiguously distinguish between a nebular gas/solid fractionation and a planetary solid/liquid or solid/solid fractionation clearly makes them an important tool in understanding conditions in the early solar system. It is important now to consider what these studies have really told us about conditions in the early solar system.

20

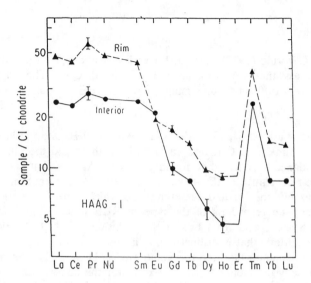

Figure 7. REE abundances in this CAI from Allende are similar in the rim and the interior. The rim REE are uniformly enriched relative to the interior except for the most volatile REE, Eu, which is depleted. These Group II patterns, with Lu more enriched than Ho and Dy, are unlike any other found in CAI. The uniqueness of this pattern suggests the rim was derived from the interior. Data are from Boynton and Wark (198).

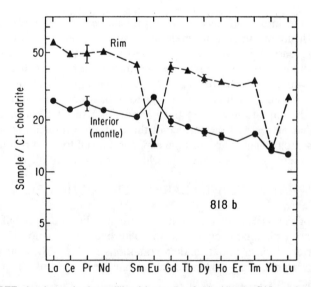

Figure 8. REE abundances in the melilite-rich mantle of this Allende CAI are fractionated by an igneous process as the CAI crystallized from an originally molten droplet; the complementary pattern is found in the pyroxene-rich core of this CAI. Again, the rim pattern appears to be derived from the interior. An intense, but brief, heating event is postulated that melted and partially vaporized the surface of CAI making rims with a uniform REE enrichment relative to the interior. The most volatile elements, Eu and Yb, were apparently also lost during this vaporization.

High temperatures were achieved in the solar nebula

It was mentioned in the discussion of the ultra-refractory component that this component provided evidence for high temperatures, if just briefly. The case was not made that these high temperatures were really in the solar nebula as opposed to in some other astrophysical environment. The only other known astrophysical setting in which temperatures fall through the range near 1500K with a sufficient density for condensation is in outflows from stars. In principle, the ultra-refractory condensates could have formed in such an environment. The Group II inclusions, on the other hand, had to form from the gas left behind after an ultra-refractory component formed, condensation stopped, and the ultra-refractory component was isolated from the gas. Since the gas density is monotonically decreasing in a stellar outflow, once condensation stops, presumably because the density is too low, it will be impossible for it to resume. The solar nebula appears to be the only environment in which, perhaps through turbulence, it may be possible to cycle material through zones of different temperatures to achieve multiple episodes of condensation.

A very efficient mechanism for gas/dust separation existed in the solar nebula

The Group II CAI are remarkable in the fact that they can have La/Lu ratios as high as 100, implying as much as 99% of the most refractory REE, Lu, is absent (Fig. 6). The Group II inclusions are usually aggregates up to a centimeter in size made up of many micrometer-sized grains. Therefore, when these grains accreted to form the centimeter-sized aggregates, they must have done so in an environment nearly devoid of any of the ultra-refractory-component grains, since they managed to avoid up to 99% of this component.

To be honest, it is hard to imagine, even in the solar nebula, how one could physically separate the gas from the ultra-refractory grains that established its composition, and then begin condensation again to form the Group II grains. The author is aware, of course, that this statement weakens the argument made in the previous section that, because we do not know how to form the Group II inclusions outside the solar nebula, they must, therefore, have formed *in* the solar nebula. Even though we also do not know how to form the Group II inclusions in the solar nebula, it seems clear that this environment offers a far richer set of conditions in which to perform these complicated processes. In addition, the accretion arguments made in this section apply to forming the Group II material outside the nebula because they would also need to accrete to centimeter-sized objects without acquiring a substantial component of normal interstellar material. Nevertheless, although we cannot describe the mechanism that formed the Group II inclusions, we can say with confidence that their formation required an efficient gas/grain separation.

The high nebular temperatures existed for a long time

This conclusion is based on the requirement that the gas/grain separation, condensation and accretion that formed the Group II aggregates must take a reasonable length of time to occur. It is difficult to be quantitative, of course, without a mechanism to model, but clearly seconds to minutes, and probably hours, are not adequate. This constraint eliminates several transient heat sources such as lightning or nebula flares, and requires that astrophysicists need to find a longer duration heat source.

A very intense, very brief heat, source also existed

The presence of the CAI rims argues for a very intense heat source to vaporize the material. Boynton (1988) calculated the minimum energy needed both to generate the thin melt zone that formed the rim and to vaporize the melt for a factor of three enrichment in REE; 300 J/cm^2 were required. He also used thermal diffusivity data to calculate the maximum time the rim could have been hot without melting the interior and found a limit of about one second. The resulting heat input of 300 W/cm^2 is very intense. If heated by photons, it would require the equivalent of an ambient medium at 2800K.

Finding a mechanism to generate such an intense heat source and a means to turn it on and off rapidly is also challenging the astrophysicists, but some progress is being made. Nebular lightning looks like it may have the required characteristics (Morfill and Levy, personal communication).

The solar nebula was a chaotic environment

Before 1973, the solar nebula was envisaged by many to be a quiescent system in which the temperatures started out high and monotonically decreased until all the condensible material had formed. It was known that some material could be isolated from the nebula at high temperatures since CAI's had been described (Marvin et al., 1970) and their mineralogy had been shown to be consistent with that expected of early condensates (Grossman, 1972). It was known that the presence of chondrules required a brief, high-temperature event, but it was not clear that this was a nebular process. In large part because of REE data such as those described in this chapter, it has now become clear that the solar nebula provided a complex environment in which multiple episodes of heating, cooling and dust/gas separation must have occurred on a variety of timescales (Boynton, 1978b).

SUMMARY

The REE are a unique group of elements that have similar properties that vary as a smooth function of size in geochemical systems, making them a very powerful tool in studying igneous processes, to which the other chapters in this volume will attest. The REE also have the interesting property that their volatilities are neither similar nor a smooth function of size. This latter property is as important to the cosmochemist as the former is to the geochemist. It permits an unambiguous discrimination between nebular and planetary fractionation processes and allows us to put important constraints on the complex processes that formed our solar system.

ACKNOWLEDGEMENTS

This work benefitted from numerous suggestions by David A. Kring. It was supported in part by NASA Grant NAG 9-37.

REFERENCES

Anders, E. and Grevesse, N., 1989, Abundances of the elements: Meteoritic and solar. Geochim. Cosmochim. Acta, 53:197-214.

Boynton, W.V., 1975, Fractionation in the solar nebula: condensation of yttrium and the rare earth elements. Geochim. Cosmochim. Acta 39:569-584.

Boynton, W.V., 1978a, Rare-earth elements as indicators of supernova condensation. Lunar and Planet. Sci. IX:120-122.

Boynton, W.V., 1978b, The chaotic solar nebula: evidence for episodic condensation in several distinct zones. In: Protostars and Planets, ed. T. Gehrels, Tucson: The University of Arizona Press, pp. 427-438.

Boynton, W.V., Frazier, R.M. and Macdougall, J.D., 1980, Identification of an ultra-refractory component in the Murchison meteorite. Lunar and Planet. Sci. XI:103-105.

Boynton, W.V., 1983, Cosmochemistry of the rare earth elements: meteorite studies. In: Rare Earth Element Geochemistry. Developments in Geochemistry, 2, ed. P. Henderson, Elsevier Science Publishers B.V., Amsterdam, pp. 63-114.

Boynton, W.V. and Wark, D.A., 1984, Trace element abundances in rim layers of an Allende type A coarse-grained, Ca-Al-rich inclusion. Meteoritics 19: 195-197.

Boynton, W.V. and Wark, D.A., 1985, Refractory rims: evidence for high temperature events in the post-formation history of Ca-Al-rich inclusions. Meteoritics 20: 613-614.

Boynton, W.V. and Wark, D.A., 1987, Origin of CAI rims--I: The evidence from the rare earth elements.Lunar and Planet. Sci. XVIII: 117-118.

Boynton, W.V., 1988, Nebular processes associated with CAI rim formation, Meteoritics 23:259.

Clayton, R.N., 1981, Isotopic variations in primitive meteorites. Phil. Trans. R. Soc. Lond. A 303:339-349.

Conard, R., 1976, A study of the chemical composition of Ca-Al rich inclusions from the Allende meteorite. M.S. Thesis, Oregon State University, Corvallis, 129 pp.

Consolmagno, G.J. and Drake, M., 1977, Composition and evolution of the eucrite parent body: evidence from rare earth elements. Geochim. Cosmochim. Acta, 41:1271-1282.

Davis, A.M. and Grossman, L., 1979, Condensation and fractionation of rare earths in the solar nebula. Geochim. Cosmochim. Acta 43:1611-1632.

Davis, A.M. and MacPherson, G.J., 1988, Rare earth elements in a hibonite-rich Allende fine-grained inclusion. Lunar and Planet. Sci. XIX, Part I, pp. 249-250.

Dodd, Robert T., 1981, Meteorites A petrologic-chemical synthesis. Cambridge: Cambridge University Press, 368 pp.

Drake, M.J. and Boynton, W.V., 1988, Partitioning of rare earth elements between hibonite and melt and implications for nebular condensation of the rare earth elements. Meteoritics 23:75-80.

Grossman, L., 1972, Condensation in the primitive solar nebula. Geochim. Cosmochim. Acta 36:597-619.

Johnston, W.D., 1965, Oxidation-reduction equilibriums in molten Na_2O - $2SiO_2$ glass. Jour. Amer. Ceramic Soc. 48:184-190.

Kant, I., 1755, Allgemeine Naturgeschichte und Theorie des Himmels.

Kring, D.A., 1988, The petrology of meteoritic chondrules: evidence for fluctuating conditions in the solar nebula, Ph.D. Thesis, Harvard University, Cambridge, 346 pp. (available from University Microfilms International, Ann Arbor, Michigan).

24

Larimer, J.W., 1975, The effect of C/O ratio on the condensation of planetary material. Geochim. Cosmochim. Acta 39:389-392.

Larimer, J.W. and Bartholomay, M., 1979, The role of carbon and oxygen in cosmic gases: some applications to the chemistry and mineralogy of enstatite chondrites. Geochim. Cosmochim. Acta 43:1455-1466.

Larimer, J.W. and Wasson, J.T., 1988a, Refractory lithophile elements. In: Meteorites and the Early Solar System, eds. John F. Kerridge and Mildred Shapley Matthews, Tucson: The University of Arizona Press, pp. 394-415.

Larimer, J.W. and Wasson, J.T., 1988b, Siderophile-element fractionation, In: Meteorites and the Early Solar System, eds. John F. Kerridge and Mildred Shapley Matthews, Tucson: The University of Arizona Press, pp. 416-435.

Lee, T., 1988, Implications of isotopic anomalies for nucleosynthesis. In: Meteorites and the Early Solar System, eds. John F. Kerridge and Mildred Shapley Matthews, Tucson: The University of Arizona Press, pp. 1063-1089.

MacPherson, G.J., Wark, D.A. and Armstrong, J.T., 1988, Primitive material surviving in chondrites: refractory inclusions. In: Meteorites and the Early Solar System, eds. John F. Kerridge and Mildred Shapley Matthews, Tucson: The University of Arizona Press, pp. 746-807.

Martin, P.M. and Mason, B., 1974, Major and trace elements in the Allende meteorite. Nature: 249:333-334.

Marvin, U.B., Wood, J.A. and Dickey, J.S., Jr., 1970, Ca-Al-rich phases in the Allende meteorite. Earth Planet. Sci. Lett. 7:356-350.

Mason, B. and Martin, P.M., 1977, Minor trace element distribution in melilite and pyroxene from the Allende meteorite. Earth Planet. Sci. Lett. 22:141-144.

McSween, H.Y., Sears, D.W.G. and Dodd, R.T., 1988, Thermal metamorphism. In: Meteorites and the Early Solar System, eds. John F. Kerridge and Mildred Shapley Matthews, Tucson: The University of Arizona Press, pp. 102-113.

Palme, H., Larimer, J.W. and Lipschutz, M.E., 1988a,b, Moderately volatile elements. In: Meteorites and the Early Solar System, eds. John F. Kerridge and Mildred Shapley Matthews, Tucson: The University of Arizona Press, pp. 436-461.

Stolper, E. and Paque, J.M., 1986, Crystallization sequences of Ca-Al-rich inclusions from Allende: The effects of cooling rate and maximum temperature, Geochim. Cosmochim. Acta 50:1785-1806.

Tanaka, T. and Masuda, A., 1973, Rare-earth elements in matrix, inclusions and chondrules of the Allende meteorite. Icarus 4:523-530.

Tanaka, T., Davis, A.M., Grossman, L., Lattimer, J.M., Allen J.M., Lee, T. and Wasserburg, G.J., 1979. Chemical study of an isotopically-unusual Allende inclusion. Lunar and Planet. Sci. X:1203-1205.

Thiemens, M. H., 1988, Heterogeneity in the nebula: evidence from stable isotopes. In: Meteorites and the Early Solar System, eds. John F. Kerridge and Mildred Shapley Matthews, Tucson: The University of Arizona Press, pp. 899-923.

Wark, D.A. and Boynton, W.V., 1987, Origin of CAI rims--II: The evidence from refractory metals, major elements and mineralogy. Lunar and Planet. Sci. XVIII: 1054-1055.

Wark, D.A. and Lovering, J.F., 1977, Marker events in the early solar system: evidence from rims on Ca-Al-rich inclusions in carbonaceous chondrites. Proc. Lunar Sci. Conf. 8:95-112.

Wasserburg, G.J., Lee, T. and Papanastassiou, D.A., 1977, Correlated and Mg isotopic anomalies in Allende inclusions, II. Magnesium. Geophys. Res. Lett. 4:299-302.

Wasson, J.T., 1974, Meteorites. New York: Springer-Verlag, 316 pp.

Wood, J.A. and Morfill, G.E., 1988, A review of solar nebula models. In: Meteorites and the Early Solar System, eds. John F. Kerridge and Mildred Shapley Matthews, Tucson: The University of Arizona Press, pp. 329-347.

Zolensky, M. and McSween, H.Y., 1988, Aqueous alteration. In: Meteorites and the Early Solar System, eds. John F. Kerridge and Mildred Shapley Matthews, Tucson: The University of Arizona Press, pp. 114-143.

RADIOGENIC ISOTOPE GEOCHEMISTRY

OF RARE EARTH ELEMENTS

INTRODUCTION

Long-lived radioactive isotopes of Rare Earth Elements

Radiogenic isotope systematics involving Rare Earth Elements (REE), particularly the Sm-Nd system, have made major contributions to understanding of planetary evolution processes. This is primarily because of the systematic and well-understood chemical fractionations among REE that occur in nature, addressed elsewhere in this volume. The utility of all long-lived radioactive isotopes and their products has been summarized in the book by Faure (1986). De Paolo (1988) has discussed Nd isotope geochemistry in a recent book. The purpose of the present chapter is to summarize, in the context of this volume, the contributions made by studies of Sm-Nd, Lu-Hf and La-Ce decay schemes to geochemistry and geochronology.

Three main parent-daughter pairs have been of use in geochronology and geochemistry: ^{138}La-^{138}Ce, ^{147}Sm-^{143}Nd and ^{176}Lu-^{176}Hf. The single extinct radionuclide that was long enough lived in the early Solar System to have produced detectable isotopic effects is ^{146}Sm (half life 10^8 years), and probable resulting deviations in ^{142}Nd have been observed in some meteorites (e.g., Lugmair et al., 1975a). Other naturally-occurring isotopes that have long lived decays, which form a single α-decay chain ^{152}Gd - ^{148}Sm - ^{144}Nd - ^{140}Ce, are of no interest geologically because the half-lives are all $>10^{14}$ years. The decays are thus more than 1000 times slower than those of ^{138}La, ^{147}Sm and ^{176}Lu, and resulting isotopic variations would be undetectable by present instrumentation. Essentials of the La-Ce, Sm-Nd and Lu-Hf decay systems are summarized in Table 1.

^{138}La-^{138}Ce decay. ^{138}La is an odd Z-odd N nucleus that has a branched decay, by electron capture or positron decay to ^{138}Ba and by β-decay to ^{138}Ce. ^{138}Ba is the most abundant isotope of an abundant element, making ^{138}Ba variations inherently hard to detect. La-Ba geochronology is not discussed further in this chapter. For both of the ^{138}La decays, the decay constants are poorly known. For the β-decay to ^{138}Ce, values recently measured and in use vary by 20% (Table 1; see also Sato and Hirose, 1981). The means of fractionation-correcting and reporting Ce isotope data have definitely not yet settled (Table 1). The main analytical problem in La-Ce studies arises because ^{138}La is a low-abundance isotope (0.1%) and the decay is quite slow, so that small differences in ^{138}Ce abundance must be measured. ^{138}Ce is only 0.25% of Ce, and the isotopes ^{136}Ce, ^{138}Ce and ^{142}Ce must be measured in the presence of an extremely large ^{140}Ce beam, which can saturate amplifier circuits as well as lead to significant peak tails in the spectrum. Ce isotope data are included in this review because they reinforce the elegant correlation between REE geochemical behavior and covariation of Nd, Hf, Sr and Pb isotopes.

^{147}Sm-^{143}Nd decay. Based on the most recent determinations, the α-decay constant is well known (De Paolo, 1988 reviews). The main ambiguity in Nd isotopic data arises from the two main bases for correcting out mass fractionation during measurement (Table 1). As is the case for Ce isotopes, use of the epsilon notation, explained in a later section, obviates this problem for general readers. Nd isotopic composition is relatively easy to measure provided an adequate chemical separation from neighboring REE can be achieved. It is especially important to separate Sm, which interferes with Nd at masses 144, 148 and 150.

^{176}Lu-^{176}Hf decay. Theoretically ^{176}Lu should decay like ^{138}La, being an odd Z-odd N nuclide. In fact, the presumed electron-capture or positron decay to ^{176}Yb has never been observed or documented. Determinations of the β-decay constant vary by tens of percent, but the most recent value of 1.93 ($\pm$0.03).10^{-11} a^{-1} (Sguigna et al., 1982) agrees well with the 1.94 ($\pm$ 0.07).10^{-11} a^{-1} value currently in use by geologists. This was based on a total-rock isochron for a series of 4.55 Ga eucrite meteorites (Patchett and Tatsumoto, 1980; Tatsumoto et al., 1981). While the assumption of a cogenetic origin for all the eucrites used is certainly open to question, the agreement between the meteoritic and physical determinations, coupled with the uncertainties in the Lu-Hf system as a whole, render a switchover to the 1.93.10^{-11} value not worthwhile. Lu is a moderately easy element to measure in a thermal ionization mass spectrometer, although like Rb and La there are only two isotopes, so that no internal correction for mass fractionation can be made. Hf is difficult to measure because of a high ionization potential for this high-melting-temperature element. At the temperature limit of platinum-group metal filaments, Hf can only be ionized around 100 times less efficiently than Nd.

Chemical variations of La/Ce, Sm/Nd and Lu/Hf ratios

REE chemical variations in depleted peridotite, mid-ocean ridge basalts and the upper continental crust relative to chondrites are shown in Figure 1. The general and specific aspects of REE behavior are covered in other sections of this volume. Here it only needs to be pointed out that La/Ce fractionations occur in the opposite direction to Sm/Nd and Lu/Hf, so that geochemical behavior predicts positive correlation of Nd and Hf isotopic variations, but negative correlation of either of these with Ce isotopes. Residual (depleted) mantle acquires high Sm/Nd, high Lu/Hf but low La/Ce; continental crust has opposite characteristics (Fig. 1). Hafnium is not a rare earth element, but several generations of studies on abundance in oceanic basalts have established that in mantle melting events, Hf behaves like the missing naturally-occurring element between Nd and Sm (e.g., Patchett et al., 1981; Hofmann, 1988). In terms of magnitude of fractionation during mantle melting events, it is clear that Sm/Nd fractionation would exceed La/Ce fractionation because La and Ce are adjacent REE, while Sm and Nd are not. Lu/Hf fractionations are greatest, and this is reflected in the isotopic variations of mantle-derived rocks described later.

All indications are that Hf behaves coherently with REE during mantle melting events, which involve mineral phases such as olivine and pyroxenes. In the continental crust, however, much Zr and Hf are present in the mineral zircon,which has a complex behavior in melting events (Watson and Harrison, 1983). This certainly leads to special fractionations of the Lu/Hf ratio between, for instance, granitoids and their crustal sources. These potential crustal Lu/Hf fractionations remain to be documented, although an important fractionation due to zircons in the sedimentary system is described later in this chapter.

GEOCHRONOLOGICAL STUDIES

La-Ce and Lu-Hf chronology

The contribution of La-Ce and Lu-Hf to dating has been relatively insignificant. Studies have mostly been directed at decay constant determination or feasibility demonstration. Tanaka and Masuda (1982) gave La-Ce results for the Bushveld complex. Dickin (1987a) determined a La-Ce total rock age for Archean rocks, with an error of 15%, while Masuda et al. (1988) determined mineral ages for other Precambrian samples. Patchett and Tatsumoto (1980) gave a total-rock isochron for eucrite meteorites by the Lu-Hf method. Pettingill and Patchett (1981) determined a Lu-Hf age for Amitsoq Gneisses of West Greenland based on total rocks and zircon separates.

Sm-Nd chronology

In contrast to the above systems, the Sm-Nd method has made major contributions to extraterrestrial chronology. The method was first developed in the context of lunar and meteoritic chronology (Lugmair et al., 1975b), and since then has stacked up a number of significant advances in lunar chronology. These include further Sm-Nd internal isochrons

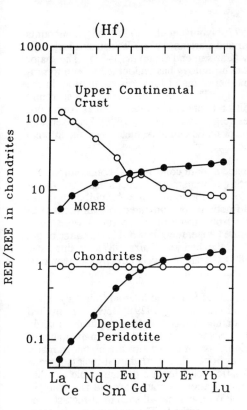

Figure 1. Typical rare-earth element abundance patterns for major reservoirs in the Earth. MORB = mid-ocean ridge basalts.

Table 1. La-Ce, Sm-Nd, and Lu-Hf decay schemes and reference parameters.

System	Radioactive isotope	Radiogenic product	Decay constant λ (a^{-1})	Parent/product ratio, CHUR value[1]	Product isotope ratio, CHUR value[1]
La-Ce[2]	^{138}La	^{138}Ce	$2.30 \cdot 10^{-12}$	$^{138}La/^{136}Ce$, 0.178	$^{138}Ce/^{136}Ce$, 1.32857
			$2.77 \cdot 10^{-12}$	$^{138}La/^{142}Ce$, 0.003089	$^{138}Ce/^{142}Ce$, 0.0225722
Sm-Nd[3]	^{147}Sm	^{143}Nd	$6.54 \cdot 10^{-12}$	$^{147}Sm/^{144}Nd$, 0.1966	$^{143}Nd/^{144}Nd$, 0.512638
				$^{147}Sm/^{144}Nd$, 0.1967	$^{143}Nd/^{144}Nd$, 0.511847
Lu-Hf	^{176}Lu	^{176}Hf	$1.94 \cdot 10^{-11}$	$^{176}Lu/^{177}Hf$, 0.0334	$^{176}Hf/^{177}Hf$, 0.28286

[1]CHUR values given are for an undifferentiated "chondritic uniform reservoir" today, taken to be representative for bulk terrestrial planets.

[2]Dual listing for La-Ce reflects different decay constants and data reporting schemes accepted by the two groups that have produced data: East Kilbride (upper line) and Tokyo (lower line).

[3]Dual listing for Sm-Nd reflects the two main bases for mass fractionation correction in Nd isotopic data that are in use. The upper line has the basis $^{146}Nd/^{144}Nd = 0.7219$, while the lower line corresponds to $^{146}Nd/^{142}Nd = 0.636151$. Approximately 75% of researchers use the basis given in the upper line.

for lunar basalts (e.g., Papanastassiou et al., 1977; Nyquist et al., 1979, 1981), constraints on KREEP chronology and evolution (Lugmair and Carlson, 1978; Carlson and Lugmair, 1979), and lunar crustal chronology generally (Carlson and Lugmair, 1981). The major contribution of Sm-Nd isotopic studies in lunar chronology has undoubtedly been to provide reliable ages for lunar crust formation and hence major lunar differentiation, in the 4.2-4.4 Ga interval (Carlson and Lugmair, 1981). The Sm-Nd method has also found wide application in chronological studies of differentiated meteorites, such as Angra dos Reis (Lugmair and Marti, 1977) and eucrites (e.g., Hamet et al., 1978; Jacobsen and Wasserburg, 1984). For well-preserved basaltic or mafic cumulate meteorites, the internal Sm-Nd isochrons are often the most reliable age.

In terrestrial chronology, Sm-Nd has made a more controversial contribution. An apparently straightforward application was dating of metamorphic rocks using two-point lines between garnet, with a high Sm/Nd ratio, and the total rock. Amphibolite facies garnets did not have sufficiently great Sm/Nd fractionation compared to total rocks (van Breemen and Hawkesworth, 1979). Garnets from granulite-grade rocks were more successful, but there are indications that the closure temperature for Sm-Nd exchange might be as low as 600°C (Humphries and Cliff, 1982; Patchett and Ruiz, 1987). In this case, the garnet Sm-Nd ages would not date the 700-800°C granulite event directly, but would be a point on the uplift and cooling curve for a terrane.

There have been straightforward applications of Sm-Nd to internal (mineral) dating of well-preserved mafic rocks (e.g., De Paolo and Wasserburg, 1979). Because of limited Sm/Nd ratio variation and the slow decay of ^{147}Sm, the method would be applicable mainly in the Precambrian; and is limited by the fact that rocks with well-preserved mafic mineralogy are not very common in Precambrian regions. As Sm/Nd ratios are higher in mafic rocks compared to felsic rocks (see Fig. 1), it was hoped that Sm-Nd total-rock isochrons could be used routinely to date Precambrian mafic rocks that are not amenable to other geochronologic techniques. However, the demonstration of biased Sm-Nd isochrons in Archean mafic volcanics (e.g., Chauvel et al., 1985) means that all such total-rock studies are suspect unless the age is corroborated by another method. The problem results from heterogeneous mantle magma sources and/or different degrees of crustal contamination in the mafic magmas. Thus the utility of Sm-Nd in terrestrial dating appears finite but quite limited.

DEFINING BULK PLANETARY ISOTOPIC EVOLUTION

For satisfactory interpretation of isotopic data from old differentiated reservoirs of planets, it is necessary to define an isotopic evolution for the total planet. This is also the evolution of any undifferentiated part of the planet, that would have the same La/Ce, Sm/Nd and Lu/Hf ratios as primary undifferentiated material. For refractory elements like REE and Hf, it is clear that even if only a small fraction of the Solar System matter were solid, it would contain those elements in the same proportion as they were present in the whole system (see chapter by Boynton in this volume). Consequently, isotope geochemists have felt confident in using carbonaceous and/or ordinary chondrites as a model for La/Ce, Sm/Nd and Lu/Hf ratios in planets like the Earth and Moon. The isotopic evolution lines constructed from chondrites are called CHUR evolution curves, based on the first letters of the words "chondritic uniform reservoir."

Table 1 has been constructed so that currently-used CHUR curves can be written for variants of Sm-Nd and Lu-Hf decay systems. The example given is Sm-Nd, where Nd isotopic data have been corrected to ^{146}Nd/^{144}Nd = 0.7219; this is the upper line given for Sm-Nd in Table 1:

$$(^{143}\text{Nd}/^{144}\text{Nd})^t_{\text{CHUR}} = 0.512638 - 0.1966 (e^{\lambda t} - 1) \qquad (1)$$

Note that it is necessary to calculate backwards from the present day and that t is always the age in years. Equivalent equations can be written for Sm-Nd and Lu-Hf variants of Table 1 by substituting appropriate parameters.

Sm-Nd CHUR evolution was defined by Jacobsen and Wasserburg (1980, 1984). The $^{143}Nd/^{144}Nd$ was an average of several chondrites that were all consistent with a 4.55 Ga age. The Sm/Nd ratio was that of a cluster of chondrite analyses within a spread of 2%. For most times in the past, the Sm-Nd CHUR curve is uncertain by about one epsilon unit (explained below).

The La-Ce CHUR evolution is poorly defined at this point. The upper line of Table 1 contains $^{138}Ce/^{136}Ce$ derived from the Ce-Nd isotopic correlation in oceanic basalts (Fig. 4; Dickin, 1987b). The lower line for La-Ce contains present-day CHUR $^{138}Ce/^{142}Ce$ ratios for meteorites (Shimizu et al., 1984, 1988). The La/Ce ratio of chondritic material is well known. However, the large uncertainty in both decay constants of ^{138}La precludes construction of a reliable CHUR curve. Until the ^{138}La decay constants are better known, studies of old samples and planetary evolution using Ce isotopes can only yield ambiguous results. Ce isotopic variations are discussed here because they reaffirm the coherence of REE abundance patterns and isotopic evolution in a general way. Because of the branched decay of ^{138}La, the equation for Ce isotopic evolution is more complex than equation (1) here. Should the ^{138}La decay constants become better defined, the equation for Ce isotopic evolution given by Shimizu et al. (1988) will be useful in geologic studies.

Lu-Hf CHUR evolution is better known than La-Ce, but less well defined than Sm-Nd. The $^{176}Hf/^{177}Hf$ of Solar System material was derived from an isochron for eucrite meteorites, because chondrites have low Hf concentrations and yielded $^{176}Hf/^{177}Hf$ with large uncertainty. Even the eucrite isochron has an uncertainty of three epsilon units on its initial $^{176}Hf/^{177}Hf$, because very low Lu/Hf rocks or mineral phases were not available. The Lu/Hf ratio was selected from chondrite data that varied by 10% (see Patchett, 1983). It is probably a good choice, however, in that it represents C1, C2 and C3 carbonaceous chondrites to less than 1% (Beer et al., 1984). Overall, the Lu-Hf CHUR curve has an uncertainty at all times of three epsilon units, due to the error on the eucrite meteorite initial $^{176}Hf/^{177}Hf$. To a large extent, mutual corroboration of the Sm-Nd and Lu-Hf curves by old and present-day isotopic variations on the Earth suggests that 3 ε-units is a very conservative error estimate for the Lu-Hf CHUR curve, at least as far as the Earth is concerned.

Because of the difficulty of different data presentations and the existence of well-defined undifferentiated reservoir evolution, it is common to quote Nd, Ce and Hf isotopic data as epsilon values, which are deviations in parts in 10,000 from CHUR evolution. The example given is again Nd:

$$\varepsilon^t_{Nd} = 10^4 \times \left(\frac{(^{143}Nd/^{144}Nd)^t_{SAMPLE} - (^{143}Nd/^{144}Nd)^t_{CHUR}}{(^{143}Nd/^{144}Nd)^t_{CHUR}} \right) \qquad (2)$$

The CHUR value at t comes from Equation 1. Clearly the same basis must be used for ε_{Nd} calculation as was used for the original Nd data, and if this is done, the different numerical bases of reporting Nd results become unimportant to the general reader. Deviations from bulk planetary evolution are in any case more informative as to chemical evolution. There is an exactly similar definition of ε_{Ce} and ε_{Hf}, with the proviso that ε_{Ce} can presently only be sensibly calculated at the present day, because of the uncertain decay constant.

30

ISOTOPIC STUDY OF PLANETARY INTERIORS

Both on the Earth and the Moon, Nd and Hf isotopes have made substantial contributions to understanding of the evolution of planetary interior mafic and ultramafic rocks. Although some ultramafic rocks have been analyzed directly, when available as xenoliths or upthrusted massifs, the main principle has been to use basalts as samplers of the isotopic composition of their sources. At melting temperatures, diffusion data suggest that equilibration between solid and melt is quite rapid.

The Moon

Apart from the contribution of Sm-Nd to lunar chronology discussed earlier, Nd and Hf isotopes have been extremely important in corroborating and developing models of lunar differentiation (Nyquist et al., 1979, 1981; Unruh et al., 1984; Fujimaki and Tatsumoto, 1984). In particular, it has become clear that mare basalts were derived by partial melting of layered cumulates produced from a magma ocean or series of major pools, that also gave rise to the lunar crust 4.2-4.4 Ga ago. The Sm/Nd and Lu/Hf data can only be satisfied, as is the case for other chemical parameters, such as Ti content, if the mare basalts represent low-degree partial melts of cumulates derived from only the later stages of magma ocean crystallization. Although several problems, above all the question of elemental mass balance in the differentiated Moon, remain to be evaluated, the isotopic data appear to point to a self-consistent picture of lunar evolution.

The Earth

Nd, Hf and Ce isotope data for mantle-derived terrestrial basalts are summarized in Figures 2, 3 and 4. Nd data come from many sources, and have been reviewed by White (1985) and Zindler and Hart (1986). Hf isotope data have been reviewed by Patchett (1983), and Figure 3 contains additional data from Stille et al. (1983) and White and Patchett (1984). Ce isotope data are from Dickin (1987b) and Tanaka et al. (1987).

The demonstration of a reasonable negative correlation of Nd and Sr isotopes in mantle-derived samples (Richard et al., 1976; De Paolo and Wasserburg, 1976; O'Nions et al., 1977) was extremely important for understanding mantle evolution. This was because Sr and Pb isotopes are poorly correlated in oceanic basalts, and it had never been clear how the isotope geochemistry related in a simple way to progressive chemical differentiation of the planet. In particular, the depleted mantle (DM) source of mid-ocean ridge basalts (MORB) could now be modeled as the high Sm/Nd-low Rb/Sr residue of continental crust extraction (O'Nions et al., 1979; Jacobsen and Wasserburg, 1979; De Paolo, 1979). This modeling resulted in the demonstration of a "mean age" of around 2 Ga for differentiation of the primitive mantle into complementary continental crust and DM reservoirs. This result would mean that approximately half of the Earth's continental crust was produced before 2 Ga, and half after 2 Ga. The result seems to be approximately consistent with the actual Nd crust extraction age pattern observed in some continents, discussed in the next section of this chapter (see Fig. 6). Another important result of modeling continental crust and depleted mantle reservoirs was the inference that the continental crust was produced from less than the whole mantle. The calculation is rather uncertain because whereas element ratios, such as Sm/Nd, are quite well known for the primitive mantle from CHUR, absolute concentrations of elements like Sm and Nd are only defined within a factor of two or so. Other uncertainties, in the continental crustal abundances for example, mean that the proportion of the mantle involved in producing the crust could vary from 30% to 90% of the total (Allègre et al., 1983). Isotope geochemists nevertheless accept that some less-than-total proportion of the mantle produced the continental crust. The remaining non-DM fraction of the mantle could be primitive mantle, or mantle that is less depleted by virtue of having had crustal materials recycled back into it.

Thus the interpretation of the MORB-producing DM source appears straightforward. It would seem that the nature and chemical history of mantle sources other than DM

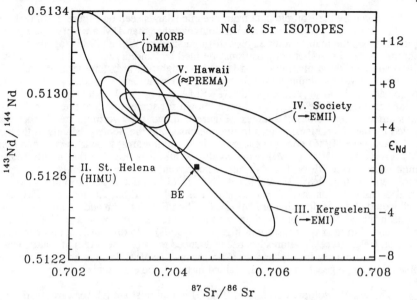

Figure 2. The mantle array for Nd-Sr isotopes in oceanic basalts. Roman numerals refer to source types of White (1985). In parentheses are shown the approximately corresponding end members of Zindler and Hart (1986). The acronyms used by White (1985) and Zindler and Hart (1986) are explained in the text. BE = bulk Earth based on chondrites for Sm/Nd, and an inferred $^{87}Sr/^{86}Sr$ ratio.

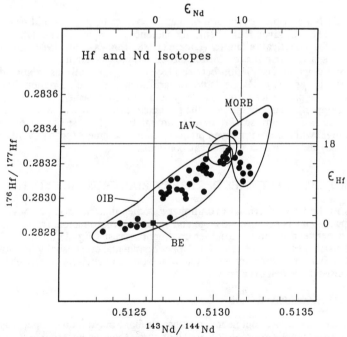

Figure 3. The mantle array for Hf-Nd isotopes in oceanic basalts, with data points. MORB = mid-ocean ridge basalts; OIB = all ocean island basalts, not divided as in Figure 2; IAV = intra-oceanic island-arc volcanics. IAV data points are not individually shown. Data from Patchett (1983) and references therein, White and Patchett (1984) and Stille et al. (1983). BE = bulk Earth based on chondrites.

would be given by the isotopic data from oceanic islands seen on Figure 2. In fact, however, the nature and origin of all mantle sources other than DM are very controversial. Figure 2 shows the mantle source types distinguished by White (1985) together with the groupings or "end-member compositions" defined by Zindler and Hart (1986). Only the MORB source (DMM of Zindler and Hart) is straightforward. Zindler and Hart (1986) distinguish a Nd-Sr isotopic signature, typified by Hawaii and even many MORB, which they consider to be a "prevalent mantle" (PREMA on Fig. 6). For the high $^{87}Sr/^{86}Sr$, low $^{143}Nd/^{144}Nd$ end of the array, it has been known for some time that there are diverging trends among oceanic basalt sources. Figure 6 shows the two main trends that occur, typified by the Kerguelen samples in the one case, and the Society Islands in the other. Zindler and Hart (1986) consider these trends as mixing arrays towards enriched mantle (EM) reservoirs EMI and EMII respectively. While it is possible that undifferentiated mantle sources (see BE on Fig. 6) figure as contributors to ocean islands, most workers regard trends to high $^{87}Sr/^{86}Sr$ and low $^{143}Nd/^{144}Nd$ as reflecting contributions from subducted oceanic or continental crustal materials (White, 1985; Zinder and Hart, 1986), which would lie in the direction of EMI and EMII on Figure 6. An additional complexity in the ocean island Nd-Sr isotopic data are the St. Helena-type results (called HIMU by Zindler and Hart after their history of high U/Pb ratios). No fully satisfactory explanation for the St. Helena type signatures has been advanced to date. The origin of mantle isotopic variations, particularly for higher-$^{87}Sr/^{86}Sr$, lower-$^{143}Nd/^{144}Nd$ sources, remains a major problem in isotope geochemistry, highlighted further at the end of this chapter.

The Nd-Hf isotopic correlation (Fig. 3) is the most straightforward, testifying to coherent Sm/Nd versus Lu/Hf fractionation in mantle processes. There does appear to be an excess variation for Hf isotopes in MORB sources, which is not understood. Whatever extra information, about subducted components for example, is given by the discordances in plots involving Sr isotopes is lost in the Nd-Hf diagram, where a single mantle array is seen.

The Ce-Nd isotopic covariation of mantle sources (Fig. 4) should probably be at least as good as Nd-Hf, given REE chemical coherence. Samples that deviate for ε_{Ce} have been disputed for analytical reliability (Dickin, 1988; Tanaka et al., 1988). Figure 4 also shows the ε_{Nd}-ε_{Ce} variation for continental crustal samples, where it is clear that: (1) there is a very good negative correlation between ε_{Nd} and ε_{Ce}, exactly as predicted by REE patterns (Fig. 1), and (2) that Ce isotope variations are much smaller than for Nd, consistent with the properties of the La-Ce system described earlier. The Ce-Nd isotopic correlation is important corroboration of the predicted relationships between chemical fractions of different REE in terrestrial reservoirs through time. The use of Ce isotopes beyond this general level will require improved precision in measurement (probably difficult to the degree required) and improved ^{138}La decay constants.

Nd ISOTOPES IN STUDIES OF TERRESTRIAL CRUSTAL EVOLUTION

It is in the field of crustal evolutionary studies that Nd isotopes have made the greatest impact. This is because the large and somewhat reproducible fractionation of the Sm/Nd ratio between mantle and upper continental crust (Fig. 1) allows rudimentary "chemical crustal ages" to be determined. Upper continental crust usually has Sm/Nd only 55% that of CHUR, and probably ~40% that of the upper mantle, much of which has been depleted by crust extraction over time.

Model Nd ages of continental crust

Because most upper continental crust has similar Sm/Nd ratios, upper crust tends to evolve away from mantle sources at a constant angle towards negative ε_{Nd} (see Fig. 5). This allows determination of model chemical ages for crust. Such a model Nd chemical age could represent the time that a segment of crust was separated from the mantle. For more complex histories, it would just be an indication of the average mantle-separation age

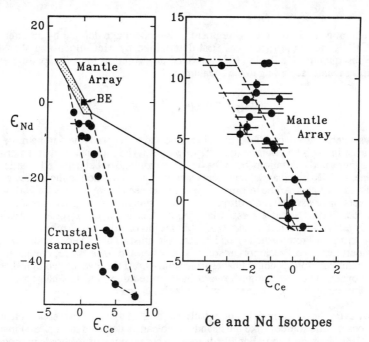

Figure 4. Ce isotopic data for oceanic and continental rocks plotted against Nd isotopes. Data from Dickin (1987b) and Tanaka et al. (1987). BE = bulk Earth based on chondrites.

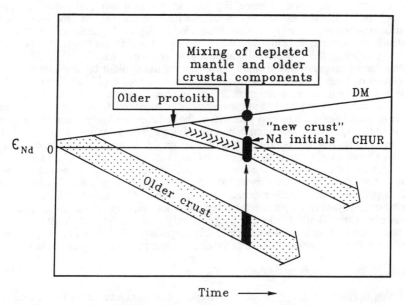

Figure 5. Two competing interpretations for ε_{Nd} values that lie below the evolution of depleted, high-Sm/Nd, upper mantle (DM). Such values could have been derived by producing crustal protoliths some 200-400 Ma earlier than the final crustal stabilization age, or by mixing a small component of much older crust into the terrane. Time scale is arbitrary. CHUR = chondritic uniform reservoir, or bulk Earth isotopic evolution, defined as $\varepsilon_{Nd} = 0$ at all times.

of all the components in the crustal segment. The model age defined by an intersection with the CHUR curve (T_{CHUR}) was first determined by McCulloch and Wasserburg (1978). Subsequently, De Paolo (1981) defined an evolution for incompatible-element depleted upper mantle according to the equation:

$$\varepsilon_{Nd}^{DM,t} = 0.25t^2 - 3t + 8.5 \tag{3}$$

This allows determination of a T_{DM} model chemical age, generally held to be more meaningful, at least in post-2.5 Ga time, than T_{CHUR} ages. The rationale is that the crust was produced by repeated extractions from the depleted mantle, so that depleted mantle, rather than pristine, CHUR-like mantle, represents a better approximation to the source of continental crust in any post-Archean time. T_{DM} ages exceed T_{CHUR} ages by 200-400 Ma. Other workers have used slightly different models, such as T_{CR} ("crustal residence") of Goldstein et al. (1984), that are essentially equivalent to T_{DM}. T_{DM} ages distinguish crustal ages easily where differences >500 Ma are involved, as in Archean versus mid-Proterozoic crust, or Proterozoic versus Phanerozoic crust. This is very important because granitoids derived from Precambrian sources may be, and usually are, chemically very similar to granitoids from younger Phanerozoic sources. At the level of geological eras, the T_{DM} age concept is quite straightforward in interpretation, and has had an enormous impact. Some examples will be described later in this chapter.

However, the T_{DM} model is only truly valid as an age for crust created from contemporaneous mantle sources (like the island arc volcanics depicted in Figure 3) in a series of events <100 Ma in duration. For all situations where mixing of materials in orogenesis, or intermediate-age crustal melting events may be involved, ambiguities arise (Arndt and Goldstein, 1987). Extrapolating the Sm/Nd ratio of a granitoid through the time of magmatism to some distinctly older T_{DM} age is an elementary example of what is not acceptable, because the intracrustal melting process probably involved Sm/Nd change; usually the source will have had higher Sm/Nd than the granitoid magma produced. There are additional problems of interpretation, illustrated in Figure 5. Initial ε_{Nd} values for newly-stabilized crust that lie below the DM curve (Fig. 5) could arise from either (a) older protoliths, predating the crust stabilization by some time, usually 100-500 Ma, or (b) mixing during orogenesis of a small proportion of much older crust together with a major fraction of juvenile arc-like material. Both models, seen on Figure 5, seem highly plausible in a variety of cases. Older protoliths make sense in terms of primitive island arc-like terranes accreted to continents and then reprocessed/remelted during orogeny. Mixing makes sense because of the transport potential of the sedimentary system, and the delivery of pelagic sediments to orogenic continental margins by plate tectonics. Other evidence must be used to resolve the interpretation. For example, a small proportion of 2.9 Ga zircons in sediments and igneous rocks of a 1.9 Ga crustal terrane would argue for mixing in of Archean crustal material. On the other hand, the presence of rocks in the same terrane with zircon ages of 2.2 Ga would argue for older Proterozoic protoliths. For most data sets, this important discriminating evidence is not available. Consequently, although T_{DM} and related model ages are very powerful at the 500 Ma level, they are difficult to interpret unequivocally at the 50-300 Ma level. Unfortunately, it is the 50-300 Ma time scale in which most plate-tectonic cycles take place. T_{DM} ages nevertheless are often given as a means of comparing samples, even though the detailed interpretation may be unclear.

Growth curves for the continental crust

Using Nd isotopic data it becomes possible to construct a curve for the cumulative age of separation of the continental crust from the mantle. This is superior to a curve for terrane age alone from conventional geochronology, because orogenic terranes often incorporate major volumes of older rocks. These older rocks can be distinguished by Nd isotopic study. Sufficient information is only available for North America and Europe (Nelson and De Paolo, 1985; Patchett and Arndt, 1986) and Australia (McCulloch, 1987).

These data are shown on Figure 6. The curves are not comparable in detail because Patchett and Arndt used the orogenic mixing model for T_{DM} ages intermediate between known events, whereas McCulloch used the older protolith model. The curves would also be modified a little by data obtained since their publication. At a gross level, the curves are similar, but there are also major differences. There was more pre-3.0 Ga crust in Australia, but a greater addition 3.0-2.7 Ga in North America-Europe. There was major Proterozoic growth in both continental masses, but more late Proterozoic and Phanerozoic growth in North America-Europe. Beyond this general level, the interpretation problem detailed in the previous section intervenes, and comparisons at any greater detail are not possible. The differences between the North America-Europe and Australia curves certainly underline that age patterns for continents cannot be assumed or extrapolated from these data, but must be constructed individually with as much data as possible.

Origin of granitoids

Nd isotopic studies have made a major impact on granitoid petrogenesis. This is because granitoids of widely differing origin may have very similar geochemical and petrologic characteristics. Sr and Pb isotopes gave petrogenetic information, but Nd yields more information about true rock masses involved in the source region. Classic cases where the origin of granitoids have been studied involve a worldwide survey (Allègre and Ben Othman, 1980), the Scottish Caledonian (Halliday, 1984; Frost and O'Nions, 1985), the Lachlan Belt of Australia (McCulloch and Chappell, 1982), and the western USA (De Paolo, 1981; Farmer and De Paolo, 1983). Data from Farmer and De Paolo (1983) are depicted in Figure 7, where all samples are of Mesozoic or Tertiary age. Only certain granitoids in the most western USA ("eugeocline") are like intraoceanic island arcs in Nd and Sr isotopes. More inboard plutons have lower ε_{Nd} and higher $^{87}Sr/^{86}Sr$, and many samples could easily represent 100% Proterozoic crust, or the presence of Archean crustal components. A distinctive signature is obtained for samples from further inboard, "craton" on Figure 7. These have Precambrian Nd isotopic signatures, but low $^{87}Sr/^{86}Sr$; this is thought to be characteristic of old, Rb-depleted lower continental crust (Farmer and De Paolo, 1983).

Nd isotopes and the sedimentary system

A fruitful field of Nd isotopic study has been sediments 3.8-0 Ga in age. Key papers are Goldstein et al. (1984), Allègre and Rousseau (1984), Miller and O'Nions (1984) and Michard et al. (1985). From these and other studies it has become clear that Archean and early Proterozoic sediments often have model Nd ages close to their stratigraphic age. On the other hand, later sediments deviate increasingly so that the average mantle separation Nd age of today's sedimentary system is 1.5-1.8 Ga. Figure 8 plots stratigraphic age against the T_{CR} age used by O'Nions and co-workers, which is similar in concept to T_{DM}. It is therefore clear that as geologic time passed, the average age of exposed crust became progressively greater, as shown by the increasing separation between the sediment data and the $T_{CR} = T_{STRAT}$ line of Figure 8. The data probably also indicate substantial cannibalistic recycling of older sediments into younger sediments.

Characterization of whole crustal terranes

Constructing growth curves for whole continents requires that individual orogenic terranes that make up continents be characterized isotopically. This involves analysis of representative rock types making up the whole terrane. As such, this type of study is a natural integration of granitoid and sediment studies already mentioned.

The utility of Nd isotopes in interpretation of orogenic belts and rock terranes can be illustrated by some straightforward examples. In a study of French sediments, Michard et al. (1985) found generally negative ε_{Nd} values, corresponding to Proterozoic T_{DM} ages, just like the average sediments of Figure 8. There were brief maxima of ε_{Nd}, however, up to near zero, that corresponded in time to major European orogenic events at 550, 450 and

36

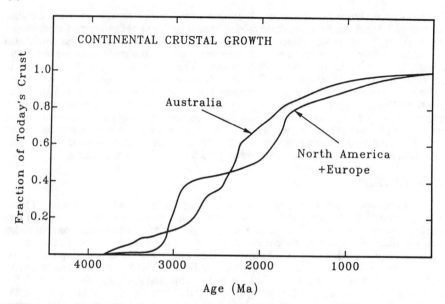

Figure 6. Approximate crustal growth curves for two large continental masses based on Nd isotopic data for crustal age and origin (Patchett and Arndt, 1986; McCulloch, 1987).

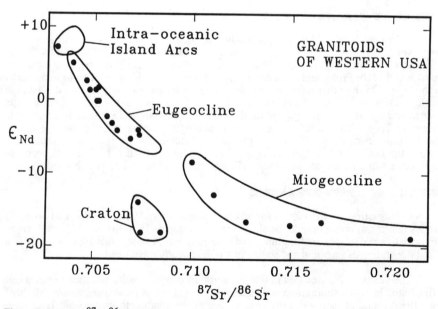

Figure 7. ϵ_{Nd} and $^{87}Sr/^{86}Sr$ values observed in granitoid plutons of Mesozoic and Tertiary age in the western U.S.A. (Farmer and De Paolo, 1983).

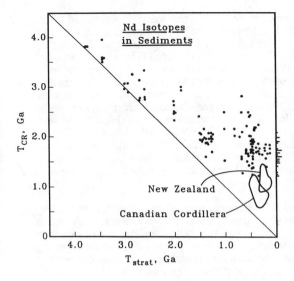

Figure 8. Stratigraphic age versus Nd crustal residence age for worldwide sediments, adapted after Frost and Coombs (1989). Two fields occupied by sediments from young circum-Pacific orogenic terranes are shown (Frost and Coombs, 1989; Samson et al., 1989).

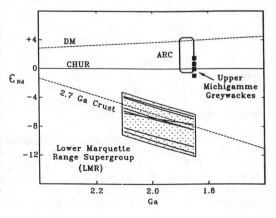

Figure 9. Nd isotopes used to discriminate sediment sources and constrain tectonic/ accretionary events. The lower Marquette Range Supergroup (LMR) sediments were derived from Archean basement to the north of Lake Superior. The ϵ_{Nd} values for the LMR sediments are shown only as evolutions through time, because the depositional age could lie anywhere between 2.1 and 1.85 Ga. All the evolution lines for LMR lie quite close to the evolution of typical Superior Province 2.7 Ga crust. In contrast, the overlying and overthrust Michigamme greywackes were derived from a juvenile arc-like terrane of 1.85-1.90 Ga age that accreted from the south around 1.85 Ga ago. The field labeled "ARC" shows the distribution of ϵ_{Nd} values for the arc-like rocks, known as the Wisconsin Magmatic Terrane. All data from Barovich et al. (1989).

300 Ma. These would a priori be expected to be times when material was added to the crust from the mantle. In a study of Proterozoic crust in the Lake Superior region, Barovich et al. (1989) found mainly Archean signatures for Marquette Range Supergroup sediments on the southern margin of the Archean Superior Craton (Fig. 9). To the south lies the juvenile island-arc-like 1.9 Ga Wisconsin Magmatic Terrane ("ARC" on Figure 9), also characterized in the study. The remarkable aspect of the sediment data was that the uppermost sediments resting on the Superior Craton, a widespread turbidite sequence called the Michigamme Formation, had Nd isotopic signatures totally different to underlying sediments. They had positive epsilon Nd at the time of deposition (Fig. 9), and were exactly similar to the Proterozoic volcanic rocks of the Wisconsin Magmatic Terrane ("ARC" on Figure 9). Clearly the results show that the Michigamme turbidites must have come from the arc to the south after it had been accreted to the Archean Superior continent (Fig. 9). The northward transport probably includes a tectonic thrusting component in addition to sediment movement.

The above examples, coupled with the granitoid studies like Figure 7, pave the way for detailed use of Nd isotopes in characterizing individual terranes as a whole, as well as documenting terrane accretion events. Several studies characterizing Precambrian terranes have been referenced already. Another developing area of research is study of young orogenic belts to discover how much juvenile crust they contain. An additional advantage over Precambrian studies is that low metamorphic grades, preserved tectonic contacts, etc. enable the Nd isotopic data and tectonics to be much better integrated. Figure 8 shows two sets of sediment data from young orogenic belts, South Island, New Zealand (Frost and Coombs, 1989) and the Alexander and Stikine Terranes of the Canadian Cordillera (Samson et al., 1989). Both sediment data sets are part of terrane-characterization studies. Both data sets show substantially lower Nd model ages than average worldwide sediment (Fig. 8), documenting Phanerozoic crustal additions in these regions. Analysis of terranes in young orogenic belts will eventually yield much information on tectonic/sedimentary processes and Phanerozoic crustal growth.

CRUSTAL Lu-Hf ISOTOPIC STUDIES

The Lu-Hf system differs from Sm-Nd in two ways: (1) Lu/Hf fractionations exceed those of Sm/Nd by a factor of more than two (Fig. 1), and (2) there are special Lu/Hf fractionations in crustal environments due to trace minerals, particularly zircon with 1% Hf. As far as the increased Lu/Hf fractionation in mantle melting is concerned, it must be stated that Lu-Hf isotopic analyses are much more difficult to carry out than Sm-Nd analyses. Furthermore, the Hf-Nd isotopic correlation in oceanic basalts is the simplest of all mantle isotopic arrays, with the only possible anomaly being the wider variation in MORB sources for ε_{Hf} (Fig. 3). For these reasons, Hf isotopic data are definitely redundant in many situations, because a good correlation with Nd data could be assumed based on existing correlations. In any situation where Hf isotopic data would yield the same conclusion as Nd data, the Lu-Hf determinations would clearly be unnecessary.

Lu-Hf analyses would be worthwhile in cases where some additional or alternative information were available that could not be obtained from Sm-Nd results. The potential for extraordinary Lu/Hf fractionations in crustal melting due to variable behavior of zircon remains largely uninvestigated. Two research areas where Hf can make a special contribution have been developed however.

The first of these is direct analysis of zircons from igneous rocks, that have been dated by U-Pb. The availability of zircon fractions from U-Pb studies was exploited by Patchett et al. (1981) and Smith et al. (1987) to obtain well-dated initial ε_{Hf} for the igneous host rocks. There are risks in this approach associated with addition of radiogenic Hf to the zircon during later overgrowths. This means that only U-Pb concordant or near-concordant zircons are acceptable samples. Both the cited papers added substantial information on Precambrian ϵ_{Hf} values for well-dated samples that were relatively easy to analyze for Lu-Hf.

   Another special application of Lu-Hf is shown in Figure 10. Patchett et al. (1984) found that whereas sedimentary processes do not fractionate Sm/Nd by more than a few percent in major sediment types, Lu/Hf varies by a factor of up to 60. This is due to the high mechanical strength and chemical resistance of zircon, with 1% Hf, but much lower REE. While REE-bearing minerals are broken down and the REE carried to the ocean in particulates or solution, Hf is retained in sandy sediments as zircon. The REE are precipitated into clays and manganese nodules, causing high REE/Hf ratios in many deep-sea sediments. The Hf finds its way to the deep ocean only as turbidite sands (Fig. 10). This lack of crustal Hf in the deep marine environment enabled White et al. (1986) to identify a component in Hf of Mn nodules that had been derived from hydrothermal activity at mid-ocean ridges. Patchett et al. (1984) also used the spectacular REE versus Hf fractionation in the sedimentary system, coupled with the coherent Hf-Nd mantle array to set constraints on possible sediment subduction into the mantle. They argued that if the ocean island Hf-Nd mantle array was due to mixing of ancient subducted sediments with depleted mantle, then average pelagic sediment would lead to an array with lower ϵ_{Nd} and would not be a possible end member. Only a roughly 1:1 mixture of average pelagic sediment with turbidite would satisfy the observed mantle array. Essentially the fractionated sediments would have to be mixed back together again before becoming part of any mantle basalt sources. This might be achievable for small degrees of sediment involvement, but is highly implausible for massive amounts of sediment because major accumulations of turbidite in today's oceans are at very different locations from where pelagic sediments might be undergoing subduction. The implication is that major subduction of sediments does not contribute to oceanic basalt magma sources. Further studies connected with Hf and zircons in the sedimentary system are in progress. Lu-Hf studies of crustal rocks will continue to concentrate on areas where a special contribution can be made over and above the information available from Sm-Nd studies.

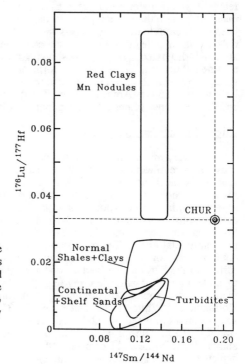

Figure 10. Lu/Hf versus Sm/Nd variation in the sedimentary system. Whereas Sm/Nd displays only minor changes, Lu/Hf has been fractionated by up to a factor of 60. This is caused by the behavior of the mineral zircon, which contains 1% Hf, is highly resistant and retained in sandy sediments (data from Patchett et al., 1984).

MAJOR UNSOLVED PROBLEMS

This section concentrates on three areas of uncertainty in terrestrial isotope geochemistry that represent major unsolved problems which will require solution if understanding of Earth evolution is to improve.

Continental crustal growth curve

Further data are needed to establish the age pattern of continental crust production, both to improve the existing North American, European and Australian curves, and develop curves for other continents. This will be achieved through Nd isotopic studies carefully integrated with geologic and geochronologic studies. Apart from global geochemical modeling, there is a major spin-off from this information into tectonics, and in particular how the plate-tectonic process has operated through geologic time.

Abundance of Archean continental crust

Modeling of crustal extraction from the mantle has certainly reached the limit of the available data base (Allègre et al., 1983; Albarède and Brouxel, 1987). It will be updated by improved crustal growth curves as they become available. Ultimately, however, an underlying question is whether there were in the past major amounts of continental crust that have been subducted into the mantle (Armstrong, 1981). In particular this question takes the form of how much pre-3.0 Ga and 3.0-2.5 Ga crust ever existed. Further calculations will not resolve this problem, and real data constraints must be found. Recent discussions have centered around two issues. One is the fact that Nd isotopes in mantle-derived rocks 3.8-2.5 Ga in age show that they were derived from depleted-mantle-like sources. This was thought to argue a priori that large amounts of long-lived continental crust must have existed 4.0-2.5 Ga ago, in order to have produced the complementary mantle depletion. Chase and Patchett (1988), however, claim that continental crust is not required, and that storage of oceanic crust is a much more likely candidate to cause upper mantle depletion than is storage of continental crust. A second major line of discussion has been the attempt to disprove large-scale subduction of sediments or any continental materials into the mantle by pointing out anomalies in the geochemistry of sediments (Patchett et al., 1984, discussed above), or by showing how certain incompatible-element ratios limit continental material in the mantle (Hofmann et al., 1986). These arguments, while urged strongly by the authors referenced, can probably be circumvented in various ways, particularly for pre-2.5 Ga (Archean) time. The questions of massive crust-to-mantle recycling and the abundance of pre-2.5 Ga continental crust must be regarded still as open issues.

Origin of mantle isotopic variations

A very much related question is the origin of mantle isotopic variations. Figure 2 shows how reviewers of the database have identified several components, possibly representing chemical end-members in mantle mixing processes. To keep the discussion simple, Figure 11 shows the most straightforward mantle isotopic correlation, Nd versus Hf, together with all the hypotheses advanced to account for such arrays. The possible explanations of mantle isotopic variation are essentially the same, whether Nd, Hf, Sr or Pb isotopes are considered. In all cases, the depleted mantle is the chemical complement of the continental crust. The discussion centers around the origin of the array of isotope compositions for ocean island sources, and how much material recycled from near the surface of the Earth is involved in producing them. Origin (1) was the first to be advanced to explain the Nd-Sr mantle array (e.g., De Paolo, 1979): here no recycling occurs, and the lowest-ε_{Nd} sources are dominated by primitive mantle. Origin (2) was developed by Hofmann and White (1981) and Chase (1981): it envisages that most oceanic island sources are dominated by subducted oceanic crust (hence "recycled MORB" of Fig. 11), stored in the mantle for up to 2 Ga. In origin (3), which has been discussed by several authors (e.g., White and Hofmann, 1982), the oceanic island array is caused by mixing of depleted mantle signatures with small amounts of subducted oceanic sediment ("S" on Fig.

41

ORIGIN OF MANTLE ISOTOPIC VARIATIONS

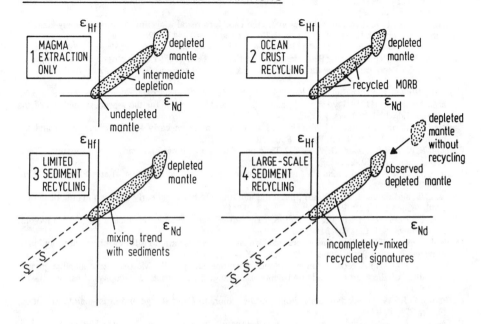

Figure 11. The range of possible hypotheses to explain the Hf-Nd isotopic mantle array, and therefore all other related arrays. See text for discussion. S = location of isotopic composition of continent-derived sediments.

11). Only 2% sediment is needed to produce the whole array. Origin (4), developed by Armstrong (1981), differs in that sediments equivalent to the volume of the continental crust are subducted into the mantle over geologic time. In this case, not only is the oceanic-island array the result of mixing, but the depleted mantle would have been severely affected (Fig. 11). Origin (4) is the mantle corollary of the steady-state continental crust hypothesis of Armstrong (1981), where crustal growth is balanced by loss of crust to the mantle. Clearly the problem of crustal growth history and mantle isotopic variations are very much related. All four of these radically different hypotheses (Fig. 11) are, in one form or another, still possible explanations of mantle isotopic variation. Obviously, diagrams that produce more spread in components, such as Nd-Sr or Sr-Pb, may discriminate better, but the same set of explanations and arguments apply ultimately to all these mantle isotopic arrays. if consensus is worth anything, then a combination of origin (2) plus minor amounts of (1) and (3) are most likely. None have been finally disproved, however. Thus central questions of crust-mantle evolution on Earth remain to be resolved by future work.

REFERENCES

Albarède, F. and Brouxel, M. (1987) The Sm/Nd secular evolution of the continental crust and the depleted mantle. Earth Planet. Sci. Lett. 82, 25-35.
Allègre, C.J. and Ben Othman, D. (1980) Nd-Sr isotopic relationship in granitoid rocks and continental crust development: a chemical approach to orogenesis. Nature 286, 335-342.
Allègre, C.J. and Rousseau, D. (1984) The growth of the continent through geological time studied by Nd isotope analysis of shales. Earth Planet. Sci. Lett. 67, 19-34.

42

Allègre, C.J., Hart, S.R. and Minster, J.-F. (1983) Chemical structure and evolution of the mantle and continents by inversion of Nd and Sr isotopic data. II. Numerical experiments and discussion. Earth Planet. Sci. Lett. 66, 191-213.

Armstrong, R.L. (1981) Radiogenic isotopes: the case for crustal recycling on a near-steady-state no-continental-growth Earth. Phil. Trans. R. Soc. London A301, 443-472.

Arndt, N.T. and Goldstein, S.L. (1987) Use and abuse of crust-formation ages. Geology 15, 893-895.

Barovich, K.M., Patchett, P.J., Peterman, Z.E. and Sims, P.K. (1989) Nd isotopes and the origin of 1.9-1.7 Ga Penokean continental crust of the Lake Superior region. Geol. Soc. Am. Bull. 101, 333-338.

Beer, H., Walter, G., Macklin, R.L. and Patchett, P.J. (1984) Neutron capture cross sections and solar abundances of $^{160,161}Dy$, $^{170,171}Yb$, $^{175,176}Lu$ and $^{176,177}Hf$ for the s-process analysis of the radionuclide ^{176}Lu. Phys. Rev. C 30, 464-478.

Carlson, R.W. and Lugmair, G.W. (1979) Sm-Nd constraints on early lunar differentiation and the evolution of KREEP. Earth Planet. Sci. Lett. 45, 123-132.

Carlson, R.W. and Lugmair, G.W. (1981) Time and duration of lunar highlands crust formation. Earth Planet. Sci. Lett. 52, 227-238.

Chase, C.G. (1981) Ocean island Pb: two-stage histories and mantle evolution. Earth Planet. Sci. Lett. 52, 277-284.

Chase, C.G. and Patchett, P.J. (1988) Stored mafic/ultramafic crust and early Archean mantle depletion. Earth Planet. Sci. Lett. 91, 66-72.

Chauvel, C., Dupré, B. and Jenner, G.A. (1985) The Sm-Nd age of Kambalda volcanics is 500 Ma too old! Earth Planet. Sci. Lett. 74, 315-324.

De Paolo, D.J. (1979) Implications of correlated Nd and Sr isotopic variations for the chemical evolution of crust and mantle. Earth Planet. Sci. Lett. 43, 201-211.

De Paolo, D.J. (1981) A neodymium and strontium isotopic study of the Mesozoic calc-alkaline granitic batholiths of the Sierra Nevada and Peninsular Ranges, California. J. Geophys. Res. 86, 10470-10488.

De Paolo, D.J. (1981) Neodymium isotopes in the Colorado Front Range and crust-mantle evolution. Nature 291, 193-196.

De Paolo, D.J. (1988) Neodymium isotope geochemistry—an introduction. Springer-Verlag, New York, 187 pp.

De Paolo, D.J. and Wasserburg, G.J. (1976) Nd isotopic variations and petrogenetic models. Geophys. Res. Lett. 3, 249-252.

De Paolo, D.J. and Wasserburg, G.J. (1979) Sm-Nd age of the Stillwater complex and the mantle evolution curve for neodymium. Geochim. Cosmochim. Acta 43, 999-1008.

Dickin, A.P. (1987a) La-Ce dating of Lewisian granulites to constrain the ^{138}La β-decay half-life. Nature 325, 337-338.

Dickin, A.P. (1987b) Cerium isotope geochemistry of ocean island basalts. Nature 326, 283-284.

Dickin, A.P. (1988) Mantle and crustal Ce/Nd isotope systematics—comment. Nature 333, 403.

Farmer, G.L. and De Paolo, D.J. (1983) Origin of Mesozoic and Tertiary granite in the western United States and implications for pre-Mesozoic crustal structure. 1. Nd and Sr isotopic studies in the geocline of the northern Great Basin. J. Geophys. Res. 88, 3379-3401.

Faure, G. (1986) Principles of isotope geology, 2nd. ed. Wiley, New York, 589 pp

Frost, C.D. and Coombs, D.S. (1989) Nd isotope character of New Zealand sediments: implications for terrane concepts and crustal evolution. Am. J. Sci. 289, 744-771.

Frost, C.D. and O'Nions, R.K. (1985) Caledonian magma genesis and crustal recycling. J. Petrol. 26, 515-544.

Fujimaki, H. and Tatsumoto, M. (1984) Lu-Hf constraints on the evolution of lunar basalts. Proc. 14th Lunar Planet Sci. Conf., B445-B458.

Goldstein, S.L., O'Nions, R.K. and Hamilton, P.J. (1984) A Sm-Nd isotopic study of atmospheric dusts and particulates from major river systems. Earth Planet. Sci. Lett. 70, 221-236.

Halliday, A.N. (1984) Coupled Sm-Nd and U-Pb systematics in late Caledonian granites and the basement under northern Scotland. Nature 307, 229-233.

Hofmann, A.W. (1988) Chemical differentiation of the Earth: the relationship between mantle, continental crust, and oceanic crust. Earth Planet. Sci. Lett. 90, 297-314.

Hofmann, A.W. and White, W.M. (1982) Mantle plumes from ancient oceanic crust. Earth Planet. Sci. Lett. 57, 421-436.

Hofmann, A.W., Jochum, K.P., Seufert, M. and White, W.M. (1986) Nd and Pb in oceanic basalts: new constraints on mantle evolution. Earth Planet. Sci. Lett. 79, 33-45.

Humphries, F.J. and Cliff, R.A. (1982) Sm-Nd dating and cooling history of Scourian granulites, Sutherland, NW Scotland. Nature 295, 515-517.

Jacobsen, S.B. and Wasserburg, G.J. (1979) The mean age of mantle and crustal reservoirs. J. Geophys. Res. 84, 7411-7427.

Jacobsen, S.B. and Wasserburg, G.J. (1980) Sm-Nd isotopic evolution of chondrites. Earth Planet. Sci. Lett. 50, 139-155.

Jacobsen, S.B. and Wasserburg, G.J. (1984) Sm-Nd isotopic evolution of chondrites and achondrites, II. Earth Planet. Sci. Lett. 67, 137-150.

Lugmair, G.W. and Carlson, R.W. (1978) The Sm-Nd history of KREEP. Proc. Lunar Planet Sci. Conf. 9th, 689-704.

Lugmair, G.W. and Marti, K. (1977) Sm-Nd-Pu timepieces in the Angra Dos Reis meteorite. Earth Planet. Sci. Lett. 35, 273-284.

Lugmair, G.W., Scheinin, N.B. and Marti, K. (1975a) Search for extinct [146]Sm: 1. The isotopic abundance of [142]Nd in the Juvinas meteorite. Earth Planet. Sci. Lett. 27, 79-84.

Lugmair, G.W., Scheinin, N.B. and Marti, K. (1975b) Sm-Nd age and history of Apollo 17 basalt 75075: evidence for early differentiation of the lunar exterior. Proc. Lunar Sci Conf. 6th, 1419-1429.

Masuda, A., Shimizu, H., Nakai, S., Makashima, A. and Lahti, S. (1988) [138]La β-decay constant estimated from geochronological studies. Earth Planet. Sci. Lett. 89, 316-322.

McCulloch, M.T. (1987) Sm-Nd isotopic constraints on the evolution of Precambrian crust in the Australian continent. In: Proterozoic Lithospheric Evolution, A. Kröner, ed. Am. Geophys. Union Geodyn. Ser. 17, 115-130.

McCulloch, M.T. and Chappell, B.W. (1982) Nd isotopic characteristics of S- and I-type granites. Earth Planet. Sci. Lett. 58, 51-64.

McCulloch, M.T. and Wasserburg, G.J. (1978) Sm-Nd and Rb-Sr chronology of continental crust formation. Science 200, 1003-1011.

Michard, A., Gurriet, P., Soudant, M. and Albarède, F. (1985) Nd isotopes in French Phanerozoic shales: external vs. internal aspects of crustal evolution. Geochim. Cosmochim. Acta 49, 601-610.

Miller, R.G. and O'Nions, R.K. (1985) Source of Precambrian chemical and clastic sediments. Nature 314, 325-330.

Nelson, B.K. and De Paolo, D.J. (1985) Rapid production of continental crust 1.7 to 1.9 b.y. ago: Nd isotopic evidence from the basement of the North American mid-continent. Geol. Soc. Am. Bull. 96, 746-754.

Nyquist, L.E., Shih, C.-Y., Wooden, J.L., Bansal, B.M. and Wiesmann, H. (1979) The Sr and Nd isotopic record of Apollo 12 basalts: implications for lunar geochemical evolution. Proc. Lunar Planet. Sci. Conf. 10th, 77-114.

Nyquist, L.E., Wooden, J.L., Shih, C.-Y., Wiesmann, H. and Bansal, B.M. (1981) Isotopic and REE studies of lunar basalt 12038: implications for petrogenesis of aluminous mare basalts. Earth Planet. Sci. Lett. 55, 335-355.

O'Nions, R.K., Hamilton, P.J. and Evensen, N.M. (1977) Variations in [143]Nd/[144]Nd and [87]Sr/[86]Sr in oceanic basalts. Earth Planet Sci. Lett. 34, 13-22.

O'Nions, R.K., Evensen, N.M. and Hamilton, P.J. (1979) Geochemical modeling of mantle differentiation and crustal growth. J. Geophys. Res. 84, 6091-6101.

Papanastassiou, D.A., De Paolo, D.J. and Wasserburg, G.J. (1977) Rb-Sr and Sm-Nd chronology and genealogy of mare basalts from the Sea of Tranquility. Proc. Lunar Sci. Conf. 8th, 1639-1672.

Patchett, P.J. (1983) Importance of the Lu-Hf isotopic system in studies of planetary chronology and chemical evolution. Geochim. Cosmochim. Acta 47, 81-91.

Patchett, P.J. and Arndt, N.T. (1986) Nd isotopes and tectonics of 1.9-1.7 Ga crustal genesis. Earth Planet. Sci. Lett. 78, 329-338.

Patchett, P.J. and Ruiz, J. (1987) Nd isotopic ages of crust formation and metamorphism in the Precambrian of eastern and southern Mexico. Contrib. Mineral. Petrol. 96, 523-528.

Patchett, P.J. and Tatsumoto, M. (1980) Lu-Hf isochron for the eucrite meteorites. Nature 288, 571-574.

Patchett, P.J., Kouvo, O., Hedge, C.E. and Tatsumoto, M. (1981) Evolution of continental crust and mantle heterogeneity: evidence from Hf isotopes. Contrib. Mineral. Petrol. 78, 279-297.

Patchett, P.J., White, W.M., Feldmann, H., Kielinczuk, S. and Hofmann, A.W. (1984) Hafnium/rare earth element fractionation in the sedimentary system and crustal recycling into the Earth's mantle. Earth Planet. Sci. Lett. 69, 365-378.

Pettingill, H.S. and Patchett, P.J. (1981) Lu-Hf total-rock age for the Amitsoq Gneisses, West Greenland. Earth Planet. Sci. Lett. 55, 150-156.

Richard, P., Shimizu, N. and Allègre, C.J. (1976) [143]Nd/[146]Nd, a natural tracer: an application to oceanic basalts. Earth Planet. Sci. Lett. 31, 269-278.

Samson, S.D., McClelland, W.C., Patchett, P.J., Gehrels, G.E., and Anderson, R.G. (1989) Evidence from neodymium isotopes for mantle contributions to Phanerozoic crustal genesis in the Canadian Cordillera. Nature 337, 705-709.

Sato, J. and Hirose, T. (1981) Half-life of [138]La. Radiochem. Radioanal. Lett. 46, 145-152.

Sguigna, A.P., Larabee, A.J. and Waddington, J.C. (1982) The half-life of [176]Lu by γ-γ coincidence measurement. Can. J. Phys. 60, 361-364.

44

Shimizu, H., Nakai, S., Tasaki, S., Masuda, A., Bridgwater, D., Nutman, A.P. and Baadsgaard, H. (1988) Geochemistry of Ce and Nd isotopes and REE abundances in the Amitsoq Gneisses, West Greenland. Earth Planet. Sci. Lett. 91, 159-169.

Shimizu, H., Tanaka, T. and Masuda, A. (1984) Meteoritic $^{138}Ce/^{142}Ce$ ratio and its evolution. Nature 307, 251-252.

Smith, P.E., Tatsumoto, M. and Farquhar, R.M. (1987) Zircon Lu-Hf systematics and the evolution of the Archean crust in the southern Superior Province, Canada. Contrib. Mineral. Petrol. 97, 93-104.

Stille, P., Unruh, D.M. and Tatsumoto, M. (1983) Pb, Sr, Nd and Hf isotopic evidence of multiple sources for Oahu, Hawaii basalts. Nature 304, 25-29.

Tanaka, T. and Masuda, A. (1982) The La-Ce geochronometer: a new dating method. Nature 300, 515-518.

Tanaka, T., Shimizu, H., Kawata, Y. and Masuda, A. (1987) Combined La-Ce and Sm-Nd isotope systematics in petrogenic studies. Nature 327, 113-117.

Tanaka, T., Shimizu, H., Kawata, Y. and Masuda, A. (1988) Mantle and crustal Ce/Nd isotope systematics—reply. Nature 333, 404.

Tatsumoto, M., Unruh, D.M. and Patchett, P.J. (1981) U-Pb and Lu-Hf systematics of Antarctic meteorites. Proc. 6th Symp. Antarctic Meteorites, pp. 237-249, Nat'l Inst. Polar Res., Tokyo.

Unruh, D.M., Stille, P., Patchett, P.J. and Tatsumoto, M. (1984) Lu-Hf and Sm-Nd evolution in lunar mare basalts. Proc. 14th Lunar Planet. Sci. Conf., B459-B477.

van Breemen, O. and Hawkesworth, C.J. (1980) Sm-Nd isotopic study of garnets and their metamorphic host rocks. Trans. R. Soc. Edinburgh Earth Sci. 71, 97-102.

Watson, E.B. and Harrison, T.M. (1983) Zircon saturation revisited: temperature and composition effects in a variety of crustal magma types. Earth Planet. Sci. Lett. 64, 295-304.

White, W.M. (1985) Sources of oceanic basalts: radiogenic isotopic evidence. Geology 13, 115-118.

White, W.M. and Hofmann, A.W. (1982) Sr and Nd isotope geochemistry of oceanic basalts and mantle evolution. Nature 296, 821-825.

White, W.M. and Patchett, P.J. (1984) Hf-Nd-Sr isotopes and incompatible element abundances in island arcs: implication for magma origins and crust-mantle evolution. Earth Planet. Sci. Lett. 67, 167-185.

White, W.M., Patchett, P.J. and Ben Othman, D. (1986) Hf isotope ratios of marine sediments and Mn nodules: evidence for a mantle source of Hf in seawater. Earth Planet. Sci. Lett. 79, 46-54.

Zindler, A. and Hart, S.R. (1986) Chemical geodynamics. Ann. Rev. Earth Planet. Sci. 14, 493-571.

PARTITIONING OF RARE EARTH ELEMENTS BETWEEN MAJOR SILICATE MINERALS AND BASALTIC MELTS

INTRODUCTION

A major geochemical application of rare earth elements (REE) is in the mathematical modelling of igneous petrogenetic processes, the fundamentals of which are discussed by Hansen elsewhere in this volume. An essential conceptual tool in such modelling is the mineral/melt partition coefficient[1], generally defined as the concentration of an element in a mineral divided by its concentration in the melt in equilibrium with that mineral. More formally, the partition coefficient of element E between a solid mineral S and the molten silicate liquid in equilibrium with that mineral is defined as $D_{E,S/L} = C(E_S)/C(E_L)$, where $C(E_S)$ is the concentration of element E in mineral S and $C(E_L)$ is its concentration in the coexisting liquid.[2] Thus, because the concentration of E in the solid is normalized to its concentration in the liquid, it is generally assumed that D will be independent of the concentration of E. Elements whose concentration in a mineral are lower than those in the coexisting melt have $D_{S/L}<1$, and are referred to as being "incompatible" in that mineral, while elements whose concentrations in a mineral are higher than that in the melt have $D_{S/L}>1$, and are often called "compatible." The rare earth elements are generally incompatible in most petrogenetically important silicate minerals, but can be compatible in a few major minerals, particularly in highly evolved systems.

Usefulness of the REE for petrogenetic modelling

The usefulness of rare earth elements for petrogenetic modelling stems primarily from four factors. First, although they are generally incompatible, the values of their partition coefficients for any particular mineral vary systematically with atomic number (some by orders of magnitude), while the shapes of the partition coefficient patterns (plots of distribution coefficient versus atomic number) differ strikingly from one mineral to another. Thus, each mineral which fractionates during the series of events culminating in the formation of an igneous rock leaves its own signature on the REE pattern of that rock. By working backwards from the REE pattern of a sample, it is often possible to identify specific minerals which have participated in the petrogenesis of that sample.

Second, a useful exception to the systematic variation of REE partition coefficients with atomic number is the well-known europium anomaly. Because a significant proportion of Eu in an igneous system is present in the divalent oxidation state, rather than the trivalent state typical of other REE in igneous systems[3], values for Eu partition coefficients typically differ from those of neighboring Sm or Gd (see discussion below). The magnitude and direction of this difference varies widely from mineral to mineral, so that anomalous Eu

---

[1]The terms "partition coefficient" and "distribution coefficient" are often used interchangeably.

[2]Some workers prefer to represent mineral/melt partition coefficients with the notation K_D, while others (including me) prefer to reserve the K_D notation for the ratio of partition coefficients for two elements, i.e., the exchange coefficient for those elements.

[3]Cerium is often present as a quadrivalent ion in oxidizing marine or sedimentary environments, but is primarily trivalent at the oxygen fugacities typical of igneous systems.

abundances can be a strong diagnostic for the involvement of certain minerals in petrogenetic processes.

Third, additional usefulness of the REE, particularly in modelling the petrogenesis of large- or planetary-scale processes, stems from the fact that they minimize one of the free parameters in most petrogenetic models, the relative elemental abundances in the starting material. The REE are highly refractory, and hence the relative abundances of these elements in most undifferentiated materials from our solar system are generally assumed to be proportional to their relative abundances in the sun and in chondritic meteorites. (See the chapter by Boynton for the rationale behind this assumption, and exceptions to the rule.) This means that when we are dealing with processes involving primitive starting materials (such as the Earth's bulk mantle or the parent bodies of basaltic meteorites), we at least have a good idea of the relative REE concentrations in the starting materials. Models must explain any changes in relative REE concentrations from primitive chondritic values, but do not need to guess at relative abundances in the starting materials.

A final characteristic of REE which lends utility to their use in petrogenetic modelling is that they are involved in three important isotopic systems, Sm-Nd, Lu-Hf, and La-Ce. As a result, they can provide information not only about the nature of the mineral separations which occurred during a petrogenetic process, but also the times at which these separations occurred. REE isotopic systematics are discussed in the chapter by Patchett in this volume.

Scope of this chapter

Rather than serve as a comprehensive literature review of REE partitioning data, this chapter is intended to (1) illustrate the methods by which REE partition coefficients are measured and some of the difficulties involved in such measurements, (2) illustrate the principles which govern the partitioning of REE, (3) give examples of some geologic applications of partition coefficients, and (4) give a brief compilation of representative values for petrogenetically important minerals to illustrate the basic shapes of REE distribution coefficient patterns (plots of distribution coefficient versus atomic number). Emphasis will be on minerals important for the petrogenesis of basaltic rocks. Partitioning for late-stage minerals, which are often quite REE-enriched, and can be important in the petrogenesis of highly evolved rocks, will not be stressed.

Caveat

As other authors in this volume also emphasize, although the REE are extremely useful in petrogenetic models, they generally should not be considered alone, in the absence of other major, minor, and trace elements. A petrogenetic model can be judged completely successful only if it explains the behavior of *all* analyzed elements. To be truly credible, models must predict abundances of elements with widely varying compatibilities.

HOW PARTITION COEFFICIENTS ARE MEASURED

There are two basic approaches to measuring mineral/melt partition coefficients. One approach involves analyzing natural samples, while the other involves analyzing synthetic samples produced in laboratory crystallization experiments. Each approach has its advantages and disadvantages, and both have been widely used and are still in use today.

Phenocryst/matrix studies of natural samples

Measurement of trace element partition coefficients from natural samples has generally been done using the phenocryst/matrix approach (e.g., Onuma et al., 1968; Schnetzler and Philpotts, 1968, 1970; Philpotts and Schnetzler, 1970). In this technique, phenocrysts in volcanic rocks are mechanically separated from the surrounding matrix, and both are analyzed for trace elements using standard bulk analysis methods such as instrumental neutron activation analysis or isotope dilution mass spectrometry. To insure clean separation, this approach requires that the sample have fairly large phenocrysts set in a fine-grained or preferably glassy matrix.

Several assumptions are implicit in the phenocryst/matrix approach. First, one must assume that the mineral separations have been effective, so that the phenocryst fraction is not contaminated to a significant degree with matrix or minor trace element-rich phases (or, in the case of trace element-rich minerals, that the matrix is not contaminated with phenocrysts). For highly incompatible elements such as light REE (LREE, usually La-Sm) in olivine, it is extremely difficult to produce mineral separates sufficiently pure that the analyses are not affected by contamination (e.g., Stosch, 1982; McKay, 1986; Michael, 1988). If similar partition coefficient values are obtained for chemically diverse elements such as Ba, Sr, La, and Zr, it is likely that the mineral separates are affected by contamination with matrix or melt inclusions.

It is sometimes possible to overcome contamination. For example, Phinney et al. (1988) recently used an approach that allowed them to overcome problems with contamination of plagioclase megacryst separates by alteration products. This approach involved analysis of multiple aliquants of the same mineral separate, each having differing levels of contamination, and extrapolation back to a contamination-free composition, as determined by the Fe content of plagioclase measured by electron microprobe analysis of clean crystals. The approach was successful in this case largely because the contaminating material had high Fe/REE, so that it was possible to determine the proportion of contaminant fairly accurately from the Fe content of the mineral separate analysis. For contaminants with low Fe/REE, the Fe content of the host phase would generally not be known well enough to make an accurate extrapolation.

Another implicit assumption is that the phenocrysts were actually in equilibrium with the coexisting melt at the time of eruption. There are two situations where this assumption might be wrong. (1) The minerals could be xenocrysts, totally unrelated to the magma in which they are found. This situation is rare, but happens occasionally. It can usually be detected by comparing the major element composition of the minerals (e.g., the Fe/Mg ratio) with that predicted for crystals in equilibrium with the observed melt based on known major element partition coefficients (e.g., $^{Fe/Mg}K_{D(OL/L)}$, Roeder and Emslie, 1970). If the mineral appears to be seriously out of equilibrium for major elements, the trace element partition coefficients should be treated with skepticism. (2) The minerals could have grown from the coexisting melt too quickly to maintain bulk equilibrium, so that the cores are out of equilibrium with the matrix, even though the rims might be in equilibrium. Because bulk analytical techniques measure the average composition of the entire crystal, erroneous results will be obtained for zoned minerals. Again, this situation can often be detected by looking for major element zoning in the crystals with the electron microprobe. In some cases, especially for highly compatible trace elements whose concentrations change quickly during fractional crystallization (see Hansen, this volume) it is possible that the phenocrysts could be zoned for trace elements, but homogeneous for major elements. In this case the disequilibrium would be very difficult to detect. Alternatively, it is possible that the phenocrysts could be zoned for highly compatible major elements such as Mg in olivine, but not significantly zoned (or seriously out of equilibrium) for incompatible trace elements, because the concentration of these elements changes much more slowly than that of compatible elements during fractional crystallization. Thus, care must be taken in any phenocryst/matrix study to evaluate the validity of the assumption of equilibrium.

Because trace element partition coefficients depend on melt and mineral compositions, as discussed below, it is important that the major element compositions of these phases be measured in any phenocryst/matrix study in order that maximum benefit can be derived from the resulting partition coefficients. Partition coefficients also depend on temperature, pressure, and, for some elements, redox conditions. Unfortunately, one disadvantage of the phenocryst/matrix approach is that the temperature, pressure, and oxygen fugacity of most natural systems are generally not known, thus making it difficult to apply distribution coefficients measured for one system to other systems with a high degree of confidence that their values are applicable.

There are also some important advantages to the phenocryst/matrix approach, however. First, partition coefficients are actually measured for natural trace element concentration levels, thus making academic the question of Henry's law behavior, the lack of dependence of D_E upon the concentration of element E. This is a very important issue for experimentally determined partition coefficients as we shall see presently. Second, if partition coefficients are measured for the same rocks whose petrogenesis is being investigated, there can be little doubt that the partition coefficients are directly applicable. Issues such as the extrapolation of partition coefficient values to other compositions, pressures, temperatures, etc., are not of concern in this case. Third, using modern analytical techniques, it is possible to accurately determine partition coefficients for a large number of elements in a reasonably short time, provided the assumptions above are justified for the samples under study.

Experimental measurement of partition coefficients

Although the phenocryst/matrix approach has provided many valuable sets of partition coefficients, there are some questions it is not well suited to answer. Because of uncertainties in the temperature, pressure, and oxygen fugacity of last equilibration of phenocrysts and matrix, it is difficult to gain an understanding of the effects of these variables from phenocryst/matrix studies. Consequently, because those effects cannot be sorted out, it is nearly impossible to understand the effects of phase composition. Moreover, for some geologic problems, such as the petrogenesis of lunar mare basalts, eucritic meteorites, or calcium- and aluminum-rich inclusions in chondritic meteorites, there are no samples suitable for measuring partition coefficients using the phenocryst/matrix technique. For these and other reasons, partitioning behavior is often studied experimentally under controlled laboratory conditions.

The early 1970's saw an explosion of work in this field. Results from many of those pioneering studies are presented in the proceedings of an international conference on experimental trace element geochemistry, published in *Geochimica et Cosmochimica Acta* (1978, Vol. 42, No. 6A), while results from many other studies are described in the excellent review by Irving (1978) in the same volume.

Basic experimental approach. Although several techniques have been employed to study partition coefficients experimentally, most approaches have several features in common. Starting materials usually consist of powdered natural samples or synthetic mixtures of reagents, possibly fused to a glass or dried to a gel. These starting mixtures are placed in furnaces of various types and held under conditions of controlled temperature, pressure (often 1 atmosphere), and occasionally $f(O_2, H_2O, CO_2, SO_2,$ etc.). Experimental conditions are chosen so that a portion of the sample is present as one or more minerals, with the remainder as a melt. It is usually best to select conditions under which only a small fraction (<10%) of the sample is crystalline, because this minimizes problems associated with zoning. The sample is held under these conditions for times ranging from a few minutes to many days, in hopes that the minerals will equilibrate with the melt, then cooled quickly to "quench in" the equilibrium phase compositions. Glass and crystals in the quenched samples are analyzed by a variety of techniques to obtain partition coefficients.

Equilibrium. One of the most important concerns which must be addressed in all such studies is whether the minerals and melt were actually in equilibrium at the time the sample was quenched. It can be argued that under conditions of normal, orderly crystal growth, minerals and melts generally approach equilibrium at the interface between the melt and the growing crystal (see McKay et al., 1986a). In this case, the two major sources of disequilibrium are zoning within the crystal as a result of changes in the melt composition caused by crystal growth, and zoning within the melt, caused by formation of a boundary layer at the interface with the growing crystal. In real situations, it is impossible to prevent either of these effects from occurring, because neither the volume of crystal formed nor the growth rate of that crystal are infinitesimally small. Thus, a certain degree of disequilibrium is inevitable, and it is the experimenter's job to minimize it.

The approach I currently favor is to grow a small number of fairly large crystals (for ease of analysis; see below) but to grow them very slowly, so that the boundary layer in the melt around the growing crystal has time to equilibrate diffusively with the bulk melt. I also attempt to keep the proportion of crystals below a few percent, so that the melt composition changes very little as a result of crystal growth, thereby minimizing zoning in the crystals. Achieving these goals often involves fairly elaborate experimental run thermal histories (e.g., McKay et al., 1986a). It is desirable to have enough nuclei of the desired mineral present when cooling and crystal growth begins so that supersaturation nucleation and rapid growth are avoided, but not so many as to result in growth of many small crystals. Final cooling and crystal growth are typically at cooling rates <2°C/hr, leading to growth rates of a few μm/hr or less.

Another commonly used approach is to accept disequilibrium crystals upon initial growth, but to rely on diffusion to restore equilibrium during the course of the run. This approach works best if crystal size is small, minimizing the distances for diffusive equilibration of bulk crystals with melt, and hence the time required for such equilibration. One problem with this approach is that small crystals can cause analytical difficulties, especially for elements with partition coefficients differing markedly from 1 (McKay, 1986). Another problem is that diffusion is so slow in some minerals (e.g., plagioclase and diopside) that they will not approach equilibrium in reasonable experimental times, e.g., days or weeks.

Regardless of which approach is used to achieve equilibrium, it is important for the investigator to assess how well it has been achieved. In traditional phase equilibria studies, "reversal" experiments, where phase transformations are shown to occur in both directions in response to a slight change in experimental conditions, are generally performed to demonstrate equilibrium. For partitioning experiments, which are generally crystal growth experiments, traditional reversals (i.e., crystal dissolution) do not help. Some workers (e.g., McKay, 1977, 1986; Lindstrom and Weill, 1978; Ray et al., 1983) have used a "partitioning reversal" technique in which the equilibrium trace element concentration in the crystal is approached from both directions, by starting with crystals having both higher and lower initial trace element concentrations. Results of such an experiment by McKay (1986) are shown in Figure 1. With increasing run duration, the initial two populations of olivine, distinguished by their different apparent distribution coefficients, converge on a single population. In this case, the partition coefficient value towards which these populations converged was the same as values obtained from normal crystallization experiments of a few days duration, suggesting that olivine crystals in the normal synthesis experiments initially grew with near-equilibrium Yb concentrations.

Such experiments are probably the most rigorous demonstration of equilibrium available for trace element partitioning studies, but require solid state diffusive equilibration in the crystals, which is not practical to achieve in laboratory timescales for most minerals. Other lines of evidence often used to support close approach to equilibrium include consistent partition coefficient values for experimental runs of varying duration, reproducibility of results in several crystals in a single run and in replicate runs, and general consistency of results. When all is said and done, it is very difficult to prove equilibrium in the absence of partitioning reversal experiments, and consistency of results from one experiment to another, although not conclusive, is probably the best indication of the absence of serious problems with disequilibrium.

Percent level doping technique. The major difference among experimental partitioning techniques involves the method used to measure trace element concentrations in experimental run products. Much of the experimental partition coefficient data in the literature has been obtained through a technique commonly referred to as "percent level doping," pioneered by Drake (1972). This technique involves adding trace elements to the experimental starting composition in sufficient amounts (usually a few weight percent) that their concentrations in the synthetic melt and minerals produced during experimental runs can be measured with an electron microprobe. Because microprobe analyses are performed *in situ* on polished mounts, this approach eliminates the need for mineral separations, which can be quite difficult for the small samples and fine grain sizes typical of most experimental charges. Moreover, it permits an assessment of the degree of zoning, thus providing information about the degree of equilibrium or disequilibrium as discussed above. Finally, it allows a means of determining the substitution mechanisms by which the trace element is incorporated into the mineral, a topic which will be considered below.

Several cautions must be exercised when using this technique. Because the x-ray lines of some trace elements interfere with others, one must be careful about which elements are combined in a single sample. This is a particular problem for the REE, for which there are many overlapping x-ray lines (Fig. 2). This problem can be overcome by breaking the REE up into four groups of non-interfering elements: (Eu,Gd,Tb,Tm), (Nd,Sm,Yb,Lu), (Y,La,Ce,Pr), and (Tb,Dy,Ho).[4]

---

[4]This approach was taken by Drake and Weill (1972) in preparing a set of synthetic REE electron microprobe standards.

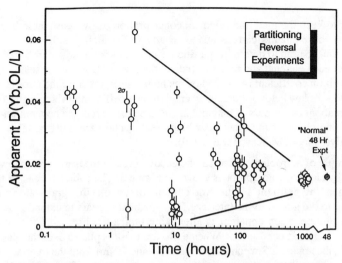

Figure 1. Reversal experiments for Yb olivine/liquid distribution coefficient. Convergence of apparent D with run duration, indicated by "trend lines," reflects approach to equilibrium through diffusion of Yb into undersaturated olivines and out of oversaturated ones. Run durations were 0.25, 2, 10, 40, 100, 200, and 1000 hrs, but points at each discreet time were spread along horizontal axis for clarity. Agreement between the average value from the 1000 and 48 hr runs indicates close approach to equilibrium in crystallization experiments. Data are from McKay (1986).

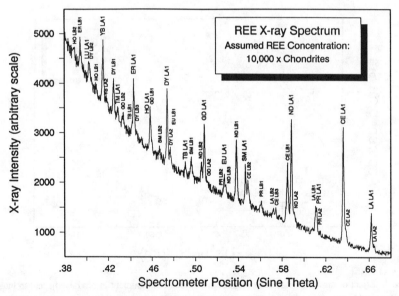

Figure 2. Synthetic X-ray spectrum of a hypothetical sample having REE concentrations of 10,000 x chondrites, as it would appear on a wavelength-dispersive microprobe spectrometer with an LiF crystal. Note the many overlapping lines, requiring that REE be added to starting materials in groups of non-interfering elements, in order for experimental run products to be analyzed by electron microprobe. Figure is adapted from McKay (1989).

In order to keep the total trace element concentration below reasonable limits, only a small number of elements can be added to any individual sample, so that multiple experiments must be run to determine partition coefficients for a large number of elements. Moreover, for each element and mineral studied, the investigator must assess the degree to which the partition coefficient is affected by the high concentration of the trace element (see section on Henry's law below). Finally, when partition coefficients which differ from unity by more than about two orders of magnitude are to be measured, the concentration will be very low in one phase or the other, in which case special care must be taken to obtain accurate analyses (e.g., McKay, 1986, 1989).

An example of a problem which can be encountered in measuring very low partition coefficients by electron microprobe is shown in Figure 3. The solid triangles in this figure show an apparent gradient in the Sm content of an olivine grain along a profile perpendicular to the glass-olivine interface. This gradient appears to extend more than 100 μm into the olivine. Such apparent gradients are typically observed for elements with extremely low partition coefficients. When a Mo foil mask is placed over the glass adjacent to the olivine, the apparent Sm concentrations at 50 and 70 μm from the interface are within uncertainty of zero, and no measurable gradient is present. These data demonstrate conclusively that the apparent gradient is an analytical artifact resulting from detection of Sm x-rays generated in glass adjacent to the olivine. It is not clear whether these x-rays are excited by secondary fluorescence, by a spray of unfocused electrons striking the glass, or by some other unidentified cause. Whatever the source of excitation, the data of Figure 3 indicate that in order to avoid resulting analytical errors when measuring low distribution

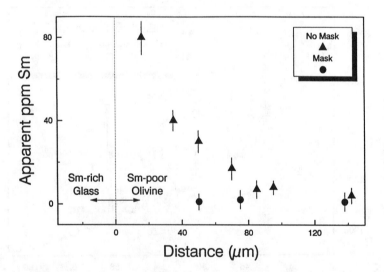

Figure 3. Effect of masking glass on apparent Sm concentration gradient in olivine adjacent to interface with glass. Triangles show apparent concentration gradient observed in olivine near interface with glass. Dots show concentration profile in olivine after glass was masked with Mo foil. Reduction in apparent Sm concentration at 50 μm from interface demonstrates that apparent concentration gradients are analytical artifacts, the Sm x-rays being generated in the glass. Olivine analyses must be performed at least 100 μm from the nearest glass to be free from this effect. Data are from McKay (1986).

coefficients an electron microprobe, it is necessary to analyze points which are more than 100 μm from the nearest glass, and workers studying very low values should take great care to avoid problems arising from this phenomenon.

Beta-track mapping technique. Another commonly used experimental approach, pioneered by Mysen and Seitz (1974), involves the determination of partition coefficients through auto-radiography of synthetic charges to which a short-lived beta-emitting radio-isotope of the element under study was added prior to the experimental run. This method allows measurement of partition coefficients at or below natural concentration levels for elements having appropriate radio-isotopes, including many elements of geochemical interest (Mysen and Seitz, 1974). Among the REE, [141]Ce, [151]Sm, and [171]Tm have received the most attention with this technique.

In principle, the technique is rather simple. The radio-isotope is usually added as a solution to the powdered starting material. After the experimental run, a polished sample of the quenched run product is held in contact with a beta-sensitive emulsion long enough to produce sufficient beta-tracks for adequate statistical measurement, but not so long as to saturate the emulsion with tracks. For partition coefficients differing strongly from 1, two exposures are often required, a short one for the high concentration phase, and a longer one for the low concentration phase. The emulsion is then developed, and the density of tracks is measured in minerals and melt, either by optical counting of tracks (e.g., Mysen and Seitz, 1974), or by microprobe analysis of the emulsion for silver (e.g., Holloway and Drake, 1977). Track densities are corrected for differences in exposure time between mineral and melt, and then ratioed to obtain partition coefficient values.

In practice, however, this technique is not as straightforward as it appears on the surface, primarily because of subtleties associated with the response of the photographic emulsion. For example, emulsion can be exposed by pressure during contact with the sample (Rogers, 1973). Also, latent images on undeveloped emulsion can fade during the course of long exposures, resulting in non-linear response to concentration (e.g., Jones and Burnett, 1981). Finally, like electron microprobe analysis, this technique is subject to adjacent phase interference when partition coefficients differing strongly from one are measured. I attempted to use the technique to investigate Henry's law behavior in olivine (McKay, 1986). I found, however, that any emulsions exposed long enough to give sufficient track density for counting in olivine showed distinct gradients in exposure density extending several tens of μm into both olivine and epoxy mounting medium, despite great care taken to ensure that sample surfaces were flat and held firmly against the emulsion. It is clear these apparent concentration gradients are artifacts, with the emulsion in olivine and epoxy being exposed by radiation originating in glass. These gradients suggest that limitations to the spatial resolution of the beta-track mapping technique are comparable to those for the electron microprobe, and that erroneously high values might result if the beta mapping technique is applied to crystals smaller than ~100 μm for measuring very low distribution coefficients. This is in agreement with a theoretical discussion of beta ranges found in Jones (1981).

Despite the analytical difficulties, the beta-track mapping technique has contributed much valuable partitioning data, and has been used to shed light (and occasionally smoke and heat) on the Henry's law problem discussed below (e.g., Drake and Holloway, 1978, 1981; Harrison, 1981; Harrison and Wood, 1980; Hoover, 1978; Mysen, 1978, 1979; Watson, 1985, Jones and Burnett, 1987).

Other experimental approaches. Although most experimental partitioning data have been produced using the percent level doping or beta-track mapping techniques, several other approaches have also been used. Traditional physical separation and bulk analysis of melt and minerals has been applied successfully in a few cases, although it has been used less commonly than *in situ* analytical techniques because of the fine-grained nature of most experimental run products. In some studies of this type, minerals were separated by conventional use of heavy liquids and hand picking (e.g., Tanaka and Nishizawa, 1975; Nagasawa et al., 1980). A slightly different approach, the differential dissolution technique of Shimizu (1974), makes use of the faster rate of solution of glass than silicate minerals in hydrofluoric acid. This technique has yielded apparently good results, although possible leaching of trace elements from the crystals during the HF treatment cannot be ruled out. A combination of physical and chemical separation techniques was used successfully by Nakamura et al. (1986) to measure REE and other partition coefficients for ilmenite. In this study, magnetically separated ilmenite crystals were subjected to additional purification by treatment with HF to dissolve any remaining glass, and final purification by careful hand-picking. Provided that successful separations of minerals and glass can be performed, mineral separation techniques have the advantage over most *in situ* techniques that partition coefficients can be measured for many elements at once, and at natural concentration levels. However, zoning of trace elements in minerals cannot be detected.

The ion probe, with its ability to analyze many elements simultaneously, *in situ*, at natural concentration levels, as well as to detect trace element zoning, has been used successfully in some studies (e.g., Shimizu, 1974; Ray et al., 1983). That use of this instrument for partitioning studies is not more widespread probably reflects lack of general accessibility more than anything else.

Another fundamentally different approach has occasionally been used to study trace element partitioning. This approach involves equilibration of crystalline and melt phases separately with aqueous vapor or molten salts, and analysis by conventional bulk techniques (e.g., Cullers et al., 1970, 1973; and Zielinski and Frey, 1974). Although the technique eliminates the problems associated with phase separations, it must rely on solid state diffusion in the minerals to achieve equilibrium, an inherently slow process for most elements and minerals.

Henry's law: The applicability of percent-level doping results. Workers employing the beta-track mapping technique (e.g., Mysen, 1978; Hoover, 1978; Harrison and Wood, 1980; Harrison, 1981; see review by Watson, 1985) have observed that distribution coefficients are constant over a wide range of trace element concentrations, extending up to the percent range. However, below a certain concentration, which seems to vary from element to element and mineral to mineral, distribution coefficients in some studies tend to increase with decreasing concentration[5]. At even lower concentrations, distribution coefficients commonly become constant again, at higher values than those observed for

---

[5]It should be noted, however, that attempts by Drake and Holloway (1981) to duplicate "non-Henry's Law" behavior of Ni partitioning in olivine reported by Mysen (1979) were unsuccessful. Jones and Burnett (1987) also failed to observe any differences in $D_{Sm,Diopside/L}$ between three sets of experiments with very different Sm concentrations: Grutzeck et al. (1974, % level doping), Ray et al. (1983, ~500-5000 ppm) and Jones and Burnett (1987, ~50 ppm). The above results, especially those of Drake and Holloway, suggest that deviations from Henry's Law behavior at low trace element concentrations reported by some workers might be influenced by subtle and poorly understood factors, or might even be an obscure experimental artifact.

greater bulk trace element concentrations. This behavior has been interpreted as indicating that entry of trace elements in minerals at very low concentrations involves defect sites, while at higher concentrations the defect sites become saturated, and the trace elements enter normal crystallographic sites (e.g., Navrotsky, 1978; Harrison and Wood, 1980).

Regardless of the reasons for apparent non-Henry's Law behavior at low-concentrations, a critical question involves whether it is the higher "Henry's law" concentration range (extending to percent levels) which is important in mineral/melt partitioning in natural systems, or the concentration range below this "Henry's law region." If it is the higher concentration region, then partition coefficients from percent level doping experi-ments are applicable to partitioning in natural systems; otherwise, they might not be.

Watson (1985) has argued convincingly that it is indeed the higher concentration Henry's law region that is applicable to partitioning in most natural systems. He reasons as follows: (1) Deviations from Henry's law behavior occur at distinctly different concentrations for different elements, even among chemically similar elements (such as the REE) in a single mineral. (2) If Henry's law were to fail for some REE, for example, but not for others, this phenomena would manifest itself in systematic variations of REE partitioning patterns for natural minerals with concentration, as opposed to the general parallelism of patterns for minerals of similar composition expected if Henry's law holds. (3) The general parallelism actually observed in natural minerals, even for elements as chemically unlike as Ba and Rb, suggests that significant deviations from Henry's law behavior do not typically occur at natural concentrations. If Watson's conclusion is correct, then the extensive body of partitioning data obtained from percent level doping experiments is applicable to partitioning at natural concentrations, and the breakdown of Henry's law behavior at low concentrations reported by some investigators, while an interesting geochemical curiosity, is not relevant to partitioning in natural samples.

FACTORS GOVERNING MINERAL/MELT PARTITIONING

This section will cover several factors which have a strong influence on the partitioning of trace elements (including the REE) between minerals and melts. In the next section, many of these factors will be considered in terms of quasi-thermodynamic treatments of partitioning which can lead to approaches for predicting partitioning behavior "from first principles".

Ionic size and charge of trace element

It has long been recognized that ionic size and charge have major effects on element substitution in crystals (e.g., Goldschmidt, 1937). One property of the trivalent REE which makes them so useful in petrogenetic modelling is that their ionic radii vary systematically with atomic number (Fig. 4)[6], while the configuration of their outer electron shells remains the same. Hence, with respect to substitution in minerals, they are virtually identical except for their ionic radii. Consequently, because of the strong influence of ionic radius on partitioning, as discussed below, their partition coefficients also vary smoothly with atomic number. The ionic radius of divalent Eu, also shown in Figure 4, is much larger than that for trivalent Eu. This distinction, combined with the difference in charge, results in large

---

[6]The effective ionic radius for any element increases with the coordination number, or number of oxygen atoms with which that ion is surrounded. Figure 4 shows ionic radii only for 8-fold coordination.

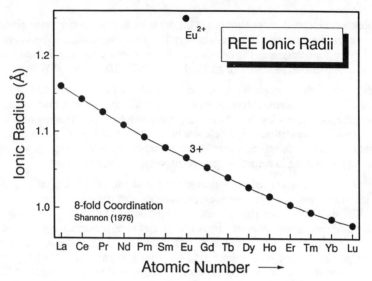

Figure 4. Ionic radius vs. atomic number for the REE. Ionic radii vary smoothly from La to Lu. Note that ionic radius depends on the coordination number; values are shown for 6-, 8-, and 10-fold coordination. Ionic radii in this and subsequent figures are from Shannon and Prewitt (1970) and Shannon (1976).

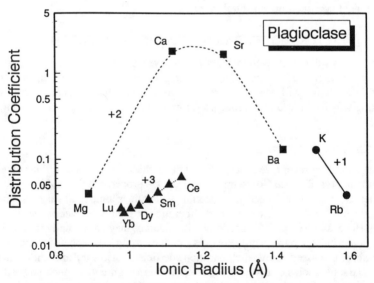

Figure 5. Plagioclase/melt partition coefficient vs. ionic radius ("Onuma" diagram) for REE and other geochemically important elements. Partition coefficients form smooth convex-upwards curves, with a different curve for each cation valence. Maxima are centered between the ionic radii of Ca (1.12 Å) and Na (1.18 Å), suggesting that large trace cations primarily substitute in the (Ca,Na) site. Partition coefficients are from Philpotts and Schnetzler (1970) and Schnetzler and Philpotts (1970).

differences between Eu^{2+} and Eu^{3+} distribution coefficients for many minerals, an important point which will be discussed below.

In one of the earliest studies of mineral/melt partition coefficients, Onuma et al. (1968) presented their data as plots of partition coefficients against ionic radii. An example of such an "Onuma diagram" for plagioclase is shown in Figure 5. Onuma et al. noted that for any particular mineral, partition coefficient values in such plots tend to lie along smooth, convex-upwards curves, with a different curve for each valence, and that each curve tends to parallel those for ions of other valences. Moreover, they noted that the maximum in these curves occur at the same ionic radius for each valence. Curves for some minerals show more than one maxima, which they interpreted as corresponding to different crystallographic sites. They concluded from their observations that (1) trace elements occupy lattice sites rather than occurring in defects or as inclusions; (2) for trace elements of the same charge, the difference between the ionic radius of the trace ion and that of the major element being substituted for is the most important factor in determining the partition coefficient; and (3) the effect of ionic charge on partition coefficients appears to be independent of the effect of ionic radius. Moreover, it is generally the case in such diagrams that the major elements occupying the site the trace elements are most likely to enter have ionic radii close to the maxima in the curves.

From the above relationships, it follows that trace element partition coefficients can be qualitatively predicted merely from inspection of a plot of valence versus ionic radius. Such a plot is shown in Figure 6, for two different coordination numbers. (Note that while there is an increase in ionic radius with coordination number, it affects most elements nearly proportionally, so that there are only minor shifts in the relative ionic radii for most elements.) Inspection of this plot suggests, for example, that partition coefficients for Sc in olivine should be greater than those for REE, because Sc is closer in size to Fe or Mg, the major structural constituents for which these elements might substitute in olivine[7]. Moreover, REE partition coefficients should decrease from Lu to La, because Lu, the smallest REE, "fits" better in olivine than any of the others. Finally, the partition coefficient for Co would be expected to be larger than that for Sc or Zr, because, although these elements all have ionic radii similar to Fe and Mg, the substitution of Co is homovalent, while those of Sc or Zr are altervalent, and require charge coupling substitutions which are energetically unfavorable. These qualitative predictions are in agreement with observed partition coefficient values, as discussed later.

Crystal field effects. An additional effect may influence the partitioning of some elements, particularly transition metals. This effect comes from the splitting of electron energy levels as a result of crystal field effects (e.g., Burns, 1970). However, the effect is not important for REE, which do not exhibit energy level splitting, and its importance in influencing the crystal/liquid partitioning behavior of even transition elements has been questioned (e.g., Philpotts, 1978; Lindstrom, 1976).

---

[7]Note, however, that sizes of trace cations relative to the major structural constituent ion are not always good indicators of relative compatibilities. In high-Ca pyroxene, for example, LREE are more incompatible than HREE, despite their ionic radii being closer to that of Ca. As discussed in a later section, this suggests that the size of the M2 site in high-Ca pyroxene does not correspond exactly to the radius of the Ca ion.

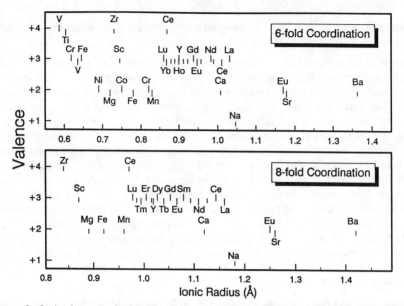

Figure 6. Ionic charge vs. ionic radius for geochemically important elements, including the REE. Partition coefficient values can be qualitatively predicted from this plot, provided the major cation for which the trace element substitutes is known. (Cr^{2+} is shown because, although uncommon in terrestrial samples, it is important in lunar samples and many meteorites.)

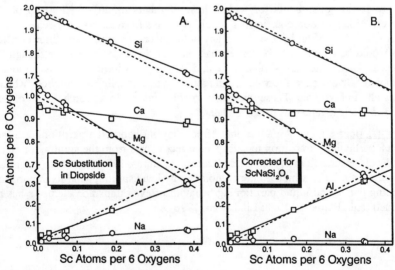

Figure 7. Variation of diopside structural formula as Sc is added in successive doping experiments. The dashed lines have slopes of ±1, and indicate the variation expected if Sc substitutes solely for Mg in the M1 site, with charge compensation through substitution of Al for Si in a tetrahedral site. (A) Actual composition as analyzed. (B) Composition after subtraction of a $ScNaSi_2O_6$ component amounting to 14.4% of the Sc present. Data are from Lindstrom (1976).

Crystallographic versus defect sites: the Henry's law question

Onuma et al. (1968) concluded that the trace elements for which they measured partition coefficients reside in crystallographic sites. This is of fundamental importance in understanding and applying thermodynamic analysis to partitioning behavior. Evidence supporting their conclusion comes both from experimental studies and from study of natural samples.

To demonstrate that partition coefficients are independent of trace element concentration in studies using the percent level doping technique, most workers conduct a series of experiments over a range of trace element "doping" levels. In most studies, for most elements and minerals, Henry's Law behavior is observed at least up to "trace" element concentrations of several percent in the melt (e.g., Drake and Weill, 1975; Lindstrom and Weill, 1978; McKay, 1986; Grutzeck et al., 1974; McKay et al., 1986b).

In addition to providing information about the upper concentration limit of Henry's Law behavior, such studies commonly provide information concerning the mechanisms by which trace elements enter minerals. By looking at variations in the major element composition of a mineral as trace element concentration increases, it may be possible to deduce the substitutions by which the trace elements enter the mineral. This approach is illustrated in Figure 7, which shows the variation in composition of synthetic diopsides as increasing amounts of Sc are added in a series of successive doping experiments (Lindstrom, 1976).

Figure 7a shows the compositions as actually analyzed. The dashed lines indicate the variations which would result from substitution of Sc for Mg in the M1 site with a coupled substitution of Al for Si in a tetrahedral site to preserve charge balance. While the variation of Mg follows the trend expected for this substitution very closely, the slopes for Si and Al are shallower than expected, and there is an also an increase in Na and a corresponding decrease in Ca with increasing Sc, suggesting that some of the charge balancing is occurring through substitution of Na for Ca in the M2 site, as a $NaScSi_2O_6$ component. When the analyses are corrected by subtracting this component in an amount sufficient to hold the Na content constant (this amount corresponds to 14.4% of the Sc present), the variations follow the expected trends much more closely (Fig. 7b).

Thus, it is clear that at the concentrations in Lindstrom's experiments, Sc predominantly enters the M1 site in diopside, while charge compensation takes place primarily through coupled substitution of Al for Si on tetrahedral sites, and, to a lesser degree of, Na for Ca on the M2 site. Similar studies of other elements and minerals have yielded analogous results: Where substitution mechanisms can be deduced from mineral analyses, trace elements invariably substitute for one or more major cations in a crystallographic site (e.g., Drake, 1972; Grutzeck et al., 1974; Drake and Weill, 1975; Colson et al., 1989a; Gallahan and Nielsen, 1989). This approach to deducing substitutions can only work, of course, if the concentration of the trace element in the mineral is significantly higher than the analytical precision for the major elements involved in the substitution. Hence, concentrations of at least a few tenths of a percent are generally required. An important question is whether these same substitutions (of trace elements for major lattice-occupying cations) occur at natural concentration levels.

As discussed above, apparent deviations have been observed from Henry's law behavior at very low concentrations (e.g., Mysen, 1978; Hoover, 1978; Harrison and

Wood, 1980; Harrison, 1981). This behavior has been interpreted as indicating that entry of trace elements in minerals at very low concentrations involves defect sites, while at higher concentrations the defect sites become saturated, and the trace elements enter normal crystallographic sites (e.g., Navrotsky, 1978; Harrison and Wood, 1980). Hence, if partitioning behavior observed for the concentration range below the high-concentration "Henry's law region" is important in mineral/melt partitioning in natural systems, then the information about trace element siting and substitution mechanisms gained from percent level doping experiments is not applicable to partitioning in natural systems. However, if Watson's (1985) conclusion is correct, that it is the higher "Henry's law" concentration range (extending to percent levels) which is important in natural systems, then this information is applicable.

In this case, (1) we can infer that trace elements generally enter crystallographic sites, making the regular dependence on ionic radius noted above understandable; (2) substitutions and charge coupling mechanisms deduced from percent level doping experiments are applicable, and this information may be used to predict the effects of mineral and melt compositional variations on partition coefficients (see subsequent sections). Finally, as a corollary of Watson's arguments, the similarity of REE partition coefficient patterns from percent level doping experiments with those from natural samples, including the location of maxima in "Onuma" diagrams at the same general ionic radius, strongly supports entry of trace elements into crystallographic sites.

Phase compositions

Because trace elements appear to enter crystallographic sites, it is reasonable to expect that mineral composition should have a strong influence on the values of the partition coefficients, especially when the size of the sites varies with composition. Among major minerals, correlations of REE partition coefficients have been reported with anorthite content in plagioclase (Drake, 1972; Drake and Weill, 1975), with Fe/Mg in olivine (McKay, 1986, Colson et al., 1988) and orthopyroxene (Colson et al., 1988), and with wollastonite content in sub-calcic pyroxene (McKay et al., 1986a).

Perhaps the most striking of these correlations is that with wollastonite in sub-calcic pyroxene. An example of this correlation is shown in Figure 8, which is based on data from the study by McKay et al. (1986a) of pyroxene/melt partitioning in a synthetic system similar in composition to the Shergotty meteorite.[8] The partition coefficients correlate strongly with wollastonite content of the pyroxene (Fig. 8a), but the variation is much steeper for the light REE than for the heavy REE. Values for La increase by a factor of about 30 as the mole fraction of wollastonite in varies from 0.1 to 0.4, while values for Lu also increase, but only by a factor of about 3. These differences in sensitivity to Ca content between LREE and HREE result in changes in the shape of the distribution coefficient pattern as pyroxene composition varies (Fig. 8b). The pattern is very steep for low-Ca pyroxenes, but becomes much flatter, especially for the HREE, at higher Ca contents. (Note, however, that Gallahan and Nielsen, 1989, report no correlation of D_{REE} with Ca content for pyroxenes of Wo > 30.)

We interpret the partition coefficient variations in Figure 8 in terms of substitution of REE for Ca in the M2 sites (Grutzeck et al., 1974), and the principle that the larger the

---

[8]This is one of the group of meteorites widely regarded as martian by most meteoriticists (e.g., Bogard and Johnson, 1983).

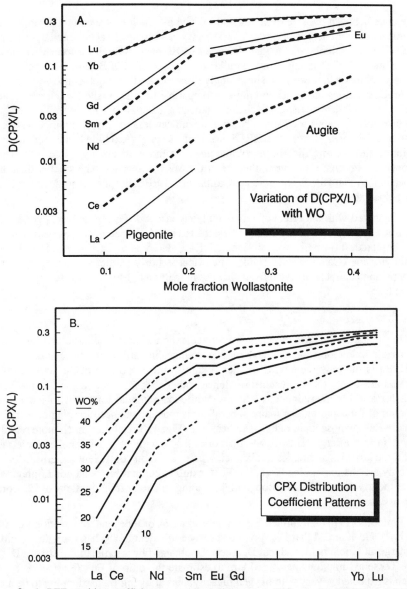

Figure 8. A. REE partition coefficients vs. wollastonite content (WO = molar Ca/(Ce+Fe+Mg)) for synthetic subcalcic pyroxenes. Lines represent least-squares fits to data points (not shown). Breaks in Lines correspond to proposed transition from pigeonite to augite structure, based solely on composition. Partition coefficients increase markedly with increasing Ca content, and increase is much steeper for light REE than for heavy REE. Insufficient pigeonite data were obtained to constrain variation of D_{Eu} at low WO. B. Variation of REE partition coefficient pattern with WO content for sub-calcic pyroxenes. Partition coefficients are much lower for low-Ca pyroxene than for high-Ca pyroxene. Moreover, the pattern for high-Ca pyroxene is much flatter, particularly from Gd-Lu, than for low-Ca pyroxene. Eu partition coefficients were measured at oxygen fugacities near the quartz-fayalite-magnetite buffer curve. Data are from McKay et al. (1986a).

difference between the size of an ion and the site in which it is accommodated, the more incompatible is the ion in that site. The M2 sites can obviously accommodate the smaller HREE ions much more readily than the larger LREE ones. At high Ca content, the average size of the M2 sites is apparently similar to the size of HREE ions, and these are not strongly excluded, nor does the site strongly discriminated among them. However, it does discriminate among the larger and more incompatible LREE. As the Ca content of the site decreases, and it is increasingly occupied by much smaller Fe and Mg ions, the average site size probably becomes progressively smaller, and can less readily accommodate even the smaller REE ions. Hence, the HREE are more strongly excluded than for higher Ca contents, and the site discriminates among them more efficiently. Note that even for pyroxenes with high Ca content, the size of the M2 site appears to be smaller than that of the Ca ion, so that HREE are more compatible than LREE, despite the ionic radii of LREE being closer to that of Ca.

Compared with the effect of Ca content, any effect of Fe/Mg ratio on pyroxene/melt partition coefficients was too small to be detected by McKay et al. (1986a). Such an effect was detected, however, for olivine (McKay, 1986 and Colson et al., 1988) and orthopyroxene (Colson et al., 1988). The effect is fairly subtle, and changes in crystal composition cannot be isolated from changes in melt composition because of their mutual dependence.

<u>Oxidation state</u>

Inasmuch as the oxidation state of a trace cation affects its charge and ionic radius, changes in oxidation state can have marked changes in partitioning behavior. Among the REE, Eu is the only element for which a significant proportion of the ions in igneous systems are likely to be present in valences other than 3+. This leads to "anomalous" behavior of Eu in many igneous systems. While the potential for anomalous partitioning behavior of Eu is present for many minerals, it is most pronounced for plagioclase. This is because Eu^{2+}, whose ionic radius is rather large (Fig. 4), substitutes much more readily for Ca or Na in the large feldspar site than does Eu^{3+}, with its much smaller ionic radius, leading to large Eu anomalies in plagioclase/melt partition coefficient patterns. Moreover, not only are the anomalies large, but Eu^{2+} exhibits compatible behavior in plagioclase, while the trivalent REE are incompatible, leading to strong fractionations of Eu from the other REE.

Philpotts (1970) interpreted Eu behavior in terms of independent partitioning of Eu^{2+} and Eu^{3+}. He assumed that $D_{Eu^{2+}}$ will be nearly equivalent to D_{Sr} based on their similarity of charge and ionic radii, and that $D_{Eu^{3+}}$ can be obtained by interpolation between D_{Sm} and D_{Gd}. These assumptions permitted him to estimate the ratio of Eu^{2+}/Eu^{3+} in each phase, which is expected to vary with oxygen fugacity, thus allowing Eu partitioning to be used as an oxygen fugacity barometer.

The variation of $D_{Eu,PL/L}$ with oxygen fugacity has been experimentally measured by several workers (e.g., Weill and Drake, 1973; Drake, 1975; Weill and McKay, 1975). Following Philpotts (1970), these workers considered D_{Eu} as the sum of $D_{Eu^{2+}}$ and $D_{Eu^{3+}}$, the contributions due to Eu^{2+} and Eu^{3+} respectively, weighted according to the relative amounts of these two ions present in the coexisting liquid:

$$D_{Eu} = D_{Eu}2+ \cdot W(Eu^{2+}) + D_{Eu}3+ \cdot W(Eu^{3+}) \tag{1}$$

where $W(Eu^{2+})$ and $W(Eu^{3+})$ are weight concentrations of each species normalized to a sum of 1 (Weill et al., 1974). For magmatic melts, the Eu^{2+}/Eu^{3+} ratio may be approximated by

$$logK(T) = log[(Eu^{2+}/Eu^{3+}) \cdot fO_2^{1/4}] = C/T + A \qquad (2)$$

where C and A are constants over a restricted range of temperature and composition.

Equations 1 and 2 may be combined to obtain

$$D_{Eu}(T,fO_2) = [D_{Eu}^{2+} \cdot K + D_{Eu}^{3+} \cdot fO_2^{1/4}]/[K + fO_2^{1/4}] \qquad (3)$$

which predicts that at constant temperature, D_{Eu} varies along an "S-shaped" curve which approaches $D_{Eu^{2+}}$ at low oxygen fugacities and $D_{Eu^{3+}}$ at high fugacities. Experimental plagioclase/melt partition coefficients follow the predicted variation quite closely (Fig. 9). Similar variations of D_{Eu} with fO_2 have been observed experimentally for diopside (Grutzeck et al., 1974), augite (Sun et al., 1974), fassaitic[9] clinopyroxene (McKay et al., 1989a) and pigeonite (McKay et al., 1989b). In a later section, an application of these variations to infer oxidation conditions during crystallization of a magmatic meteorite will be discussed.

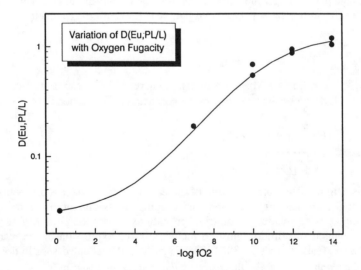

Figure 9. Variation of $D_{Eu,PL/L}$ with oxygen fugacity. Curve is based on model described in text. Experimental data are from Weill and McKay (1975).

---

[9]Al- and Ti-rich high-Ca pyroxene

THERMODYNAMIC RELATIONSHIPS: DEPENDENCE OF PARTITIONING ON TEMPERATURE AND COMPOSITION

Many workers have discussed the thermodynamic treatment of trace element mineral/melt partitioning in order to relate partitioning to temperature and composition (e.g., Banno and Matsui, 1973; Drake and Weill, 1975; Watson, 1977; Navrotsky, 1978; Mysen and Virgo, 1980; Ford et al., 1983; Jones, 1984; Nielsen, 1985, 1988; Colson et al., 1988). It is possible to consider the entry of a trace element into a mineral either as an exchange reaction or a synthesis reaction. Expressing partitioning as an exchange reaction decreases pressure dependence (Colson et al., 1988) and minimizes the dependence of partitioning on unknown melt structure, assuming the trace cation occupies the same site in the melt as the major element for which it exchanges. Several steps are necessary in formulating an expression for the dependence of the partition coefficient on temperature and phase composition: (1) Formulate an exchange or crystallization reaction for incorporating the trace cation in the mineral. (2) Formulate an equilibrium constant for that reaction. (3) Choose relations to express activities of the reaction components in terms of compositional parameters. (4) Rearrange the equilibrium constant expression to isolate the partition coefficient, thereby expressing it in terms of temperature and "non-coefficient" compositional parameters.

Colson et al. (1988) have recently considered trace element partitioning between olivine or orthopyroxene and melt in terms of exchange reactions and their treatment of orthopyroxene partitioning will be used as an example for this discussion. They expressed the equilibrium partitioning of trace cations between orthopyroxene and melt by the general exchange reaction

$$Mg^{2+}{}_{OPX} + Tr^{n+}{}_{L} \Leftrightarrow Tr^{n+}{}_{OPX} + Mg^{2+}{}_{L}$$

where $Mg^{2+}{}_{OPX}$ denotes a Mg^{2+} ion in orthopyroxene, Tr^{n+} denotes a trace cation of valence n and L denotes melt. The equilibrium constant for this reaction is

$$K = \frac{a(Tr^{n+}{}_{OPX}) \cdot a(Mg^{2+}{}_{L})}{a(Tr^{n+}{}_{L}) \cdot a(Mg^{2+}{}_{OPX})}$$

where a is activity. The numeric value of the equilibrium constant is a function of the standard states chosen for trace and major components and of the accompanying composition-activity relations. For trivalent cations substituting for Mg^{2+}, a charge balancing coupled substitution must be included in the expression of the reaction. In orthopyroxene, Colson et al. (1988, 1989a) found that this coupled substitution is $Al^{3+} \Leftrightarrow Si^{4+}$ in the tetrahedral site. For this reaction, the equilibrium constant is

$$K = \frac{a(MgTr^{3+}AlSiO_6) \cdot a(Mg^{2+}{}_{L}) \cdot a(Si^{4+}{}_{L})}{a(MgMgSi_2O_6) \cdot a(Tr^{3+}{}_{L}) \cdot a(Al^{3+}{}_{L})} \qquad (4)$$

where $a(MgTr^{3+}AlSiO_6)$ is the activity of the $MgTr^{3+}AlSiO_6$ component in orthopyroxene and $a(Mg^{2+}{}_{L})$ is the activity of Mg^{2+} in the melt. Colson et al. used the following activity-composition relations for orthopyroxene, based on Kerrick and Darken (1975), and melt, based on a modified version of the two-lattice model of Bottinga and Weill (1972):

$$a(MgMgSi_2O_6) = X(En)^2$$

$$a(MgTr^{3+}AlSiO_6) = \gamma \cdot X(En) \cdot [C(Tr_{OPX})/[C(Mg_{OPX}) + C(Fe_{OPX}) + C(Tr_{OPX})]]$$

$$a(Mg_L) = C(Mg_L/[\Sigma C(cations_L) - C(Al_L) - C(Si_L)]$$

$$a(Tr_L) = C(Tr_L/[\Sigma C(cations_L) - C(Al_L) - C(Si_L)]$$

$$a(Si_L) = C(Si_L)/[C(Al_L) + C(Si_L)]$$

$$a(Al_L) = C(Al_L)/[C(Al_L) + C(Si_L)]$$

where $X(En)$ is the mole fraction of enstatite in orthopyroxene, the coefficient γ is a nonideality term interpreted as describing ordering of Fe and Mg around trace cations (Colson et al., 1989b), and $C(Mg_{OPX})$ is the atomic concentration of Mg in orthopyroxene. The partition coefficient D_M (expressed in molar proportions) is

$$D_M = [C(Tr_{OPX})/[C(Mg_{OPX}) + C(Fe_{OPX}) + C(Tr_{OPX})]]/a(Tr_L) \tag{5}$$

The temperature dependence of any reaction is given by

$$\ln K = \Delta S/R - \Delta H/RT \tag{6}$$

where K is the equilibrium constant, R the gas constant, ΔS the enthalpy change of the reaction, ΔH the entropy change of the reaction, and T the temperature (Kelvin).

The variation of D_M with temperature and compositional parameters not included in the partition coefficient is obtained by combining equations 4, 5 and 6 to yield

$$\ln(D_M) = \Delta S/R - \Delta H/RT - \ln \gamma + \ln[X(En) \cdot a(Al_L)] - \ln[a(Mg_L) \cdot a(Si_L)] \tag{7}$$

Up to this point, except for the choice of activity composition relations, which varies widely from author to author, the treatment by Colson et al. (1988) is similar to that of many other authors. However, Colson et al. next performed simultaneous multiple regressions on a large body of experimentally determined partition coefficients to obtain values for ΔS and ΔH, allowing prediction of partition coefficients as functions of temperature and phase composition for the elements studied. Moreover, Colson et al. then related the values of ΔH and ΔS to the ionic size of the trace cation by semi-empirical expressions, allowing successful prediction of partition coefficients for elements not studied experimentally.

Other predictive approaches

While the approach of Colson et al. (1988) to predicting partitioning behavior is unusual, both for the large data set employed in the regressions and for the correlation with ionic size which allows predictions for unstudied elements, several other approaches have also been quite successful. For example, Jones (1984) noted the strong linear correlation of molar olivine/melt partition coefficients for Mg, Fe, Mn, and Ni. Because the stoichiometry of olivine is simple, it is possible to use these correlations to predict D values for Mg, Fe, Mn, and Ni if the bulk composition of a basalt in equilibrium with olivine is known. Jones concluded from these regressions that temperature rather than melt composition is the most important control on elemental partitioning. Jones and Burnett (1987) extended this approach to Sm partitioning between diopside and liquid.

Nielsen (1985) also devised a useful approach for extrapolating partition coefficients to different compositions. He emphasized the proper formulation of the activity of the trace component in the melt, and showed that use of a modification of the Bottinga-Weill (1972) two-lattice melt model significantly reduced the compositional dependence of partition coefficients. According to the Bottinga-Weill model, the melt consists of two independent quasi-lattices, the network formers, composed primarily of the components $NaAlO_2$, $KAlO_2$, $CaAl_2O_4$, $MgAl_2O_4$, $FeAl_2O_4$ and SiO_2, and the network modifiers, mainly Mg, Fe, and Ca. All Al is assumed to be present as tetrahedrally coordinated network formers complexed with the cations K, Na, Ca, Mg, and Fe^{2+}, in order of preference. Melt components are assumed to mix ideally within their own quasi-lattice, with no mixing between quasi-lattices. In this case, the activity of a component in the melt is assumed to be its mole fraction normalized to the sum of the components within its quasi-lattice. For example, the activity of Si is the mole fraction of Si in the melt divided by the sum of the mole fractions of Si, NaAl, KAl, $CaAl_2$, $MgAl_2$, $FeAl_2$.

Ray et al. (1983) demonstrated that the original Bottinga-Weill model is not fully successful in accounting for the effects of composition on partitioning of Sc, Ti, Sr, and Sm between diopside and liquid in the system diopside-albite-anorthite. In particular, they observed a variation in partition coefficient values along isotherms in the system, which are not predicted by the Bottinga-Weill model. The isotherms in this system are nearly parallel to the albite-anorthite join. According to the Bottinga-Weill model, as long as the proportion of diopside remains constant, there is no change in the mole fraction of the network formers and modifiers as the albite/anorthite ratio changes, simply a substitution of NaAl for $CaAl_2$. Thus no variation the activity of melt components and hence in partition coefficients should be observed along the isotherms. Nielsen (1985) demonstrated that a modification of the Bottinga-Weill model (Nielsen and Drake, 1979) reduces the compositional dependency of the Ray et al (1983) data below analytical error. The modification consists of assigning all Ca, Mg, Fe, and Al above that which can be charge balanced by Na or K to be network modifiers. This has the effect of increasing the calculated activity of network modifying components in alkali-rich melts and thus reducing the distribution coefficients and the effect of composition. Nielsen (1988) demonstrated that this approach reduces compositional dependence in several other systems as well.

The above approaches have proven useful in extrapolating partition coefficients to compositions and temperatures differing from those for which they were measured. However, except for Colson et al.'s (1988) approach, they require that at least some experimental data be obtained for the elements of interest. Although the approach of Colson et al. allows prediction of partitioning behavior for elements for which no data exist, the uncertainties for such elements are rather large because of the approximations involved in the approach. Moreover, it remains to be demonstrated that any of these approaches are successful in extrapolating to melt compositions differing considerably from the basaltic compositions for which the partitioning data are most commonly obtained. Hence, to date, no fully reliable, generally applicable method is available for confidently predicting partition coefficients for temperatures and compositions differing considerably from those for which they were measured.

SPECIAL APPLICATIONS

For many applications involving petrogenetic modelling, the absolute values of partition coefficients are not as important as the shape of the pattern for each mineral (e.g., Hansen, this volume). Hence, it is not always essential to have partition coefficients which are appropriate for the specific phase compositions and temperatures of the system being modelled. However, for some other applications, appropriate absolute values are required. The following are examples from my own research of the use of partition coefficients in ways which differ from the usual modeling of melt evolution, and in some cases require more accurate partition coefficient values.

Eu as an oxygen fugacity indicator

Several workers have used the variation of Eu distribution coefficients with oxygen fugacity to infer redox conditions under which magmas have crystallized (e.g., Drake, 1975). In this section I will describe application of this relationship to understanding redox conditions during the crystallization of the antarctic achondrite Lewis Cliff (LEW) 86010, an unusual refractory-rich meteorite with close chemical and petrologic affinities to the very ancient and primitive achondrite Angra dos Reis (e.g., McKay et al., 1988). Both of these meteorites have phase assemblages which include Fe metal and an Fe-Ti oxide whose microprobe analysis suggests the presence of Fe^{3+}, thereby giving conflicting signals about redox conditions during crystallization.

In an attempt to discover the oxygen fugacity under which these samples crystallized, McKay et al. (1989a) measured partitioning of REE among fassaitic[10] pyroxene, anorthite, and a synthetic analog melt of LEW 86010 (at ~1170°C and 1 atm), focusing on variations in the ratio of D_{Eu}/D_{Gd} with changing fO_2. Resulting partition coefficient ratios are plotted against oxygen fugacity (relative to the iron-wüstite buffer) in Figure 10a. D_{Eu}/D_{Gd} for plagioclase decreases with increasing oxygen fugacity, as we would predict from Figure 9, while that for pyroxene increases with oxygen fugacity, indicating that $D_{Eu^{2+},PX} < D_{Eu^{3+},PX}$. Figure 10b shows the ratio of D_{Eu}/D_{Gd} for plagioclase to that for pyroxene as a function of oxygen fugacity. Individual values were obtained from average values for each oxygen fugacity from Figure 10a. The ratio plotted in Figure 10b varies by nearly an order of magnitude over the 4 log-unit range in oxygen fugacity studied.

Although the ratios of mineral/melt distribution coefficients were used to construct Figure 10b, trace element abundances in the melt cancel out of the expression, leaving mineral/mineral distribution coefficients as the quantity plotted. Hence, if Eu and Gd abundances are known for plagioclase and pyroxene which crystallized in mutual equilibrium, the redox conditions during crystallization may be obtained from Figure 10b (provided that the natural phase compositions and crystallization temperature are similar to those of the experiments). Equilibrium ratios of trivalent REE between core plagioclase and pyroxene (Crozaz and McKay, 1989) suggest equilibration between these phases. The Eu/Gd ratio between these minerals (Crozaz and McKay, 1989) is shown in Figure 10b,

---

[10] Al-, and Ti-rich

68

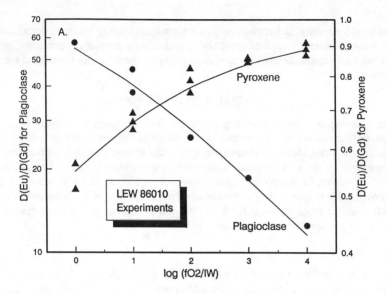

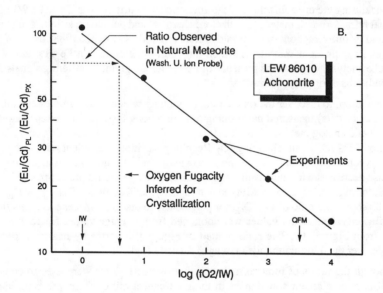

Figure 10. (A) Variation of D_{Eu}/D_{Gd} with oxygen fugacity for plagioclase and fassaitic pyroxene in equilibrium with a synthetic analog melt of achondritic meteorite LEW 86010 at ~1170°C and 1 atm. Curves are calculated using Equation 3. (B) Variation of $(Eu/Gd)_{PL}/(Eu/Gd)_{PX}$ with fO_2 for LEW 86010 experimental charges. Curve was obtained by combining curves in A. Ratio observed in ion probe analyses of mineral cores in natural sample of LEW 86010 is also shown, and indicates crystallization at oxygen fugacities slightly above those of the iron-wüstite buffer, but well below those of the quartz-fayalite-magnetite buffer. Experimental data are from McKay et al. (1989a), and data for natural sample are from Crozaz and McKay (1989).

and suggests that crystallization occurred under relatively reducing conditions, closer to the iron-wüstite buffer than to the quartz-fayalite-magnetite buffer. Hence the partitioning of Eu provides clear evidence concerning the redox conditions under which this sample crystallized.

Origin of the Eu anomaly in lunar mare basalts

A long-held paradigm of lunar science is that the complementary REE patterns and Eu anomalies of the lunar crust and mare basalts (Fig. 11) reflect an early differentiation event of global scale resulting from the crystallization of a lunar magma ocean (see chapter by Haskin in this volume). The positive Eu anomaly of the crust is generally thought to result from plagioclase enrichment, while the negative Eu anomaly in mare basalts is thought to be inherited from an evolved magma ocean in which plagioclase removal had produced a negative Eu anomaly prior to crystallization of the mafic cumulates that form the mare basalt source region (e.g., Taylor, 1982)[11].

The need for prior plagioclase removal has recently been re-examined by Brophy and Basu (1989) and Shearer and Papike (1989). These authors explicitly addressed the question of whether prior plagioclase removal is required to produce Eu anomalies of the magnitude observed in mare basalts. However, using different partition coefficient values, they arrived at opposite conclusions.

Part of the uncertainty in this issue is a result of inadequate mineral/melt partition coefficient data, especially for Eu at lunar oxygen fugacities. The situation is most critical for low-Ca pyroxene (a major carrier of REE among magma ocean crystallization products throughout much of its crystallization sequence). Orthopyroxene distribution coefficients of Weill and McKay (1975), the only available coefficients determined specifically for lunar systems, indicate only a very minor Eu anomaly. However, those results were obtained before the difficulty of measuring very low distribution coefficients on small crystals was appreciated (McKay, 1986) so the magnitude of the anomaly is likely to be unreliable. To provide reliable values for the partitioning of REE between low-Ca pyroxene and melt under near-lunar fO_2 and thus better constrain the origin of the mare basalt Eu anomaly, we recently studied partitioning of REE and Sr between pigeonite and mare basalt melts at oxygen fugacities near the iron-wüstite buffer (McKay et al., 1989b).

In addition to noting a correlation of D_{Gd} with the Wo content of pyroxene, as in our study of Shergotty pyroxenes (McKay et al., 1986), we also observed a correlation of D_{Eu} with fO_2. Values of D_{Eu}/D_{Gd} and D_{Sr}/D_{Gd} from these experiments are plotted against oxygen fugacity in Figure 12, and the observed correlation is indicated by the solid line. The correlation predicted according to reasoning discussed above for feldspar is shown as a dotted line. The predicted curve is calculated using the value interpolated between D_{Sm} and D_{Gd} for $D_{Eu^{3+}}$, D_{Sr} for $D_{Eu^{2+}}$, and assuming that Eu^{2+}/Eu^{3+} varies as $fO_2^{-1/4}$ and that the correct value for Eu^{2+}/Eu^{3+} is given by the data at higher oxygen fugacity. The predicted correlation is not as steep as the observed one, but it could be brought into coincidence by using a slightly lower value for $D_{Eu^{2+}}$ and/or a slightly higher value for $D_{Eu^{3+}}$, both within uncertainty. Hence the proposition of Philpotts (1970) that Sr serves as a proxy for Eu^{2+} appears valid for low-Ca pyroxene.

---

[11]Note, however, that Korotev and Haskin (1988) have presented evidence against a positive Eu anomaly in at least the upper portion of the lunar crust.

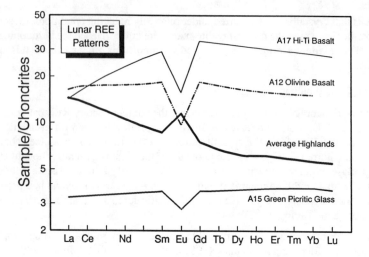

Figure 11. REE patterns of the lunar highlands crust and lunar mare basalts and picritic pyroclastic glass. Note the complementary nature of the Eu anomalies in the average crust and mare basaltic materials. Data are from Taylor (1982).

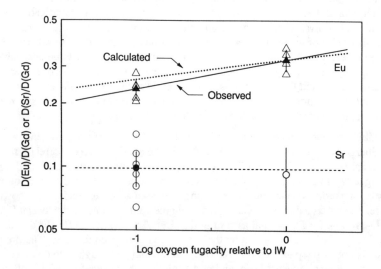

Figure 12. D_{Eu}/D_{Gd} and D_{Sr}/D_{Gd} for pigeonite vs oxygen fugacity relative to the iron-wüstite buffer. Open symbols are values from individual analyses, and filled symbols are averages. Error bars are ±2 σ. D_{Eu}/D_{Gd} shows a positive correlation with fO_2, while the uncertainties for D_{Sr}/D_{Gd} at the higher oxygen fugacity so large that a correlation for this ratio cannot be ruled out. However, no correlation is expected for Sr, as indicated by the dashed line through the error-weighted mean of all Sr determinations. The theoretically expected variation of D_{Eu}/D_{Gd} (see text) is indicated by the dotted line, and, within uncertainties in the assumptions, is in agreement with the observed variation.

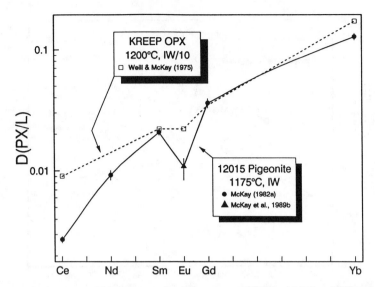

Figure 13. Distribution coefficient patterns for orthopyroxene (Weill and McKay, 1975) and pigeonite (McKay, 1982a, 1989b). The pigeonite pattern displays a significant Eu anomaly. The orthopyroxene pattern has a minimal Eu anomaly, slopes much less steeply for the LREE, but these differences may be due to smaller crystal size in the earlier experiments, and the ensuing difficulties for measuring low partition coefficients (McKay, 1986).

Figure 13 compares the partition coefficient pattern obtained for pigeonite in this study (McKay et al., 1989b) with that obtained by Weill and McKay (1975) for orthopyroxene. The pigeonite pattern has a much larger Eu anomaly, and slopes more steeply from Ce to Sm. It is likely that these differences, rather than resulting from differences in pyroxene structure, instead are the result of analytical difficulties caused by the small crystal size in the earlier experiments. These results show that low-Ca pyroxene has a much larger capacity to generate a europium anomaly in mare basalts than the earlier data suggest. Whether the Eu anomaly in low-Ca pyroxene partition coefficients is large enough to explain the Eu anomaly in mare basalt without prior plagioclase removal awaits detailed modelling studies.

REE PARTITION COEFFICIENT PATTERNS FOR THE MAJOR MINERALS

For many modelling applications, the shape of REE partition coefficient patterns is more important than the absolute values of the coefficients. Shapes of the distribution coefficient patterns of the major minerals involved in basalt petrogenesis are illustrated in Figure 14, and the features of these patterns are discussed below. The absolute values are generally appropriate for basaltic magma systems, but are likely to vary to some degree with temperature and phase composition.

Plagioclase

Figure 14 shows plagioclase/melt partition coefficients for a lunar highlands basaltic system (McKay, 1982b). With the exception of Eu, values are generally low, and the pattern drops off sharply from LREE to HREE. Except for very small fractions of melt, partial melting of a plagioclase-rich source region does not strongly fractionate the trivalent

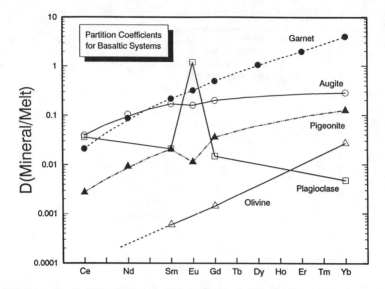

Figure 14. Mineral/melt REE partition coefficients for plagioclase (McKay, 1982b), olivine (McKay, 1986), clinopyroxene (McKay et al., 1986), pigeonite (McKay, 1982a; McKay et al., 1989b), and garnet (Shimizu and Kushiro, 1974). Plagioclase and pigeonite patterns were measured at oxygen fugacities near iron-wüstite, augite near QFM, and garnet under unknown but probably oxidizing conditions.

REE, because their partition coefficient values are quite low, and they are nearly quantitatively excluded from the plagioclase residuum[12]. A significant proportion of Eu, however, will be retained in the plagioclase residuum, producing a melt with a negative Eu anomaly. The magnitude of this anomaly is inversely correlated with the oxidation fugacity prevailing during melting.

Olivine

Figure 14 also shows olivine/melt partition coefficients for a synthetic lunar mare basalt (McKay, 1986). The pattern slopes very steeply upwards from LREE to HREE. However, the values are very low, so olivine is not capable of producing significant REE fractionation in partial melts. The magnitude of the Eu anomaly in the olivine distribution coefficient pattern is not well known, but is not significant for most situations because the partition coefficient values are so low that the REE will be dominated by almost any other phase in the system.

Pyroxene

Pyroxene partition coefficient patterns are strongly dependent on composition, as discussed above.

---

[12]Note that while partial melting produces little fractionation among highly incompatible elements, other petrogenetic models, such as the cumulate remelting model for the origin of lunar mare basalts (see Haskin, can produce large fractionations among even highly incompatible elements.

Low-Ca pyroxene. Figure 14 also shows pigeonite/melt partition coefficients for a lunar mare basalt (McKay et al., 1989b). The pattern slopes steeply upwards from Ce to Sm, has a distinct negative Eu anomaly at near-lunar oxygen fugacities (discussed above), and slopes upwards somewhat less steeply from Gd to Yb. A smaller but still distinct Eu anomaly should be present at terrestrial oxygen fugacities, based on extrapolation of the predicted variation shown in Figure 12. Again, values for all elements are so low that little REE fractionation is expected during partial melting of a low-Ca pyxroxene-rich source region except at very low degrees of melting. Distribution coefficient patterns for orthopyroxene are quite similar to that for pigeonite (e.g., Weill and McKay, 1975).

High-Ca pyroxene. Figure 14 also shows augite/melt partition coefficients for a ferrobasaltic system similar in composition to the Shergotty meteorite (McKay et al., 1986a). The pattern slopes upwards somewhat from Ce to Sm, and is fairly flat from Gd to Yb. This pattern was measured under oxidation conditions appropriate for terrestrial magmatic systems, and shows a small negative Eu anomaly. A larger anomaly would be expected under more reducing conditions. Absolute values are higher than for most other minerals involved in basalt petrogenesis, especially for the heavy REE. Significant fractionation of REE would be expected in a partial melt of a high-Ca pyroxene-dominated source region for small to moderate degrees of melting.

Garnet

Finally, Figure 14 shows garnet/melt partition coefficients for a hydrous synthetic diopside-pyrope system at 30 kbar pressure (Shimizu and Kushiro, 1975). The pattern slopes quite steeply upwards from LREE to HREE and is very similar in slope to olivine, another orthosilicate. The pattern shows no discernable Eu anomaly, but the experiments were run under conditions much more oxidizing than those experienced by terrestrial magmatic systems. However, phenocryst/matrix partition coefficients for pyrope garnet commonly show little Eu anomaly, suggesting that $D_{Eu^{2+}}$ and $D_{Eu^{3+}}$ are fairly similar for pyrope. The partition coefficient values span a very large range for garnet, with the LREE being incompatible and the HREE being compatible. A partial melt of a source region containing even a small amount of residual garnet would be significantly LREE-enriched.

FUTURE DIRECTIONS

Despite the large volume of high quality partition coefficient data, there is much work yet to be done. One of the areas in which additional study is needed is in systematizing the variation of partition coefficients with phase composition and temperature. Progress has been made, particularly for olivine and low-Ca pyroxene, but it is still impossible to predict partition coefficients with the accuracy required for some applications. Another area where progress is needed is understanding the effect of volatiles on partitioning. As in understanding the effects of temperature, pressure, and phase composition, progress in understanding volatiles will not come easily because of the intimate inter-relationships among all these variables. For example, changes in composition affect liquidus temperatures, yielding different partition coefficient values, and it is difficult to separate the combined effects. Until predictive models improve to the point where they can be used with confidence, many scientific problems will require the measurement of partition coefficients for compositions and temperatures specifically applicable to that problem.

An additional exciting area for future work, which is just beginning to be explored, is the measurement of partition coefficients at the ultra-high pressures of the earth's lower mantle or the center of Mars. As an example of the scientific frontiers to be opened by newly developed high pressure technology, Kato et al. (1988a,b) have recently reported Mg-perovskite/melt and majorite garnet partition coefficients in a modified chondritic bulk composition for a number of key elements at pressures up to ~250 kbar and 2300°C. Their results provide strong constraints upon hypotheses which maintain that the mantle experienced extensive melting during formation of the earth followed by fractional crystallization and differentiation involving majorite garnet and/or Mg-perovskite. The partition coefficients show that fractionation of garnet would be accompanied by sharp decreases of Al/Ca and Sc/Sm ratios in residual liquids, and fractionation of Mg-perovskite would cause marked variations of Lu/Hf, Sc/Sm and Hf/Sm ratios in residual liquids. The observed near-chondritic relative abundances of these elements as well as Ca, Al, Yb, and Zr appear to exclude models which propose that the bulk mantle once had chondritic abundances of Mg, Si, Al, Ca and other lithophile elements but has differentiated to form a perovskitic lower mantle and a peridotitic upper mantle. Thus, these partition coefficients suggest that the mantle did not experience extensive melting during the formation of the earth.

ACKNOWLEDGMENTS

This paper benefitted from reviews by Bruce Lipin and John Jones. Contributions of D.F. Weill to the then-fledgling field of experimental trace element partitioning are gratefully acknowledged. W. Gates and J. Garcia provided technical support.

REFERENCES

Banno S. and Matsui Y. (1973) On the formulation of partition coefficients for trace elements distribution between minerals and magma. Chem. Geol. 11, 1-15.

Bogard D. D. and Johnson P. (1983) Martian gases in an antarctic meteorite. Science 221, 651-654.

Bottinga Y. and Weill D. F. (1972) The viscosity of magmatic silicate liquids: A model for calculation. Amer. J. Sci. 272, 438-475.

Brophy J. G. and Basu A. (1989) Europium Anomalies in mare basalts as a consequence of mafic cumulate fractionation from an initial lunar magma. Proc. Lunar Planet. Sci. Conf. 20th, in press.

Burns, R. G. (1970) Mineralogical Applications of Crystal Field Theory. Cambridge Univ. Press, London.

Colson R. O., McKay G. A., and Taylor L. A. (1988) Temperature and compositional dependence of trace element partitioning: olivine/melt and orthopyroxene/melt. Geochim. Cosmochim. Acta 52, 539-553.

Colson R. O., McKay G. A., and Taylor L. A. (1989a) Charge balancing of trivalent trace elements in olivine and low-Ca pyroxene: A test using experimental partitioning data. Geochim. Cosmochim. Acta 53, 643-648.

Colson R. O., McKay G. A., and Taylor L. A. (1989b) Partitioning data pertaining to Fe-Mg ordering around trace cations in olivine and low-Ca pyroxene. Contrib. Mineral. Petrol. (1989) 102, 242-246.

Crozaz G. and McKay G. (1989) Minor and trace element microdistributions in Angra dos Reis and LEW 86010: Similarities and differences (abstract). Lunar Planet. Sci. XX, 208-209.

Cullers R. L., Medaris L. G., and Haskin L. A. (1970) Gadolinium: distribution between aqueous and silicate phases. Science 169, 580-583.

Cullers R. L., Medaris L. G., and Haskin L. A. (1973) Experimental studies of the distribution of rare earths as trace elements among silicate minerals and liquid and water. Geochim. Cosmochim. Acta 37, 1499-1512.

Drake M. J. (1972) The distribution of major and trace elements between plagioclase feldspar and magmatic silicate liquid: an experimental study. Ph.D. dissertation, University of Oregon, Eugene, OR.

Drake M. J. (1975) The oxidation state of europium as an indicator of oxygen fugacity. Geochim. Cosmochim. Acta 39, 55-64.

Drake M. J. and Holloway J. R. (1978) "Henry's law" behavior of Sm in a natural plagioclase/melt system: importance of experimental procedure. Geochim. Cosmochim. Acta 42, 679-683.

Drake M. J. and Holloway J. R. (1981) Partitioning of Ni between olivine and silicate melt: the 'Henry's law problem' reexamined. Geochim. Cosmochim. Acta 45, 431-437.

Drake M. J. and Weill D. F. (1972) New rare earth element standards for electron microprobe analysis. Chem. Geol. 10, 179-181.

Drake M. J. and Weill D. F. (1975) Partition of Sr, Ba, Ca, Y, Eu^{2+}, Eu^{3+}, and other REE between plagioclase feldspar and magmatic liquid: An experimental Study. Geochim. Cosmochim. Acta 39, 689-712.

Ford C. E., Russell D. G., Craven J. A., and Fisk M. R. (1983) Olivine-liquid equilibria: Temperature, pressure and compositional dependence of the crystal/liquid cation partition coefficients for Mg, Fe^{2+}, Ca and Mn. J. Petrol. 24, 256-265.

Gallahan W. E. and Nielsen R. L. (1989) Experimental determination of the partitioning of Sc, Y, and REE between high-Ca clinopyroxene and natural mafic liquids (abstract). EOS, Trans. Amer. Geophys. Union., in press.

Goldschmidt V. M. (1937) The principle of distribution of chemical elements in minerals and rocks. J. Chem. Soc., 655-672.

Grutzeck M. W., Kridelbaugh S. J., and Weill D. F. (1974) The distribution of Sr and the REE between diopside and silicate liquid. Geophys. Res. Lett. 1, 273-275.

Harrison W. J. (1981) Partition coefficients for REE between garnets and liquids: implications of non-Henry's law behavior for models of basalt origin and evolution. Geochim. Cosmochim. Acta 45, 242-259.

Harrison W. J. and Wood B. J. (1980) An experimental investigation of the partitioning of REE between garnet and liquid with reference to the role of defect equilibria. Contrib. Mineral. Petrol. 72, 145-155.

Hoover J. D. (1978) The distribution of samarium and thulium between plagioclase and liquid in the systems An-Di and Ab-An-Di at 1300°C. Carnegie Inst. Washington Yearbook 77, 703-706.

Holloway J. R. and Drake M. J. (1977) Quantitative microautoradiography by x-ray emission micro-analysis. Geochim. Cosmochim. Acta 41, 1295-1297.

Irving A. J. (1978) A review of experimental studies of crystal/liquid trace element partitioning. Geochim. Cosmochim. Acta 42, 743-770.

Jones J. H. (1981) Studies of the geochemical similarity of plutonium and samarium and their implications for the abundance of ^{244}Pu in the early solar system. Ph.D. dissertation, California Inst. of Technology, Pasadena, CA.

Jones J. H. (1984) temperature- and pressure-independent correlations of olivine/liquid partition coefficients and their application to trace element partitioning. Contrib. Mineral. Petrol. 88, 126-132.

Jones J. H. and Burnett D. S. (1981) Quantitative radiography using Ag X-rays. Nuclear Instruments and Methods 180, 625-633.

Jones J. H. and Burnett D. S. (1987) Experimental geochemistry of Pu and Sm and the thermodynamics of trace element partitioning. Geochim. Cosmochim. Acta 51, 769-782.

Kato T., Ringwood A. E., and Irifune T. (1988a) Experimental determination of element partitioning between silicate perovskites, garnets and liquids: Constraints on early differentiation of the mantle. Earth Planet. Sci. Lett. 89, 123-145.

Kato T., Ringwood A. E., and Irifune T. (1988b) Constraints on element partition coefficients between $MgSiO_3$ perovskite and liquid determined by direct measurements. Earth Planet. Sci. Lett. 90, 65-68.

Kerrick D. M. and Darken L. S. (1975) Statistical thermodynamic models for ideal oxide and silicate solid solutions, with application to plagioclase. Geochim. Cosmochim. Acta 39, 1431-1442.

Korotev R. L. and Haskin L. A. (1988) Europium mass balance in polymict samples and implications for plutonic rocks of the lunar crust. Geochim. Cosmochim. Acta 52, 1795-1813.

Lindstrom D. J. (1976) Experimental study of the partitioning of the transition metals between clinopyroxene and coexisting silicate liquids. Ph.D. dissertation, University of Oregon, Eugene, OR.

Lindstrom D. J. and Weill D. F. (1978) Partitioning of transition metals between diopside and coexisting silicate liquids. I. Nickel, cobalt, and manganese. Geochim. Cosmochim. Acta 42, 817-831.

McKay G. (1977) Crystal/liquid equilibration in trace element partitioning experiments (abstract). In papers presented to the International Conference on Experimental Trace Element Geochemistry, 81-82.

McKay G. (1982a) Experimental REE partitioning between pigeonite and Apollo 12 olivine basaltic liquids (abstract). EOS, Trans. Amer. Geophys. Union 63, 1142.

McKay G. (1982b) Partitioning of REE between olivine, plagioclase, and synthetic basaltic melts: Implications for the origin of lunar anorthosites (abstract). Lunar Planet. Sci. XIII, 493-494.

McKay G. (1986) Crystal/liquid partitioning of REE in basaltic systems: Extreme fractionation of REE in olivine. Geochim. Cosmochim. Acta 50, 69-79.

McKay G. (1989) Analysis of Rare Earth Elements by Electron Microprobe. Microbeam Analysis - 1989, 549-553.

76

McKay G., Wagstaff J., and Yang S.-R. (1986a) Clinopyroxene REE distribution coefficients for shergottites: The REE content of the Shergotty melt. Geochim. Cosmochim. Acta 50, 927-937.

McKay G., Wagstaff J., and Yang S.-R. (1986b) Zirconium, hafnium, and rare earth element partition coefficients for ilmenite and other minerals in high-Ti lunar mare basalts: An experimental study. Proc. Lunar. Planet. Sci. Conf. 16th, J. Geophys. Res. 91, D229-D237.

McKay G., Lindstrom D., Yang S.-R., and Wagstaff J. (1988) Petrology of Unique Achondrite Lewis Cliff 86010 (abstract). Lunar Planet. Sci. XIX, 762-763.

McKay G., Le L., and Wagstaff J. (1989a) Redox conditions during the crystallization of unique achondrite LEW 86010 (abstract). Lunar Planet. Sci. XX, 677-678.

McKay G., Wagstaff J., and Le L. (1989b) REE distribution coefficients for pigeonite: constraints on the origin of the mare basalt europium anomaly (abstr.). Workshop on Lunar and Volcanic Glasses, Lunar and Planetary Institute.

Michael P. J. (1988) Partition coefficients for rare earth elements in mafic minerals of high silica rhyolites: The importance of accessory mineral inclusions. Geochim. Cosmochim. Acta 52, 275-282.

Mysen B. O. (1978) Limits of solution of trace elements in minerals according to Henry's: Review of experimental data, Geochim. Cosmochim. Acta 42, 871-885.

Mysen B. O. (1979) Nickel partitioning between olivine and silicate melt: Henry's law revisited. Amer. Mineral. 64, 1107-1114.

Mysen B. O. and Seitz M. G. (1974) Trace element partitioning determined by beta track mapping: An experimental study using carbon and samarium as examples. J. Geophys. Res. 80, 2627-2635.

Mysen B. O. and Virgo (1980) Trace element partitioning and melt structure: An experimental study at 1 atm pressure. Geochim. Cosmochim. Acta 44, 1917-1930.

Nagasawa H., Schreiber H. D., and Morris R. V. (1980) Experimental mineral/liquid partition coefficients of the rare earth elements (REE), Sc, and Sr for perovskite, spinel, and melilite. Earth Planet. Sci. Lett. 46, 431-437.

Nakamura Y., Fujimaki H., Nakamura N., Tatsumoto M., McKay G., and Wagstaff J. (1986) Hf, Zr, and REE partition coefficients between ilmenite and liquid: Implications for lunar petrogenesis. Proceedings Lunar Planet. Sci. Conf. 16th, J. Geophys. Res. 91, D239-D250.

Navrotsky A. (1978) Thermodynamics of element partitioning; (1) Systematics of transition metals in crystalline and molten silicates and (2) Defect chemistry and "the Henry's Law problem." Geochim. Cosmochim. Acta 42, 887-902.

Nielsen R. H. (1985) A method for the elimination of the compositional dependence of trace element distribution coefficients. Geochim. Cosmochim. Acta 49, 1775-1779.

Nielsen R. H. (1988) the calculation of combined major and trace element liquid lines of descent I. The temperature dependence of compositionally independent partition coefficients. Geochim. Cosmochim. Acta 52, 27-38.

Nielsen R. H. and Drake M. J. (1979) Pyroxene-melt equilibria. Geochim. Cosmochim. Acta 43, 1259-1272.

Onuma N., Higuchi H., Wakita H., and Nagasawa H. (1968) Trace element partition between two pyroxenes and the host lava. Earth Planet. Sci. Lett. 5, 47-51.

Philpotts J. (1970) Redox estimation from a calculation of Eu^{2+} and Eu^{3+} concentrations in natural phases. Earth Planet. Sci. Lett. 9, 257-268.

Philpotts J. (1978) The law of constant rejection. Geochim. Cosmochim. Acta 42, 909-920.

Philpotts J. A. and Schnetzler C. C. (1970) Phenocryst-matrix partition coefficients for K, Rb, Sr, and Ba, with applications to anorthosite and basalt genesis. Geochim. Cosmochim. Acta 34, 307-322.

Phinney W. C., Morrison D. A., and Maczuga D. E. (1988) Anorthosites and related megacrystic units in the evolution of archean crust. J. Petrol. 29, 1283-1323.

Ray G. L., Shimizu N., and Hart S. R. (1983) An ion microprobe study of the partitioning of trace elements between clinopyroxene and liquid in the system diopside-albite-anorthite. Geochim. Cosmochim. Acta 47, 2131-2140.

Roeder P. L. and Emslie R. F. (1970) Olivine-liquid equilibrium. Contrib. Mineral. Petrol. 29, 275-289.

Rogers A. W. (1973) Techniques of Autoradiography, 2nd edition, 372 pp. Elsevier, Amsterdam.

Schnetzler C. C. and Philpotts J. A. (1968) Partition coefficients of rare-earth elements and barium between igneous matrix material and rock-forming mineral phenocrysts--I. In: Origin and Distribution of the Elements, L.H. Ahrens, ed., p. 929-938. Pergamon, New York.

Schnetzler C. C. and Philpotts J. A. (1970) Partition coefficients of rare-earth elements between igneous matrix material and rock-forming mineral phenocrysts--II. Geochim. Cosmochim. Acta 34, 331-340.

Shannon R. D. (1976) Revised effective ionic radii and systematic studies of interatomic distances in halides and chalcogenides. Acta Cryst. A32 751-767.

Shannon R. D. and Pruitt C. T. (1970) Revised values of effective ionic radii. Acta Cryst. B26, 1046-1048.

Shearer C. K. and Papike J. J. (1989) Is plagioclase removal responsible for the negative Eu anomaly in the source regions of mare basalts? Geochim. Cosmochim. Acta, in press.

Shimizu N. (1974) An experimental study of the partitioning of K, Rb, Cs, Sr and Ba between clinopyroxene and liquid at high pressure. Geochim. Cosmochim. Acta 38, 1789-1798.

Shimizu N. and Kushiro I. (1975) The partitioning of rare earth elements between garnet and liquid at high pressures: Preliminary experiments. Geophys. Res. Lett. 2, 413-416.

Stosch H.-G. (1982) Rare earth element partitioning between minerals from anhydrous spinel peridotite xenoliths. Geochim. Cosmochim. 46, 793-811.

Sun C-O., Williams R. J., and Sun S-S. (1974) Distribution coefficients of Eu and Sr for plagioclase-liquid and clinopyroxene-liquid equilibria in oceanic ridge basalt: An experimental study. Geochim. Cosmochim. Acta 38, 1415-1433.

Tanaka T. and Nishizawa O. (1975) Partitioning of REE, Ba, and Sr between crystal and liquid phases for a natural silicate system at 20 kb pressure. Geochem. J. 9, 161-166.

Taylor, S. R. (1982) Planetary science: A lunar perspective. Lunar and Planetary Institute, Houston.

Watson E. B. (1977) Partitioning of manganese between forsterite and silicate liquid. Geochim. Cosmochim. Acta 41, 1363-1374.

Watson E. B. (1985) Henry's law behavior in simple systems and in magmas: Criteria for discerning concentration-dependent partition coefficients in nature. Geochim. Cosmochim. Acta 49, 917-923.

Weill D. F. and Drake M. J. (1973) Europium anomaly in plagioclase feldspar: Experimental results and semiquantitative model. Science 180, 1059-1060.

Weill D. F. and McKay G. A. (1975) The partitioning of Mg, Fe, Sr, Ce, Sm, Eu, and Yb in lunar igneous systems and a possible origin of KREEP by equilibrium partial melting. Proc. Lunar Sci. Conf. 6th, 1143-1158.

Weill D. F., McKay G. A., Kridelbaugh S. J., and Grutzeck M. (1974) Modeling the evolution of Sm and Eu abundances during lunar igneous differentiation. Proc. Lunar Sci. Conf. 5th, 1337-1352.

Zielinski R. A. and Frey F. A. (1974) An experimental study of a rare earth element in the system diopside/water. Geochim. Cosmochim. Acta 38, 545-565.

AN APPROACH TO TRACE ELEMENT MODELING
USING A SIMPLE IGNEOUS SYSTEM AS AN EXAMPLE

INTRODUCTION

Approaching a set of trace element data to evaluate the petrogenesis of a suite of igneous rocks can be an intimidating experience even if one has developed an understanding of how trace elements are distributed in igneous systems. A common approach is to propose a hypothesis and to then use the batch melting or fractional crystallization equations used for trace elements to **forward** calculate from selected mineralogies and chemical compositions. The analytical data are then compared to the calculated data and the closeness of fit is evaluated. This is a time consuming, hit-or-miss approach that often leads to erroneous conclusions, because the solutions are not unique. Also, there are large uncertainties in the selected compositions of the sources, the proportions of minerals involved, and the distribution coefficients to be used.

Presented here is an **inverse** approach to evaluating the processes responsible for the trace element characteristics in a suite of igneous rocks. Using element-element scatter plots constraints can be placed on the processes, compositions, mineralogy and bulk distribution coefficients before doing the necessary **forward** calculations (Sun and Hanson, 1975). In some cases petrogenetic models can be eliminated simply by inspection.

A common criticism of forward modeling approaches concerns the validity of the values for the mineral-melt distribution coefficients selected. In the inverse modeling approach the bulk distribution coefficients[1] for the elements of interest are calculated for a given process. Then one considers whether the mineral proportions and distribution coefficients for a proposed suite of minerals are appropriate.

Generally in an igneous system we are most interested in knowing what process or processes are responsible for the production of the suite of rocks. This usually includes wanting to know whether a given mineral or suite of minerals is involved in any one of the processes, because the mineral assemblage can be used to place constraints on the temperature, pressure and composition of the system. We are usually less interested in knowing the exact proportions of minerals involved or what the exact mineral-melt distribution coefficients might be.

The rare earth elements (REE) play an important role because the changes in the shapes of the REE patterns within a suite of rocks or melts are directly related to the shape of the mineral-melt distribution coefficient patterns of the minerals involved. Changes in the shapes of REE patterns are best evaluated by normalizing the REE data for the whole rocks to one of the more primitive rocks in the sequence. Even though there may be rather large variations in the absolute values for the mineral-melt distribution coefficients, the relative shapes of the patterns of the mineral-melt distribution coefficients for the REE for a given mineral do not usually change much.

When approaching a set of data, it is best to assume that the rocks within one suite of apparently closely related rocks are not the products of one process or one set of mineral assemblages. Using one process to forward model a set of data without a preliminary evaluation of the range of possible processes and mineral assemblages involved can be

---

[1]The bulk distribution coefficient, D_i, for a given element i is the sum of the weight proportions of each mineral, X, in the residue for melting or in the crystallizing assemblages for fractional crystallization multiplied times its mineral-melt distribution coefficient, K_{di}, i.e., $D = \text{sum} (X\ K_{di})$.

80

frustrating. While you may be able to model from one rock to another, the whole suite of rocks may not be explained by the selected model. As a result one keeps trying different models, but not one of them explains the complete suite of rocks. This is because the samples may have been derived from multiple sources with different mineralogies and trace and major element characteristics. These sources may have undergone a range in extents of melting. Each primary melt may have differentiated under different physical conditions involving a separate mineral assemblage, and the melts may have been contaminated by various melts, fluids or country rocks.

REVIEW OF TRACE ELEMENT EQUATIONS

In order to do inverse modeling it is essential to understand how trace element pairs will appear on element-element scatter plots as a function of processes, bulk distribution coefficients and extents of processes. In this section we will calculate how trace elements appear on such plots for melting and fractional crystallization.

We will use the definition of a trace element as presented in Hanson and Langmuir (1978). A trace element is an element: (1) which has such a low abundance in a system that changing its abundance does not affect the stability of any phase in the system[2], (2) which follows Henry's Law in all phases in the system, and (3) therefore, over the total range of abundances considered, its distribution between any two phases in the system can be calculated using a mineral-melt or mineral-mineral distribution coefficient (K_{di}) without consideration of the element's abundance in any other phase in the system.

A number of studies have considered the complexities involved in melting and fractional crystallization. We will consider the simplest conditions for equilibrium or batch melting and fractional crystallization. We are not able to consider more complex models within the confines of this paper. In any case it is more appropriate to first apply simple models. If the simple models do not work for a suite of rocks, one can use the same inverse modeling approaches incorporating more complex models. *Table 1 summarizes the symbols used in this paper.*

Table 1. List of symbols used

C_{li}	Weight concentration of element i in liquid or melt
C_{oi}	Weight concentration of element i in system
C_{si}	Weight concentration of element i in solid phase or phases
D_i	Bulk distribution coefficient
ESC	Essential structural constituent
F	Fraction of liquid or melt in system
i	Element of interest
j	Element of interest
K_{di}	Mineral-melt distribution coefficient for element i
REE	Rare earth element(s)
X	Weight fraction of mineral in residue or mineral assemblage

[2]This is a subjective criterion depending on how well the stabilities of the various phases in the system are known. Or, how well you want to know their stabilities. For example, is it important in the description of the system if changing the concentration of an element by a factor of two decreases or increases the occurrence or disappearance of a phase by 1°C? 10°C? Or 100°C?

Melting

For equilibrium melting the abundance of a given element i in the melt, C_{li}, relative to its abundance in the original parent, C_{oi}, for a given fraction of melting, F, and a given bulk distribution coefficient, D_i, is given by Schilling (1966) as:

$$\frac{C_{li}}{C_{oi}} = \frac{1}{(1-F)D_i + F} \quad . \tag{1}$$

Because this is an equilibrium process the value of C_{li} is independent of the melting path and is dependent only on the bulk distribution coefficient for the minerals in the residue at the time of melt removal. In Figure 1a C_{li}/C_{oi} is plotted against F for a range of D_i's. If D_i is 0 for a given element, $C_{li}/C_{oi} = 1/F$. *Also if D_i is not equal to 0, but is much less than F, $C_{li}/C_{oi} = 1/F$. If for a given suite of rocks, there is an incompatible element, i.e., $C_{li}/C_{oi} = 1/F$, it is possible to place constraints on the relative extents of melting for each of the rocks even without knowing C_{oi}.*

If the extent of melting is less than about 50%, it can be seen from Figure 1 that elements with D_i's near one or larger will show only small ranges in abundance during melting. Elements with D_i's much less than one will show relatively large ranges in abundances.

For a suite of samples the relative compatibilities of the elements in the system can be determined by the use of element-element scatter plots. On a scatter plot of the concentrations of the elements i and j, i.e., C_{li} against C_{lj}

$$C_{li} = \frac{C_{lj}(C_{oi})(D_j(1-F)+F)}{(C_{oj})(D_i(1-F)+F)} \quad . \tag{2}$$

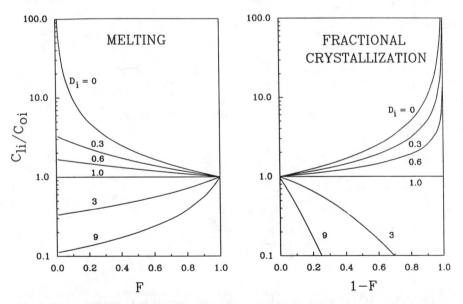

Figure 1. The concentration of a trace element in the liquid C_{li} relative to the trace element in the original parent C_{oi} versus F the fraction of melting and 1-F the fraction of fractional crystallization. D_i is the bulk distribution coefficient.

82

If $D_i = D_j$, or if D_i and D_j are much less than F,

$$C_{li} = \frac{C_{oi}}{C_{oj}} \, C_{lj} \, . \tag{3}$$

This is the equation of a line whose slope is positive and equal to the ratio of the abundances of elements i and j in the source and whose intercept is the origin. Figure 2a is a plot of two elements, C and D, both with $D_i = 0$ from a calculated suite of melts in which the extent of melting varies from 5% to 30%. It can be seen that the data lie along a line which is collinear with the origin and has a slope of 0.333, the ratio of the abundances of the two elements in the parent. *Thus, if two elements are incompatible or have the same D_i's for a suite of melts, the ratio of concentrations in any sample from the suite of melts is the ratio of these elements in the source. On an element-element scatter plot samples from that suite of melts will lie along a line collinear with the origin.*

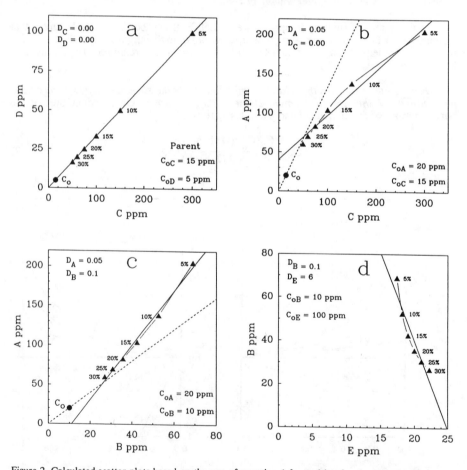

Figure 2. Calculated scatter plots based on the use of equation 1 for melting for the hypothetical trace elements A, B, C, D and E with a variety of bulk distribution coefficients, D_i, and source compositions, C_{oi}. The source composition is the filled circle labeled C_{oi}. The solid line is a regression line through the "data". The dashed line is a line between the source composition, C_{oi}, and the origin. The dotted line is the path for melting with per cents of melting noted for each data point.

In Figure 2b, element A with $D_A = 0.05$ from the same suite of melts is plotted against element C with $D_C = 0$. The data lie about a regression line which intersects the axis along which A is plotted. A similar relation is shown in Figure 2c in which A is plotted against B with $D_B = 0.1$ from the same suite of melts. The data lie about a regression line which intersects the axis along which the abundance of B is plotted. Thus, it is theoretically possible to distinguish relatively minor differences in D_i. In Figure 2d element E with $D_E = 6$ is plotted against element B with $D_B = 0.1$. In this case there is a relatively small range in abundances in E. Also note that for elements with $D_i > 1$ plotted against a nearly incompatible element, the regression line through the data has a negative slope.

From Figure 1a and Figure 2 it has been shown that for a given suite of melts an element with a larger D_i will have a smaller range in abundances than an element with smaller D_i, especially if the extent of melting is less than about 50%. *For elements with $D_i < 1$, a regression line through the data on an element-element plot will have a positive slope and intersect the axis of the more compatible element, i.e., the one with the larger Di. If an element known to have $D_i < 1$ is plotted against an element with a $D_i > 1$, a line through the data will have a negative slope. A negative slope can also be found if the D_i for the more compatible element is less than 1 and decreases as melting proceeds. See Figure 8b and discussion in the text.*

In order to evaluate the relative compatibilities of the elements, it is essential that the data for both elements are plotted from 0 ppm to just slightly above the highest abundance for each element. It is easier to evaluate the data if the plots are square, i.e., the two axes have the same length (in centimeters, not ppm). Using this approach one systematically plots one element against another, ranking the relative compatibilities of the elements within the suite. For example, we have considered five elements in Figure 2. If one looks at the slopes of the lines and intersections on the plots, D and C are equally compatible, A is more compatible than C, B is more compatible than A, and E is much more compatible than B. Thus, D and C are the least compatible elements, and E is the most compatible.

If $F = 0$, Equation 1 becomes

$$\frac{C_{li}}{C_{oi}} = \frac{1}{D_i} \tag{4}$$

For elements with D_i less than one, the maximum enrichment in a melt relative to the source occurs at $F = 0$, but the extent of enrichment is never larger than $1/D_i$. Only those elements with D_i's much less than 1 will show a large range in abundances for a range of F's. For example the maximum possible enrichment for an element with $D_i = 0.02$ is 50, but for an element with a D_i of 0.5 the maximum enrichment is 2. Thus the total range in abundances for an element in a suite of rocks related by melting places constraints on the maximum value of D_i. For example, if the highest abundance is twice that of the lowest, D_i must be less than 0.5. Probably D_i will be much less than 0.5, because rocks derived under similar conditions will most likely have undergone similar extents of melting, thereby greatly restricting the range in abundances.

Because of the large range in abundances of A, B, C and D in Figure 2, we would suspect that D_i for these elements is much less than 1. The six-fold variation in concentrations in D and C would suggest that D_D and D_C are much less than 0.17 and the extent of melting varied by at least a factor of 6. In this case the increase in concentration in E with the decrease in concentration in B in Figure 2d suggests that D_E may be significantly greater than 1.

In a real situation the bulk distribution coefficients may change during melting due to changes in the mineralogy of the residue, temperature or composition of the melt. In many cases it appears that mineral-melt distribution coefficients decrease with increasing temperature. Thus, a decrease in D_i might be expected during static melting where heat is introduced into the rock undergoing melting, for example, near the contact of an igneous intrusion. However, if the proportion of a mineral with a large distribution coefficient increases with extent of melting, D_i would increase during melting.

During adiabatic melting, for example in a rising mantle diapir, temperature decreases during melting because the latent heat of fusion is supplied adiabatically. Thus, for adiabatic melting D_i may increase during melting as a result of decreasing temperature.

It is, of course, always necessary to consider possible changes in the mineral assemblage and its effect on D_i during melting. If melting is an equilibrium process as presented here, changes in mineralogy before the melt separates from residual minerals will have no effect on the trace element composition of the melt. This is because equilibrium processes are path independent. So, the trace element composition of the melt will depend only on the C_{oi}'s and the D_i's for the mineral assemblage at the time of separation.

Fractional crystallization

For fractional crystallization the abundance of a given element in the melt, C_{li}, relative to the abundance in the initial melt, C_{oi}, as a function of the fraction of the melt remaining, F, given by Neumann et al. (1954) is

$$C_{li}/C_{oi} = F^{(D_i - 1)} \tag{5}$$

The extent of fractional crystallization is 1-F. As D approaches 0, $C_{li}/C_{oi} = 1/F$. This is the same relationship as given for melting. If one has an incompatible element in given suite of rocks, it is possible to place constraints on the extents of fractional crystallization. Figure 1b is a plot of C_{li}/C_{oi} against $1-F$ for fractional crystallization. For D_i's less than 1 the curves in Figure 1a and 1b are essentially mirror images, except that for high extents of fractional crystallization, C_{li}/C_{oi} is not limited to being less than $1/D_i$ but can increase indefinitely. *If D_i is greater than 1, there are large changes in C_{li}/C_{oi} for small changes in the extents of fractional crystallization.*

On an element-element plot

$$C_{li} = C_{lj}(C_{oi}/C_{oj})F^{(D_i - D_j)} . \tag{6}$$

If $D_i = D_j$, then $C_{li} = C_{lj}(C_{oi}/C_{oj})$ and the data will lie along a line collinear with the origin, and the slope will be the ratio of the abundances of the elements in the parent composition. If D_i and D_j are both significantly less than 1 for less than about 50% crystallization, the ratio of C_{li}/C_{lj} will be close to C_{oi}/C_{oj}. (For example, for 30% fractional crystallization with $D_i = 0.0$ and $D_j = 0.2$, $C_{li}/C_{lj} = 1.07(C_{oi}/C_{oj})$.) However, it is still possible to determine the relative compatibilities of the elements on a scatter plot using the same approach as was used for melting.

In Figure 3 the effects of up to 50% fractional crystallization are compared to those for melting using the same elements and data as used in the scatter plots from Figure 2. The parental melt for fractional crystallization is the melt which was derived by 15% melting in Figure 2. In Figure 3a for elements D and C in which D_D and D_C are 0, the data from both melting and fractional crystallization lie along the same line and are collinear with the origin. In Figure 3b and 3c regression lines through the fractional crystallization data (open triangles) intersect the axis of the element with the larger D_i. In Figure 3d a regression line through the data has a negative slope. This occurs when $(D_i - D_j)$ has an absolute value greater than 1.

Melting versus fractional crystallization

The discussion to this point has assumed that either melting or fractional crystallization was the operating process. With real data sets the operating process is not known. It is thus necessary to develop criteria for distinguishing these processes. *On plots of elements with D_i less than 1, it is not easy to distinguish between melting and fractional crystallization. However, on a plot of an element with a D_i much less than 1 and an element with D_i greater than 1, melting may be distinguished from fractional crystallization.* This is shown in Figure 3d where for melting there is a relatively greater range in the abundances of element D ($D_D = 0$) relative to element E ($D_E = 6$). The inverse relationship is found for fractional crystallization.

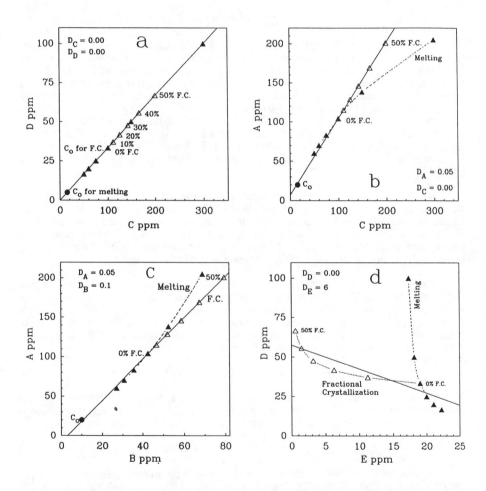

Figure 3. Calculated scatter plots based on the use of equation 5 for fractional crystallization for the hypothetical trace elements A, B, C, D and E superposed on the data for melting from figure 2. The data resulting from the melting calculations are filled triangles. The data resulting from fractional crystallization calculations are unfilled triangles. The parental melt composition, C_{oi}, for fractional crystallization, labeled 0% F.C., is the result of 15% melting (see Fig. 2). The filled circle is the source composition for melting. The dashed line is the path of melting. The dotted line is the path of fractional crystallization. The solid line is a regression line through the fractional crystallization data.

Essential structural constituents

Sun and Hanson (1975) classified an element that virtually filled one site of any mineral in a mineral-melt system as an **essential structural constituent** (ESC). Examples would be Zr in zircon or P in apatite. If zircon or apatite or any other phases in which Zr or P fill a site were not present, Zr or P may be treated as trace elements. The concentration of an ESC in the melt is not dependent on the bulk distribution coefficient but is dependent only on the mineral-melt distribution coefficient for the mineral which has its site filled with the ESC. This is because from the definition of a mineral-melt distribution coefficient, K_{di}, where the element i is the ESC

$$K_{di} = C_{si}/C_{li}, \tag{7a}$$

$$C_{li} = C_{si}/K_{di}. \tag{7b}$$

For a given set of T, P, X conditions K_{di} is a constant. Also, because the site is filled, the concentration of the element in the mineral (C_{si}) cannot be made greater or less. If the concentration of the ESC in the site were to become less, the mineral would no longer be stable. The concentration of the ESC in the site cannot be greater, because the site is already filled. Therefore, C_{si} is fixed as long as the mineral is present. C_{li} is given by Equation 7. C_{li} will vary only if K_{di} varies. For an ESC, C_{li} is not dependent upon the proportion of the mineral containing the ESC either in the residue or among the crystallizing phases. The concentration of the ESC in all other phases in the system is also fixed by Equation 7, i.e., $C_{si} = C_{li}K_{di}$. For a given set of T, P, X conditions the K_{di}'s for the ESC in other phases are constants, and C_{li} is a constant; therefore, C_{si} for each of the other phases is a constant.

For an ESC, as for any other element, K_{di} will be a function of T, P and X. For example, the P and Zr concentrations in a range of silicate melts saturated with apatite and zircon, respectively, can be used to calculate apatite and zircon saturation temperatures (Watson and Harrison, 1984; Harrison and Watson, 1984). For a given composition as temperature decreases, the P or Zr concentration of a melt saturated in apatite or zircon, respectively, decreases because the K_{di}'s increase with decreasing temperature.

The path for Zr in a melt with a zircon-melt K_d of 2,500 is shown for both melting and fractional crystallization in Figure 4. For these examples, the other minerals in the system produce a bulk distribution coefficient for Zr (D_{Zr}) of 0.1 for melting and 0.05 for fractional crystallization. Stoichiometric zircon has 545,000 ppm Zr. Accounting for limited substitution of Hf, U, etc. for Zr, zircon has approximately 500,000 ppm Zr. When zircon is present the concentration of Zr in the melt is given by

$$C_{lZr} = C_{ZirZr}/K_{dZr} = (500,000 \text{ ppm})/2,500 = 200 \text{ ppm}. \tag{8}$$

For melting, with 100 ppm Zr in the original source, zircon will be present in the residue and the melt will have 200 ppm Zr until the last infinitesimal amount of zircon is left. The fraction of melting at which the last zircon melts is given from Equation 1 by:

$$C_{lZr} = C_{oZr}/(D_{Zr}(1-F)+F) = 200 = 100/(0.1(1-F)+F). \tag{9}$$

Solving for F gives 44.4% melting. After zircon is no longer a phase in the residue, the melt has less than 200 ppm Zr. It could also be stated that if the melt has less than 200 ppm Zr, zircon is not stable. From Equation 9, if D_{Zr} for the minerals other than zircon in the residue is equal to or greater than C_{oZr}/C_{lZr}, F has a negative value, showing that the other minerals have incorporated enough Zr so that zircon is not a stable phase in the residue at any time during the melting.

For fractional crystallization, the parental melt has 125 ppm Zr, and the other minerals crystallizing have a D_{Zr} of 0.05. Zircon will not crystallize until the melt has 200 ppm Zr, which will occur when

$$C_{lZr} = C_{oZr}F^{(D_{Zr} - 1)} = 200 = 125F^{-0.95} . \qquad (10)$$

Solving for F gives 0.609. 1-F, the extent of fractional crystallization, is 39.1%.

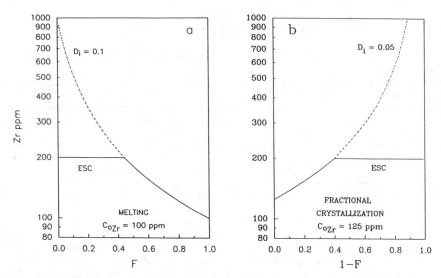

Figure 4a and 4b. Calculated paths on a concentration versus F the fraction of melting and 1-F the fraction of fractional crystallization for Zr. For melting the parent has 100 ppm Zr. The mineral melt distribution coefficient for Zr in zircon is 2500 and for this calculation does not change during melting. The solid line labeled ESC is the path of the liquid when zircon is present. The solid curve is the path of the liquid when zircon is not present in the residue. For melting the minerals in the residue besides zircon have a $D_i=0.1$. So when zircon is no longer in the residue the path follows the extension of the dashed $D_i=0.1$ path. For fractional crystallization the parent liquid has 125 ppm Zr, the crystallizing minerals other than zircon have a $D_i=0.05$, so before zircon begins to crystallize the liquid follows the curved path of $D_i=0.05$. When zircon is present the Zr abundance of the melt is fixed by the zircon-melt distribution coefficient for Zr (see equation 7 and discussion).

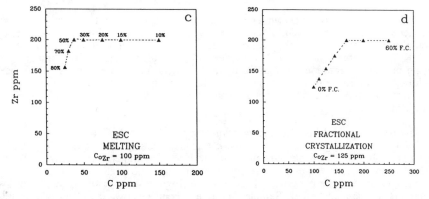

Figure 4c and 4d. Scatter plots showing Zr from Figure 4a and b versus incompatible element C from Figures 2 and 3 for melting and fractional crystallization. In a scatter plot when zircon is present, Zr is an ESC and has a constant abundance. When zircon is not present, Zr has the properties of a slightly compatible element.

In summary, if an element occurs as an essential structural constituent (ESC) in any phase in the system, its concentration in the melt for either batch melting or fractional crystallization is not dependent on the proportion of that mineral in the residue or among the crystalizing phases. It is dependent only on the concentration of the ESC in the solid phase and the mineral-melt distribution coefficient (K_{di}) for the ESC in that phase. In real systems, the concentration of the ESC in the melt varies with the extent of melting or crystallization while the phase is present, because K_{di} is a function of T, P and X, which are changing.

EXAMPLE OF PETROGENETIC APPROACH

In this section we will place constraints on the processes responsible for the trace element composition of a suite of volcanics. Rather than using actual data, we will use a set of calculated data for different extents of melting and fractional crystallization. To keep things simple, we will consider only batch or equilibrium melting and fractional crystallization. It is assumed that the reader has a basic knowledge of trace element abundances and REE patterns for rocks and potential mantle sources and appropriate mineral-melt distribution coefficients. See, for example, Hanson (1977, 1980) and Taylor and McLennan (1981). The following data are calculated for a sequence of nearly aphyric volcanic rocks, i.e., with negligible fractions of phenocrysts. We will define the volcanics as occurring on an oceanic island, where there is no possible contamination by components of the continental crust. Also, all of the volcanics will have been ultimately derived from one source. The data have an analytical uncertainty of +1% of the amount present.

While it would be useful to also consider the major elements, we will restrict our considerations to the trace elements. We will limit our data base to six whole-rock samples. The trace element data are presented in Table 2. Samples A, B and C are basalts with only olivine phenocrysts. Sample D is a basaltic andesite with phenocrysts of olivine and clinopyroxene. Samples E and F are andesites with phenocrysts of plagioclase and clinopyroxene. Because the fractions of phenocrysts are negligible and the phenocrysts represent liquidus phases, the mineral to whole rock ratios for the trace elements given in Table 3 approximate mineral-melt distribution coefficients. The data have been calculated such that samples A, B and C are the results of different extents of melting of identical sources. Samples D, E and F are on the liquid line of descent from sample B. Sample D is derived by 11% fractional crystallization of only olivine from sample B. Sample E is derived by 16% fractional crystallization of equal amounts of olivine and clinopyroxene. Sample F is derived from sample E by 12% fractional crystallization of equal amounts of plagioclase and clinopyroxene.

We will now consider what constraints the trace element data place on the petrogenesis of these rocks. The chondrite normalized REE patterns for the six samples are parallel or sub-parallel and light REE enriched (Fig. 5). We will need to test whether the light REE enrichment or heavy REE depletion is a result of garnet involvement either during melting or fractional crystallization. All of the samples except sample F have slightly positive Eu anomalies. Is the positive Eu anomaly a source characteristic, or does it result from a mineral such as garnet or clinopyroxene with negative Eu anomalies in their K_{di} REE patterns?

Normalizing the REE abundances of the samples to sample C, which has the lowest REE abundances in Figure 6a, allows a more magnified view of the variations from rock to rock than is possible on a chondrite normalized plot. The mineral-to-rock ratios for the REE in clinopyroxene and plagioclase are shown in Figure 6b to allow a comparison with the changes seen in the patterns in Figure 6a. In Figure 6a the patterns for samples A through D are flat and parallel. Sample E is slightly concave upward with a slight positive Eu anomaly. Sample F is more concave upward with a distinct negative Eu anomaly. These characteristics suggest that sample E is derived from a melt that had undergone removal of clinopyroxene relative to samples A, B, C and D. These characteristics also

Table 2. Trace element data for rocks and phenocrysts in ppm or chondrite normalized values (cn).

Sample	wr A	ol A	wr B	wr C	ol C	wr D	ol D	cpx D	wr E	plag E	wr F
Rock Type	Basalt		Basalt	Basalt		Basaltic Andesite			Andesite		Andesite
Phenocrysts	ol		ol	ol		ol+cpx			plag+cpx		plag+cpx
Element											
Ni ppm	270	2700	265	275	2780	93	1230	178	25.2		23.9
Zr ppm	195		224	152		252		63	294	4.89	328
Ce cn	110		127	85.8		143		28.6	167	26.7	186
Nd cn	74.2		85.7	57.9		96.5		36.4	110.3	13.02	121
Sm cn	50.1		57.9	39.1		65.2		34.9	73.5	6.91	80.2
Eu cn	42.8		49.4	33.4		55.7		27.7	63.1	60.7	65.3
Gd cn	34.2		39.5	26.7		44.5		27.4	49.9	3.89	54.2
Dy cn	23.7		27.4	18.5		30.8		18.1	34.5	2.17	37.6
Er cn	17.44		20.1	13.6		22.67		12.2	25.7	1.336	28.1
Yb cn	14.36		16.58	11.2		18.67		7.95	21.3	0.959	23.5
Rb ppm	41.4		47.6	32.3		53.5		0.053	63.6	4.45	71.9
Sr ppm	652		750	509		843		59	997	2850	940
K %	1.66		1.90	1.29		2.14		0.0021	2.54	0.490	2.88
K/Rb	401		399	399		400			399		401
Rb/Sr	0.0635		0.0635	0.0635		0.0635			0.0638		0.0765
Ce/Yb	7.64		7.65	7.66		7.69			7.84		7.91

Table 3. Phenocryst to whole rock distribution coefficients

Element	ol A/wr A	ol C/wr C	ol D/wr D	cpx D/wr D	plag E/wr E
Ni	10.0	10.1	13.2	1.91	—
Zr				0.250	0.0166
Ce				0.200	0.160
Nd				0.377	0.118
Sm				0.536	0.094
Eu				0.498	0.962
Gd				0.616	0.078
Dy				0.587	0.063
Er				0.538	0.052
Yb				0.426	0.045
Rb				0.001	0.070
Sr				0.070	2.86
K				0.001	0.193

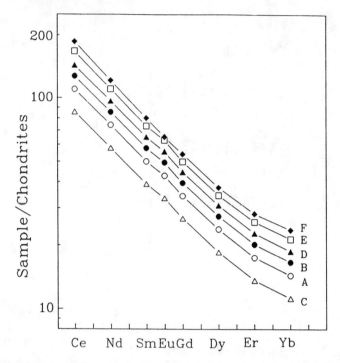

Figure 5. Chondrite normalized rare earth element plot of whole rock samples A, B, C, D, E and F (data from Table 2).

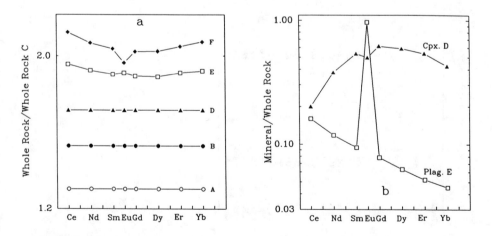

Figure 6a. Rare earth element plot of whole rocks A, B, D, E and F normalized to whole rock sample C.
Figure 6b. Mineral to whole rock plot of plagioclase in whole rock sample E and clinopyroxene in whole rock sample D (data from Table 3).

suggest that sample F had undergone removal of plagioclase and clinopyroxene relative to the other samples. These hypotheses will need to be considered in other element-element plots and then quantitatively evaluated.

To place constraints on the relative importance of the possible processes involved, the extents of any process, and the minerals involved, it is necessary to determine the relative compatibility of the trace elements considered using element-element plots. For a sequence of basic rocks, Ce would be expected to be the most incompatible of the REE. K and Rb might be similarly incompatible to Ce. For the plots involving K, Rb, Ce, Zr, and Yb (Fig. 7), there is a large range in abundances and the data are collinear with an intersection near the origin. This suggests that these elements have low and similar compatibilities. For such elements it is difficult to distinguish whether the variations are a result of different extents of melting or crystallization.

The plot of Ni and Rb (Fig. 7f), however, shows divergent trends which is useful for discriminating between melting and fractional crystallization (compare with Figure 3d). We know from Table 3 that the olivine-liquid K_{dNi}'s are 10 to 13 in samples A, C and D. We are also suspicious that the basaltic rocks may have been derived from an olivine-rich peridotitic mantle with about 2000 ppm Ni (see Taylor and McLennan, 1981). Basalt samples A, B and C show little variation in Ni abundance, about 270 ppm, and a rather large variation in Rb abundance. In comparison, samples D, E and F have much lower abundances of Ni. Compare Figure 7f with Figure 3d. A hypothesis that we will want to consider is that samples A, B and C represent near primitive melts from a mantle, and that samples D, E and F represent liquids whose parents have undergone olivine removal from melts with compositions similar to samples A, B or C.

In Figure 7a, Rb is plotted against K. A regression line through the data intersects the origin, suggesting that K and Rb are equally compatible (or incompatible) and the K/Rb ratio (Table 2) is the K/Rb ratio for the parent or source. The range in Rb and K concentrations greater than a factor of two suggests that D_K and D_{Rb} are small. In Figure 7b, Rb is plotted against Ce. A regression line through the data intersects the Ce axis showing that if this sequence of rocks is related, Rb is more incompatible than is Ce. Samples A, B and C may represent different extents of melting of a common source. A regression line (dashed) through samples A, B and C shows that they are collinear with the origin and with sample D. Samples E and F are slightly off the line.

Figure 7c is a plot of Rb against Zr in which a regression line intersects the Zr axis, showing that Zr is more compatible than Rb. A regression line through samples A, B and C is collinear with the origin and sample D. Samples E and F are slightly off the line. We would suggest the large range of Zr abundances suggests that zircon was not a crystallizing phase (compare with Fig. 4).

We will try to resolve whether the strong depletion of the heavy REE's is a result of a mineral such as garnet with relatively small K_{di}'s for the light REE and relatively large K_{di}'s for the heavy REE or whether it is a characteristic of the source. In Figure 7d, Ce is plotted against Yb. A regression line through the data suggests that Ce is more incompatible, but the large range in Ce and Yb abundances suggests that both elements have relatively small D_i's. A regression line through samples A, B and C is also collinear with the origin and sample D. Samples E and F are slightly off the line.

For comparison with these data, Figure 8a and 8b show the calculated effects of garnet on the REE patterns and Ce-Yb plots during melting from 1.6% to 22% of garnet peridotite. The original mantle has a flat REE pattern with two times chondritic abundances and a mineralogy consisting of 10% garnet, 10% clinopyroxene, 20% orthopyroxene, and 60% olivine. The K_{di}'s are those for mantle-melting in Hanson (1980) and plotted in Figure 9. The proportions of minerals going into the melt are given by Mysen (1982, p. 513). Using this relationship, garnet is present in the residue in decreasing proportions until 16% melting when it is used up. Clinopyroxene is used up by 22% melting at which

92

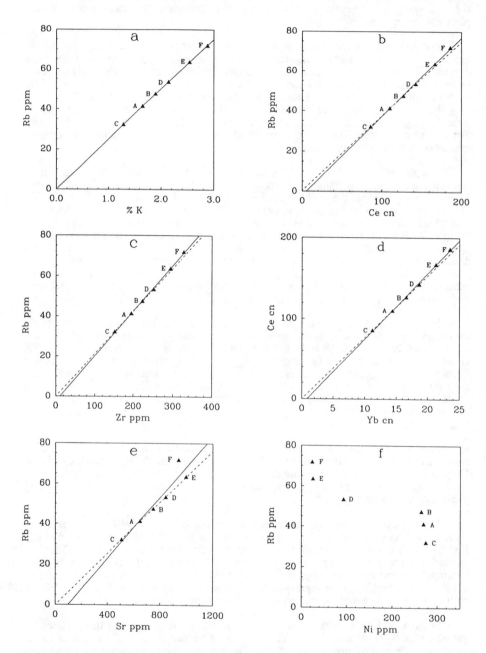

Fig. 7. Selected element-element scatter plots for the data in Table 2. The solid line is a regression through all of the samples. The dashed line is a regression through only samples A, B and C.

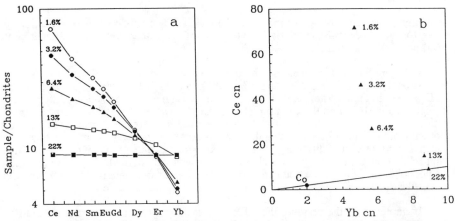

Figure 8a. Calculated chondrite normalized plot of different extents of melting of a garnet peridotite with a flat REE pattern with an original abundance of 2 times chondrites and 10% garnet, 10% clinopyroxene, 20% orthopyroxene and 60% garnet. The minerals involved in melting follow the relations of Mysen, 1982. The mineral-melt distribution coefficients for the REE's are shown in Figure 9 They are taken from Table 2 (mantle melting) in Hanson, 1980. The percentages on the curves are the percents of melting. At 16% melting garnet is used up. At 22% melting only olivine and orthopyroxene are in the residue. The key feature for REE patterns for melts derived from different extents of melting with garnet in the residue is the fanning of the REE patterns. That is, there is a wider range in abundances for the light REE than for the heavy REE. For this example the difference between the bulk D_i's for the heavy and light REE's is great enough to allow crossing of the REE patterns. Compare this plot with that of Figure 5 where the REE patterns are also strongly enriched in the light REE and depleted in the heavy REE, but the patterns are parallel, i.e they do not fan out from the heavy REE much less cross.

Figure 8b. Scatter plot of chondrite normalized Ce versus Yb for the REE patterns shown in Figure 8b. The percents of melting are given for each of the melt. In Figure 7d Ce and Yb have about the same relatively large rang abundances, suggesting both are relatively incompatible. In Figure 8b the smaller variation in abundance of Yb compared to that for Ce suggests that Yb is much more compatible than Ce in this plot for melting of a garnet peridotite.

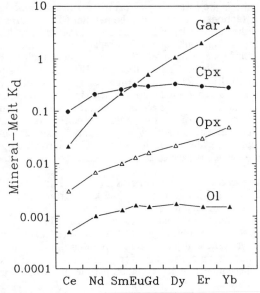

Figure 9. Mineral-Melt distribution coefficients (K_d) for garnet (Gar), clinopyroxene (Cpx), orthopyroxene (Opx) and olivine (Ol) used for calculating the REE abundances for the melts derived from a garnet peridotite shown in Figure 8. The distribution coefficients are from Table 2 (mantle melting) in Hanson, 1980.

time only olivine and orthopyroxene are in the residue. The bulk distribution coefficients D_i for Ce and Yb decrease regularly from 0.012 and 0.41, respectively, at 1.6% melting to 0.001 and 0.01 at 22% melting.

In Figure 8a it can be seen that there is a rotation and crossing of the REE patterns from strongly light REE enriched at 1.6% melting to flat at 22% melting. In contrast, the REE patterns in Figure 5 are parallel to each other. The effect of the rotation of the patterns in Figure 8a is shown on the Ce-Yb plot in Figure 8b where Ce can be seen to decrease with greater extents of melting, but the Yb abundance increases. Comparing this Ce-Yb plot with that in Figure 7d and noting the relationships of samples A, B and C in Figure 7d which may be a result of different extents of melting of a mantle source, it can be seen that samples A, B and C lie on a line which is collinear with the origin. In Figure 8b the melts are not collinear. If a line were drawn through the data, it would intersect the Yb axis with a negative slope, suggesting that Yb has a much larger D_i than does Ce.

If garnet is in the residue, the REE patterns for a sequence of melts would be expected to not vary as much in the heavy REE's as in the light REE's. Depending on the proportion of garnet in the residue and the K_{di}'s, i.e., the value of D_i, for the heavy REE's, the patterns may cross, as they do in Figure 8a, or may not cross if the D_i's for the heavy REE's were smaller. In any case, the REE patterns would fan out from the heavy REE and on a Ce-Yb plot Yb would appear to be much more compatible than Ce.

The parallel REE patterns and the collinear array of the data on the Ce-Yb plot with an intersection near the origin suggest that a mineral with large distribution coefficients for the heavy REE is not the cause of the heavy REE depletion in samples A through F. Instead, the heavy REE depletion of these samples is most likely a characteristic of their source.

In Figure 7e, Rb is plotted against Sr in which the data for samples A, B, C, D and E are collinear with the origin, but sample F lies significantly off the line, showing that for sample F the Sr abundances are depleted relative to Rb.

If samples A, B and C are essentially direct melts from the mantle, perhaps having undergone only minor fractionation of olivine, the parallel REE patterns for samples A, B and C in Figure 6a and the collinearity of the elements in Figure 7a to 7e for samples A, B and C suggest that these elements are equally incompatible, i.e., either they have the same D_i's or, more likely, their D_i's are much less than the extent of melting F. In either case, the ratios for the elements K, Rb, Sr, Zr and all of the REE are the ratios in the source. The heavy REE depletion or light REE enrichment and the positive Eu anomaly are characteristics of the source and are not due to a mineral in the residue such as garnet. This also means that we cannot use the trace element data for these elements to give us any information about the mineralogy of the source during melting, other than to say that the D_i's are much less than the extent of melting, F. The only constraint we can place on the extent of melting is that it is <66%[3] and that sample C is derived by greater extents of melting than is sample A, which is derived by greater extents of melting than is sample B.

We are now going to consider that samples D, E and F may be differentiates from parent liquid(s) similar to samples A, B or C. We know from the phenocrysts that the crystallizing phases are only olivine in samples A, B and C, olivine and clinopyroxene in sample D and plagioclase and clinopyroxene in samples E and F. From other studies Rb is known to have very low olivine-liquid K_{di}'s and the calculated K_{di}'s based on the trace element data for clinopyroxene and plagioclase show that the K_{dRb} for this sequence of rocks is 0.001 for clinopyroxene and 0.07 for plagioclase (see Table 3). As a first approximation we will assume that Rb is incompatible and can be used to constrain

---

[3]The maximum concentration of the near incompatible elements in the source is the abundance in sample C. Thus, the ratio of the abundances in sample C to sample B i equal to F, which is a maximum for the extent of melting.

possible values of the extent of fractional crystallization (1-F). If sample D is derived from sample A and D_{Rb} is essentially 0, from Equation 5, $F = C_{ORb}/C_{IRb} = 41.4/53.5 = 0.774$ or 22.6% fractional crystallization. If sample D is derived from sample B, there is 11% fractional crystallization. If sample D is derived from sample C, there is 40% fractional crystallization. Considering the Ni data, the D_{Ni} for derivation from sample A is given by Equation 5 as $C_{INi}/C_{oNi} = F^{(DNi-1)}$. So

$$D_{Ni} = \frac{Log(C_{INi}/C_{oNi})}{Log\ F} + 1 \qquad (11)$$

solving $D_{Ni} = 5.2$. For derivation from sample B, $D_{Ni} = 10$. For derivation from sample C, $D_{Ni} = 3.1$.

If sample D is derived from sample A, assuming only olivine and clinopyroxene as possible crystallizing phases, the olivine-melt K_{dNi} would be between 10 and 13 based on the phenocryst data for samples A and D. The clinopyroxene-melt K_{dNi} value would be about two based on the clinopyroxene phenocryst data for sample D. The proportions of olivine and clinopyroxene as crystallizing phases will be given by

$$D_{Ni} = X_{cpx}\ ^{cpx/l}K_{dNi} + X_{ol}\ ^{ol/l}K_{dNi} . \qquad (12)$$

$X_{ol} = (1- X_{cpx})$, so we can solve for X_{ol} and X_{cpx} using the phenocryst data. X_{cpx} is 0.6 for olivine-liquid K_{dNi} of 10 and 0.71 for K_{dNi} of 13. For this amount of clinopyroxene and 22.6% fractional crystallization, we would expect to see a depletion of the middle REE in Figure 5a. We do not. This suggests that Sample A is probably not the parent.

Solving Equation 12 for sample B as the parent with $D_{Ni} = 10$ gives X_{cpx} of 0 if the olivine-melt K_{dNi} is 10, and 27% if K_{dNi} is 13. If the K_{dNi} is closer to 10 and the proportion of clinopyroxene in the crystallizing liquid is low, the middle REE would not be expected to be relatively depleted in sample D, as seen in Figure 5a.

Fractionating from sample C to D would require more clinopyroxene in the crystallizing phases than fractionating from sample A. So sample C is even less likely as a parent. We would thus conclude that a melt with trace element characteristics such as those of sample B could be the parent for sample D.

Sample E has a rock-normalized REE pattern (Fig. 6a) which is bowed down in the middle REE and has a positive Eu anomaly suggestive of the removal of clinopyroxene. It is also strongly depleted in Ni, suggestive of the removal of olivine. Although sample E has plagioclase phenocrysts, it has the same Rb/Sr ratio as samples A, B, C and D, suggesting that if it is related to those samples, it has not had significant plagioclase removal. If it were derived from a liquid similar in composition to sample D, use of Rb as an incompatible element gives 16% fractional crystallization.

The plagioclase-liquid $K_{dSr} = 2.9$ based on the plagioclase phenocrysts in sample E. To have less than a 2% effect on the Rb/Sr ratio, i.e., twice the analytical uncertainty, use of Equation 6 shows that D_{Sr} would need to be less than 0.12. Because the plagioclase-melt K_{dSr} is about 2.9, plagioclase makes up less than 4% of the crystallizing phases. $D_{Ni} = 8.5$. Such a large D_{Ni} requires that olivine be one of the crystallizing phases. Assuming that the olivine-liquid $K_{dNi} = 13$ and the only other phase crystallizing is clinopyroxene with $K_{dNi} = 2$, this gives 41% clinopyroxene and 59% olivine in the crystallizing phases.

Sample F has a rock normalized REE pattern (Fig. 6a) that is bowed down in the middle REE with a significant negative Eu anomaly. Sr is depleted relative to Rb (Fig. 7d). Plagioclase and clinopyroxene are phenocryst phases. Ni has a similar although slightly higher abundance than it does in sample E. If sample F is related to this sequence of rocks, the negative Eu anomaly and the Sr depletion suggest plagioclase removal. The bowed down nature of the middle REE is suggestive of the removal of clinopyroxene.

If sample F is derived from a melt with a chemistry similar to that of sample E, the extent of fractional crystallization using Rb as an incompatible element is 11.5%. $D_{Sr} = 1.5$. Assuming that the plagioclase involved is similar to that analyzed in sample E, this would require about 50% plagioclase among the crystallizing phases. $D_{Ni} = 1.4$. The K_{dNi} for clinopyroxene is 1.9 in sample D and may be even greater in the more evolved melts represented by samples E and F. If clinopyroxene were a crystallizing phase with a K_{dNi} similar to that in sample D, this would require about 50% clinopyroxene in the crystallizing phases. If olivine were a crystallizing phase, the low value of D_{Ni} in going from sample E to F would restrict the proportion of olivine in the crystallizing phases to significantly less than 10%, because the more bowed down nature of the REE in sample F relative to sample E suggests that significant clinopyroxene would be crystallizing. Using a similar approach, one can evaluate the bulk distribution coefficients for each of the REE and Zr. Such an analysis would show that fractional crystallization involving subequal fractions of clinopyroxene similar to that in sample D and plagioclase similar to that in sample F would be consistent with derivation of sample F from a melt with trace element characteristics similar to those of sample E.

DISCUSSION AND SUMMARY

The approach presented here has been to look for trends in data in order to evaluate the relative importance of batch melting and fractional crystallization as processes. Igneous systems can be complicated by the effects of ranges in composition of the source for the melts, cumulate mineralogy, and contamination of the melt or rock. The approach presented here allows one to test to see if the simple systems can explain the data. If they do not, then it is quite possible to add to the complexity. In adding to the complexity one should place special emphasis on those elements which may be highly affected by the complexities as well as those that are relatively insensitive to them.

This approach is most effective if applied to relatively large numbers of samples that are closely related to each other, minimizing the gaps in the extents of the processes that have affected them. It is best to assume that a simple liquid line of descent will not be found and that any one sample may have had a somewhat unique history relative to other samples, i.e., it is not directly related to any other sample studied. A major goal is to understand the processes and minerals involved which can be evaluated from the trends for the variations in extents and conditions of melting and fractional crystallization.

Such an evaluation requires using a range of compatible and incompatible elements with highest quality analytical data, i.e., low analytical uncertainties. For example, 10% fractional crystallization results in large changes in the abundances of elements with large bulk distribution coefficients. Elements with bulk distribution coefficients less than 1 which are needed to place constraints on the extent of a process, however, will change in abundance by at most 10%. If the analytical uncertainties for these elements are on the order of 10%, it will be difficult to evaluate which rock is more evolved and impossible to evaluate the extent of the process relating the rocks and their bulk distribution coefficients.

For a suite of samples the relative compatibilities of the elements in the system can be determined by the use of element-element scatter plots. Some of the evaluations that can be made by inspection include:

(1) If two elements are incompatible or have the same D_i's for a suite of melts, the ratio of concentrations in any sample from the suite of melts is the ratio of these elements in the source. On an element-element scatter plot samples from that suite of melts will lie along a line collinear with the origin.

(2)   The relative compatibilities of the elements can be evaluated. For elements with $D_i < 1$ a regression line through the data on an element-element plot will intersect the axis of the most compatible element, i.e., the one with the largest D_i. If an element known to have $D_i < 1$ is plotted against an element with a $D_i > 1$, a line through the data will have a negative slope. However, a negative slope does not mean that D_i for the more compatible element is greater than one. A negative slope can also occur if the D_i for the more compatible element with a $D_i < 1$ is decreasing during melting.

(3)   On plots of elements with D_i equal to or less than 1 it is not easy to distinguish between melting and fractional crystallization. However, on a plot of an element with a D_i much less than 1 and an element with D_i greater than 1, melting may be distinguished from fractional crystallization.

(4)   If an element occurs as an essential structural constituent (ESC), i.e., it essentially fills a site in any phase in the system, its concentration in the melt for either batch melting or fractional crystallization is determined by the concentration of the ESC in the solid phase and the mineral-melt distribution coefficient (K_{di}) for that phase. Examples of ESC's include Zr in zircon, and P in apatite.. In real systems the concentration of the ESC in the melt varies with the extent of melting or crystallization while the phase is present because K_{di} is a function of T, P and X.

(5)   If for a given suite of rocks, there is an incompatible element, i.e., $C_{li}/C_{oi} = 1/F$, it is possible to place constraints on the relative extents of melting or fractional crystallization.

ACKNOWLEDGMENTS

Many of the ideas presented here were first developed at Stony Brook with J.G. Arth and S.S. Sun and have evolved since the mid-1970's through interactions with many graduate students and colleagues who have done their petrogenetic research at Stony Brook. Members of the Isotope Laboratory as well as the editors of this volume reviewed the original manuscript. The latest grants supporting this presentation are NSF EAR-8607973, NSF INT-8814308 and NASA NAG985.

REFERENCES

Hanson, G.N. (1977) Geochemical evolution of the suboceanic mantle. J. Geol. Soc. 134, 235-253.

Hanson, G.N. (1980) Rare earth elements in petrogenetic studies of igneous systems. Ann. Rev. Earth Planet. Sci. 8, 371-406.

Hanson, G.N. and Langmuir, C.H. (1978) Modelling of major and trace elements in mantle-melt systems using trace element approaches. Geochim. Cosmochim. Acta 42, 725-741.

Harrison, T.M. and Watson, E.B. (1984) The behavior of apatite during crustal anatexis: equilibrium and kinetic considerations. Geochim. Cosmochim. Acta 48, 1467-1477.

Mysen, B.O. (1982) The role of mantle anatexis. In: R.S. Thorpe, ed., Andesites: Orogenic Andesites and Related Rocks. John Wiley and Sons, New York., p. 489-522.

Neumann, H., Mead, J. and Vitaliano, C.J. (1954) Trace element variation during fractional crystallization as calculated from the distribution law. Geochim. Cosmochim. Acta 6, 90-99.

Schilling, J.-G. (1966) Rare earth fractionation in Hawaiian volcanic rocks. Unpublished Ph.D. thesis, Mass. Inst. Tech., Cambridge, MA.

Sun, S.S. and Hanson, G.N. (1975) Origin of Ross Island basanitoids and limitations upon the heterogeneity of mantle sources for alkali basalts and nephelinites. Contrib. Mineral. Petrol. 52, 77-106.

Taylor, S.R. and McLennan S.M. (1981) The composition and evolution of the continental crust: rare earth element evidence from sedimentary rocks. Phil. Trans. R. Soc. London A301, 381-399.

Watson, E.B. and Harrison, T.M. (1984) Accessory minerals and the geochemical evolution of crustal magmatic systems: a summary and prospectus of experimental approaches. Physics Earth Planet. Interiors 35, 19-30.

RARE EARTH ELEMENTS IN UPPER MANTLE ROCKS

INTRODUCTION

Abundances of rare earth elements (REE) in upper mantle rocks and minerals were previously reviewed by Frey (1984). In this chapter we build upon this earlier review and emphasize ideas, models and interpretations that are based on REE data published since 1982. This chapter is not a thorough review of the data base in the literature, but it is a comprehensive review of how REE abundance data can be used to understand and constrain upper mantle processes. Specifically, our discussion focuses on:

I.   MASSIVE PERIDOTITES
      Massive peridotites: dominantly lherzolite.
            Western Alps - Lanzo.
            Western Alps - Baldissero, Balmuccia.
            Eastern Liguria, Italy.
            Western Liguria, Italy.
            Eastern Pyrenees - France.
            Ronda, Spain.
            Effects of late stage alteration on REE.
      What can be inferred about the melting process and the segregated melts?
      Massive peridotites: pyroxenite layers and veins and their wall rocks.
            Amphibole-bearing pyroxenite veins.
            Anhydrous pyroxenite layers.
            How were the pyroxenite layers created? Evidence for multistage processes.
            Implications for mantle enrichment processes (metasomatism) and genera-
                tion of different primary magmas.
      Massive peridotites: dominantly harzburgite.
      Oceanic peridotites.

II. ULTRAMAFIC XENOLITHS
      Group 1 spinel peridotites.
      Garnet peridotites.
      Pyroxenite and related xenoliths.
      Models for REE abundance trends in peridotite xenoliths.

III. MEGACRYSTS, MINERALS IN XENOLITHS AND DIAMOND INCLUSIONS
      Megacrysts.
      Minerals in peridotites and pyroxenites.
      Inclusions in diamonds.

IV. SUMMARY: COMPARISON OF PERIDOTITES FROM MASSIFS AND XENO-
      LITHS AND IMPLICATIONS OF REE DATA FOR UPPER MANTLE
      COMPOSITION

We emphasize that in addressing problems related to mantle processes, REE should not be used alone. A major strength of geochemistry is the ability to use data for many elements with different geochemical characteristics. REE are clearly a very useful geochemical group, but even here where we focus on REE, it will be apparent that the strongest interpretations result from evaluation of how REE abundances correlate with abundances of other trace elements and major elements.

In the broadest sense REE abundances in upper mantle rocks and minerals have two general applications. (1) A recorder of mantle processes; e.g., the mechanisms of partial melting and melt segregation; the range in degree of melting, and the role of melt and fluid migration in causing mantle enrichment[a] and metasomatism. (2) Definition of compositional heterogeneity within the upper mantle; i.e., how are the sources of compositionally different primary magmas distributed in the mantle; and how much compositional variation is there on a variety of scales ranging from centimeters to kilometers?

MASSIVE PERIDOTITES

Geochemical studies of massive peridotites are more amenable to interpretation than geochemical studies of mantle xenoliths in volcanic rocks because field relations can be used to constrain hypotheses. In the past, the more highly altered nature of massive peridotites (e.g., the ubiquitous presence of serpentine) led many geochemists to concentrate their efforts on the less altered mantle xenoliths found in volcanic rocks. However, the problems created by alteration have become less important because: (1) several studies have shown that major element and REE abundance trends are not obviously affected by serpentinization (e.g., Frey et al., 1985; Bodinier et al., 1988a); and (2) recently developed techniques to separate and analyze pure minerals and use of the ion microprobe for "in situ" analyses of unaltered minerals minimize the need for analyses of whole-rocks which contain minerals formed during late-stage alteration.

In our discussion of massive peridotites we divide these rocks into two groups: massifs where lherzolite is dominant and massifs, where harzburgite is dominant, e.g., the ultramafic tectonite forming the basal section of ophiolites. This is a logical distinction because these groups tend to have very different chondrite-normalized REE patterns. The following discussion is focussed on identifying the igneous processes that have affected these massive peridotites. Like most mantle rocks these peridotites have metamorphic textures developed at subsolidus conditions; it is important to determine if we can see through this metamorphism and understand higher temperature processes that affected these peridotites.

Early studies of REE abundances in massive peridotites were limited to a few samples (usually <10) from individual massifs (Frey, 1984). However, in the last few years several massive peridotites have been studied in much more detail. With geochemical data for several peridotites and pyroxenites from a single body it is now possible to utilize the observed range in REE abundances to identify and constrain the processes that determined the compositions of these upper mantle rocks. For example, most of these peridotites have (LREE/HREE)$_N$ <1 (the subscript "N" indicates a chondrite-normalized ratio)[b]. If these peridotites initially had relative abundances of REE similar to chondritic meteorites, the relative LREE depletion is consistent with an origin as residues from partial

---

[a] We use "enrichment" and "depletion" to describe increases and decreases, respectively, in abundances of incompatible elements. The terms "depleted" and "enriched" designate mantle that has lower and higher, respectively, abundances of incompatible elements than inferred primitive mantle (for estimates of primitive mantle composition, see Hart and Zindler, 1986; Sun and McDonough, 1989). As an example, depleted mantle has LREE/HREE abundances ratios less than chondritic ratios; primitive mantle has chondritic LREE/HREE ratios; and enriched mantle has LREE/HREE ratios greater than chondritic ratios. (LREE = light rare earth elements, La to Nd; MREE = middle rare earth elements, Sm to Tb; HREE = heavy REE, Er to Lu.)

[b] In this chapter we use several REE abundance ratios as a measure of slopes on a chondrite-normalized REE diagram; e.g., La/Yb reflects fractionation between the typically highly incompatible LREE and relatively more compatible HREE (see REE partitioning chapter by McKay). (Yb is used rather than Lu because Yb is more abundant and determined with a higher precision; Ce/Yb is used when La data are absent). Tb/Yb ratios are used because, unlike clinopyroxene, spinel and plagioclase, garnet is effective in changing the abundance ratio of these two REE (McKay, this volume). Some peridotites are preferentially depleted in the intermediate atomic number REE, Sm to Tb. A measure of this depletion is given by Tb/Yb and La/Sm (or Ce/Sm) ratios. Sm/Nd ratios are also used because this is the parent/daughter ratio that controls the abundance of ^{143}Nd (see, Patchett, this volume).

melting. If this is an appropriate explanation, we need to evaluate if REE abundance data can be used to answer the following more detailed questions: (1) what was the range and extent of melting; (2) how did the melts form, accumulate and segregate; e.g., can we distinguish between batch melting, fractional melting[c] and porous flow models; (3) what were the residual phases that equilibrated with the melt; e.g., were recrystallized plagioclase or spinel peridotites originally formed as garnet peridotite residues; (4) what was the composition of the segregated melt; e.g. tholeiitic or alkalic?

Despite the dominance of peridotites with $(LREE/HREE)_N$ <1 in massive peridotites, it is definitely established that some peridotites in these massifs have $(LREE/HREE)_N$ >1. Does this reflect pre-melting compositions, low-temperature alteration or high temperature melt/fluid migration in the mantle? Can the spatial constraints provided by these massive peridotites be used to constrain enrichment and metasomatic processes in the mantle?

Also important is the diversity of rock types, e.g., lherzolite, harzburgite, dunite and pyroxenite, in massive peridotites (see Fig. 1 for nomenclature). Pyroxenites typically occur as dikes and ~1 cm to 2 m thick tabular bodies traceable for several hundred meters within the peridotite. In the peridotite literature, the latter are commonly referred to as mafic or pyroxenite layers. These pyroxenites have a wide range of REE abundances ranging from relative LREE depletion to enrichment. Do REE abundances in these pyroxenites provide constrains on the formation, migration, segregation and compositional evolution of melts in the mantle?

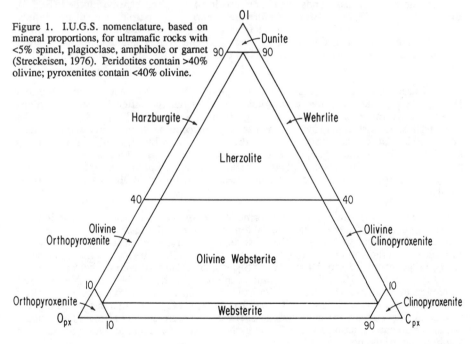

Figure 1. I.U.G.S. nomenclature, based on mineral proportions, for ultramafic rocks with <5% spinel, plagioclase, amphibole or garnet (Streckeisen, 1976). Peridotites contain >40% olivine; pyroxenites contain <40% olivine.

---

[c] Two simple end-member melting models are: (1) batch melting where a single batch of melt is created in equilibrium with the bulk residual solid; and (2) fractional melting where instantaneous melts are formed in equilibrium with the bulk residual solid, but they are immediately segregated from this residue. As a result, a series of melts of varying composition are formed and incompatible elements are removed from the residues very efficiently. In both models solid state diffusion is assumed to be rapid with respect to melt segregation.

We first present a summary of REE data for the massive peridotites containing abundant lherzolite. The most intensively studied massive peridotites are in Europe. They include the Ronda peridotite in southern Spain, several massifs, including Lherz, in the eastern Pyrenees (France), several massifs including Lanzo in the western Alps, and peridotites in the northern Apennines of Liguria, Italy. Following this data summary, we address the questions posed above. First, we address questions that can be constrained with data for peridotites unaffected by pyroxenite layers. Secondly, we address questions that can be constrained with data for pyroxenite layers and adjacent peridotites (those within 1 meter of the layers).

<u>Massive peridotites: dominantly lherzolite</u>

<u>Western Alps - Lanzo</u>. The structural and petrological characteristics of the Lanzo massif have been studied for many years. The dominant rock type is plagioclase lherzolite with the plagioclase occurring in thin veinlets which are interpreted to reflect magma flow (Nicolas, 1986). Harzburgite is less abundant, but discontinuous bands of dunite are well exposed. In addition, the peridotites contain pyroxenite layers. A detailed trace element study of the Lanzo massif (29 peridotites and 13 pyroxenites, Bodinier, 1988) showed that the plagioclase lherzolites which contain 2.2-3.6% CaO and Al_2O_3 with Mg numbers (molar ratios of 100 $Mg/(Mg+\Sigma Fe)$) from 89 to 90 have very systematic REE abundances with LREE at 0.1 to 0.7 x chondrites and HREE at 1-2 x chondrites (Fig. 2). Abundances of REE, especially HREE, are positively correlated with abundances of CaO and Al_2O_3, and similar well-defined trends are characteristic of most massive peridotites (Fig. 3). Because CaO, Al_2O_3 and REE are preferentially enriched in partial melts of peridotites, such correlations are expected in residues from varying degrees of a melting of a common source. Harzburgite and dunite with lower CaO and Al_2O_3 abundances have lower contents of all REE; however, the Ce/Sm in a dunite is higher than in the peridotites (Fig. 2).

Based on major and trace element, including REE, abundance trends, Bodinier (1988) inferred that the Lanzo peridotites formed as residues after extraction of melts created over a range in degree of melting, ~0 to 12% for lherzolites and ~18-20% for harzburgites. He also concluded that melt segregation occurred in the spinel stability field. For example, Tb/Yb ratios are close to chondritic in the lherzolites (Fig. 4) which precludes a major role for residual garnet. Also the absence of positive Eu anomalies (Fig. 2) provides no evidence for residual plagioclase. The segregated melts were inferred to range in $(Ce/Yb)_N$ from 0.4 to ~1; i.e., from depleted to transitional MORB (mid-ocean ridge basalt).

Massive peridotites provide information on spatial compositional variations on scales ranging from cm to several km. Bodinier found that the Lanzo peridotite massif is zoned in Al_2O_3 and Ce/Yb with the least depleted peridotites occurring in the north. Specifically, in the north $(Ce)_N$ ranges from 0.7 to 1 with $(Ce/Yb)_N$ >0.25 and in the south $(Ce)_N$ ranges from 0.15 to 0.4 and $(Ce/Yb)_N$ <0.25. Also in the north there is a region with Al_2O_3-rich (3.5-3.9%) lherzolite which contrasts with the relatively low Al_2O_3 (<3%) lherzolites in the south. However, in detail, the spatial zonation in Ce/Yb does not correlate with Al_2O_3 abundance variation (see Bodinier, 1988, Fig. 3). Bodinier suggested that the zonation in Ce/Yb may reflect an enrichment event that preferentially affected the northern portion or an early melting episode in the garnet stability field that primarily affected the southern part of the body.

Dunites in the Lanzo massif do not have higher Mg/(Mg+Fe) than the peridotites; they have relatively low Tb/Yb (Figs. 2 and 4), high $(Ce/Sm)_N$, ~0.81, and occur in the southernmost depleted part of the massif. Because they lack clinopyroxene, the dunites are highly depleted rocks (<0.9% Al_2O_3 and CaO contents) but they do not have relatively high Mg/Fe or low LREE/HREE ratios. Two hypotheses for the origin of dunites are that they

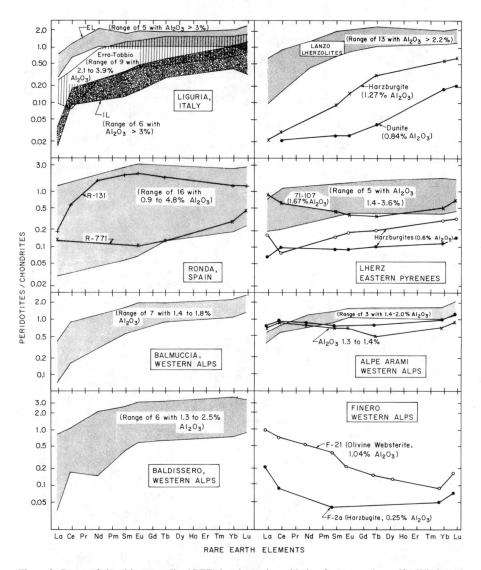

Figure 2. Range of chondrite-normalized REE abundances in peridotites from several massifs. Whole-rock Al₂O₃ (wt %) contents are given as a measure of the basaltic component in the peridotite. The majority of samples are relatively depleted in LREE; however a few samples (plotted individually) have more complex patterns. Data for massive peridotites in this and subsequent figures are from Ernst, 1978; Ernst and Piccardo, 1979; Ottonello et al., 1979, 1984a, 1984b; Frey et al., 1985; Bodinier et al., 1988a; and Bodinier, 1988. In this and all subsequent figures REE abundances are normalized to ordinary chondrites (Boynton, 1984, Table 3.3, 'Recommended chondrite abundances').

104

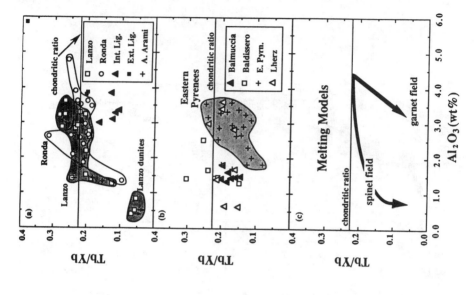

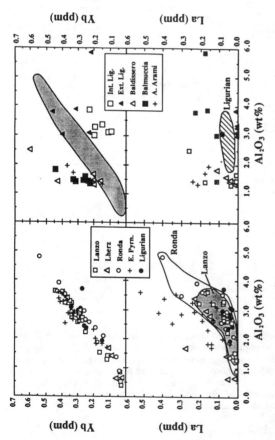

Figure 3 (left). Upper: Whole-rock Yb and Al₂O₃ contents for peridotites from several massifs. The left panel shows the very good correlation defined by peridotites from Lanzo, Lherz and other eastern Pyrenean bodies, Ronda and Erro-Tobbio, Liguria, and these data are shown as the shaded field in the upper right panel. In contrast the right panel illustrates the very poor correlation defined by peridotites from the Internal and External Ligurides, Balmuccia and Baldissero. Lower: Whole-rock La and Al₂O₃ contents for the same peridotites plotted in the upper panels. Note the poor correlation between La, a light REE, and Al₂O₃. However, the outlined fields for Ronda and Lanzo show a broad positive correlation. See the caption of Figure 2 for data sources.

Figure 4 (right). Tb/Yb versus Al₂O₃. The upper panel shows that peridotites from Lanzo and Ronda define a field centered near the chondritic Tb/Yb ratio. However, peridotites from the Internal Ligurides and Lanzo dunites have (Tb/Yb)N <1. The middle panel shows that most peridotites from Balmuccia and the eastern Pyrenees, including Lherz, have (Tb/Yb)N <1. As illustrated in the bottom panel Tb/Yb ratios are sensitive to the presence of garnet as a residual mineral. The arrows indicate trends for residues from increasing degrees of melting. See Figure 8b of Bodinier et al. (1988a) for quantitative trajectories of Tb/Yb during partial melting of spinel and garnet peridotites.

are cumulates segregated from a magma or residues from high degrees of melting. Bodinier favored a hybrid model where dunites represent residual peridotites which subsequently reacted with a genetically unrelated melt.

Western Alps - Baldissero, Balmuccia. A study (Ottonello et al., 1984a) of several other peridotite massifs from the western Alps, but with fewer samples per massif showed that like the Lanzo peridotites, spinel peridotites from Baldissero and Balmuccia have $(La/Yb)_N$ <1 (Fig. 2) with all Balmuccia samples having $(Tb/Yb)_N$ <1 (Fig. 4). Contrary to most peridotites, Baldissero and Balmuccia samples define poor Yb-Al_2O_3 trends (Fig. 3); they also have atypically low Al_2O_3/CaO (<1). Either the Al_2O_3 data for these samples are erroneous or these samples have been depleted in Al_2O_3 by a process that did not affect Lanzo peridotites.

In contrast, peridotites from two other western Alp massifs have very different REE abundance characteristics. Garnet peridotites from Alpe Arami equilibrated at 40±10 kbar (Ernst, 1978). Three garnet peridotites from Alpe Arami contain hornblende and have whole-rock $(La/Yb)_N$ <1; however, two other garnet peridotites lack hornblende and have lower HREE contents with $(La/Yb)_N$ ~1 (Fig. 2). Ottonello et al. (1984a) interpreted that these higher La/Yb ratios reflect the presence of low temperature alteration minerals; however, one of the samples with lower La/Yb contains similar amounts of these alteration minerals. Despite the presence of garnet the Alpe Arami peridotites with $(La/Yb)_N$ <1 have nearly chondritic Tb/Yb (Fig. 4); i.e., there is no REE evidence that garnet was a residual phase during a melting process.

The two spinel peridotite samples analyzed from the Finero body contain less than 2% alteration minerals, but both are significantly enriched in LREE. One sample contains 8% hornblende and phlogopite, but these phases are absent in the other sample which contain only 0.25% CaO and 0.13% Al_2O_3 (Ernst, 1978). These REE patterns are atypical for massive peridotites. Cawthorn (1975) previously suggested that this peridotite recrystallized in a hydrous environment. Thus, it is plausible that LREE were preferentially introduced by late-stage fluids.

In order to use trace element abundances to evaluate if these peridotites represent residues after partial melting Ottonello et al. (1984a) used triangular diagrams showing the relative abundances of Ni-Co-Sc and La-Sm-Lu. They favor these figures because there is no dependence upon absolute abundances; consequently, this approach avoids uncertainties in choosing absolute values of partition coefficients. Based on these plots, their choice of relative partition coefficients and source abundances, they concluded that peridotites from Lanzo, Balmuccia and Baldissero are residues after incipient melting.

Eastern Liguria, Italy. The ultramafic rocks associated with Ligurian ophiolites (northern Apennines) are divided into the Internal and External Ligurides (IL and EL). The IL are interpreted as being derived from the internal part of an ocean basin whereas the EL were derived from an oceanic-continental boundary region (e.g., Ottonello et al. 1984b). In both terrains spinel lherzolites are abundant, but subsolidus recrystallization is reflected by plagioclase exsolution from pyroxenes and plagioclase reaction rims around spinels. Based on mineral compositions Ottonello et al. concluded that the IL peridotites are more depleted in basaltic constituents, e.g., TiO_2, Al_2O_3 and Na_2O, than EL peridotites. However, the whole-rock peridotites (11 IL and 6 EL samples) in both groups are relatively Al_2O_3-rich (>3%, Fig. 3).

REE abundances define two distinct groups with the IL having lower REE contents, especially LREE (Fig. 2). None of these Ligurian peridotites are relatively enriched in LREE. Even two samples with very low CaO (~0.5%) have the typical LREE-depleted patterns. These peridotites range widely in Al_2O_3/CaO (~1.2 to 7) and unlike most suites of peridotites, these Ligurian samples do not define systematic positive correlations between HREE, Al_2O_3 and CaO abundances (Fig. 3); in fact, Al_2O_3 and CaO

106

are not correlated with MgO. If this discordance reflects late-stage alteration, this process did not result in LREE enrichment. Alternatively, as with the Baldissero and Balmuccia data, the Al_2O_3 data may be erroneous.

Based on trends in a La-Sm-Lu triangular diagram, Ottonello et al. (1984b) concluded that EL peridotites formed as residua from incipient melting, and they postulated that EL peridotites were associated with tholeiite lavas erupted during continental rifting. In contrast, the IL peridotites are interpreted as residues from higher degrees of melting. Although Ottonello et al. found that the La-Sm-Lu trajectory in a triangular diagram is intermediate to their calculated trends for equilibrium and fractional melting, the relatively high Al_2O_3 contents coupled with unusually low LREE contents in the IL peridotites (Fig. 2) are similar to trends expected during fractional melting. That is, fractional melting residues become very highly depleted in incompatible elements as the degree of melting increases (e.g. Figs. 5.3 and 5.4 from Frey 1984 and Fig. 8b from Prinzhofer and Allègre, 1985). Among all the massive peridotite samples studied, these IL peridotites with their very low La contents are the most similar to residues from fractional melting. Alternatively, prior to the melting event the IL peridotites may have had low LREE/HREE ratios. Also their low Tb/Yb (Fig. 4) may reflect a source characteristic or residual garnet.

Western Liguria, Italy. The Erro-Tobbio lherzolite in western Liguria is the largest ultramafic complex in the Alps. The studied rocks are spinel lherzolites with trace amounts of plagioclase in reaction rims. All samples have $(La/Yb)_N$ <1, but several samples have negative Ce anomalies; (Ottonello et al., 1979). The degree of relative LREE depletion is intermediate to the depletion of LREE in the two peridotite groups from eastern Liguria (Fig. 2), and there is a positive correlation between HREE and Al_2O_3 contents (Fig. 3). Based on REE melting models, Ottonello et al. (1979) concluded that these lherzolites are not residues remaining after generation of tholeiitic melts; but they may have formed as residues from 1 to 5% partial melting. These peridotites, like most massive peridotites have the relative LREE depletion that is expected in the source rock of MORB.

Eastern Pyrenees, France. Bodinier et al. (1988a) studied 45 peridotite samples (collected >1 m from pyroxenite layers) from nine massifs, the largest being the well known Lherz locality. Typically these are spinel peridotites with intercalated bands (up to tens of meters) of harzburgite and layers (3 to 20 cm) of pyroxenite. The majority of the whole-rocks have $(La/Yb)_N$ <1 with positive correlations between abundances of HREE, CaO and Al_2O_3 but much poorer correlations between LREE and CaO or Al_2O_3 (Fig. 3). Moreover, at Lherz, a lherzolite with 1.67% Al_2O_3 is clearly relatively enriched in LREE $(La/Yb)_N$ = 1.72. This enrichment also occurs in clinopyroxene separated from this rock. This LREE enrichment is not characteristic of all Lherz peridotites with low CaO and Al_2O_3 because two harzburgites with <0.7% Al_2O_3 have low REE contents with $(La/Yb)_N$ <1, although one sample has an anomalously high La/Ce ratio (Fig. 2).

Bodinier et al. concluded that samples with $(Tb/Yb)_N$ <1 (Fig. 4) contained small amounts of residual garnet which was eliminated during subsequent recrystallization. On the basis of Tb/Yb-Yb trends (see Fig. 8 in Bodinier et al., 1988a), they concluded that these samples formed as residues from batch melting of peridotites with garnet or spinel as the aluminous phase. By choosing residual modes with 2 to 4% garnet, they calculated that the equilibrium melts had $(Ce/Yb)_N$ of 2 to 5; i.e., similar to continental tholeiites and significantly higher than in the melts which segregated from the Lanzo peridotite.

Ronda, Spain The Ronda peridotite is the largest (300 km^2) peridotite massif that has been studied in detail. It is zoned systematically from garnet peridotite in the west to plagioclase peridotite in the east (Obata, 1980). All of the 17 peridotites studied have $(La/Yb)_N$ <1. The relative LREE depletion in these whole-rock Ronda peridotites also characterizes the constituent clinopyroxenes (Salters and Shimizu, 1988). However, one harzburgite (R-771) has anomalously high La/Sm with a REE pattern similar to Lanzo dunites (Fig. 2). In addition, a sample with anomalous Fe and Ti contents (R-131) also has a distinctive convex upwards chondrite-normalized pattern (Fig. 2). This sample was located near pyroxenite layers; as discussed later, there is abundant evidence for complex interactions between pyroxenite layers and the surrounding peridotite.

Based on major and trace element, including REE, abundances, Frey et al. (1985) concluded that the Ronda peridotites formed as residues created by variable degrees of batch melting. The extracted melts were inferred to be picritic with relative REE abundances similar to MORB. The peridotites do not have the highly depleted LREE content expected in residues of fractional melting. Although $(Tb/Yb)_N$ ratios are near unity (Fig. 4), they concluded that the residual peridotites ranged from garnet to spinel peridotites.

Among the Ronda peridotites there is an excellent correlation between abundances of REE, CaO and Al_2O_3 (Fig. 3). An important inference from this bulk rock study of Ronda peridotites is that there is a west to east decrease in bulk rock CaO, Al_2O_3 and REE content which resulted from a systematic gradient in degree of melting (Frey et al., 1985). However, a surprising result is that clinopyroxenes from some plagioclase peridotites in the northeast have $(Sm/Nd)_N$ <1 and $^{143}Nd/^{144}Nd$ ratios less than bulk earth estimates (Reisberg and Zindler, 1986, 1987). These peridotites do not contain evidence of modal metasomatism. Peridotites with $(LREE/HREE)_N$ >1 were not found by Frey et al. (1985, see Fig. 2, this chapter), but they did not study the eastern portion of the massif in detail. The additional isotopic data of Reisberg et al. (1989a) show that there is a west to east isotopic zonation with higher $^{87}Sr/^{86}Sr$ and lower $^{143}Nd/^{144}Nd$ in the east. This trend for "enriched" isotopic characteristics in the east is contrary to the element abundance trend which shows that there is a decrease in basaltic constituents from west to east across the massif; i.e., higher Mg/(Mg+Fe) and lower incompatible element abundances in the east. Reisberg and Zindler (1986/87) proposed that the lower Nd contents in the eastern part of the massif made this region more susceptible to isotopic perturbation by incorporation of an exotic component. Strong supporting evidence for this interpretation is the recent finding that $^{187}Os/^{186}Os$ in Ronda peridotites correlates negatively with Mg/(Mg+Fe); i.e., as expected the most depleted peridotites have high Mg/(Mg+Fe) and low Re/Os which led to low $^{187}Os/^{186}Os$ (Reisberg et al., 1989b). This result is consistent with the systematic west to east compositional changes which Frey et al. (1985) interpreted as reflecting a gradient in degree of melting from the interior (higher degree of melting) to the exterior (lower degree of melting) of an ascending diapir with the eastern portion of Ronda representing the interior of the diapir. The significant Os isotopic variation requires that the melting event(s) was ancient. Also, the absence of a correlation between $^{143}Nd/^{144}Nd$ and $^{187}Os/^{186}Os$ is consistent with the interpretation that the Nd isotopic system was disturbed by post-melting events, a mantle enrichment event in the east and tectonic disaggregation of mafic layers in the west (Reisberg et al., 1989b).

Effects of late stage alteration on REE. Although most massive peridotites contain secondary minerals, such as serpentine, formed at low temperatures, the effects of this alteration on REE abundances have not been thoroughly documented. Relative LREE enrichment has been attributed to alteration phases (e.g., some Alpe Arami peridotites, Ottonello et al., 1984a). However, the relative LREE enrichment in many peridotites reflects LREE enrichment in clinopyroxene (e.g., Frey and Green, 1974; Stosch, 1982 and Bodinier et al., 1988a); therefore, relative LREE enrichment can be created by high temperature processes. Moreover, when several samples from an individual massif are studied the same coherent major and trace element abundance trends are defined by samples with widely different amounts of serpentine (e.g., Frey et al., 1985; Bodinier et al. 1988a).

Negative Ce anomalies have been reported for several peridotites. Few of these have been confirmed with replicate analyses. Ottonello et al. (1979) investigated the thermodynamics of Ce in hydrous environments and concluded that Ce could be preferentially leached from a rock during hydrothermal alteration. Recently, Neal and

Taylor (1989) showed that clinopyroxene separated from a spinel peridotite xenolith in the Malaita alnoite (Solomon Islands) has a well-defined negative Ce abundance anomaly. They speculate that this anomaly reflects recycling of oceanic crust into the mantle.

What can be inferred about the melting process and the segregated melts?

Most of the major and trace element abundance trends in these lherzolite-dominated massive peridotites have been interpreted as characteristic of residues formed by varying degrees of melting. The systematic correlation of HREE with major oxide abundances (e.g., Fig. 3) have been important in reaching this conclusion. However, to realistically infer the compositions of the segregated melt and the residual mineralogy, it is necessary to be aware of several likely complexities. For example, have the peridotites been a closed system since the partial melting event? The poor correlation between LREE abundances and Al_2O_3 (Fig. 3) may reflect post-melting addition of highly incompatible elements; in this case, LREE abundances and LREE/HREE ratios can not be used to understand the melting process. If a closed system and a simple batch melting model are assumed the parameters required to infer the REE content of the equilibrium melt are the mineral/melt partition coefficients, the residual minerals and their composition all at the pressure and temperature of melting. Complexities that must be considered are that the mode of these recrystallized peridotites may differ considerably from the mode during the partial melting process; e.g., the proportions of clinopyroxene and orthopyroxene change significantly with temperature, and the aluminous phase changes with pressure. Moreover, even if the whole-rock has been a closed chemical system, the present mineral compositions have been affected by subsolidus re-equilibration; e.g., changes in major element contents, such as wollastonite content in clinopyroxene, will affect the cpx/melt partition coefficients for REE (McKay et al., 1986), and changes in the REE content of the residual minerals will directly affect the inferred REE content of the equilibrium melt. For more complex melting processes such as fractional melting, multistage melt segregation and melt percolation models the problems are much greater. In these cases, the final residue was in equilibrium only with the last melt created and details of the melting process must be known before the melt characteristics can be inferred.

Systematic compositional changes in a suite of peridotites have been used to infer the range in degree of melting. One approach is to use variations in major oxide content (e.g., Frey et al., 1985) or variations in modal proportions (e.g., Dick et al., 1984). Typically, a wide range from <5% to ~30% melting is inferred. If a source with uniform abundances of incompatible elements (a major assumption that may not be generally valid on a scale of 10's of kilometers) is melted to varying degrees, abundances of highly incompatible elements, such as LREE, reflect the degree of melting with abundances in the residues decreasing with increasing degrees of melting. However, abundances of highly incompatible elements in residual rocks are also very sensitive to post-melting enrichment processes and the melt segregation processes; e.g., at the same total degree of melting, residues from fractional melting are much more depleted in incompatible elements than residues from batch melting (see e.g., Figs. 5.3 and 5.4 of Frey, 1984, Fig. 8b of Prinzhofer and Allègre, 1985). In fact, few peridotites have the very low LREE contents expected in residues from >5% fractional melting. The closest examples are the peridotites from the Internal Ligurides which have significant Al_2O_3 contents (>3%), (reflecting only small amounts of melting) but very low LREE, especially La contents (Fig. 2). Many of the lherzolites from massive peridotites contain more than 2% Al_2O_3 and CaO and have (LREE/HREE)N >0.1 (Figs. 2 and 3); these peridotites probably formed by relatively low extents of melting, typically <5%. Almost all of the massive peridotites have REE contents that are consistent with residues formed by small to moderate amounts (1-25%) of batch melting (e.g., Frey, 1984, Fig. 5.3).

Massive peridotites: pyroxenite layers and veins and their wall rocks

A characteristic feature of most massive peridotites are tabular pyroxenites commonly referred to as mafic or pyroxenite layers. Typically these pyroxenites are parallel to foliation within the peridotites. These layers range from a few cm to 3 m in thickness with some thicker layers symmetrically zoned in modal proportions; e.g., from spinel orthopyroxenite at the peridotite contact to spinel websterite to garnet pyroxenite in the interior of the layer. Locally these pyroxenite layers are more abundant than the host peridotite. In addition, at a few localities such as Lherz and Freychinède in the French Pyrenees there are thin (a few mm to <30 cm) amphibole-containing pyroxenite veins which cross-cut the anhydrous pyroxenite layers.

Pyroxenite layers in peridotite are important indicators of mantle processes. Several researchers have concluded that these layers reflect pathways for migration of accumulated melts within the upper mantle; (for example, Bodinier et al., 1987a, 1987b; Frey et al., 1985; Suen and Frey, 1987 and Wilshire, 1987). However, based on isotopic data, Polvé and Allègre (1980) interpreted the associated pyroxenites and lherzolites as "sandwiches" of mantle with different histories coupled together by mechanical stirring in a convection cell. In a further development of this model Allègre and Turcotte (1986) proposed that the pyroxenite layers represent elongated strips of former oceanic crust injected into the mantle after subduction at convergent plate margins. Regardless of their origin, kilometer-size portions of upper mantle peridotite with abundant pyroxenite layers are an example of chemical and physical heterogeneities in the upper mantle. Consideration of how a heterogeneous mantle would melt has led to hypotheses for explaining coeval and spatially associated but geochemically heterogeneous basalts (e.g., Hanson, 1977; Fitton and Dunlop, 1985; Allègre and Turcotte, 1986; Prinzhofer et al., 1989).

Amphibole-bearing pyroxenite veins. Perhaps the simplest pyroxenites to understand are the relatively young, cross-cutting veins of amphibole-bearing pyroxenite. Loubet and Allègre (1982) found at Lherz that amphibole-bearing veins were relatively enriched in LREE, i.e. $(La/Yb)_N >1$, very much unlike most of the anhydrous pyroxenites (Figs. 5a and 6). Loubet and Allègre (1982) concluded that these amphibole pyroxenites are cumulates crystallized from melts that were highly enriched in LREE. Bodinier et al. (1987a) showed in more detail that amphibole pyroxenite and hornblendite veins have convex upwards chondrite-normalized REE patterns (Fig. 5a) which are consistent with a cumulate origin. Using REE abundances in clinopyroxene and amphibole, they calculated that these pyroxenites formed as upper mantle cumulates segregated from alkalic basalts. Obviously, the emplacement of these veins is an effective way of locally enriching the upper mantle in incompatible elements such as K, Ti and LREE.

Anhydrous pyroxenite layers. Although the anhydrous pyroxenite layers are approximately basaltic in major element composition, the early papers on REE abundances clearly showed that many pyroxenites have had a complex history. Relative to the host peridotite they typically have higher HREE contents (Figs. 5 and 6) but more variable LREE/HREE abundance ratios; in fact, many pyroxenites have lower La/Yb than the host peridotites (Fig. 6). Almost none of the pyroxenites analyzed have trace element abundances similar to erupted basalts. On the basis of compatible and incompatible trace element contents, the simplest model is an origin as mineral segregations from a migratory melt. However, more complex models have been proposed, such as a late-stage partial melting of a layer that had formed earlier as a cumulate (e.g., see Fig. 5.10 of Frey, 1984; Figs. 2 and 3 of Loubet and Allègre (1982)).

A cumulate origin has been proposed for the anhydrous pyroxenites at Ronda, Spain, Lanzo (western Alps) and Lherz, Freychinède and Prades in the French Pyrenees (Suen and Frey, 1987; Bodinier, 1988 and Bodinier et al., 1987b). Several compositional features are used to support this interpretation. (1) The bulk-rock pyroxenite compositions do not lie along extrapolations of the well-defined major element abundance trends formed

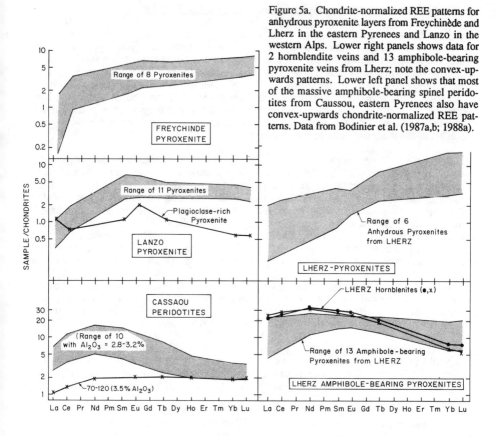

Figure 5a. Chondrite-normalized REE patterns for anhydrous pyroxenite layers from Freychinède and Lherz in the eastern Pyrenees and Lanzo in the western Alps. Lower right panels shows data for 2 hornblendite veins and 13 amphibole-bearing pyroxenite veins from Lherz; note the convex-upwards patterns. Lower left panel shows that most of the massive amphibole-bearing spinel peridotites from Caussou, eastern Pyrenees also have convex-upwards chondrite-normalized REE patterns. Data from Bodinier et al. (1987a,b; 1988a).

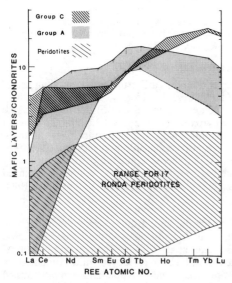

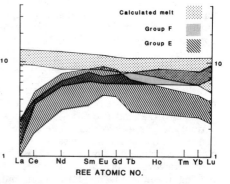

Figure 5b. Chondrite-normalized REE abundances of mafic layers (Groups A,C, E and F) from Ronda. These layers were divided into Groups A-F by Suen and Frey (1987). Also shown are the range of 17 Ronda peridotites (left) and a 15-20% equilibrium melt of the least depleted Ronda peridotite (right).

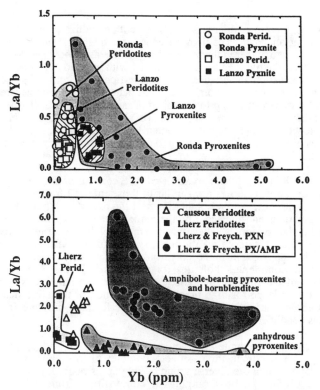

Figure 6. (a) La/Yb vs Yb showing that pyroxenite layers from Lanzo and Ronda have higher HREE contents than their host peridotites but the pyroxenites and peridotites <u>overlap</u> in LREE/HREE ratios.
(b) The anhydrous pyroxenite layers from Lherz also have higher HREE but trend to lower LREE/HREE than the host peridotites; in contrast, the amphibole-bearing pyroxenites and hornblendites from Lherz have higher HREE and trend to higher La/Yb.

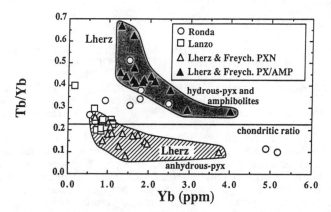

Figure 7. Tb/Yb vs Yb for pyroxenite layers showing that most anhydrous pyroxenites from Ronda have $(Tb/Yb)_N$ >1 but two Ronda samples (Group C in Fig. 5B) and most of the Lherz and Freychinede anhydrous pyroxenites have $(Tb/Yb)_N$ <1. High HREE contents and $(Tb/Yb)_N$ <1 probably reflect primary garnet. In contrast, hydrous pyroxenites have $(Tb/Yb)_N$ >1.

by the peridotites; therefore, the pyroxenites are not crystallized melts derived from the peridotites. (2) The abundances of elements concentrated in pyroxenes, such as Sc, V, and Cr, and Cr/Ni ratios are higher in the pyroxenites than the associated peridotites; moreover, these abundances in the pyroxenites are correlated with variation in mineral abundances. These are geochemical characteristics of pyroxene-rich cumulates. (3) The pyroxenites have a wide variety of chondrite-normalized REE patterns which are interpreted to reflect different cumulus minerals; e.g., primary spinel in samples with $(Tb/Yb)_N$ ~1 but primary garnet in several pyroxenites from Lherz with $(Tb/Yb)_N$ <1 (Fig. 7); primary plagioclase in a Lanzo sample with atypically low REE contents and a positive Eu anomaly. Finally, the low abundances of incompatible elements in these pyroxenites require very small amounts of trapped melts.

Bodinier (1988) found that the calculated REE contents of melts in equilibrium with the anhydrous Lanzo websterites are similar to those inferred for the partial melts that segregated from peridotites in the southern part of the massif. As a result, he speculated "...that the magmatic event registered by the pyroxenites is related to the partial melting event observed in the peridotites". In contrast to the amphibole-bearing pyroxenites which apparently equilibrated with an alkalic melt, the calculated REE contents of melts in equilibrium with clinopyroxene in the anhydrous pyroxenites are typical of tholeiitic lavas (interpreted to be continental tholeiites for the massifs in the French Pyrenees and transitional to depleted MORB for Lanzo). Therefore, the anhydrous pyroxenites are interpreted to be evidence for high pressure fractionation of tholeiitic magmas and the younger amphibole-bearing pyroxenites as evidence for high pressure fractionation of alkalic magmas (Bodinier et al., 1987a,b).

How were the pyroxenite layers created? Evidence for multistage processes. In general, the pyroxenite compositions reflect a complex history. For example, Suen and Frey (1987) proposed that the diverse group of anhydrous mafic layers at Ronda, garnet pyroxenites to olivine gabbros ranging in composition from 9.2 to ~22% MgO ($100Mg/(Mg+\Sigma Fe)$ = 58 to 89), formed at high pressures (>19 kbar) as cumulates (mainly clinopyroxene, + orthopyroxene + spinel and clinopyroxene + garnet) along the walls of conduits which are now represented by the mafic layers. However, in detail the incompatible element abundances in these mafic layers do not define simple coherent trends (e.g., the wide variety of REE patterns in Fig. 5b). It is evident from compositional and isotopic data that the Ronda layers reflect a complex series of processes which probably include multi-stage crystal/melt segregation from compositionally distinct melts (Suen and Frey, 1987), mixing of pyroxenites and peridotites accompanying deformation and sub-solidus recrystallization (Reisberg et al., 1989a,b) and contamination by continentally-derived components (Zindler et al., 1983).

As a whole the radiogenic isotopic data for the peridotites and pyroxenites define a large range that includes the field for recent MORB and OIB (ocean island basalts). For example, clinopyroxenes from Ronda peridotites range in $^{143}Nd/^{144}Nd$ from 0.51228 (ε_{Nd} = -6) to 0.51363 (ε_{Nd} = +20) and in $^{87}Sr/^{86}Sr$ from 0.70205 to 0.70406 (Reisberg et al., 1989b); $^{187}Os/^{186}Os$ ranges from 0.99 to 1.11 in Ronda peridotites but from 1.7 to 47.9 in the mafic layers (Reisberg et al., 1989a); $^{206}Pb/^{204}Pb$ in lherzolites from Lherz-Freychinède range from 17.16 to 18.52 while associated pyroxenites range only from 18.33 to 19.01 (Hamelin and Allègre, 1988). These large isotopic heterogeneities in massive peridotite bodies led Allègre and Turcotte (1986) to propose a marble cake mantle. They proposed that the layering in the oceanic lithosphere which was formed by igneous processes is deformed, primarily by elongation of layers, as oceanic lithosphere is subducted and incorporated into the convecting asthenosphere by streamline mixing. In this model the pyroxenite layers represent discrete elongated layers of subducted oceanic lithosphere. These layers are different in age, composition and isotopic characteristics and on average the oldest layers should be more elongated and thinner. The mantle thus has the appearance of a marble cake comprised of enriched former oceanic crust, the mafic layers, which may have been partially depleted by subduction zone melting, embedded in highly depleted peridotite.

However, to understand the isotopic heterogeneity in massive peridotites it is necessary to identify how much of this isotopic variation results from relatively young events such as mantle metasomatism and crustal contamination during emplacement, and how much can be attributed to primary characteristics resulting from ancient melting events or mixing of isotopically disparate components.

Implications for mantle enrichment processes (metasomatism). Pyroxenite layers and veins in peridotite are examples of how upper mantle can be locally enriched in incompatible elements. If these pyroxenites formed from mantle-derived melts, then the peridotite-pyroxenite contact is the focal point of postulated mantle metasomatic processes (e.g., Irving, 1980; Wilshire, 1984, 1987; Menzies et al., 1985). Bodinier et al. (1988a) studied 22 samples of peridotites that were adjacent, within 1 to 5 cm, to pyroxenite layers. They found significant differences between the wall rocks adjacent to hydrous and anhydrous pyroxenites.

Compared to peridotites distant (>1 m) from veins, peridotites adjacent to amphibole-bearing veins have increased Fe/Mg and Ce/Yb and higher abundances of Sr, LREE, Hf, and Ti (e.g., Fig. 8). These data are consistent with partial equilibration of the wall-rock peridotite with the alkalic magma that was parental to the amphibole-pyroxenites. Relative to peridotites distant from layers, the peridotites adjacent to anhydrous pyroxenites also have increased Fe/Mg; however, they have lower LREE/HREE (Fig. 8) and higher abundances of Al, Ca, Ti, Sc, Hf and HREE. Thus these wall rock peridotites have several geochemical characteristics intermediate between massive peridotites and the anhydrous pyroxenites. Bodinier et al. (1988a) noted that these wall rocks have geochemical characteristics that are consistent with solid state mixing of the anhydrous pyroxenites and host peridotite, perhaps during deformation and recrystallization. Alternatively, they proposed that the enhanced clinopyroxene signature in the wall rocks resulted from a reaction between melt and wall rock peridotite which resulted in the crystallization of clinopyroxene. This interpretation is now preferred by Bodinier (pers. comm., 1989).

Important results bearing on REE mobility at peridotite/pyroxenite contacts have recently been reported by Bodinier et al. (1988b,1989).[d] They determined REE abundances in harzburgites from a 65 cm traverse perpendicular to an amphibole pyroxenite vein at Lherz. Close to the contact (within 15 cm) the wall rock harzburgites contain interstitial amphibole and have normalized-REE patterns with slight upwards convexity in the LREE region; specifically, the patterns are similar to those of the wall rocks in the upper part of Figure 8, but REE abundances are considerably lower reflecting the low clinopyroxene content in these wall rock harzburgites. However, at distances of 15 to 25 cm from the contact REE patterns of the harzburgites are convex-downwards (V-shaped with LREE and HREE at ~0.2 to 0.5x chondrites and Tb~0.08 to 0.2x chondrites). From 25 to 65 cm the harzburgite wall rocks are enriched in LREE (0.6 to 1x chondrites), depleted in HREE (0.05 to 0.2x chondrites) with $(La/Yb)_N$ >1, and amphibole is absent. Thus, the REE abundance variations in these wall-rock harzburgites, which presumably result from a single enrichment/metasomatic event, encompass the wide range of LREE/HREE variations found in upper mantle peridotites. Bodinier et al. (1988b, 1989) propose that these variations in REE abundances as a function of distance from the peridotite/pyroxenite contact result from porous flow and diffusion processes.

There is also considerable evidence that subsequent sub-solidus re-equilibration can remove the compositional differences between the layers and wallrocks. For example,

---

[d] Because of time constraints we could not include these "in press" data in our figures.

114

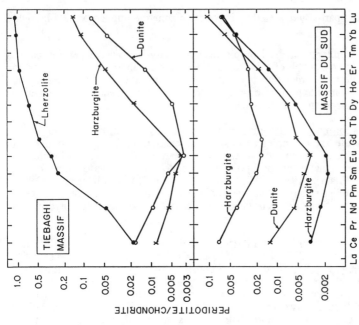

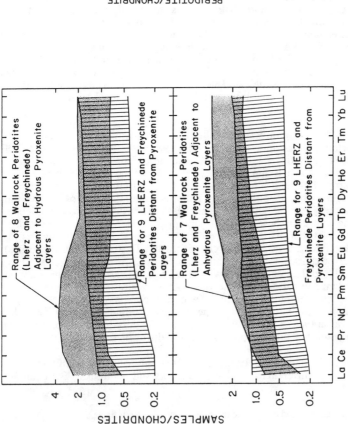

Figure 8 (left). Chondrite-normalized REE abundances in wall-rock peridotites (within 1 m of a pyroxenite layer or vein) adjacent to hydrous (amphibole-bearing) pyroxenite veins (upper) and anhydrous pyroxenite layers (lower). Note that wallrocks adjacent to hydrous pyroxenites have been increased in LREE/HREE and a convex-upwards pattern has developed from La to Tb. In contrast, the wallrocks adjacent to anhydrous pyroxenite layers do not have higher LREE/HREE than the massive peridotites >1 m from the pyroxenites.

Figure 9 (right). Chondrite-normalized REE patterns in peridotites from New Caledonian ophiolites. Note the very low REE contents in the dunites and harzburgites and the V-shaped pattern. Data from Prinzhofer and Allègre (1985).

isotopic differences between lherzolites and pyroxenites are typically absent when the layers are closely spaced (<2 cm); therefore Polvé and Allègre (1980) and Hamelin and Allègre (1988) suggested that isotopic equilibration occurred via solid-state diffusion. Also, Bodinier et al. (1987a) found that the $Mg/(Mg+Fe)$ of amphibole-bearing veins was inversely correlated with layer thickness; apparently, the thinnest veins have re-equilibrated with the surrounding peridotite. Finally, Allegre and Turcotte (1986) proposed that many thin (<2 cm) pyroxenite layers have been destroyed by recrystallization and diffusion.

In contrast to these wall-rock peridotites which Bodinier et al. (1988a,b; 1989) describe as evidence for "wall-rock metasomatism", the clinopyroxene/amphibole-rich, ilmenite-bearing peridotites from Caussou, a 60x100 m outcrop in the northeast Pyrenees, have high $Mg/(Mg+Fe^{+2})$ and high $(La/Yb)_N$ with convex upward chondrite-normalized REE patterns (Fig. 5a). Similar convex upward REE patterns occur in hydrous pyroxenites (Fig. 5a) and their wall rocks; however, the Caussou peridotites are not intimately associated with pyroxenites. Bodinier et al. (1988a) and Fabriès et al. (1989) postulated that these peridotites were pervasively metasomatized by percolation of an alkalic basalt melt, and that precipitation of clinopyroxene and amphibole occurred as the peridotite and alkalic basalt reacted and equilibrated. Thus, the LREE enrichment of Caussou lherzolites is interpreted to have resulted from interaction with a pervasive, percolating alkali basalt melt whereas in the nearby Lherz and Freychinède peridotites relative enrichment of LREE occurred only locally in the peridotite wall rocks of pyroxenite veins.

Massive peridotites: dominantly harzburgite

Frey (1984) summarized the literature which showed that very low REE contents and convex-downward, chondrite-normalized REE patterns are characteristic of the dunites and harzburgites forming the basal section of ophiolites (e.g., Fig. 9). REE analyses of clinopyroxene are not available from the dunites and harzburgites of depleted ophiolites. Therefore, the v-shaped patterns (Fig. 9) could reflect a post-melting enrichment. Recently, three quantitative models have been proposed to explain convex-downwards REE patterns in residual peridotites:

(1) Mixing of light REE-enriched melts with depleted residues (e.g., see Fig. 11 of Song and Frey, 1989).
(2) Differential ascent rates of REE as ascending melts interact (ion-exchange) with peridotite wallrocks (Navon and Stolper, 1987, and see Fig. 10 of Song and Frey, 1989).
(3) Formation of residues during a sequential process of disequilibrium melting (Prinzhofer and Allègre , 1985). In this model phases in the source are melted but the melt does not equilibrate with residual minerals. Therefore, the melt inherits the REE abundances of the minerals that contribute to the melt; i.e., the residue preferentially loses heavy REE when garnet melts and Eu when plagioclase melts. Thus harzburgite and dunite with convex-downward REE patterns and a negative Eu anomaly are residues that began melting in the garnet stability field and continued to melt as the earlier residues ascended into the plagioclase stability field. This model is effective in creating convex-downwards REE patterns in peridotites containing no clinopyroxene, but it can not explain the dominance of peridotites with relative LREE depletion (Fig. 2).

Oceanic peridotites

The samples of upper mantle that are most likely related to MORB petrogenesis are the peridotites obtained from the ocean floor by dredging or drilling. Because they contain alteration minerals formed at low temperature, bulk rock compositional data are difficult to interpret. However, Johnson et al. (1987,1988,1989) used an ion microprobe to obtain in situ REE abundances of clinopyroxene in peridotites (2-12 vol. % cpx) dredged from the south Atlantic and southwest Indian Ocean. Almost all of these clinopyroxenes are markedly depleted in LREE; e.g., all have $(Ce/Yb)_N$ <1 with only 4 having $(Ce/Nd)_N$ >1

(Fig. 10). Although Johnson et al. avoided samples containing plagioclase, hydrous phases and veins, the most significant result is that clinopyroxene from these ocean floor peridotites have the relative LREE depletion that is characteristic of MORB. Moreover, the very low $(Ce/Yb)_N$, ~0.002 to 0.5, in these clinopyroxenes from abyssal peridotites is much lower than whole-rock $(Ce/Yb)_N$ in most massive peridotites; e.g., the highly LREE depleted peridotites from the Internal Ligurides (Fig. 2) have $(Ce/Yb)_N$ ~0.15, and more commonly $(Ce/Yb)_N$ ratios are ~0.4 in clinopyroxenes from massive peridotites exposed on continents (Fig. 10). Only the highly LREE depleted clinopyroxenes from the Lizard peridotite have similarly low $(Ce/Yb)_N$ (Frey, 1969). Clinopyroxenes from abyssal peridotites also have unusually low Ti, Sr and Zr contents (Johnson et al., 1989). If a MORB source with constant REE abundances (LREE = 1.5x, HREE = 2.5x) is assumed, it is not possible to relate the REE in MORB and these peridotites in a batch melting process. However the trend to very low LREE abundances in these clinopyroxenes is consistent with residues formed by fractional fusion of spinel and garnet peridotites (Johnson ct al., 1989).

Mantle peridotites also occur as subaerial exposures on some oceanic islands. For example, St. Paul's rocks, a subaerial, mylonitized ultramafic complex in the equatorial Atlantic Ocean near the axis of the mid-Atlantic Ridge, contains spinel peridotite ± primary amphibole with kaersutitic hornblendite layers up to 10 m in thickness. All samples have $^{143}Nd/^{144}Nd$ exceeding bulk earth estimates with $(Sm/Nd)_N$ <1 and relative enrichment in LREE with $(La/Ce)_N$ up to 4.4 (Fig. 11). Roden et al. (1984) noted that as a whole this complex has the geochemical characteristics of a source for alkalic basalt.

Unlike amphibole-rich rocks from Lherz, some of the St. Paul's hornblendites have REE abundances similar to alkalic basalts (Fig. 11). Roden et al. (1984) concluded that (1) the hornblendites formed from metasomatic fluids segregated from the spinel peridotite; and (2) subsolidus equilibration of depleted (i.e., with negligible clinopyroxene) wall-rock spinel peridotites with amphibole-rich regions led to the relative LREE-enrichment of the spinel peridotites. This model is satisfactory only if olivine has La/Sm larger than that of coexisting clinopyroxene and amphibole. Relatively higher La/Sm in olivine is not expected from crystal chemical constraints and has not been found in olivine/melt experimental studies (McKay, 1986). However, ol/cpx REE partition coefficients determined from mineral separates of spinel peridotite xenoliths typically have V-shaped patterns (e.g., Ottonello, 1980; Stosch, 1982). This anomalous enrichment of LREE in olivine has been attributed to inclusions in olivine (e.g., Kurat et al., 1980; Stosch, 1982; Zindler and Jagoutz, 1988). If LREE-rich inclusions in olivine are important during equilibration of olivine with other phases, the model of Roden et al. (1984) can explain the relative LREE enrichment in St. Paul's peridotites.

Zabargad island, interpreted as a fragment of sub Red Sea lithosphere, is composed of spinel lherzolite, plagioclase peridotite and amphibole peridotite (e.g., Bonatti et al., 1986). These peridotites are not representative of mature oceanic mantle and they may represent subcontinental mantle lithosphere exposed during opening of the Red Sea. The spinel peridotites have $(La/Yb)_N$ <1 (Fig. 11) similar to massive lherzolites exposed in orogenic regions. These Zabargad peridotites are interpreted as residues from small degrees of melting (Bonatti et al., 1986; Piccardo et al., 1988). There is no evidence for the enrichment process that affected spinel peridotites from St. Paul's Rocks.

Chondrite-normalized REE patterns in the Zabargad amphibole peridotites are convex-upward similar to the amphibole-bearing pyroxenites from Lherz (Figs. 5 and 11). An exceptionally amphibole-rich Zabargad peridotite contains >10% amphibole and minor apatite and has high P_2O_5 and LREE abundances. Bonatti et al. (1986) proposed that the amphibole-bearing Zabargad peridotites formed as ascending H_2O-rich fluids interacted with peridotite; specifically, that with decreasing pressure the solubility of REE in these fluids decreased and LREE were deposited in the wall rocks along fracture systems.

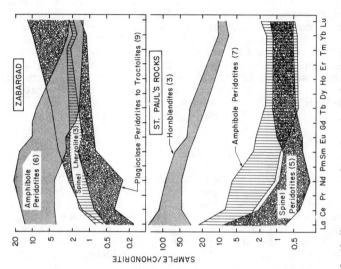

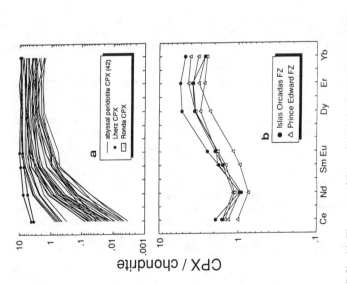

Figure 10 (left). (a) The group of lines with relative LREE depletion, (Ce)$_N$ <0.7, are for 42 clinopyroxenes from abyssal oceanic peridotites (dredged samples from the South Atlantic and Southwest Indian Ocean; data from Johnson et al., 1989). These clinopyroxenes have much lower LREE contents than clinopyroxenes from Ronda and Lherz peridotites (upper shaded field and filled circles, data from Salters and Shimizu, 1988; Bodinier et al., 1988a). (b) Unusual REE abundances patterns, (Ce/Nd)$_N$ >1, have been found in clinopyroxenes from only 4 abyssal peridotites. As seen in the upper panel, the majority have systematically decreasing normalized-abundances from Sm to Ce (data from Johnson et al., 1989).

Figure 11 (right). Chondrite-normalized REE abundances in massive peridoites from oceanic islands: St. Paul's Rocks in the equatorial Atlantic Ocean and Zabargad in the Red Sea. In both occurrences amphibole-bearing samples have (LREE/HREE)$_N$ >1. Even anhydrous spinel peridotites from St. Paul's Rocks have (La/Ce)$_N$ and (La/Sm)$_N$ >>1. In contrast, spinel lherzolites, plagioclase peridotites and troctolites from Zabargad have (LREE/HREE)$_N$ <1 with patterns similar to other massive peridotites (cf. Fig.2). Data from Frey (1970), Roden et al. (1984) and Bonatti et al. (1986).

The plagioclase (>2%) peridotites from Zabargad have igneous textures and occur as discrete bodies and as small stringers and blebs. They range from peridotites with CaO and Al_2O_3 <3% to troctolitic gabbros and have a wide range of REE abundances; all with $(La/Yb)_N$ <1 and many with La/Yb less than the spatially associated spinel peridotites (Fig. 11). Despite the trend to lower LREE/HREE, Bonatti et al. (1986) concluded that the plagioclase peridotites are mixtures of residues and an exotic melt. This interpretation is not obvious from the REE data and requires either residues more depleted in LREE than the spinel peridotites or incorporation of fractional melts with very low LREE content. Brueckner et al. (1988) recognized this difficulty and hypothesized that the troctolites are accumulates formed by processes similar to those forming the pyroxenite layers in most massive peridotites. Another alternative is that the plagioclase-bearing rocks have equilibrated at subsolidus conditions with the surrounding spinel peridotites. Because plagioclase is a poor host for REE and these rocks have low contents of clinopyroxene (<10%), atypically low REE contents could develop; however, this process does not predict that La/Yb should be lower in the plagioclase peridotites (Fig. 11).

ULTRAMAFIC XENOLITHS

Analyses of ultramafic xenoliths hosted in basalts and kimberlites provide an additional source of data on upper mantle composition and processes. An advantage to studying peridotite xenolith suites is that they occur in a diversity of crustal environments. For example, peridotite xenoliths hosted in continental basalts and kimberlites may represent fragments of the continental lithosphere, and thus they may be the only source of samples for this portion of the mantle. Similarly, peridotite xenoliths in oceanic basalts (e.g., Hawaii and Tahiti) may represent fragments of the oceanic lithosphere and these pieces of mantle can be compared with peridotites from the ocean floor and ophiolites to obtain a more comprehensive view of oceanic mantle. In addition, basalt-hosted xenolith suites are less altered than kimberlite-hosted suites and often are fresher than massive peridotites which contain varying amounts of serpentine.

Compared to massive peridotites, there are several disadvantages to peridotite xenoliths; e.g., their small size, and the possibility that they have interacted with the host magma. Probably the most frustrating aspect of studying a peridotite xenolith suite is the difficulty in determining the relative positions of individual xenoliths in the mantle prior to sampling by the host magma. Precise equilibration pressures of the spinel peridotites are especially difficult to infer. Also it is likely that the sampling process does not provide a representative sampling of upper mantle.

An important aspect of studying peridotite xenoliths hosted in continental volcanics is the possibility that they represent continental lithospheric mantle of varying age. For example, Richardson et al (1984) suggested that garnet-clinopyroxene inclusions in diamonds have been stored in the southern African lithospheric mantle since the Archean; Walker et al. (1989) concluded that Os isotopic ratios in peridotite xenoliths from South Africa reflect isolation in continental lithospheric mantle for more than 2 Ga. Others (e.g., Stosch and Lugmair, 1986; Menzies and Halliday, 1988) have argued for long term isolation of peridotite xenoliths in the lithospheric mantle based on 1.0 Ga and greater Nd model ages. In contrast, Nicolas et al (1987) suggested that many of the Massif Central spinel-bearing peridotite xenoliths, particularly those with deformation microstructures, have remained in the lithosphere for relatively short time periods (e.g., <10 m.y.).

The REE geochemistry of peridotite xenoliths will be discussed in four sections: (1) spinel peridotites, (2) garnet peridotites, (3) pyroxenites and (4) upper mantle minerals, including minerals in peridotite xenoliths, mineral inclusions in diamonds and megacrystic mineral suites. The spinel peridotite and pyroxenite xenoliths hosted in alkali basalts are commonly divided into Group 1 and Group 2 peridotite xenoliths. This classification is according to Frey and Prinz (1978), with the former including the Cr-rich spinel-bearing lherzolite and harzburgite xenoliths, and the latter including the Al-Ti-rich pyroxenitic and related xenoliths (Wilshire and Shervais, 1975). The most obvious difference in the REE geochemistry between peridotite xenoliths and other peridotite samples (e.g., massive, ophiolite and ocean floor peridotites) is the much greater diversity of relative REE abundances, especially the common occurrence of relative LREE enrichment in xenoliths.

Group 1 spinel peridotites

Spinel-bearing lherzolites and harzburgites are the most common ultramafic xenoliths found in alkalic basalts. They typically have Mg-numbers > 85, $TiO_2 \leq 0.25$ wt % and CaO and Al_2O_3 contents < 4.5 wt % and display trends of decreasing CaO, Al_2O_3, TiO_2 and increasing Cr and Ni with increasing MgO contents (Maaløe and Aoki, 1975). As with the massif peridotites, these major element and compatible trace element trends are consistent with the xenoliths representing residues from variable degrees of melting. However, at each locality Sr and Nd isotopic ratios of the spinel peridotite inclusions span a larger range than the host lavas. This isotopic diversity precludes simple one-stage models for derivation of the xenoliths and establishes that these spinel peridotite inclusions are accidental xenoliths which are not genetically related to the host lavas.

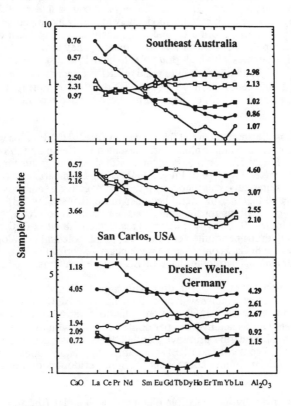

Figure 12. Representative chondrite-normalized REE abundances for spinel peridotite xenoliths. Also included are the respective wt % CaO and Al_2O_3 of the peridotite xenoliths. The relatively fertile peridotites with high CaO and Al_2O_3 have higher Yb and generally lower La than the depleted peridotite xenoliths. Data sources: Frey and Green (1974), Frey and Prinz (1978), Stosch and Seck (1980) and Stosch and Lugmair (1986).

Since the earlier review of Frey (1984) there have been several new studies focussed on REE abundances in peridotite xenolith suites. Distinguished by their continental location they include European (Stosch and Lugmair, 1986; Downes and Dupuy, 1987; Dupuy et al., 1987; Menzies and Halliday, 1988), African (Dupuy et al., 1986; Dautria and Girod, 1987), Asian (Preβ et al., 1986; Stosch et al., 1986; Song and Frey, 1989), North American (Roden, et al., 1984, 1988, 1989; Menzies et al., 1985; Wilshire et al., 1988; Galer and O'Nions, 1989) and Australian localities (O'Reilly and Griffin, 1988; Stolz and Davis, 1988, 1989; Chen et al., 1989). Other important publications present general models of mantle metasomatism and global coverage of xenolith localities (e.g., Roden and Murthy, 1985; Menzies and Hawkesworth, 1987; Nixon, 1987).

Peridotite xenoliths have a wide range of REE patterns varying from relative LREE-enrichment to LREE-depletion (Fig. 12). Xenoliths with more than 2 wt % CaO and Al_2O_3may have either LREE-depleted or LREE-enriched patterns; however, most samples with less than 2 wt % CaO and Al_2O_3 have LREE-enriched patterns (Fig. 13). Because LREE are highly incompatible in a peridotite mineral assemblage (see chapter by McKay), this LREE enrichment in depleted peridotites, i.e., <2% CaO and Al_2O_3, is not consistent with a simple melting model. Frey and Green (1974) first recognized this paradox in peridotite xenoliths and proposed that major elements and compatible trace elements record an early partial melting event, whereas the moderately to highly incompatible trace elements (e.g., LREE, P, Sr, K, etc.) record a second stage event involving the introduction of a melt and/or fluid component enriched in these elements. They also noted that there was a general tendency for samples with high $(La/Yb)_N$ to have low, less than chondritic, HREE abundances (Fig. 12). In addition, despite relative LREE enrichment, e.g., $(Sm/Nd)_N <1$, these spinel peridotite xenoliths have $^{143}Nd/^{144}Nd$ ratios requiring time integrated $(Sm/Nd)_N >1$; therefore the LREE enrichment has been created recently (e.g., Menzies et al., 1987).

In Figure 13 we illustrate the trend for increasing LREE/HREE ratios with decreasing CaO and Al_2O_3 contents. Similar plots were presented by Kempton (1987) and Nielson and Noller (1987). Figure 13 should be compared with similar diagrams for xenolith suites from a single locality (see Fig. 5.19 of Frey, 1984). Such a comparison shows that several individual xenolith suites have reasonably well defined negative trends; however, when all of the data are plotted together the trend becomes more diffuse (Fig. 13). Nevertheless, there are negative correlations for the anhydrous peridotite xenoliths with chondritic to LREE-depleted samples, $(La/Yb)_N<1$, having greater than 1.8 wt % CaO and Al_2O_3. In contrast, La/Yb in hydrous peridotite xenoliths, i.e., amphibole and/or phlogopite bearing shows no correlation with CaO content and a weak negative correlation with Al_2O_3 content. Note also that most hydrous peridotites have LREE-enriched patterns.

Frey (1984) noted that all peridotites with high CaO and Al_2O_3 contents have $(LREE/HREE)_N <1.0$. For samples with more than 4.5 wt % Al_2O_3, a reasonable estimate of the earth's primitive mantle Al_2O_3 content, this observation is still valid (Fig. 13b); however, several anhydrous and hydrous peridotite xenoliths with more than about 3.6 wt % CaO, an estimate of the earth's primitive mantle CaO content, have $(La/Yb)_N >1.0$ (Fig. 13a). These high CaO contents and LREE-enrichments probably reflect the addition of a clinopyroxene and/or amphibole component to these peridotites during a metasomatic enrichment process.

Abundances of mildly incompatible elements, such as HREE, Sc and V correlate positively with CaO and Al_2O_3 abundances. In fact, data for most massive peridotites (Fig. 3) and anhydrous peridotite xenoliths define the same HREE-Al_2O_3-CaO correlation trend (Frey, 1984). Moreover, abundance ratios such as CaO/Yb, Sc/Yb and V/Yb, are systematically correlated with Yb content (Fig. 14). If the trend from high to low Yb contents reflects increasing degrees of melting, these four mildly incompatible elements have similar partition coefficients during melting. For example, there is about a factor of

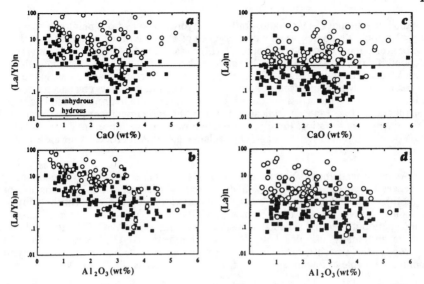

Figure 13. The variation in (a) (La/Yb)$_N$ vs CaO (wt %) and (b) (La/Yb)$_N$ vs Al$_2$O$_3$ (wt %) for anhydrous and hydrous (i.e. containing amphibole) spinel peridotite xenoliths. The hydrous peridotite xenoliths show the highest degree of LREE-enrichment. There is a general negative trend in these diagrams for the anhydrous xenoliths, whereas the hydrous peridotites show only a negative trend in the Al$_2$O$_3$ versus La/Yb diagram. (c) and (d) Chondrite normalized La vs CaO (wt %) and Al$_2$O$_3$ (wt %). No trends are obvious except that on average hydrous xenoliths contain more La than anhydrous xenoliths. Data from references in text, Frey (1984) and Nixon (1987).

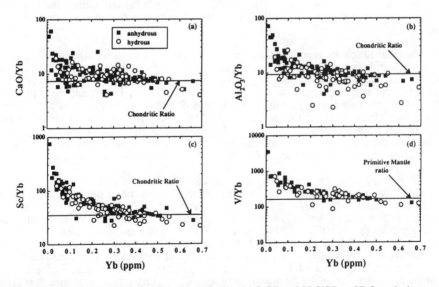

Figure 14. The variation in (a) CaO/Yb, (b) Al$_2$O$_3$/Yb, (c) Sc/Yb and (d) V/Yb vs Yb for anhydrous and hydrous spinel peridotite xenoliths; Sc, V and Yb are given in ppm and CaO and Al$_2$O$_3$ in wt %. There are reasonably coherent trends for CaO/Yb, Sc/Yb and V/Yb in hydrous and anhydrous xenoliths, whereas some of the hydrous peridotites have lower Al$_2$O$_3$/Yb values at Yb contents greater than 0.2 ppm. Chondritic and primitive mantle ratios are from Jagoutz et al. (1979b), Anders and Ebihara (1982), Sun (1982) and Sun and McDonough (1989). Data are from the literature.

three total variation in CaO/Yb ratios for both anhydrous and hydrous peridotite xenoliths at low (<0.1 ppm) Yb contents and a convergence of CaO/Yb ratios on the chondritic ratio (7.5) between about 0.45 and 0.50 ppm Yb (Fig. 14a). The Sc/Yb ratio also converges on the chondritic ratio at ~0.4 ppm Yb, but there is almost a factor of ten variation in this ratio at low Yb concentrations (Fig. 14c). The V/Yb versus Yb graph is analogous to the Sc/Yb plot and also shows about a factor of ten variation in this ratio at low Yb concentrations. In addition, there is no clear distinction between anhydrous and hydrous peridotites in these three diagrams (Fig. 14a,c,d), indicating that melting and enrichment processes generally do not change these ratios. Finally, the relative range of variation in V/Yb, Sc/Yb and CaO/Yb ratios and their trends suggest that relative partition coefficients for these elements are Yb<CaO<Sc~V.

As mentioned above, Yb concentrations also correlate with Al_2O_3 abundances, yet differences between the trends for anhydrous and hydrous peridotite xenoliths are found (Fig. 14b). The Al_2O_3/Yb and Yb correlation is reasonably coherent for the anhydrous peridotites, showing larger variations in Al_2O_3/Yb at lower Yb concentrations. In contrast, hydrous peridotite xenoliths display a much larger variation in Al_2O_3/Yb ratios at high Yb contents and many samples have low, subchondritic Al_2O_3/Yb ratios (Fig. 14b). Apparently, the enrichment/metasomatic processes involved in forming the hydrous xenoliths were able to fractionate Al_2O_3 from CaO, as well as Yb, Sc and V.

From studies of basalts and komatiites it is established that Ti/Eu abundance ratios are nearly constant and chondritic over a wide range of concentrations, suggesting that these elements have similar partition coefficients during partial melting (Sun and McDonough, 1989). Ti/Eu ratios in anhydrous peridotite xenoliths vary over a wide range (Fig. 15), but the average value, 7850, is close to the chondritic ratio, 7700. Because Eu anomalies are rare in peridotite xenoliths, it is not likely that the Ti/Eu range reflects fractionation of Eu from other REE; however, some of the range in Ti/Eu may reflect analytical errors in whole-rock abundances of TiO_2. The hydrous peridotites are more variable in Ti/Eu and they have a much lower average value (4410). Apparently the enrichment/metasomatic processes involved in forming the hydrous xenoliths were also able to fractionate Ti from Eu.

Normalization of incompatible element abundances in peridotite xenoliths to a primitive mantle composition enables simultaneous comparison of many element abundance ratios. For example, the smooth trends from Sr to V in Figure 16 show that in these xenoliths the adjacent elements (e.g., Sr/Nd, Zr/Hf, Ti/Eu, Ho/Y, and Sc/Yb) have abundance ratios similar to primitive mantle (and chondrites). Abundances of the most compatible elements plotted (e.g., Sc, V and Mn) converge towards primitive mantle abundances, but abundances of the moderately incompatible elements, Sr to Ca, range from near to less than primitive mantle abundances. The increased scatter among the most highly incompatible elements (Th, Nb and La) may reflect mantle enrichment processes.

There may be geographic variations in the LREE contents of peridotite xenoliths. For example, histograms of La contents in spinel peridotite xenoliths from four localities on three continents display markedly different shapes (Fig. 17). Both of the North American examples (San Carlos and Kilbourne Hole) have lower average and maximum La concentrations than the West German (Dreiser Weiher) and Australian suites. The differences in absolute LREE contents of these suites result from the hydrous peridotites in Dreiser Weiher and southeast Australia suites and their virtual absence at the two North American localities. Low LREE contents are also a feature of the Assab, Ethiopia suite (Ottonello, 1980). Given the large number of samples studied at these localities, it is clear that some regions of the lithospheric mantle have relatively low LREE contents, whereas other regions have variable and relatively high LREE contents.

High LREE abundances associated with the presence of amphibole also occur in samples of oceanic mantle; for example, St. Paul's Rocks in the Atlantic Ocean and

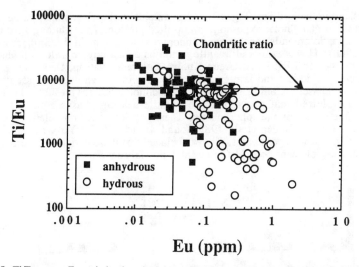

Figure 15. Ti/Eu versus Eu variation in anhydrous and hydrous spinel peridotite xenoliths; Ti and Eu are in ppm. Ratios of Ti/Eu in the anhydrous xenoliths scatter above and below the chondritic ratio and on average their ratio is near chondritic. However, most hyrous xenoliths have $(Ti/Eu)_N \leq 1$. Apparently, in some hydrous xenoliths REE enrichment was unaccompanied by Ti enrichment.

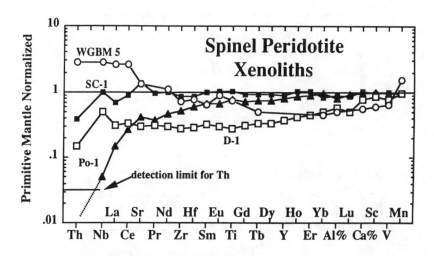

Figure 16. Incompatible element diagram for selected spinel peridotite xenoliths. The order of elements are based on trends defined by oceanic basalts (Sun and McDonough, 1989). Peridotites with LREE-enriched patterns are also enriched in other highly incompatible elements. Data for some of the highly incompatible elements (e.g., K, Rb, Ba, U) have not been plotted because their abundances may have been modified by surface weathering and infiltration of the host magma (e.g., Stosch, 1982; Zindler and Jagoutz, 1988). Data are from Griffin et al. (1988) and Jochum et al. (1989). Primitive mantle normalizing values are from Sun and McDonough (1989) and references in Figure 14.

124

Zabargad Island in the Red Sea (Fig. 10). Also, amphibole-bearing, spinel-garnet peridotite xenoliths from an alnoite exposed in an uplifted portion of the Ontong Java oceanic plateau (Malaita, Solomon Islands) contain clinopyroxene ranging in $(La/Sm)_N$ from 0.3 to 2.6 (Neal, 1988). However, not all mantle xenoliths containing volatile-rich phases are relatively enriched in LREE. At Green Knobs (Colorado Plateau, USA) the abundant Group 1 spinel peridotites and rare garnet peridotites contain an unusual assemblage of volatile-bearing phases including amphibole, titanoclinohumite, magnesite and antigonite. Smith (1979) inferred that many of these volatile-bearing phases equilibrated with anhydrous minerals within the upper mantle. These peridotites are not relatively enriched in LREE; for example, Roden et al. (1989) found whole rock $(La/Yb)_N$ ratios of 0.24 to 0.74 in four Green Knobs peridotites. They concluded that these peridotites were hydrated by an incompatible element-poor, H_2O-rich fluid, possibly seawater.

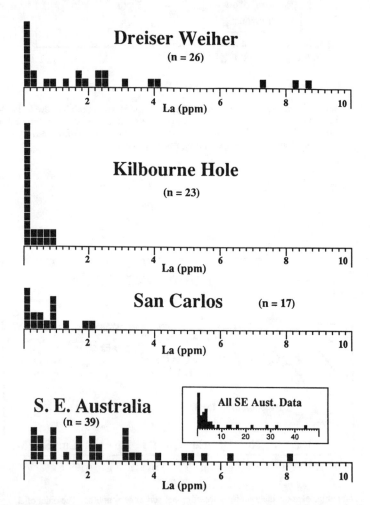

Figure 17. Histogram of La contents (ppm) for four different spinel peridotite xenoliths localities. Inset diagram labelled "All SE Aust. Data" shows the total range of La contents in spinel peridotite xenoliths from southeast Australian xenolith localities. The number of samples for each histogram are reported as n = value. Data are from Frey and Green (1974), Frey and Prinz (1978), Stosch and Seck (1980), BVSP (1981), Feigenson (1986), Stosch and Lugmair (1986), O'Reilly and Griffin (1988), Roden et al. (1988), Stolz and Davies (1988) and Chen et al. (1989).

In summary, REE abundances in spinel peridotite xenoliths can be viewed in terms of a two stage evolution model (Frey and Green, 1974). An initially undepleted peridotite experiences loss of a partial melt, resulting in depletion of incompatible elements and enrichment of compatible elements, such as MgO, Ni, Cr, Co. This stage is referred to as the development of component A, the residual component (Frey and Green, 1974). This residue is subsequently subjected to an enrichment event, addition or interaction with a mobile fluid or melt, which adds moderately to highly incompatible elements (e.g., LREE, K, P, Th, Nb). This stage is referred to as the development of component B, the enriched component (Frey and Green, 1974).

Garnet peridotites

These xenoliths include the garnet-bearing lherzolites and harzburgites hosted in alkali basalts and kimberlites. Garnet lherzolite xenoliths in alkalic lavas (e.g., Siberia (Kovalenko et al., 1988); Patagonia, South America (Skewes and Stern, 1979); Malaita, Solomon Islands, (Neal, 1988)) typically have higher CaO and Al_2O_3 contents than garnet lherzolites in kimberlites (cf. Nixon et al., 1981; BVSP, 1981). Because garnet lherzolites in alkalic basalts are compositionally more similar to spinel lherzolite xenoliths than to the garnet lherzolites and garnet harzburgites in kimberlites and associated rocks, we restrict this discussion to garnet peridotite xenoliths hosted in kimberlite and related rocks from southern Africa, Tanzania and the western U.S.

In all cases, kimberlite-hosted xenoliths have been affected to varying degrees by host infiltration and late stage, post-emplacement alteration processes. Consequently, Shimizu (1975) focused on mineral analyses and calculated bulk rock REE abundances. By analyzing garnet and clinopyroxene, the only significant hosts of REE in these xenoliths, he calculated whole rock REE patterns and identified two groups of garnet peridotites; samples with strongly fractionated LREE/HREE and high LREE abundances and samples with relatively flat REE normalized patterns and near chondritic absolute abundances. Furthermore, he correlated these characteristic REE patterns with bulk rock composition and texture thereby distinguishing two groups of garnet peridotites: (1) a sheared-fertile[e] type with near chondritic REE patterns and (2) a coarse granular-depleted type with high LREE/HREE ratios. A subsequent study (Nixon et al., 1981) of REE abundances in whole-rock garnet-bearing xenoliths from South Africa also identified these compositional groups. In general, as with the spinel lherzolite xenoliths, there is a broad spectrum of compositions ranging from fertile garnet lherzolites to highly depleted garnet harzburgites. However, unlike fertile spinel peridotites, the fertile garnet lherzolite xenoliths often have sheared or deformed textures and higher equilibration temperatures and pressures (e.g., Boyd and Nixon, 1972; Nixon et al., 1981).

Whole-rock abundances of REE (both calculated and measured) in garnet peridotite xenoliths for a selection of localities in north America and Africa are shown in Figure 18. In all of these examples, samples with the highest Al_2O_3 and CaO contents have relatively high HREE contents and $(LREE/HREE)_N \sim 1$; in contrast, samples with lower Al_2O_3 and CaO contents have lower HREE and $(LREE/HREE)_N >1$ (see also Fig. 5.2 in Frey, 1984). These trends are identical to those found in spinel peridotite xenolith suites and indicate that similar processes may have affected spinel and garnet peridotite xenoliths.

Nixon et al. (1981), following the model of Frey and Green (1974), proposed a two-stage evolutionary history for depleted peridotites whereby an initial partial melting event removes a basaltic component and creates depleted lithosphere. A subsequent second stage upward migration of melt and/or fluids from the base of the lithosphere enriches the peridotites in incompatible trace elements. In contrast with the studies of spinel peridotite

---

[e] We use fertile to indicate peridotites that can yield significant amounts (>10%) of basalt in a partial melting event; i.e., these peridotites have relatively high contents of the major elements (CaO, Al_2O_3, Na_2O, TiO_2) concentrated in partial melts.

126

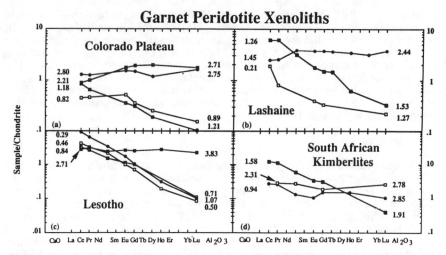

Figure 18. Representative chondrite-normalized REE abundances for garnet peridotite xenoliths from (a) Colorado Plateau, western USA, (b) Lashaine, Tanzania, (c) Lesotho, southern Africa and (d) South African kimberlite pipes: Frank Smith, Jagersfontein and Monastary. Also included are the respective CaO and Al_2O_3 contents (wt %) of the peridotite xenoliths. The relatively fertile peridotites with high CaO and Al_2O_3 contents have higher Yb contents and generally lower La contents than the depleted peridotite xenoliths. Data sources include Ridley and Dawson (1975), Nixon et al. (1981), BVSP (1981) and Ehrenburg (1982).

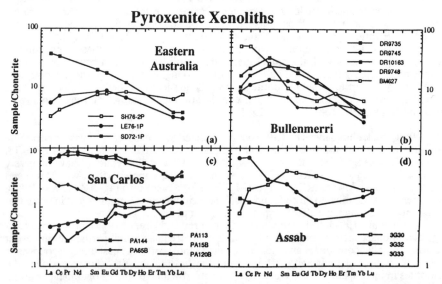

Figure 19. Representative chondrite-normalized REE abundances for clinopyroxenite, orthopyroxenite, websterite and wherlite xenoliths from (a) Eastern Australia, (b) Bullenmerri, Victoria, Australia, (c) San Carlos, western USA and (d) Assab, Ethiopia. Sample numbers given in each panel correspond to those used in the original papers. Data sources include Frey and Prinz (1978), Irving (1980), Ottonello (1980) and Griffin et al (1988).

xenoliths, Nixon et al. (1981) considered the fertile garnet peridotites xenoliths to represent asthenospheric mantle samples and suggested that their low levels of incompatible trace elements and nearly chondritic REE patterns indicated that they were not affected by a metasomatic enrichment process. More recent studies, however, have shown that the deformed or sheared fertile garnet peridotite samples, e.g., the extensively studied sample PHN 1611, have also been affected by the influx of a second stage or metasomatic component (Smith and Boyd, 1987; Bodinier et al., 1987c; Griffin et al., 1989). Evidence for the influx of a second component is documented by the systematic core to rim enrichments of incompatible trace elements in garnets. Also, based on Os isotopic data Walker et al. (1989) concluded that both types of garnet peridotite resided in continental lithosphere mantle for more than 2 Ga.

Pyroxenite and related xenoliths

A wide variety of pyroxenite xenoliths occur in alkali basalts and kimberlites. Chromium-rich diopsidic pyroxenites and orthopyroxenite xenoliths found in alkalic basalts and kimberlites generally have high Mg-numbers (>85) and are often considered with Group 1 spinel peridotites (Wilshire and Shervais, 1975; Frey and Prinz, 1978). However, the most common pyroxene-rich xenolith type in alkalic basalts are the Cr-poor and Al-Ti-enriched clinopyroxenites and wehrlites. The pyroxenites are commonly garnet-bearing and have often been mislabeled as garnet lherzolites and/or eclogites. Other pyroxenite xenoliths are spinel- and/or amphibole-bearing and these tend to be garnet-free; it is possible that some of these pyroxenite xenoliths formed at crustal pressures. In general, most pyroxenite xenoliths may be related to the passage of basaltic magmas through the mantle (Wilshire and Jackson, 1975; Frey and Prinz, 1978; Irving, 1980; Frey, 1980; Wilshire et al., 1980), although some may represent oceanic basalts which were recycled into the mantle (e.g., Allègre and Turcotte, 1986). An important question is: what is the relationship of pyroxenite and wherlite xenoliths to the contemporaneous magmatism that brought them to the surface? Isotope data can provide important constraints; e.g., some wehrlite xenoliths from SE Australia may be related to recent volcanism, but many of the pyroxenite xenoliths formed during an older magmatic episode (Griffin et al., 1988).

The REE patterns of pyroxenite xenoliths are quite variable, ranging from relatively smooth LREE-enriched to LREE-depleted patterns to others having convex-upward patterns, with a maximum among the LREE or the MREE (Fig. 19, and Fig. 5.22 in Frey, 1984). Many of the Group 2, or Al-Ti-rich, pyroxenite xenoliths are relatively enriched in LREE (e.g., Frey and Prinz, 1978); however, $(La/Ce)_N$ ratios are quite variable (Fig. 19a,b,d). In contrast, Group 1 pyroxenite xenoliths, particularly orthopyroxenites and some websterites, have lower overall abundances of the REE and LREE-depleted patterns are common (Fig. 19c).

In a detailed study of pyroxenite xenoliths from Salt Lake Crater, Hawaii, Frey (1980) used major and trace elements, in particular the convex-upward chondrite-normalized REE patterns (Fig. 5.22 of Frey, 1984), to argue for a cumulate origin for these xenoliths. Many subsequent studies of Al-Ti-rich pyroxenite xenoliths have reached similar conclusions. The La/Ce variability has been interpreted as a result of open system behavior whereby a LREE-rich fluid segregated from crystallizing minerals (e.g., Kurat et al., 1980). However, Kurat et al. (1980) proposed that a garnet-spinel websterite from Kapfenstein, Austria, which has a severe LREE-depleted pattern, represents a mafic liquid crystallized at upper mantle pressures that was subsequently subjected to partial melting. A similar model has been proposed for LREE depleted pyroxenite layers in massive peridotites (e.g., Loubet and Allègre, 1982). The diversity of compositions and isotopic ratios (e.g., Griffin et al., 1988) suggest that pyroxenite xenoliths record multiple melting and recrystallization episodes in the upper mantle. As discussed earlier, the compositions of pyroxenite layers in massive peridotites requires similar complexity. Nevertheless, there is a consensus that these pyroxenites reflect the passage of magmas through a peridotite mantle.

Models for REE abundance trends in peridotite xenoliths

REE have played a significant role in characterizing and identifying the processes that create incompatible element enrichment in mantle rocks. Dawson (1984) defined two end products of mantle enrichment processes as patent and cryptic metasomatism. Patent metasomatism, also known as modal metasomatism, is characterized by LREE- (and other incompatible elements) enrichment accompanied by the presence of hydrous minerals, whereas cryptic metasomatism is characterized by a similar style of compositional enrichment without the presence of hydrous minerals. Hydrous peridotites typically show the highest degree of LREE-enrichment (i.e., high $(La/Yb)_N$ as well as the highest La concentrations (Figs. 13 and 17; Kempton, 1987; Nielson and Noller, 1987)

Two general models have been proposed for enrichment processes within the upper mantle: (1) enrichment caused by local processes associated with intrusion and crystallization of magma in dikes and veins, and (2) widespread and pervasive enrichment of peridotite by ascending melts or fluids.

Studies of composite peridotite xenoliths, where Group 2 pyroxenite veins are in contact with Group 1 spinel peridotite (Wilshire and Jackson, 1975; Irving, 1980; Wilshire et al., 1980; Kempton, 1987), and studies of pyroxenite layers and their peridotite wall rocks in massive peridotites (Bodinier et al., 1988a,b, 1989) have been used to document the processes of magma transport through the lithospheric mantle and its effect on the wallrock peridotite. Irving (1980) suggested that the pyroxenite veins in composite xenoliths are crystals which were plated onto the walls of magma conduits; i.e., they record the passage of basaltic melt through the mantle, but their bulk compositions are not representative of melts because a melt component was segregated from the precipitated minerals now forming the pyroxenite. Interaction between the melt and peridotite wall-rocks before crystal deposition is recorded by compositional zoning in the peridotite wall rock. Alternatively, Wilshire et al. (1980), in a similar study of composite xenoliths, rejected the crystal plating hypothesis in favor of a model whereby the pyroxenite veins represent crystallized basaltic intrusions which reacted with their wall rocks. Wilshire (1984,1987) and Menzies et al. (1985,1987) proposed models for mantle enrichment (metasomatism) based on injection of a residual melt from the pyroxenite dike into the wall rock peridotite. Evolution of REE-bearing, H_2O- or CO_2-rich fluids from the cooling residual melt then form an inner aureole of hydrous metasomatism and further penetration of a CO_2-rich fluid into the peridotite produces an outer, anhydrous metasomatic aureole.

The finding by Bodinier et al. (1989) of V-shaped, chondrite-normalized REE patterns in harzburgite wall rocks 16-25 cm from an amphibole pyroxenite vein and strongly LREE enriched patterns, $(La)_N$ ~1 and $(Yb)_N$ ~0.1, in more distant harzburgites provides the important spatial control that is necessary to constrain models for REE mobility. Their preferred models are also based on interactions between melts and wall-rock peridotites, but they conclude that REE mobility was not the result of CO_2- or H_2O-rich fluids. In the model favored by Bodinier et al. (1989) the important processes causing REE mobility are reactions between melt and anhydrous minerals, diffusion in the melt and minerals near the pyroxenite-peridotite contact and chromatographic fractionation of REE during long-range (>25 cm from the contact) porous flow percolation. Because their percolation-chromatographic model does not assume instantaneous equilibrium between melt and minerals, they find that grain size is an important parameter. Specifically, the tendency for clinopyroxene to be smaller than olivine and orthopyroxene in harzburgites accentuates LREE/HREE fractionation and results in V-shaped REE patterns (chondrite-normalized) in olivine and orthopyroxene. Because all of the wall-rock harzburgites in this example have <1% CaO and <1.5% Al_2O_3, these data do not directly address the origin of the inverse trend between $(La/Yb)_N$ and CaO or Al_2O_3 (Fig. 13). Nevertheless, this study is the best example of REE mobility at a pyroxenite-peridotite contact and demonstrates that large and systematic variations in LREE/HREE and LREE/MREE abundance ratios occur in these wall rocks as a function of distance from the contact. Moreover, this example convincingly demonstrates that patent (modal) and cryptic metasomatism can result from the same igneous event.

Regional metasomatic models are not dependent on localized processes such as dike emplacement. These models appeal to a wide variety of processes including: upward melt migration from a mantle low velocity zone (e.g., Varne and Graham, 1971; Frey and Green, 1974); fluxes of incompatible element-rich fluid/melt through the lithosphere (O'Reilly and Griffin, 1988); carbonatite metasomatism (e.g., Green and Wallace, 1988), and ion exchange between ascending melts and porous wall-rock peridotite (Navon and Stolper, 1987).

The observation that peridotite xenoliths with relatively low CaO and Al_2O_3 are not preferentially depleted in LREE and tend to have high $(La/Yb)_N$ (Fig. 13) must be addressed by these general models for enrichment and metasomatism. In the following discussion, we summarize and briefly evaluate the specific models and ideas developed to explain the inverse correlations between $CaO-Al_2O_3$ abundances and LREE/HREE ratios (Fig. 13).

Frey and Green (1974) and Stosch (1982), among others have shown that the REE are dominantly in clinopyroxene in most anhydrous lherzolites. However, recent studies using careful mineral separation and acid leaching techniques (e.g., Stosch, 1982; Zindler and Jagoutz, 1988) have shown that LREE/HREE ratios in clinopyroxene may be less than the whole-rock ratios. These differences between whole rock and clinopyroxene REE patterns are commonly attributed to high concentratios of LREE on grain boundaries and in fluid inclusions. Furthermore, Ottonello (1980) and more recently Nielson and Noller (1987) noted that $(La/Sm)_N$ and $(La/Yb)_N$ ratios of olivine can be greater than unity and greater than these ratios in coexisting clinopyroxene. This unexpected finding results from REE-rich inclusions in olivine (e.g. Kurat et al., 1980; Stosch, 1982; Noller, 1986). However, these olivines have $(La)_N < 1$, i.e., much lower than the La content of associated clinopyroxene. Therefore, Nielson and Noller (1987) are correct in inferring that olivine with its inclusions can create relatively high $(La/Yb)_N$ in clinopyroxene-poor rocks such as harzburgites. However, in xenolith suites from individual areas some harzburgite xenoliths have relatively high $(La/Yb)_N$ and high LREE contents (i.e., $La_N > 1$), often exceeding those in rocks containing more clinopyroxene (see Fig. 12, this paper and the discrete anhydrous xenoliths in Fig. 9B of Kempton, 1987). As a consequence, relative LREE enrichment in olivine is not a viable explanation for harzburgites with relatively high La/Yb ratios and LREE contents.

Galer and O'Nions (1989) concluded that spinel peridotites with diverse $(La/Yb)_N$ and $CaO-Al_2O_3$ contents coexist in the upper mantle and that the $La/Yb-CaO$ or Al_2O_3 trends must reflect a local process. They propose that localized influx of fluids into peridotite causes simultaneous partial melting and relative LREE enrichment. This model requires that (1) the CO_2- and H_2O-rich fluids introduce significant amounts of LREE and (2) the LREE/HREE ratio of the resulting melt is very high; probably exceeding that in the host alkalic basalt. For example, cpx 21B analyzed by Galer and O'Nions has $(Sm/Nd)_N = 0.55$ (data for other REE are not available) and a fluid/melt in equilibrium with this clinopyroxene would presumably have an even lower Sm/Nd ratio. In contrast, the host basanites have $(Sm/Nd)_N \sim 0.6$ (Frey and Prinz, 1978; Galer and O'Nions, 1989).

O'Reilly and Griffin (1988) also proposed that relative LREE enrichment in peridotites results from an infiltrating fluid with high LREE content. They proposed that in an anhydrous mineral assemblage the degree of LREE enrichment is controlled by clinopyroxene/fluid partitioning. They stated that "Rocks with small amounts of clinopyroxene will have low initial REE contents and their REE patterns will therefore be more easily modified toward that of the fluid." If fluid/rock ratios were similar for lherzolites and harzburgites, a peridotite-fluid equilibration process can explain the higher

La/Yb ratios in samples with lower CaO contents (i.e., lower clinopyroxene contents) (Fig. 12 and 13), but this model can not explain high La contents in harzburgites (e.g., Fig. 12).

In light of these models, which call upon fluids (both H_2O- and CO_2-rich) to cause REE mobility, one needs to consider the results of recent experiments. Eggler (1987) reviewed much of the experimental data on the solubility of the REE in mantle fluids and concluded that the REE may be soluble in H_2O-rich and CO_2-rich fluids. However, Meen et al. (1989) showed that CO_2 fluid in equilibrium with diopside at 10 to 15 kbar contains insignificant amounts of Nd. They concluded that "...CO_2-rich fluids are extremely poor agents of metasomatism...".

By comparing the likely rates of melt segregation, mantle upwelling and equilibration by diffusion, Prinzhofer and Allègre (1985) and Bédard (1989) proposed that disequilibrium melting may be an important process. Disequilibrium melting can create residues with relative LREE enrichment (e.g., $(La/Sm)_N$ and $(La/Yb)_N >1$) from sources with relative LREE depletion because (1) the minerals preferentially forming the melt, clinopyroxene in spinel peridotite and garnet in garnet peridotite, have low LREE/HREE ratios; and (2) residual olivine (with its inclusions) typically has La/Sm greater than coexisting clinopyroxene or garnet. Therefore, the olivine-rich residues created by large degrees of disequilibrium melting will have v-shaped chondrite-normalized REE patterns (see Figs. 8 and 9 of Prinzhofer and Allègre, 1985) provided that the inclusions in olivine are not destroyed during the melting process. However, both Prinzhofer and Allègre, (1985) and Bédard (1989) found that all residues from disequilibrium melting have LREE abundances less than the unmelted source. Therefore, this model can not explain higher LREE abundances in harzburgites than in lherzolites. Moreover, in this model residues from small degrees of melting have $(LREE/HREE)_N >1$ and this ratio decreases as the degree of melting increases (Fig. 8c of Prinzhofer and Allègre, 1985). This theoretical trend is opposite of the trend defined by anhydrous xenoliths in Figure 13a,b. It is also unlikely that residues formed by disequilibrium melting would have the systematic and continuous trends in CaO/Yb, Sc/Yb and V/Yb found in peridotite xenoliths (Fig. 14).

Finally, Song and Frey (1989) proposed an incipient melt-residue mixing model that can explain (1) the inverse La/Yb-CaO correlation and (2) high LREE contents in harzburgites. In this model, which is a quantitative version of the two component model proposed by Frey and Green (1974), melts formed by very small degrees of melting (e.g., <1%) are mixed with residues from higher degrees of melting. For example, Figure 11d of Song and Frey shows that mixing a melt formed by 1% melting with residues formed by 3 to 13% melting in proportions of 3% melt: 97% residue will develop an inverse La/Yb-CaO trend with similarly high LREE contents in all the mixtures. A physical setting for this mixing process is an ascending and melting mantle diapir which will have a vertical gradient in pressure, temperature, and degree of melting. As melts segregate during diapiric ascent, large amounts of melt extracted from the shallow parts of the diapir create a vertical stratification with harzburgite overlying lherzolite. Subsequently, it is likely that upward migration of incipient melts from deeper parts of the diapir will ascend and mix with shallower, highly depleted harzburgite. An important aspect of this and other fluid/melt-peridotite interaction models is the determination of how fluids/melts migrate through the peridotite matrix. Recent experimental studies, for example, are investigating the importance of how the connectivity of the fluid/melt varies with the mineralogy and fluid/melt composition (e.g., Toramaru and Fujii, 1986; Fujii et al., 1986; von Bargen and Waff, 1986, 1988; Watson and Brenan, 1987).

MEGACRYSTS, MINERALS IN XENOLITHS AND DIAMOND INCLUSIONS

REE abundances in megacrysts, minerals from mantle-derived xenoliths and diamond inclusions provide information about the distribution of REE in the mantle. Mineral/mineral and mineral/melt abundance ratios for REE also provide insights into REE partitioning at upper mantle pressures. In addition, REE contents of minerals in ultramafic xenoliths and massif peridotites have been important in establishing the bulk rock REE characteristics of samples which have been extensively altered and/or affected by host magma infiltration. Finally, absolute and relative abundances of the REE and other incompatible elements in minerals from the mantle are important in identifying and constraining a variety of mantle processes. For example, recent studies have shown that the REE contents of olivine, vary directly with the amount of fluid inclusions (e.g., Stosch, 1982; Zindler and Jagoutz, 1988). Additionally, recent measurements show that minute (fluid and/or melt) inclusions in diamonds have high volatile contents and are strongly enriched in incompatible elements, including the LREE (Navon et al., 1988, 1989; Akagi and Masuda, 1988).

Megacrysts. Large discrete minerals, megacrysts, are commonly found in association with upper mantle xenoliths in alkali basalts and kimberlites and they are considered to form at upper mantle pressures. There are a large variety of minerals which occur as megacrysts in alkali basalts and kimberlites. The most common are clinopyroxene, garnet, amphibole and mica, but there are infrequent occurrences of orthopyroxene, spinel, olivine, anorthoclase, apatite, zircon, ilmenite, rutile, perovskite, titanomagnetite, sapphire, scapolite and sulphides. Some of the papers which report REE data for a variety of megacrysts include: Wass et al. (1980), Kramers et al. (1983), Liotard et al. (1983), Fujiimaki et al. (1984), Irving and Frey (1984) Jones and Wyllie (1984), Menzies et al. (1985; 1987), Dautria et al. (1987), Erlank et al. (1987), Liotard et al. (1988) and Menzies and Halliday (1988).

The REE concentrations in megacryst minerals vary considerably. For example, amphibole megacrysts from alkali basalts typically have high (LREE/HREE)$_N$; however, their chondrite normalized REE patterns are generally flat to LREE-depleted from La to about Nd or Sm (Irving and Frey, 1984; Menzies et al., 1985). In contrast, kimberlite-hosted amphibole megacrysts have LREE-enrichments and exhibit no flattening in the LREE region (Kramers et al., 1983; Erlank et al., 1987). Clinopyroxene megacrysts in alkali basalts show a range of REE patterns from LREE-enriched to LREE-depleted (Irving and Frey, 1984); however, convex upward patterns with (LREE/HREE)$_N$ >1.0 are typical. In contrast, clinopyroxene megacrysts in kimberlites have strong LREE-enrichments and a flattening of the LREE-pattern at Ce to Nd (Kramers et al., 1983; Erlank et al., 1987). Apatite and mica megacryst in alkali basalts have similar, strongly fractionated, LREE-enriched patterns, but with apatite having up to 2-3 orders of magnitude higher REE concentrations (Irving and Frey, 1984; Menzies et al., 1985). The same relative abundances and strong LREE-enrichments occur in apatites and micas from MARID (Mica-Amphibole-Rutile-Ilmenite-Diopside) nodules (Kramers et al., 1983). These nodules are xenoliths which may be samples of mantle pegmatites (Dawson and Smith, 1977). Other megacryst phases have REE contents that seem to reflect mineral/basaltic melt partition coefficients. For example, orthopyroxene and zircon have HREE-enriched patterns and anorthoclase is LREE-enriched with positive Eu anomalies (Irving and Frey, 1984).

Although the diversity of REE patterns in megacrysts is large, most of these data are consistent with crystallization of megacrysts at high pressure from magmas related to those erupted locally. For example, calculated melt compositions for megacrysts in kimberlites suggest that they crystallized from a kimberlitic or lamproitic magma (Kramers et al., 1983; Erlank et al., 1987; Menzies et al., 1987), while megacrysts from alkali basalts appear to have crystallized from a basaltic magma (e.g. Irving and Frey, 1984; Menzies et al., 1985, 1987). However, Sr and Nd isotopic ratios in megacrysts are commonly similar but not identical to isotopic ratios in their host (e.g., Irving and Frey, 1984), and in some occurrences the megacrysts are isotopically very different from their host lavas (e.g., Menzies and Wass, 1983), indicating that the megacrysts formed from earlier magmatism and/or metasomatism in the lithospheric mantle.

However, some mineral inclusions in kimberlite pipes do not appear to have crystallized from associated kimberlitic or lamproitic magmas. For example, sub-calcic garnets found in kimberlite heavy mineral concentrates have variable chondrite-normalized patterns ranging from sinuous or s-shaped to strongly LREE-enriched (Shimizu and Richardson, 1987). Shimizu and Richardson (1987) used a range of garnet/melt partition coefficients to calculate the REE patterns of a postulated equilibrium melt. They concluded that the REE patterns in these garnets reflect subsolidus partitioning rather than equilibration with melts or fluids.

Minerals in peridotites and pyroxenites. There are considerable data for REE abundances in minerals from peridotites and pyroxenites from massifs and xenoliths. Analyses of REE in clinopyroxenes are the most abundant, although REE data are also available for garnet, amphibole, mica, orthopyroxene, olivine, apatite and spinel, the other phases commonly found in these rocks. The largest amount of data on REE abundances in minerals is from spinel peridotite xenoliths (e.g., Nagasawa et al., 1969; Frey et al., 1971; Varne and Graham, 1971; Frey and Green, 1974; Frey and Prinz, 1978; Ottonello et al., 1978a, 1978b; Jagoutz et al., 1979a; Kurat et al., 1980; Ottonello, 1980; Stosch and Seck, 1980; Tanaka and Aoki, 1981; Stosch, 1982; Roden, M.F. et al., 1984; Menzies et al., 1985; Feigenson, 1986; Hutchison et al., 1986; Stosch and Lugmair, 1986; Stosch et al., 1986; Downes and Dupuy, 1987; Dupuy et al., 1987; Dautria et al., 1988; Neal, 1988; Salters and Shimizu, 1988; Stolz and Davies, 1988; Zindler and Jagoutz, 1988; Song and Frey, 1989).

In anhydrous spinel peridotites the bulk of the REE tend to reside in clinopyroxene, whereas in anhydrous garnet peridotites the HREE are dominantly contained in garnets and the LREE are dominantly in clinopyroxenes. In hydrous peridotites, amphiboles and apatite contain appreciable quantities of REE along with clinopyroxene and garnet, if present; in contrast, micas have relatively low REE contents. For the most part, orthopyroxenes and olivines have low REE contents (<0.5 ppm); however, phases with abundant fluid inclusions can be enriched 50 to 100 fold relative to the same mineral without fluid inclusions (Ottenello, 1980; Stosch, 1982; Zindler and Jagoutz, 1988). Typically, REE abundances decrease in the order amphibole > clinopyroxene > orthopyroxene > olivine > spinel.

The shape of the chondrite-normalized REE pattern in clinopyroxenes from spinel peridotites usually reflects that of the bulk rock. In detail, REE abundances in clinopyroxenes from spinel peridotite xenoliths show a wide variety of chondrite normalized REE patterns; i.e., ranging from LREE-depleted to LREE-enriched and including REE profiles that are flat, convex and concave. These REE abundances coupled with data for other incompatible elements (e.g., Ti, Zr, Hf) have been used to identify processes in the lithospheric mantle, as well as its composition and evolution (e.g., Stosch and Lugmair, 1986; Salters and Shimizu, 1988).

For garnet peridotites there are fewer data, but REE data for constituent minerals have provided important information about the bulk rock REE pattern (Philpotts et al., 1972; Ridley and Dawson, 1975; Shimizu, 1975; Mitchell and Carswell, 1976; Ehrenberg, 1982; Erlank et al., 1987; Neal, 1988) because these rocks are commonly altered, with abundant serpentine and limited or no preservation of olivine and orthopyroxene. Clinopyroxenes and phlogopites in garnet peridotites have LREE-enriched patterns and a flattening of the pattern at La-Ce. In contrast, REE patterns of garnets vary from relative enrichment in HREE to complex sinuous patterns which peak in the Nd to Gd region. Menzies et al (1987) provide a good illustration of the relative differences between clinopyroxenes in spinel and garnet peridotites (their Figs. 10 and 11); although all of these clinopyroxenes are LREE-enriched, the clinopyroxenes from garnet peridotites tend to have higher $(La/Yb)_N$ and much lower HREE than clinopyroxenes from spinel peridotites.

REE data for minerals from massif peridotites are relatively few, with REE abundances having been determined for olivine, orthopyroxene, clinopyroxene, spinel and amphibole separates (Frey, 1969, 1970; Menzies, 1976; Menzies et al., 1977; Ottenello et al., 1984a; Bodinier et al., 1988a). In general, the REE abundance characteristics of each mineral are similar in xenolith and massive peridotites.

The REE contents of garnet, clinopyroxene, amphibole and orthopyroxene from pyroxenite layers in peridotite massifs and pyroxenite xenoliths are reported in Reid and Frey (1971), Philpotts et al. (1972), White et al. (1972) and Bodinier et al. (1987a,b). Garnets and orthopyroxenes are HREE-enriched, with garnets having lower (La/Yb)$_N$ values. Amphibole REE patterns are LREE-enriched, commonly with (La/Ce)$_N$ <1.0. Like the whole-rock pyroxenites, clinopyroxenes have a wide range of REE patterns.

As discussed by Sun and McDonough (1989) and illustrated in Figure 16, abundances of incompatible elements in basalts and peridotites normalized to primitive mantle or chondritic abundances can be arranged in a sequence that reflects the relative incompatibility of these elements in a peridotite-melt system. Using such plots, Salters and Shimizu (1988) reported HFSE (high field strength element; i.e., Ti, Zr, Hf, Nb and Ta) depletions relative to the REE in clinopyroxenes from anhydrous spinel and garnet peridotites. In particular, they suggested that clinopyroxene is the dominant host for Zr and Ti, as well as the REE in these peridotites and proposed that (1) many peridotites have geochemical characteristics, (specifically, relative depletion in Ti and Zr) similar to island arc volcanics, (2) this type of mantle is widespread in both the suboceanic and subcontinental lithosphere, and (3) these peridotites "cannot generate nor be in equilibrium with mid-ocean ridge or ocean island basalts...". Stosch and Lugmair (1986) also reported high Sm/Hf in clinopyroxenes from hydrous spinel peridotites (note that Hf is analogous to using Zr because Zr/Hf ratios in peridotites (Jochum et al., 1989) and many mantle minerals (Fujimaki et al., 1984) are nearly constant and equal to the chondritic ratio). They also noted that both clinopyroxene and amphibole, the other dominant carrier of Sm and Hf in these rocks, have comparable Sm/Hf ratios; therefore Sm/Hf of clinopyroxene reflects the bulk rock Sm/Hf. Also, (Ti/Eu)$_N$ is <1 in many hydrous peridotites (Fig. 15). Stosch and Lugmair (1986) suggested that depletions of Hf, and by analogy Zr and Ti, were due to fluid phase metasomatism rather than silicate melt metasomatism.

In contrast, Jochum et al. (1989) found no evidence for HFSE depletion in several whole-rock, spinel peridotite xenoliths (e.g., Fig. 16), even in a sample (D-1) whose constituent clinopyroxene is relatively depleted in Ti and Zr. Based on the incompatible element abundance patterns in Figure 16, Jochum et al. concluded that these peridotites "...could be either the sources and/or residues from divergent margin or intraplate magmatism."

Choosing between the disparate interpretations of Salters and Shimizu (1988) and Jochum et al. (1989) requires more complete trace element analyses of peridotites and their minerals. For example, upper mantle rocks are metamorphic rocks; this is apparent from their textures and their inferred equilibration temperatures, typically <1100°C, which are considerably less than temperatures during partial melting. Therefore, in order to realistically interpret trace element data for mantle minerals in terms of igneous processes, it is necessary to evaluate the effects of subsolidus partitioning. For example, it is necessary to know how the modal mineralogy, mineral compositions and mineral/mineral trace element partition coefficients changed as temperature decreased. This information can be obtained from laboratory studies of peridotites at upper mantle pressures and temperatures. However, important data on trace element partitioning can also be obtained by analyses of minerals in carefully chosen natural samples.

Another approach for evaluating the significance of trace element abundances in whole-rock peridotites and their constituent minerals is to determine the abundances in each phase and the whole-rock. Although interpretation of whole-rock data for relatively coarse-grained metamorphic rocks with a complex history may be difficult, whole-rock and mineral data can be used to identify the principal host phases for each trace element. If a single phase, such as clinopyroxene, controls REE and HFSE abundances, it can be argued that subsolidus effects are unimportant. However, it is still possible that an important host phase at high temperature was eliminated during recrystallization. On the other hand, if more than one phase is important in the trace element mass balance, subsolidus partitioning may have significantly modified the trace element content of phases that were present in the partial melting residue.

The data plotted in Figure 20 illustrate the complexity and ambiguity of the present data set. Salters and Shimizu (1988, Fig. 1) show that in several peridotites both the constituent clinopyroxene and whole-rocks have HFSE depletions. However, in Figure 20 the whole rocks do not have the large Ti anomaly that occurs in several minerals (spinel, orthopyroxene and clinopyroxene). However, in this plot data are lacking for several important elements (Zr, Hf, Nb, Ta) and the quality of the data needs to be assessed.

Inclusions in diamonds. Richardson et al. (1984) reported Sm-Nd and Rb-Sr isotope model ages of 3.2 to 3.3 Ga for garnet inclusions in southern African diamonds. Later, Shimizu and Richardson (1987) showed that these garnet inclusions are relatively enriched in LREE (typically peaking at Nd in chondrite-normalized patterns) similar to the previously discussed sub-calcic garnets found in the heavy mineral concentrates from kimberlite. More recently Shimizu et al. (1989) reported a wide variation of LREE-enriched patterns for garnet inclusions in diamonds from the Finsch and Koffiefontien kimberlite pipes. Both sub-calcic and calcic garnets have LREE-enrichments, with sub-calcic garnets having greater degrees of LREE-enrichment ($(Nd/Yb)_N$ up to 59). It is significant that: (1) none of the garnet inclusions in diamonds are relatively depleted in LREE, and (2) there is a qualitative similarity between the LREE-enriched patterns of garnet inclusions in diamonds and the garnets in metasomatized garnet peridotites from southern Africa (Shimizu, 1975; Shimizu and Richardson, 1987).

Other recent studies of inclusions in diamonds have focussed on micro-inclusions, of fluids rich in water, carbonate, chlorine and K_2O (Bibby, 1979; Navon et al., 1988, 1989; Akagi and Masuda, 1988). These micro-inclusions are enriched in LREE and other incompatible elements; in fact, the relative LREE-enrichments in these micro-inclusions are greater than in kimberlites. Additionally, $^{87}Sr/^{86}Sr$ ratios of these micro-inclusions in Zaire diamonds are similar to those of the host kimberlite (Akagi and Masuda, 1988), indicating a probable genetic link between the kimberlitic magmatism and fluids trapped in the diamonds. However, Navon et al. (1988, 1989) argue against direct trapping of kimberlitic melt because of differences between kimberlite and micro-inclusions in LREE-enrichment and major element composition. The results of Akagi and Masuda (1988) support this conclusion in that they found the alkalies and alkaline earth elements in crushed diamonds were readily leached by water. Together these data indicate that the micro-inclusions are mostly trapped fluids with little or no trapped melt component. The exact origin of these fluids is uncertain; however, their enriched character strongly indicates the importance of volatiles in dissolving large quantities of REE and other incompatible elements, including Ti (Navon et al., 1988, 1989; Akagi and Masuda, 1988). These and other results on hydrous and anhydrous peridotites indicate that more mineral/melt, mineral/fluid and melt/fluid partition coefficient data are needed in order to understand the nature of incompatible element enriched components in the mantle.

SUMMARY: COMPARISON OF PERIDOTITES FROM MASSIFS AND XENOLITHS AND IMPLICATIONS OF REE DATA FOR UPPER MANTLE COMPOSITION

Since the earlier review of Frey (1984) there has been a significant increase in REE data for mantle rocks and minerals. Additionally, recent measurements of REE and other incompatible element abundances plus Sr and Nd isotopic ratios in mineral inclusions and micro-inclusions in diamonds have added important new data. Many of the earlier

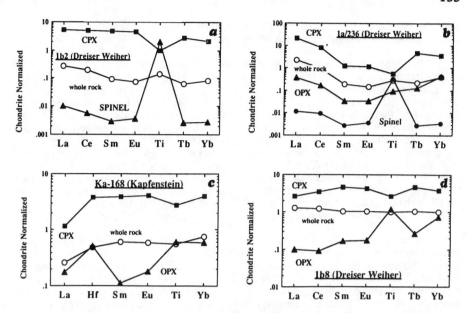

Figure 20. Incompatible element diagrams for minerals from spinel peridotite xenoliths from Dreiser Weiher, West Germany (a, b, c) and from Austria (d). Samples 1b2 (a), 1b8 (b) and Ka-168 (c) are anhydrous spinel peridotite xenoliths, whereas sample 1a/236 is an amphibole bearing spinel peridotite. The relative incompatibility of Ti is considered as being about equal to Eu. Relative enrichments and/or depletions in Ti are compared to the REE pattern for minerals and bulk rock. Note that depletions of Ti in clinopyroxenes are not necessarily reflected in the bulk rock pattern. CPX refers to clinopyroxene and OPX refers to orthopyroxene. Data are from Kurat et al. (1980), Stosch and Seck (1980), Sachtleben and Seck (1981) and Stosch (1982). Sources of chondritic normalizing values in Figure 14 caption.

Xenolith and Massif Peridotites

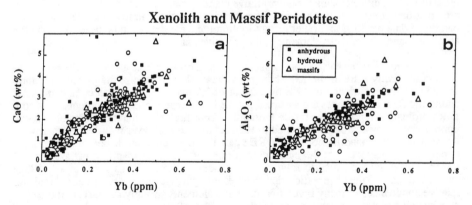

Figure 21. Variation in (a) CaO and (b) Al_2O_3 vs Yb (ppm) for anhydrous and hydrous spinel peridotite xenoliths and massif peridotites. The two sources of peridotite samples, xenoliths and massive peridotite intrusions, show similar trends. Data from references in other figure captions, Frey (1984) and Nixon (1987).

observations and conclusions in Frey (1984) are further supported with these new data; in addition, several new conclusions can be made.

Peridotites from massifs and xenoliths in alkalic basalts have a similar range of major element compositions (e.g., Maaløe and Aoki, 1975). Moreover, abundance trends for mildly incompatible elements, such as CaO and Al_2O_3 versus Yb, are identical in massif peridotites and these peridotite xenoliths (Fig. 21). These trends, in particular for the depleted lherzolites and harzburgites, indicate that these moderately incompatible elements have behaved similarly during melting in both suites of peridotites. Although massif peridotites are dominated by relative LREE depletion (Fig. 2), the recent documentation of relative LREE enrichment in some samples from massive peridotites (e.g., the Caussou peridotite (Fig. 5a) and wall-rock peridotites adjacent to hydrous pyroxenite layers (Fig. 8)) further establishes the compositional similarities between massif peridotites and peridotite xenoliths in alkalic lavas. The dominance of samples with relative LREE depletion in massif peridotites and LREE enrichment in peridotite xenoliths probably reflects the intimate association of peridotite xenoliths with regions of intense volcanism.

Another important result is that relative LREE enrichment occurs in both the oceanic and continental mantle; e.g., peridotite xenoliths in Hawaiian lavas and the massive St. Paul's peridotites in the Atlantic Ocean are as enriched in LREE as peridotite xenoliths in continental alkalic basalts (cf. Figs. 10 and 12). Thus, there is compositional (and isotopic) overlap between oceanic and continental alkalic basalts and their included peridotite xenoliths. However, abyssal peridotites from the seafloor have clinopyroxenes that are markedly depleted in LREE (Fig. 11); these may be the only mantle samples that are directly related to the tholeiitic magmatism at oceanic spreading ridges.

Fertile lherzolite samples (i.e., with CaO and Al_2O_3 >3%) occurring in massifs and as xenoliths have a limited range of HREE contents, ~1.5 to 3 times ordinary chondrites. This result can be used to constrain the composition of the earth's primitive mantle. The primitive mantle is defined as the silicate portion of the earth, after core formation and prior to the differentiation of the crust-mantle system. Knowledge of the absolute abundance of REE in the primitive mantle would provide important information concerning the bulk composition of the earth's primitive mantle and constraints on accretionary processes and core formation. The REE are refractory lithophile elements similar to Ca, Al, Ti, Zr, Hf, Nb, Ta, Th, U, Sr, Ba, Y and Sc and these elements apparently accreted in chondritic relative proportions. Thus, like the REE, a chondrite normalized diagram of all these refractory lithophile elements for the primitive mantle would be flat, and like the REE there are no strong constraints on their absolute abundance in the primitive mantle. However, combined data for REE and other refractory lithophile elements can be used to define the absolute abundance of these elements in the primitive mantle.

The mass of the earth's core is about one third of the whole earth and is assumed to contain relatively little of the REE budget of the earth. Therefore, the primitive mantle should have been enriched in REE by a factor of 1.5 over the bulk earth composition. If chondritic REE abundances are assumed for the bulk earth, the HREE contents, ~1.5 to 3 times ordinary chondrites, in fertile lherzolites from the upper mantle indicates that the upper mantle is not strongly enriched in REE relative to the lower mantle. Furthermore, the relatively flat chondrite-normalized REE patterns in the most fertile peridotites from the upper mantle suggests that the bulk compositions of the upper mantle-crust system and lower mantle have chondritic relative abundances of REE. Thus Kato et al. (1988) concluded that ... "the observation that the present upper mantle contains near-chondritic relative abundances of many involatile lithophile elements strongly suggests that the bulk compositions of the upper and lower mantles are similar."

As seen in the CaO/Yb and Sc/Yb diagrams (Fig. 14) there is a convergence of the peridotite data to chondritic or primitive mantle ratios at about 0.45 to 0.50 ppm Yb. These primitive mantle abundance ratios at a constant Yb content can be used to constrain the absolute composition of the earth's primitive mantle. A Yb concentration of 0.45 to 0.50 ppm for the primitive mantle is consistent with the $(Ce)_N$-$(Yb)_N$ trend of massive peridotites (e.g., Loubet et al., 1975; Fig. 5.3 of Frey, 1984), and the models of Jagoutz et al. (1979b), Sun (1982), and McDonough and Sun (1989); however, this estimate is higher than those of Anderson (1983), Taylor and McLennan (1985) and Hart and Zindler (1986) and is lower than the estimate of Palme and Nickel (1985). Additionally, the V/Yb versus Yb correlation (Fig. 14) and this estimate for Yb content of the primitive mantle suggests that the primitive mantle has about 82 ppm V, which is also consistent with estimates of the primitive mantle V/Yb ratio (Jagoutz et al., 1979b; Sun, 1982; Wänke et al., 1984).

If HREE contents are similar in the upper and lower mantle and abundances of refractory lithophile elements in the mantle are in chondritic proportions, a mantle Yb content of 0.45 to 0.50 ppm indicates that the primitive mantle is enriched in REE and all other refractory lithophile elements by a factor of 2.2±0.1 times ordinary chondrites. This approach was used by McDonough and Sun (1989) to determine the absolute abundance of

Table 1. Model compositions for the Earth's primitive mantle

(wt % for oxides, ppm for elements)

	C1	1	2	3	4	5	6	7	Preferred
SiO_2	21.62	45.10	44.52	45.98	45.96	49.90	47.95	46.2	44.8
TiO_2	0.0734	0.20	0.217	0.225	0.181	0.16	0.204	0.23	0.21
Al_2O_3	1.514	3.30	4.31	4.20	4.060	3.64	3.817	4.75	4.45
Cr_2O_3	0.387	0.40	0.44	0.44	0.468	0.44	0.342	0.43	0.38
FeO	23.410	8.00	8.17	7.58	7.540	8.00	7.86	7.70	8.00
MnO	0.256	0.15	0.14	0.13	0.130	0.13	0.131	0.13	0.14
NiO	1.374	0.20	0.25	0.27	0.277	0.25	0.25	0.23	0.22
MgO	15.170	38.10	38.00	36.85	37.78	35.10	34.02	35.5	37.2
CaO	1.234	3.10	3.50	3.54	3.210	2.89	3.078	4.36	3.60
Na_2O	0.674	0.40	0.39	0.39	0.332	0.34	0.275	0.40	0.34
K_2O	0.0663	0.03	0.03	0.03	0.032	0.02	0.018	–	0.028
P_2O_5	0.280	0.02	0.022	0.015	0.019	–	0.013	–	0.022
Mg#	53.6	89.5	89.2	89.6	89.9	88.7	88.5	89.1	89.2
V	56	81	87	82.1	–	77	128	–	82
Sc	5.98	–	17.0	17.0	14.8	13.0	15.0	19.0	17.1
La	0.244	–	0.720	0.52	0.603	0.551	0.57	–	0.695
Ce	0.632	–	1.86	1.73	1.56	1.436	1.40	–	1.80
Pr	0.096	–	0.283	–	0.237	0.206	–	–	0.274
Nd	0.471	–	1.39	1.43	1.16	1.067	1.02	–	1.34
Sm	0.153	–	0.451	0.52	0.378	0.347	0.32	–	0.436
Eu	0.058	–	0.171	0.188	0.143	0.131	0.130	–	0.165
Gd	0.205	–	0.605	0.74	0.506	0.459	–	–	0.584
Tb	0.0374	–	0.110	0.126	0.0924	0.087	0.090	–	0.107
Dy	0.253	–	0.746	0.766	0.625	0.572	–	–	0.721
Y	1.57	–	4.63	–	3.88	3.4	–	–	4.47
Ho	0.0566	–	0.167	0.181	0.140	0.128	–	–	0.161
Er	0.166	–	0.490	0.46	0.410	0.374	–	–	0.473
Tm	0.0255	–	0.0752	–	0.0630	0.054	–	–	0.7027
Yb	0.166	–	0.489	0.490	0.410	0.372	0.320	0.620	0.475
Lu	0.0254	–	0.0749	0.074	0.0627	0.057	0.060	–	0.0724

C1 Chondritic composition was compiled from Palme et al. (1981), Anders and Ebihara (1983), Beer et al. (1984) and Jochum et al. (1986). Model 1 = Ringwood (1979), Model 2 = Sun (1982), Model 3 = Wänke et al. (1984), Model 4 = Hart and Zindler (1987), Model 5 = Taylor and McLennan (1985), Model 6 = Anderson (1983), Model 7 = Palme and Nickel (1985). Preferred = McDonough and Sun (1989).
Mg# = 100 MgO/(MgO+FeO) on a molar basis.

138

refractory lithophile elements in Earth's primitive mantle. Using these constraints and other element correlation diagrams they also determined the absolute abundances of non-refractory lithophile elements in Earth's primitive mantle. This approach provided an internally consistent model for inferring the primitive mantle bulk composition. Their model is presented in Table 1 along with a compilation of other models. This primitive mantle model compares favorably with the pyrolite model of Ringwood (1979) and the komatiite and peridotite models of Sun (1982) and Wänke et al. (1984), respectively. However, significant differences are observed when this model is compared with other peridotite-based models (Palme and Nickel, 1985; Hart and Zindler, 1987) or the more silica-rich bulk Earth models based on carbonaceous chondrites (Anderson, 1985; Taylor and McLennan, 1985).

As we look to the future, it is apparent that advances in defining the present composition of the mantle and understanding the evolution of the mantle will require comprehensive geochemical studies of mantle samples. There are now extensive data for abundances of major elements, some trace elements, such as REE, and isotopic ratios of Sr and Nd. However, there is a lack of even survey data for other trace elements, such as the HFSE, platinum group elements, Th and U and other radiogenic isotopic systems; thus there is a need for combined geochemical studies. Individual samples must be analyzed for major elements, REE and other trace elements, plus radiogenic and stable (e.g., oxygen) isotopic ratios. Both whole-rock and mineral analyses will be needed. Analytical techniques and instruments capable of obtaining accurate and precise data for a wide variety of elements and isotopes at natural abundances are essential. Current analytical approaches are good for some elements and certainly the ion microprobe with its capability for spatial resolution and in situ analyses will be increasingly important. However, in many ways geochemical studies of mantle rocks are at the frontiers of analytical geochemistry; therefore new advances in understanding depend in part upon development of new analytical techniques and instrumentation. Also, much more emphasis on experimental laboratory studies of element partitioning between minerals, melts and fluids at mantle pressures and temperatures is essential for realistic interpretations of elemental and isotopic abundance data. Finally, geochemists must carefully choose the samples to be studied. For example, studies focussed on mantle wall rocks adjacent to dikes, veins and layers will lead to new understanding of the mantle processes, such as diffusion, melt infiltration and volatile transfer, involved in element migration within the mantle. In contrast, in order to estimate the bulk composition of the upper mantle, zones of local complexity must be avoided.

We anticipate that geochemical studies of mantle rocks and minerals will be an exciting and productive field in the next ten years. The existing data base are sufficient to identify what needs to be done next to solve important problems in understanding mantle evolution. The diverse approaches of direct geochemical studies of mantle materials and studies of mantle-derived melts of varying age combined with geophysical and petrological constraints on melt formation and migration will be utilized to understand how melts have formed, segregated and migrated in the mantle.

ACKNOWLEDGMENTS

We thank L. Kah and S. Martinez for their efforts in creating data files, R.L. Rudnick for comments on an early version of the chapter, K. Johnson for preparation of Figure 11, and D. Frank for manuscript preparation. In addition, this paper has benefitted from constructive reviews by J.L. Bodinier, L. Reisberg and N. Shimizu. Each of these has made important contributions in this field, and we appreciate their efforts in helping us to accurately present their data and interpretations.

REFERENCES

Akagi, T., and Masuda, A. (1988) Isotopic and elemental evidence for a relationship between kimberlite and Zaire cubic diamonds. Nature, 336, 665-667.

Allègre, C.J., and Turcotte, D.L. (1986) Implications of a two-component marble-cake mantle. Nature, 323, 123-127.

Anders, E., and Ebihara, M. (1982) Solar-system abundances of the elements. Geochim. Cosmochim. Acta, 46, 2363-2380.

Anderson, D.L. (1983) Chemical composition of the mantle. J. Geophys. Res. (suppl.), 88, B41-B52.

Beer, H., Walter, G., Macklin, P.J., and Patchett, P.J. (1984) Neutron capture cross sections and solar abundances of $^{160,161}Dy$, $^{170,171}Yb$, $^{175,176}Lu$ and $^{176,177}Hf$ for the S-process analysis of the radio-nuclide ^{176}Lu. Phys. Rev. C30, 464-478.

Bibby, D.M. (1979) Zonal distribution of impurities in diamond. Geochim. Cosmochim. Acta, 43, 415-423.

Bodinier, J.L. (1988) Geochemistry and petrogenesis of the Lanzo peridotite body, western Alps. Tectonophys., 149, 67-88.

Bodinier, J.L., Dupuy, C., and Dostal, J. (1988a) Geochemistry and petrogenesis of eastern Pyrenean peridotites. Geochim. Cosmochim. Acta, 52, 2893-2907.

Bodinier, J.L., Dupuy, C., Dostal, J., and Merlet, C. (1987c) Distribution of trace transition elements in olivine and pyroxenes from ultramafic xenoliths: application of microprobe analysis. Amer. Mineral., 72, 902-913.

Bodinier, J.L., Dupuy, C., and Vernieres, J. (1988b) Behaviour of trace elements during upper mantle metasomatism: evidences from the Lherz massif. Chem. Geol., 70, 152.

Bodinier, J.L., Fabriès, J., Lorand, J.P., Dostal, J., and Dupuy, C. (1987a) Geochemistry of amphibole pyroxenite veins from the Lherz and Freychinède ultramafic bodies (Ariège, French Pyrenees). Bull. Mineral., 110, 345-358.

Bodinier, J.L., Guiraud, M., Fabriès, J., Dostal, J., and Dupuy, C. (1987c) Petrogenesis of layered pyroxenites from the Lherz, Freychinède and Prades ultramafic bodies (Ariège, French Pyrénées). Geochim. Cosmochim. Acta, 51, 279-290.

Bodinier, J.L., Vasseur, G., Vernières, J., Dupuy, C., and Fabriès, J. (1989) Mechanisms of mantle metasomatism: geochemical evidence from the Lherz orogenic peridotite. J. Petrol., (in press).

Bonatti, E., Ottonello, G., and Hamlyn, P.R. (1986) Peridotites from the island of Zabargad (St. John), Red Sea: petrology and geochemistry. J. Geophys. Res., 91, 599-631.

Boyd, F.R., and Nixon, P.H. (1972) Ultramafic nodules from the Thaba Putsoa kimberlite pipe. Carnegie Inst. Washington Yearb., 71, 362-373.

Boynton, W.V. (1984) Cosmochemistry of the rare earth elements: meteorite studies. In: P. Henderson Ed., Rare earth element geochemistry, Elsevier, Amsterdam, p. 63-114.

Brueckner, H.K., Zindler, A., Seyler, M., and Bonatti, E. (1988) Zabargad and the isotopic evolution of the sub-Red Sea mantle and crust. Tectonophys., 150, 163-176.

BVSP (1981) Basaltic Volcanism on the Terrestrial Planets, Pergamon, New York, 1286pp.

Cawthorn, R.G. (1975) The amphibole peridotite-metagabbro complex, Finero, northern Italy. J. Geol., 83, 437-454.

Chen, C.Y., and Frey, F.A. (1989) Evolution of the upper mantle beneath southeast Australia: geochemical evidence from peridotite xenoliths in Mount Leura basanite. Earth Planet. Sci. Lett., 93, 195-209.

Dautria, J.M., Dostal, J., Dupuy, C., and Liotard, J.M. (1988) Geochemistry and petrogenesis of alkali basalts from Tahalra (Hoggar, Northwest Africa). Chem. Geol., 69, 17-35.

Dautria, J.M., and Girod, M. (1987) Cenozoic volcanism associated with swells and rifts. In: P.H. Nixon Ed., Mantle Xenoliths, John Wiley and Sons Ltd., Chichester, p. 195-214.

Dautria, J.M., Liotard, J.M., Cabanes, N., Girod, M., and Briqueu, L. (1987) Amphibole-rich xenoliths and host alkali basalts: petrogenetic constraints and implications on the recent evolution of the upper mantle beneath Ahaggar (Central Sahara, southern Algeria). Contrib. Mineral. Petrol., 95, 133-144.

Dawson, B.J., and Smith, J.V. (1977) The MARID (Mica-Amphibole-Rutile-Ilmenite-Diopside) suite of kimberlite xenoliths in kimberlite. Geochim. Cosmochim. Acta, 41, 309-323.

Dawson, J.B. (1984) Contrasting types of upper-mantle metasomatism? In: J. Kornprobst Ed., Kimberlites II: the mantle and crust - mantle relationships, Elsevier, Amsterdam, p. 289-294.

Dick, H.J.B., Fisher, R.L., and Bryan, W.B. (1984) mineralogic variability of the uppermost mantle along mid-ocean ridges. Earth Planet. Sci. Lett., 69, 88-106.

Downes, H., and Dupuy, C. (1987) Textural, isotopic and REE variations in spinel peridotite xenoliths, Massif Central, France. Earth Planet. Sci. Lett., 82, 121-135.

140

Dupuy, C., Dostal, J., and Bodinier, J.L. (1987) Geochemistry of spinel peridotite inclusions in basalts from Sardinia. Mineral. Mag., 51, 561-568.

Dupuy, C., Dostal, J., Dautria, J.M., and Girod, M. (1986) Geochemistry of spinel peridotite inclusions in basalts from Hoggar, Algeria. J. African Earth Sci., 5, 209-215.

Eggler, D.H. (1987) Solubility of major and trace elements in mantle metasomatic fluids: experimental constraints. In: M.A. Menzies, and C.J. Hawkesworth Eds., Mantle Metasomatism, Academic Press Inc., London, p. 21-41.

Ehrenberg, S.N. (1982) Rare earth element geochemistry of garnet lherzolite and megacrystalline nodules from minette of the Colorado Plateau province. Earth Planet. Sci. Lett., 57, 191-210.

Erlank, A.J., Waters, F.G., Hawkesworth, C.J., Haggerty, S.E., Allsopp, H.L., Richard, R.S., and Menzies, M.A. (1987) Evidence for mantle metasomatism in peridotite nodules from the Kimberley pipes, South Africa. In: M.A. Menzies, and C.J. Hawkesworth Eds., Mantle Metasomatism, Academic Press Inc., London, p. 221-312.

Ernst, W.G. (1978) Petrochemical study of lherzolitic rocks from the western Alps. J. Petrol., 19, 341-392.

Ernst, W.G., and Piccardo, G.B. (1979) Petrogenesis of some Ligurian peridotites -- I. Mineral and bulk-rock chemistry. Geochim. Cosmochim. Acta, 43, 219-237.

Fabriès, J., Bodinier, J.L., Dupuy, C., Lorand, J.P., and Benkerrou, C. (1989) Evidence for modal metasomatism in the orogenic spinel lherzolite body from Caussou (Northern Pyrenees, France). J. Petrol., 30, 199-228.

Feigenson, M.D. (1986) Continental alkali basalts as mixtures of kimberlite and depleted mantle: evidence from Kilbourne Hole Maar, New Mexico. Geophys. Res. Lett., 13, 965-968.

Fitton, J.G., and Dunlop, H.M. (1985) The Cameroon line, West Africa, and its bearing on the origin of oceanic and continental alkali basalt. Earth Planet. Sci. Lett., 72, 23-38.

Frey, F.A. (1969) Rare earth abundances in a high-temperature peridotite intrusion. Geochim. Cosmochim. Acta, 33, 1429-1447.

Frey, F.A. (1970) Rare earth abundances in alpine ultramafic rocks. Phys. Earth Planet. Interiors, 3, 323-330.

Frey, F.A. (1980) The origin of pyroxenites and garnet pyroxenites from Salt Lake Crater, Oahu, Hawaii: trace element evidence. Am. J. Sci., 280-A, 427-449.

Frey, F.A. (1984) Rare earth element abundances in upper mantle rocks,. In: P. Henderson Ed., Rare earth element Geochemistry, Elsevier, Amsterdam, p. 153-203.

Frey, F.A., and Green, D.H. (1974) The mineralogy, geochemistry and origin of lherzolite inclusions in Victorian basanites. Geochim. Cosmochim. Acta, 38, 1023-1059.

Frey, F.A., Haskin, L.A., and Haskin, M.A. (1971) Rare-earth abundances in some ultramafic rocks. J. Geophys. Res., 76, 2057-2070.

Frey, F.A., and Prinz, M. (1978) Ultramafic inclusions from San Carlos, Arizona: petrologic and geochemical data bearing on their petrogenesis. Earth Planet. Sci. Lett., 38, 129-176.

Frey, F.A., Suen, C.J., and Stockman, H.W. (1985) The Ronda high temperature peridotite: Geochemistry and petrogenesis. Geochim. Cosmochim. Acta, 49, 2469-2491.

Fujii, N., Osamura, K., and Takahashi, E. (1986) Effect of water saturation on the distribution of partial melt in the olivine-pyroxene-plagioclase system. J. Geophys. Res., 91, 9253-9260.

Fujimaki, H., Tatsumoto, M., and Aoki, K.I. (1984) Partition coefficients of Hf, Zr, and REE between phenocrysts and groundmasses. J. Geophys. Res., 89 (supp), B662-B672.

Galer, S.J.G., and O'Nions, R.K. (1989) Chemical and isotopic studies of ultramafic inclusions from the San Carlos volcanic field, Arizona bearing on their petrogenesis. Geochim. Cosmochim. Acta, (in press).

Green, D.H., and Wallace, M. (1988) Mantle metasomatism by ephemeral carbonatite melts. Nature, 336, 459-462.

Griffin, W.L., O'Reilly, S.Y., and Stabel, A. (1988) Mantle metasomatism beneath western Victoria, Australia: II. Isotopic geochemistry of Cr-diopside lherzolites and Al-augite pyroxenites. Geochim. Cosmochim. Acta, 52, 449-459.

Griffin, W.L., Smith, D., Boyd, F.R., Cousens, D.R., Ryan, C.G., Sie, S.H., and Sutter, G.F. (1989) Trace-element zoning in garnets from sheared mantle xenoliths. Geochim. Cosmochim. Acta, 53, 561-567.

Hamelin, B., and Allègre, C.J. (1988) Lead isotope study of orogenic lherzolite massifs. Earth Planet. Sci. Lett., 91, 117-131.

Hanson, G.N. (1977) Geochemical evolution of the suboceanic mantle. J. Geol. Soc., 134, 235-253.

Hart, S.R., and Zindler, G.A. (1986) In search of a bulk-earth composition. Chem. Geol., 57, 247-267.

Hutchison, R., Williams, C.T., Henderson, P., and Reed, S.J.B. (1986) New varieties of mantle xenoliths from the Massif Central, France. Mineral. Mag., 50, 559-565.

Irving, A.J. (1980) Petrology and geochemistry of composite ultramafic xenoliths in alkalic basalts and implications for magmatic processes within the mantle. Am. J. Sci., 280-A, 389-426.

141

Irving, A.J., and Frey, F.A. (1984) Trace element abundances in megacrysts and their host basalts: Constraints on partition coefficients and megacryst genesis. Geochim. Cosmochim. Acta, 48, 1201-1221.

Jagoutz, E., Lorenz, V., and Wänke, H. (1979a) Major trace elements of Al-augites and Cr-diopsides from ultramafic nodules in European alkali basalts. In: F.R. Boyd, and H.O.A. Meyer Eds., The mantle sample: inclusions in kimberlites and other volcanics, American Geophysical Union, Washington, D.C., p. 382-390.

Jagoutz, E., Palme, H., Baddenhausen, H., Blum, K., Cendales, M., Dreibus, G., Spettel, B., Lorenz, V., and Wänke, H. (1979b) The abundances of major, minor and trace elements in the earth's mantle as derived from primitive ultramafic nodules. Proc. Lunar Planet. Sci. Conf., 10th, 2031-2050.

Jochum, K.P., McDonough, W.F., Palme, H., and Spettel, B. (1989) Compositional constraints on the continental lithospheric mantle from trace elements in spinel peridotite xenoliths. Nature, (in press).

Jochum, K.P., Seufert, H.M., Spettel, B. and Palme, H. (1986) The solar-system abundances of Nb, Ta, and Y and the relative abundances of refractory lithophile elements in differentiated planetary bodies. Geochim. Cosmochim. Acta 50, 1173-1183.

Johnson, K.T.M., Dick, H.J.B., and Shimizu, N. (1987) Rare earth element composition of discrete diopsides from the oceanic upper mantle: implications for MORB genesis and processes of exotic melt infiltration. EOS Trans. Am. Gephys. Union, 68, 1541.

Johnson, K.T.M., Dick, H.J.B., and Shimizu, N. (1988) Trace element composition of diopsides in abyssal peridotites: implications for generation of mid-ocean ridge basalts. EOS Trans. Am. Gephys. Union, 69, 1516.

Johnson, K.T.M., Dick, H.J.B., and Shimizu, N. (1989) Melting in the oceanic upper mantle: an ion microprobe study of diopsides in abyssal peridotites. J. Geophys. Res., (in press).

Jones, A.P., and Wyllie, P.J. (1984) Minor elements in perovskite from kimberlites and distribution of the rare earth elements: an electron probe study. Earth Planet. Sci. Lett., 69, 128-140.

Kato, T., Ringwood, A.E., and Irifune, T. (1988) Experimental determination of element partitioning between silicate perovskites, garnets and liquids: constraints on early differentiation of the mantle. Earth Planet. Sci. Lett., 89, 123-145.

Kempton, P.D. (1987) Metasomatism and enrichment in lithosphereic peridotites. In: M.A. Menzies, and C.J. Hawkesworth Eds., Mantle Metasomatism, Academic Press Inc., London, p. 45-90.

Kovalenko, V.I., Ionov, D.A., Yarmolyak, V.V., Jagoutz, E., Lugmair, G., and Stosch, H.G. (1988) Correlation of mantle and crustal evolution for some regions of central Asia by isotope data. Composition and Processes of Deep-Seated Zones of Continental Lithosphere, Internat'l Symp. Abstracts, Novosibirsk, USSR, 136-138.

Kramers, J.D., Roddick, J.C.M., and Dawson, J.B. (1983) Trace element and isotope studies on veined, metasomatic and "MARID" xenoliths from Bultfontein, South Africa. Earth Planet. Sci. Lett., 65, 90-106.

Kurat, G., Palme, H., Spettel, B., Baddenhausen, H., Hofmeister, H., Palme, C., and Wänke, H. (1980) Geochemistry of ultramafic xenoliths from Kapfenstein, Austria: evidence for a variety of upper mantle processes. Geochim. Cosmochim. Acta, 44, 45-60.

Langmuir, C.H., Bender, J.F., Bence, A.E., Hanson, G.N., and Taylor, S.R. (1977) Petrogenesis of basalts from the FAMOUS area: Mid-Atlantic Ridge. Earth Planet. Sci. Lett., 36, 133-156.

Liotard, J.M., Boivin, P., Cantagrel, J.M., and Dupuy, C. (1983) Megacristaux d'amphibole et basaltes alcalins associes. Problemes de leurs relations petrogenetiques et geochimiques. Bull. Mineral., 106, 451-464.

Liotard, J.M., Briot, D., and Boivin, P. (1988) Petrological and geochemical relationships between pyroxene megacrysts and associated alkali-basalts from Massif Central (France). Contrib. Mineral. Petrol., 98, 81-90.

Loubet, M., and Allegre, C.J. (1982) Trace elements in orogenic lherzolites reveal the complex history of the upper mantle. Nature, 298, 809-814.

Loubet, M., Shimizu, N. and Allegre, C.J. (1975) Rare earth elements in alpine peridotites. Contrib. Mineral. Petrol., 53, 1-12.

Maaløe, S., and Aoki, K.I. (1975) The major element composition of the upper mantle estimated from the composition of lherzolites. Contrib. Mineral. Petrol., 63, 161-173.

McDonough, W.F., and Sun, S.s. (1989) Composition of the Earth's primitive mantle., (in preparation).

McKay, G.A. (1986) Crystal/liquid partitioning of REE in basaltic systems: extreme fractionation of REE in olivine. Geochim. Cosmochim. Acta, 50, 69-79.

McKay, G., Wagstaff, J., and Yang, S.R. (1986) Clinopyroxene REE distribution coefficients for shergoittites: the REE content of the Shergotty melt. Geochim. Cosmochim. Acta, 50, 927-937.

Meen, J.K., Eggler, D.H., and Ayers, J.C. (1989) Experimental evidence for very low solubility of rare-earth elements in CO_2-rich fluids at mantle conditions. Nature, 340, 301-303.

142

Menzies, M. (1976) Rare earth geochemistry of fused ophiolitic and alpine lherzolites -- I. Othris, Lanzo and Troodos. Geochim. Cosmochim. Acta, 40, 645-656.

Menzies, M., Blanchard, D., Brannon, J., and Korotev, R. (1977) Rare earth geochemistry of fused ophiolitic and alpine lherzolites II. Beni Bouchera, Ronda and Lanzo. Contrib. Mineral. Petrol., 64, 53-74.

Menzies, M.A., and Halliday, A. (1988) Lithospheric mantle domains beneath the Archean and Proterozoic crust of Scotland. J. Petrol., Special Lithosphere Issue, 275-302.

Menzies, M.A., and Hawkesworth, C.J. (1987) Mantle Metasomatism. Academic Press, London, 472pp.

Menzies, M.A., Kempton, P., and Dungan, M. (1985) Interaction of continental lithosphere and asthenospheric melts below the Geronimo volcanic field, Arizonia, U.S.A. J. Petrol., 26, 663-693.

Menzies, M.A., Rogers, N., Tindle, A., and Hawkesworth, C.J. (1987) Metasomatic and enrichment processes in lithospheric peridotites, an effect of asthenosphere-lithosphere interaction. In: M.A. Menzies, and C.J. Hawkesworth Eds., Mantle Metasomatism, Academic Press Inc., London, p. 313-361.

Menzies, M.A., and Wass, S.Y. (1983) CO_2- and LREE-rich mantle below eastern Australia: a REE and isotopic study of alkaline magmas and apatite-rich mantle xenoliths from the Southern Highlands Province, Australia. Earth Planet. Sci. Lett., 65, 287-302.

Mitchell, R.H., and Carswell, D.A. (1976) Lanthanium, Samarium and Ytterbium abundances in some southern African garnet lherzolites. Earth Planet. Sci. Lett., 31, 175-178.

Nagasawa, H., Wakita, H., Higuchi, H., and Onuma, N. (1969) Rare earths in peridotite nodules: an explanation of the genetic relationship between basalt and peridotite nodules. Earth Planet. Sci. Lett., 5, 377-381.

Navon, O., Hutcheon, I.D., Rossman, G.R., and Wasserburg, G.J. (1988) Mantle-derived fluids in diamond micro-inclusions. Nature, 335, 784-789.

Navon, O., Spettel, B., Hutcheon, I.H., Rossman, G.R., and Wasserburg, G.J. (1989) Micro-inclusions in diamonds from Zaire and Botswana. 28th Internat. Geol. Congress Extended Abstr., Workshop on Diamonds, 69-72.

Navon, O., and Stolper, E. (1987) Geochemical consequences of melt percolation: the upper mantle as a chromatographic column. J. Geol., 95, 285-307.

Neal, C.R. (1988) The origin and composition of metasomatic fluids and amphiboles beneath Malaita, Solomon Islands. J. Petrol., 29, 149-179.

Neal, C.R., and Taylor, L.A. (1989) A negative Ce anomaly in a peridotite xenolith: evidence for crustal recycling into the mantle or mantle metasomatism? Geochim. Cosmochim. Acta, 53, 1035-1040.

Nicolas, A., (1986) A melt extraction model based on structural studies in mantle peridotites. Jour. Petrol., 27, 999-1022.

Nicolas, A., Lucazeau, F., and Bayer, R. (1987) Peridotite xenoliths in Massif Central basalts, France: textural and geophysical evidence for asthenospheric diapirism. In: P.H. Nixon Ed., Mantle Xenoliths, John Wiley and Sons Ltd., Chichester, p. 563-574.

Nielson-Pike, J.E., and Noller, J.S. (1987) Processes of mantle metasomatism: evidence from xenoliths and peridotite massifs. In: E.M. Morris, and J.D. Pasteris Eds., Mantle metasomatism and alkaline volcanism, Geol. Soc. Amer. Spec. Paper 215, p. 62-78.

Nixon, P.H. (1987) Mantle Xenoliths. J. Wiley and Sons, New York, 844pp.

Nixon, P.H., Rogers, N.W., Gibson, I.L., and Grey, A. (1981) Depleted and fertile mantle xenoliths from southern African kimberlites. Ann. Rev. Earth Planet. Sci., 9, 285-309.

Noller, J.S. (1986) Solid and fluid inclusions in mantle xenoliths: an analytical dilemma. Geology, 14, 437-440.

O'Reilly, S.Y., and Griffin, W.L. (1988) Mantle metasomatism beneath western Victoria, Australia: I. Metasomatic processes in Cr-diopside lherzolites. Geochim. Cosmochim. Acta, 52, 433-447.

Obata, M. (1980) The Ronda peridotite: garnet-, spinel-, and plagioclase-lherzolite facies and the P-T trajectories of a high-temperature mantle intrusion. J. Petrol., 21, 533-572.

Ottonello, G. (1980) Rare earth abundances and distribution in some spinel peridotite xenoliths from Assab (Ethiopia). Geochim. Cosmochim. Acta, 44, 1885-1901.

Ottonello, G., Ernst, W.G., and Joron, J.L. (1984a) Rare earth and 3d transition element geochemistry of peridotite rocks: I. Peridotites from the western Alps. J. Petrol., 25, 343-372.

Ottonello, G., Joron, J.L., and Piccardo, G.B. (1984b) Rare earth and 3d transition element geochemistry of peridotite rocks: II. ligurian peridotites and associated basalts. J. Petrol., 25, 373-393.

Ottonello, G., Piccardo, G.B., and Ernst, W.G. (1979) Petrogenesis of some Ligurian peridotites -- II. Rare earth element chemistry. Geochim. Cosmochim. Acta, 43, 1273-1284.

Ottonello, G., Piccardo, G.B., Joron, J.L., and Treuil, M. (1978a) Evolution of the upper mantle under the Assab region (Ethiopia): suggestions from petrology and geochemistry of tectonic ultramafic xenoliths and host basaltic lavas. Geol. Rundsch., 67/2, 547-575.

143

Ottonello, G., Piccardo, G.B., Mazzucotelli, A., and Cimmino, F. (1978b) Clinopyroxene-orthopyroxene major and rare earth elements partitioning in spinel peridotite xenoliths form Assab (Ethiopia). Geochim. Cosmochim. Acta, 42, 1817-1828.

Palme, H., and Nickel, K.G. (1985) Ca/Al ratio and composition of the Earth's upper mantle. Geochim. Cosmochim. Acta, 49, 2123-2132.

Palme, H., Suess, H.E., and Zeh, H.D. (1981) Abundances of the elements in the solar system. In Landolt-Bornstein, Group VI: Astromomy, Astrophysic, Extension and Supplement 1, Subvolume a (ed. in chief, K.-H. Hellwege), pp.257-272, Springer-Verlag, Berlin.

Philpotts, J.A., Schnetzler, C.C., and Thomas, H.H. (1972) Petrogenitic implications of some new geochemical data on eclogitic inclusions. Geochim. Cosmochim. Acta, 36, 1131-1166.

Piccardo, G.B., Messiga, B., and Vannucci, R. (1988) The Zabargad peridotite-pyroxenite association: petrological constraints on its evolution. Tectonophys., 150, 135-162.

Polvé, M., and Allègre, C.J. (1980) Orogenic lherzolite complexes studied by ^{87}Rb-^{87}Sr: a clue to understand the mantle convection processes? Earth Planet. Sci. Lett., 51, 71-93.

Preß, S., Witt, G., Seck, H.A., Eonov, D., and Kovalenko, V.I. (1986) Spinel peridotite xenoliths from the Tariat Depression, Mongolia. I: Major element chemistry and mineralogy of a primitive mantle xenolith suite. Geochim. Cosmochim. Acta, 50, 2587-2599.

Prinzhofer, A., and Allègre, C.J. (1985) Residual peridotite and the mechanism of partial melting. Earth Planet. Sci. Lett., 74, 251-265.

Prinzhofer, A., Lewin, E., and Allègre, C.J. (1989) Stochastic melting of the marble cake mantle: evidence from local study of the East Pacific Rise at 12°50'N. Earth Planet. Sci. Lett., 92, 189-206.

Reid, J.B., and Frey, F.A. (1971) Rare earth distributions in lherzolite and garnet pyroxenite xenoliths and the constitution of the upper mantle. J. Geophy. Res., 76, 1184-1196.

Reisberg, L.C., Luck, J.M., and Allegre, C.J. (1989b) The Re-Os systematis of the Ronda ultramafic complex. EOS Trans. Am. Gephys. Union, 70, 509.

Reisberg, L., and Zindler, A. (1986/87) Extreme isotopic variations in the upper mantle: evidence from Ronda. Earth Planet. Sci. Lett., 81, 29-45.

Reisberg, L.C., Zindler, G.A., and Jagoutz, E. (1989a) Further Sr and Nd results from peridotites of the Ronda ultramafic complex. Earth Planet. Sci. Lett., (in press).

Richardson, S.H., Gurney, J.J., Erlank, A.J., and Harris, J.W. (1984) Origin of diamonds in old enriched mantle. Nature, 310, 198-202.

Richter, F.M. (1985) Models of the Archaean thermal regime. Earth Planet. Sci. Lett., 73, 350-360.

Ridley, W.I., and Dawson, J.B. (1975) Lithophile trace element data bearing on the origin of peridotite xenoliths, ankaramite and carbonatite from Lashaine volcano, N. Tanzania. Phys. Chem. Earth, 9, 559-569.

Ringwood, A.E. (1979) Origin of the Earth and Moon, Springer-Verlag, New York, 295 pp.

Roden, M.F., Frey, F.A., and Francis, D.M. (1984) An example of consequent mantle metasomatism in peridotite inclusions from Nunivak Island, Alaska. J. Petrol., 25, 546-577.

Roden, M.F., Irving, A.J., and Murthy, V.R. (1988) Isotopic and trace element composition of the upper mantle beneath a young continental rift: results from Kilbourne Hole, New Mexico. Geochim. Cosmochim. Acta, 52, 461-473.

Roden, M.F., and Rama Murthy, V. (1985) Mantle Metasomatism. Ann. Rev. Earth Planet. Sci., 13, 269-296.

Roden, M.F., Smith, D., and Rama Murthy, V. (1989) Geochemical constraints on lithosphere composition and evolution beneath the Colorado Plateau. J. Geophy. Res., (in press).

Roden, M.K., Hart, S.R., Frey, F.A., and Melson, W.G. (1984) Sr, Nd and Pb isotopic and REE geochemistry of St. Paul's Rocks: the metamorphic and metasomatic development of an alkali basalt mantle source. Contrib. Mineral. Petrol., 85, 376-390.

Sachtleben, T., and Seck, H.A. (1981) Chemical control of Al-solubility in orthopyroxene and its implications on pyroxene geothermometry. Contrib. Mineral. Petrol., 78, 157-165.

Salters, V.J.M., and Shimizu, N. (1988) World-wide occurence of HFSE-depleted mantle. Geochim. Cosmochim. Acta, 52, 2177-2182.

Shimizu, N. (1975) Rare earth elements in garnets and clinopyroxenes from garnet lherzolite nodules in kimberlites. Earth Planet. Sci. Lett., 25, 26-32.

Shimizu, N., Gurney, J.J., and Moore, R. (1989) Trace element geochemistry of garnet inclusions in diamonds from the Finsch and Koffiefontein kimberlite pipes. 28th Internat. Geol. Congress Extended Abstr., Workshop on Diamonds, 100-101.

Shimizu, N., and Richardson, S.H. (1987) Trace element abundance patterns of garnet inclusions in peridotite-suite diamonds. Geochim. Cosmochim. Acta, 51, 755-758.

144

Skewes, M.A., and Stern, C.R. (1979) Petrology and geochemistry of alkali basalts and ultramafic inclusions from the Palei-Aike volcanic field in southern Chile and the origin of the Patagonian plateau lavas. J. Volcanol. Geotherm. Res., 6, 3-25.

Smith, D. (1979) Hydrous minerals and carbonates in peridotite inclusions from the Green Knobs and Buell Park kimberlitic diatremes on the Colorado Plateau. In: F.R. Boyd, and H.O.A. Meyer Eds., The mantle sample: inclusions in kimberlite and other volcanics, Am. Gephys. Union, Washington, D.C., p. 345-356.

Smith, D., and Boyd, F.R. (1987) Compositional heterogeneities in a high-temperature lherzolite nodule and implications for mantle processes. In: P.H. Nixon Ed., Mantle Xenoliths, John Wiley and Sons Ltd., Chichester, p. 551-561.

Song, Y., and Frey, F.A. (1989) Geochemistry of peridotite xenoliths in basalts from Hannuoba, Eastern China: implications for subcontinental mantle heterogeneity. Geochim. Cosmochim. Acta, 53, 97-113.

Stolz, A.J., and Davies, G.R. (1988) Chemical and isotopic evidence from spinel lherzolite xenoliths for episodic metasomatism of the upper mantle beneath southeast Australia. J. Petrol., Special Lithosphere Issue, 303-330.

Stolz, A.J., and Davies, G.R. (1989) Metasomatised lower crustal and upper mantle xenoliths from north Queensland: chemical and isotopic evidence bearing on the composition and source of the fluid phase. Geochim. Cosmochim. Acta, 53, 649-660.

Stosch, H.G. (1982) Rare earth element partitioning between minerals from anhydrous spinel peridotite xenoliths. Geochim. Cosmochim. Acta, 46, 793-811.

Stosch, H.G., and Lugmair, G.W. (1986) Trace element and Sr and Nd isotope geochemistry of peridotite xenoliths from the Eifel (West Germany) and their bearing on the evolution of the subcontinental lithosphere. Earth Planet. Sci. Lett., 80, 281-298.

Stosch, H.G., Lugmair, G.W., and Kovalenko, V.I. (1986) Spinel peridotite xenoliths from the Tariat Depression, Mongolia. II: Geochemistry and Nd and Sr isotopic composition and their implications for the evolution of the subcontinental lithosphere. Geochem. Cosmochem. Acta, 50, 2601-2614.

Stosch, H.G., and Seck, H.A. (1980) Geochemistry and mineralogy of two spinel peridotite suites from Dreiser Weiher, West Germany. Geochim. Cosmochim. Acta, 44, 457-470.

Streckeisen, A. (1976) To each plutonic rock its proper name. Earth Sci. Rev., 12, 1-33.

Suen, C.J., and Frey, F.A. (1987) Origins of the mafic and ultramafic rocks in the Ronda peridotite. Earth Planet. Sci. Lett., 85, 183-202.

Sun, S.s. (1982) Chemical composition and origin of the earth's primitive mantle. Geochem. Cosmochem Acta, 46, 179-192.

Sun, S.s., and McDonough, W.F. (1989) Chemical and isotopic systematics of oceanic basalts: implications for mantle composition and processes. In: A.D. Saunders, and M.J. Norry Eds., Magmatism in the ocean basins, Geol. Soc. Lond. Spec. Pub., p. 313-345.

Tanaka, T., and Aoki, K.I. (1981) Petrogenic implications of REE and Ba data on mafic and ultramafic inclusions from Itinome-gata, Japan. J. Geol., 89, 369-390.

Taylor, S.R., and McLennan, S.M. (1985) The Continental Crust: its Composition and Evolution. Blackwell Scientific Publ., Oxford, 312pp.

Toramaru, A., and Fujii, N. (1986) Connectivity of melt phase in a partially molten peridotite. J. Geophys. Res., 91, 9239-9252.

Varne, R., and Graham, A.L. (1971) Rare earth abundances in hornblende and clinopyroxene of a hornblende lherzolite xenolith: implications for upper mantle fractionation processes. Earth Planet. Sci. Lett., 13, 11-18.

von Bargen, N., and Waff, H.S. (1986) Permeabilities, interfacial areas, and curvatures of partially molten systems: results of numerical computations of equilibrium microstructures. J. Geophys. Res., 91, 9261-9276.

von Bargen, N., and Waff, H.S. (1988) Wetting of enstatite by basaltic melt at 1350°C and 1.0 to 2.5 GPa pressure. J. Geophys. Res., 93, 1153-1158.

Walker, R.J., Carlson, R.W., Shirey, S.B. and Boyd, F.R. (1989) Os, Sr, Nd and Pb isotope systematics of southern African peridotite xenoliths: Implications for the chemical evolution of subcontinental mantle. Geochim. Cosmochim. Acta, 53, 1583-1595.

Wänke, H., Dreibus, G., and Jagoutz, E. (1984) Mantle chemistry and accretion history of the earth. In: A. Kröner, G. Hanson, and A. Goodwin Eds., Archaean Geochemistry, Springer-Verlag, Berlin, p. 1-24.

Wass, S.Y., Henderson, P., and Elliott, C.J. (1980) Chemical heterogeneity and metasomatism in the upper mantle: evidence from rare earth and other elements in apatite-rich xenoliths in basaltic rocks from eastern Australia. Phil. Trans. R. Soc. London, A 297, 333-346.

Watson, E.B., and Brenan, J.M. (1987) Fluids in the lithosphere, 1. experimentally-determined wetting characteristics of CO_2-H_2O fluids and their implications for fluid transport, host-rock physical properties and fluid inclusion formation. Earth Planet. Sci. Lett., 85, 497-515.

White, A.J.R., and Chappell, B.W. (1972) Coexisting clinopyroxene, garnet and amphibole from an "eclogite", Kakanui, New Zealand. Contrib. Mineral. Petrol., 34, 185-191.

Wilshire, H.G. (1984) Mantle metasomatism: The REE story. Geology, 12, 395-398.

Wilshire, H.G. (1987) A model of mantle metasomatism. In: E.M. Morris , and J.D. Pasteris Eds., Mantle metasomatism and alkaline volcanism, Geol. Soc. Amer. Spec. Paper 215, p. 47-60.

Wilshire, H.G., and Jackson, E.D. (1975) Problems in determining mantle geotherms from pyroxene compositions of ultramafic rocks. J. Geol., 83, 313-329.

Wilshire, H.G., Meyer, C.E., Nakata, J.K., Calk, L.C., Shervais, J.W., Nielson, J.E., and Schwarzman, E.C. (1988) Mafic and ultramafic xenoliths from the western United States, U. S. Geological Survey Pofessional Paper 1443, United States Government Printing Office, Washington, D.C., 179pp.

Wilshire, H.G., Nielson Pike, J.E., Meyer, C.E., and Schwarzman, E.C. (1980) Amphibole-rich viens in lherzolite xenoliths, Dish Lake and Deadman Lake, California. Am. J. Sci., 280-A, 576-593.

Wilshire, H.G., and Shervais, J. (1975) Al-augite and Cr-diopside ultramafic xenoliths in basaltic rocks from western United States. Phys. Chem. Earth, 9, 257-272.

Zindler, A., and Jagoutz, E. (1988) Mantle cryptology. Geochim. Cosmochim. Acta, 52, 319-333.

Zindler, A., Staudigel, H., Hart, S.R., Endres, R., and Goldstein, S. (1983) Nd and Sr isotopic study of a mafic layer from Ronda ultramafic complex. Nature, 304, 226-230.

RARE EARTH ELEMENTS IN METAMORPHIC ROCKS

INTRODUCTION

Recent advances in understanding the distribution and geochemistry of the rare earth elements (REE) have primarily been used to study the evolution of igneous rocks, lower crustal processes, and sedimentary provenance. Metamorphic rocks, which comprise at least 70 percent of the Earth's crust, have received comparatively little attention in regard to their REE content and especially in regard to the mineral host (residence) of REE. Where REE in metamorphic rocks have been investigated, the purpose has generally been to decipher the origin of metaigneous rocks, to determine the nature of pieces of ancient crust, or to examine deep crustal ultrametamorphic processes. Recently, Hickmott and others (1987) have demonstrated the usefulness of trace element including REE distributions in garnets to help delineate P)T)X (pressure)temperature)composition) paths in metamorphic terranes.

Metamorphism, as generally defined, excludes changes that occur during weathering and diagenesis. Furthermore, there is the presumption that the metamorphic changes are isochemical (i.e., the system is chemically closed) with the, generally tacit, understanding that the system undergoing metamorphism can lose or gain volatile components such as H_2O and CO_2. When other components are either lost or gained, the system is chemically open and the processes are termed metasomatic. In practice, there is a great deal of room for overlap in the definitions. The distinction between metamorphites and metasomatites is often ambiguous and the result of subjective judgments regarding the degree of chemical openness of the system. Throughout the following discussion a loose definition of metamorphism will be used wherein the system will be considered somewhat open.

This chapter reviews available data regarding the distribution and residence of REE in metamorphic rocks. As far as possible, the REE contents of a variety of bulk compositions is traced through progressively increasing metamorphic grades. Classic metasomatites are excluded, as are anatexites. Unless noted otherwise, the REE chondrite normalization values used are the preferred values of Boynton (1984). The accuracy and precision of the data presented have not been evaluated.

REE RESIDENCE IN METAMORPHIC ROCKS

There are very few data that pertain to the residence of REE in metamorphic rocks. Most studies have been concerned with whole rock values and only rarely (cf. Pride and Muecke, 1981; Reitan et al., 1980; Roaldset, 1975) have REE contents of individual minerals been documented. The residence of REE in metamorphic rocks depends on the minerals present in the rock, the modal abundance of those minerals, and the physical and chemical conditions in which those minerals grew (assuming that the minerals were not subsequently altered). The changes in mineral assemblages and mineral composition (especially major element chemistry) have been extensively documented for a wide variety of metamorphic terranes and are summarized in numerous textbooks and papers (cf. Winkler, 1967; Miyashiro, 1973; Ferry, 1982).

Figures 1 through 20 summarize the available data for REE contents of metamorphic minerals; where data are not available for common metamorphic minerals, values for minerals hosted by igneous and metasomatic rocks are plotted. The figures clearly demonstrate that accessory minerals (zircon, monazite, xenotime, allanite, sphene, apatite) tend to concentrate REE much more than do the major rock)forming minerals such as

148

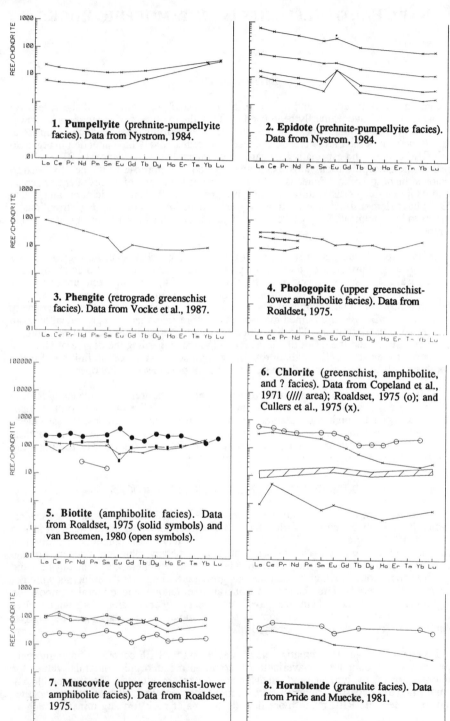

1. **Pumpellyite** (prehnite-pumpellyite facies). Data from Nystrom, 1984.

2. **Epidote** (prehnite-pumpellyite facies). Data from Nystrom, 1984.

3. **Phengite** (retrograde greenschist facies). Data from Vocke et al., 1987.

4. **Phologopite** (upper greenschist-lower amphibolite facies). Data from Roaldset, 1975.

5. **Biotite** (amphibolite facies). Data from Roaldset, 1975 (solid symbols) and van Breemen, 1980 (open symbols).

6. **Chlorite** (greenschist, amphibolite, and ? facies). Data from Copeland et al., 1971 (//// area); Roaldset, 1975 (o); and Cullers et al., 1975 (x).

7. **Muscovite** (upper greenschist-lower amphibolite facies). Data from Roaldset, 1975.

8. **Hornblende** (granulite facies). Data from Pride and Muecke, 1981.

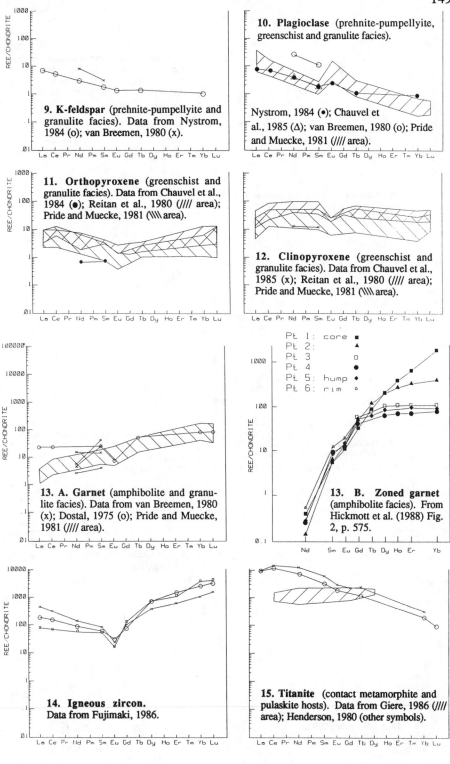

9. K-feldspar (prehnite-pumpellyite and granulite facies). Data from Nystrom, 1984 (o); van Breemen, 1980 (x).

10. Plagioclase (prehnite-pumpellyite, greenschist and granulite facies). Nystrom, 1984 (•); Chauvel et al., 1985 (Δ); van Breemen, 1980 (o); Pride and Muecke, 1981 (//// area).

11. Orthopyroxene (greenschist and granulite facies). Data from Chauvel et al., 1984 (•); Reitan et al., 1980 (//// area); Pride and Muecke, 1981 (\\\\ area).

12. Clinopyroxene (greenschist and granulite facies). Data from Chauvel et al., 1985 (x); Reitan et al., 1980 (//// area); Pride and Muecke, 1981 (\\\\ area).

13. A. Garnet (amphibolite and granulite facies). Data from van Breemen, 1980 (x); Dostal, 1975 (o); Pride and Muecke, 1981 (//// area).

13. B. Zoned garnet (amphibolite facies). From Hickmott et al. (1988) Fig. 2, p. 575.

Pt 1: core ■
Pt 2: ▲
Pt 3 □
Pt 4 ●
Pt 5: hump ◆
Pt 6: rim △

14. Igneous zircon. Data from Fujimaki, 1986.

15. Titanite (contact metamorphite and pulaskite hosts). Data from Giere, 1986 (//// area); Henderson, 1980 (other symbols).

150

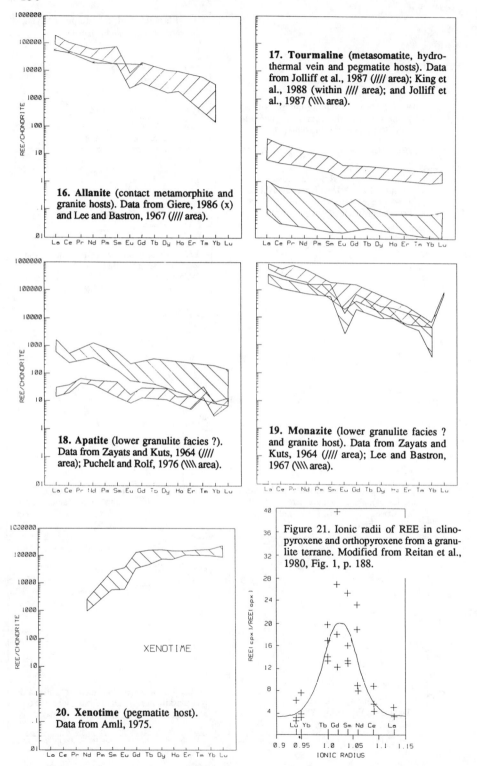

16. Allanite (contact metamorphite and granite hosts). Data from Giere, 1986 (x) and Lee and Bastron, 1967 (//// area).

17. Tourmaline (metasomatite, hydrothermal vein and pegmatite hosts). Data from Jolliff et al., 1987 (//// area); King et al., 1988 (within //// area); and Jolliff et al., 1987 (\\\\ area).

18. Apatite (lower granulite facies ?). Data from Zayats and Kuts, 1964 (//// area); Puchelt and Rolf, 1976 (\\\\ area).

19. Monazite (lower granulite facies ? and granite host). Data from Zayats and Kuts, 1964 (//// area); Lee and Bastron, 1967 (\\\\ area).

20. Xenotime (pegmatite host). Data from Amli, 1975.

XENOTIME

Figure 21. Ionic radii of REE in clinopyroxene and orthopyroxene from a granulite terrane. Modified from Reitan et al., 1980, Fig. 1, p. 188.

feldspars, micas, pyroxenes and amphiboles. Quartz, not shown, has a very low REE content. It is also obvious that the REE content of individual groups of minerals is highly variable. For example, epidotes (Fig. 2), formed under prehnite-pumpellyite facies conditions in a single metabasalt flow, show almost 2 orders of magnitude difference between their rare earth patterns. The form of the patterns within a mineral group can also be quite variable (cf. the two metamorphic hornblendes of Fig. 8).

The REE content of individual minerals and partitioning of REE between minerals are controlled by the P-T-X conditions in which the minerals grew and a variety of crystallo-chemical factors, such as valence and effective ionic radius (Adams, 1968; Fleischer, 1965; Fleischer and Altschuler, 1969; Jensen, 1973; Khomyakov, 1967; Reitan et al., 1980; Henderson,1984). The effect of cation size is shown in Figure 21 where the distribution of REE in coexisting pyroxenes is plotted against ionic radius. The middle REE (MREE) are preferentially partitioned into the M(2)-site of diopside (Reitan et al., 1980). This partitioning is also quite evident in the REE patterns of the coexisting pyroxenes (Fig. 22). The overall pattern of orthopyroxene is concave)upward while that of clinopyroxene is concave-downward, substantiating the partitioning of the MREE into clinopyroxene. The pronounced Eu anomaly in the diopside pattern is a reflection of the oxidation state of Eu during the granulite-facies conditions of pyroxene growth and Eu availability during crystal growth. The effect of variations in bulk composition of the host rock is probably reflected in the differences in the phlogopite REE patterns of Figure 4. No data are available to unequivocally demonstrate the effects of variations in P and T.

The net effects of variations of dependent variables such as P, T, and X can also be evaluated by examining the distribution coefficients for REE in coexisting (presumably, equilibrium) minerals. Figure 23A shows the variability of the distribution coefficients (D = REE concentration in phase A / REE concentration in phase B) between coexisting pyroxenes. All four samples are from plagioclase-rich, two-pyroxene granulites of intermediate composition. The samples are from two different granulite facies metamorphic terranes, but there is no consistent variation in the value of D for different REE between the two data sets. Slightly larger variations (Fig. 23B) exist in the data of Pride and Muecke (1981) for granulites from the Scourian complex (Scotland). These variations can be due to analytical error, inclusions in the minerals, alteration of the minerals, non-equilibrium growth of the minerals, growth of zoned minerals, differences in P and/or T within the sampled terranes during mineral growth, and/or non)ideal (non)Henry's Law) behaviour of REE in which the partitioning of REE between phases is a function of the composition of the phases (see Ganguly and Saxena, 1987 for a thermodynamic formulation of this effect; also see Harrison, 1981; Ottonello et al., 1984 and Watson, 1985 for discussions specifically related to REE).

The data are insufficient to permit determination of the controls on the variability of D values. Reitan et al. (1980) eliminated all samples with major element zoning within the pyroxenes, inclusions in the pyroxenes, alteration, and non-equilibrium textures. Analytical errors are not specifically stated, but the variation in D values for the individual REE tend to exceed that which would result from a 15% analytical error. They did not specifically evaluate possible P-T differences between the two sampled terranes. However, the crossing tie-lines in Figure 24A suggest that the coexisting pyroxenes grew under different P-T regimes or non)equilibrium conditions. The figure also shows the compositional vari-ability of the pyroxenes, which also could affect the D values. A similar variation of pyroxene compositions and possible disequilibrium exists in the Pride and Muecke (1981) samples (Fig. 24B). This data set has the added problem of encompassing a wide variation in whole rock compositions (felsic, mafic, ultramafic).

Different whole rock compositions (especially major element variations) can result in different mineral assemblages for two rocks that have the same metamorphic history. This is demonstrated by comparing Pride and Muecke's (1981) samples 64-12 (garnet-absent) and 67-109 (garnet-bearing). The possible effect of variations in mineral assemblage (and

Table 1.

Authors addressing REE mobility in metamorphic rocks

REE Mobile	REE Immobile

Low Temperature Processes

REE Mobile	REE Immobile
Church, 1987	Frey et al., 1968
Condie et al., 1977	Kay, et al., 1970
Frey et al., 1974	Kay & Senechal, 1976
Hellmen & Henderson, 1977	Masuda et al., 1971
Humphris et al., 1978	Menzies and Seyfried, 1979
Ludden & Thompson, 1978	Philpotts et al., 1969
_____, 1979	
Michard et al., 1983	
Seifert et al., 1985	
_____, 1987	
Stauigel & Hart, 1983	

Low-Moderate Temperature Metamorphism

REE Mobile	REE Immobile
Bartley, 1986	Cullers et al., 1974
Cerny et al., 1987	Garmann et al., 1975
Dickin, 1988	Hanson, 1975
Hellman et al., 1977	Helvaci & Griffin, 1983
_____, 1979	Herrmann et al., 1974
Jahn & Sun, 1979	Smewing & Potts, 1976
Lausch et al., 1974	
Menzies et al., 1977	
Nystrom, 1984	
Sun & Nesbitt, 1978	
Vocke et al., 1987	
Wood et al., 1976	

High Temperature Metamorphism

REE Mobile	REE Immobile
Collerson & Fryer, 1978	Bernard-Griffiths et al., 1985
Stahle et al., 1987	Dostal & Capedri, 1979
	Green et al., 1972
	O'Nions & Pankhurst, 1974
	Rollinson & Windley, 1980

Hydrothermal and Metasomatic Processes

REE Mobile	REE Immobile
Alderton et al., 1980	Hajash, 1984
Leroy & Turpin, 1988	Muecke et al, 1979
MacLean, 1988	
Martin et al., 1978	
Whitford et al., 1988	

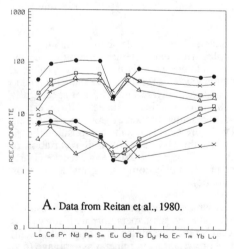

A. Data from Reitan et al., 1980.

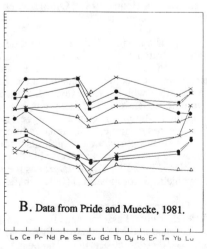

B. Data from Pride and Muecke, 1981.

Figure 22. REE patterns of coexisting clinopyroxenes and othopyroxenes from ganulite facies terranes.

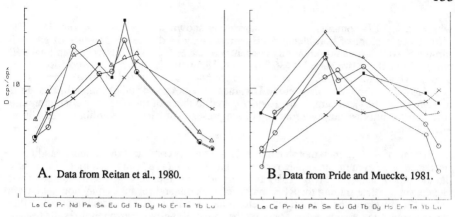

Figure 23. REE distribution coefficients for coexisting clinopyroxenes and orthopyroxenes from ganulite facies terranes. In (B), solid symbols are from sample 64-12; "x"s from sample 67-109.

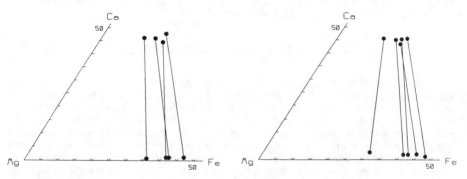

Figure 24. Major element compositions of coexisting clinopyroxenes and orthopyroxenes from granulite facies terranes. A. Data from Reitan et al., 1980. B. Data from Pride and Muecke, 1981.

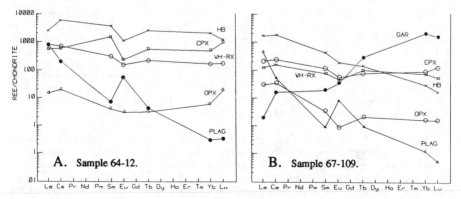

Figure 25. REE patterns of individual minerals and the whole rock value from granulite facies metamorphites. All data from Pride and Muecke, 1981.

154

whole rock REE content) on REE patterns is shown in Figures 25A and 25B. The differences in slope of hornblende REE patterns may be due to the presence of garnet in sample 67-109, which preferentially concentrates the HREE. The differences in total REE composition of the two samples (36.9 ppm vs. 18.5 ppm) is shown by the positions of the whole rock and mineral REE patterns of sample 64-12 as compared to those of sample 67-109. However, the differences in the D values for cpx/opx, or other mineral pairs (Pride and Muecke, 1981), in sample 67-109 compared to those of the other Scourian granulites (Fig. 23B) cannot be simply a function of the presence of garnet if equilibrium conditions were attained.

Zoning of REE in metamorphic minerals is a further complication in understanding the residence of REE in metamorphic rocks. Hickmott et al. (1987) and Hickmott (1988) have demonstrated that garnet (Fig. 13B) and hornblende can have zoned REE compositions. Hence, simply determining the REE content of bulk mineral separates may not yield useful values. For instance, the use of a bulk determination for garnet in determining the D values for garnet/biotite (or some other nonrefractory mineral) will give erroneous results because the composition of the garnet that grew in equilibrium with the biotite may have been only that of the rim of the garnet. Gromet and Silver (1983) have used zoned igneous allanite and titanite to help delineated the changing REE composition of their parent magma. A similar use of zoning in metamorphic minerals may be possible in unraveling changes in the composition of metamorphic fluids where P and T can be independently estimated and the metamorphic reactions are completely documented.

REE MOBILITY DURING METAMORPHISM

The REE content of a metamorphic rock (assuming closed-system conditions) will directly mimic the REE content of the protolith. This assumption has lead to a great many studies of metaigneous and metasedimentary rocks in which the evolution of the protolith was examined in detail without specifically testing for the immobility of the REE. In the extreme, some of those studies refer to amphibolite)facies amphibolite gneisses and pelitic schists as basalts and mudstones and do not acknowledge the possibility that the rock chemistry could have been altered during metamorphism, diagenesis, or weathering. The question of REE mobility during those processes has been debated for several decades and has not been resolved. What can be concluded is that under some circumstances the REE are mobile and under other circumstances the REE are immobile (Table 1). The nature of those circumstances has rarely been determined and even more rarely have mechanisms for the mobility or reasons for the immobility been examined.

Experimental data (Hajash, 1984; Flynn and Burnham, 1978; Zielinski and Frey, 1974; Cullers et al., 1973; Kosterin, 1959; Cantrell and Byrne, 1987) directly related to the mobility of REE during metamorphism are scarce. Experiments where basalt and seawater were allowed to react under hydrothermal conditions suggested little (Hajash, 1984) or no (Menzies et al., 1979) REE solubility in the temperature ranges of 500-600°C and 150-350°C, respectively. However, based on experimentally determined complexation constants, Cantrell and Byrne (1987) suggest that REE form strong carbonate complexes in seawater. This interpretation is supported by the high REE content of fluid from the East Pacific Rise hydrothermal vent field (13°N) (Michard et al., 1983).

Cullers et al. (1973) and Zielinski and Frey (1974) report relatively large REE partition coefficients between diopside, enstatite, plagioclase, forsterite and aqueous vapor in the temperature range of 550-850°C. Variable partial pressures of CO_2 in the aqueous vapor did not appreciably change the D values determined by Zielinski and Frey (1974); these results suggest that REE-carbonate complexes are not stable under the experimental conditions. These data are often cited to support arguments of REE immobility during metamorphism (cf. Rollinson and Windley, 1980). However, additional variables such as mineral assemblage, fluid compositions, and fluid/rock ratios must be evaluated before reaching such conclusions.

The study of metamorphic fluids is a relatively new and evolving science. One approach to understanding the nature of those fluids is to examine the mineralogy and chemistry of veins that result from fluid migration. Lausch et al. (1974) in an examination of alpine carbonate veins concluded that REE, Sc, Fe, and Co were removed from the host gneisses and concentrated in the veins. Furthermore, the effective partition coefficients for the elements between the veins and their hosts increase with increasing metamorphic grade. This result implies increasing REE mobility, possibly as carbonate complexes, over the temperature range of approximately 500 to 650°C.

Humphris et al. (1978) concluded that the mobility of the REE in hydrothermally altered basalts is a function of the crystallization history of the igneous rock because it controls the distribution and residence of the REE and hence their availability for subsequent alteration. They further concluded that the metamorphic mineral assemblage also plays an important role by controlling which elements are incorporated in the altered rocks and which are available for transport out of the system. With few exceptions (cf. Stahle et al., 1987; Muecke et al., 1979; Nystrom, 1984; Pride and Muecke, 1981; Condie et al., 1977) the role of individual minerals has been ignored and discussions of REE mobility have been based on whole-rock compositions.

In order to use metamorphite compositions to evaluate the mobility of elements during metamorphic processes, the composition and density of both the protolith and the metamorphite must be known. Additionally, the change in volume that resulted from the metamorphism must be known. Gresens (1967) and Grant (1986) developed sets of equations in order to quantify the chemical changes associated with a variety of alteration processes. The equations have the form $X_i = [(v^B g^B / v^A g^A) C_i^B - C_i^A] m$, where X_i is the change in the mass of component i, v is volume, g is specific gravity, C_i^B is the concentration of i in B, B is the metamorphite, A is the protolith, and m is some arbitrary mass (generally 100 grams). Associated graphical representations of composition-volume relationships permit the evaluation of assumptions, such as constant volume or immobility of some component, that are required to solve the simultaneous equations and determine the degree of mobility of the various components.

Bartley (1986) used this mass balance approach, coupled with an analysis of analytical precision to evaluate the previously collected data of Hellman et al. (1977; referred to as 1979 by Bartley). Hellman et al. examined mineralogically distinct domains of an outcrop (exposed face approximately 20 m²) of prehnite)pumpellyite facies metabasalt; the protolith of which was assumed to have been homogeneous. They concluded that the REE had been mobilized during metamorphism, but they had moved coherently. Figure 26 shows that the REE patterns for individual samples are very similar in form and slope, but have different abscissas. However, Bartley (1986) concluded that the data require that of the REE only La be mobile with the concentration differences of the other REE in Figure 26 resulting from a net volume change of about 18% brought about by major element mobility. This interpretation also requires mobility of Ti, Y, and Zr. However, if Ti were assumed to be immobile, then all of the REE had to be mobile. In order to avoid conflicting interpretations which result from the use of different assumptions, as in this case, it is necessary to constrain the mathematics with geologic data and to evaluate the conclusions in geologic terms. For instance, the 18% volume change in the above example is interpreted by Bartley as being reasonable and consistent with well preserved primary igneous textures in the metabasalts.

Petrologic data describing the reactions (both continuous and discontinuous) that occurred during metamorphism should be especially useful in determining reasonable

156

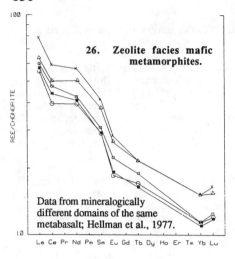

26. **Zeolite facies mafic metamorphites.**

Data from mineralogically different domains of the same metabasalt; Hellman et al., 1977.

La Ce Pr Nd Pm Sm Eu Gd Tb Dy Ho Er Tm Yb Lu

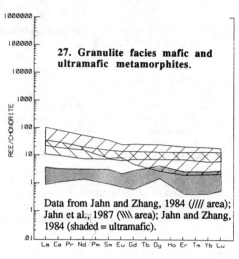

27. **Granulite facies mafic and ultramafic metamorphites.**

Data from Jahn and Zhang, 1984 (//// area); Jahn et al., 1987 (\\\\ area); Jahn and Zhang, 1984 (shaded = ultramafic).

La Ce Pr Nd Pm Sm Eu Gd Tb Dy Ho Er Tm Yb Lu

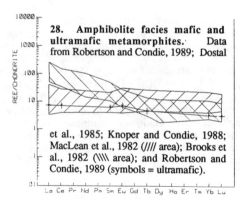

28. **Amphibolite facies mafic and ultramafic metamorphites.** Data from Robertson and Condie, 1989; Dostal et al., 1985; Knoper and Condie, 1988; MacLean et al., 1982 (//// area); Brooks et al., 1982 (\\\\ area); and Robertson and Condie, 1989 (symbols = ultramafic).

La Ce Pr Nd Pm Sm Eu Gd Tb Dy Ho Er Tm Yb Lu

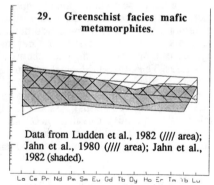

29. **Greenschist facies mafic metamorphites.**

Data from Ludden et al., 1982 (//// area); Jahn et al., 1980 (//// area); Jahn et al., 1982 (shaded).

La Ce Pr Nd Pm Sm Eu Gd Tb Dy Ho Er Tm Yb Lu

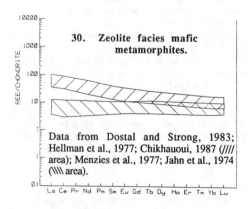

30. **Zeolite facies mafic metamorphites.**

Data from Dostal and Strong, 1983; Hellman et al., 1977; Chikhauoui, 1987 (//// area); Menzies et al., 1977; Jahn et al., 1974 (\\\\ area).

La Ce Pr Nd Pm Sm Eu Gd Tb Dy Ho Er Tm Yb Lu

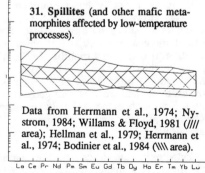

31. **Spillites** (and other mafic metamorphites affected by low-temperature processes).

Data from Herrmann et al., 1974; Nystrom, 1984; Willams & Floyd, 1981 (//// area); Hellman et al., 1979; Herrmann et al., 1974; Bodinier et al., 1984 (\\\\ area).

La Ce Pr Nd Pm Sm Eu Gd Tb Dy Ho Er Tm Yb Lu

constraints. If garnet were to disappear, the REE (especially HREE) released would have to leave the system or be incorporated in other phases. Similarly, the partial or complete disappearance of accessory minerals rich in REE could significantly alter the REE content of a rock, unless the REE are incorporated in other minerals. However, the documentation of variations in the modal abundance of accessory minerals is often difficult or impossible. In an investigation of incipient granulite formation, Stahle et al. (1987) found a decrease in P and Zr content of the gneisses as they became more charnockitic. They attributed this to a modal decrease in apatite (and possibly monazite) and zircon. The resulting availability of REE (especially HREE) in the presence of a CO_2-rich fluid provides an explanation for the also observed decrease in REE content of the charnockitic rocks. Unfortunately, they were unable to document the inferred modal decrease of the accessory minerals and they were able to constrain their mass-balance analysis of element mobility only in a negative sense by assuming that Zr, P and REE were mobile.

The question of REE mobility during metamorphism and the related question of the relationship between REE patterns (and contents) of metamorphites and those of their protoliths have not been resolved. Existing experimental data permit conflicting interpretations regarding REE mobility within metamorphic P and T ranges and the observational data has lead to confilicting interpretations (even for the same data set). In short, the hard work required to demonstrate REE mobility or immobility during metamorphism has generally not been done.

REE CONTENT OF METAMORPHIC ROCKS

As discussed in the previous sections, the REE content of a metamorphic rock is a function of the chemistry of the rock's protolith and the physical and chemical conditions under which the rock evolved. These variables are expressed in the mineralogy of the rock. Many of the minerals participate in continuous reactions through which their compositions change in response to their changing environment. The extent to which REE participate in continuous reactions has not been evaluated. But, it can be inferred from the dependence of partition coefficients on P-T-X that minor amounts of REE will be either incorporated in the participating minerals or will be given up to a metamorphic fluid phase during continuous reactions. Discontinuous reactions, which result in the appearance or disappearance of a mineral, play a major role in controlling the partitioning of trace elements into individual minerals by controlling, in part, the availability of those elements to participate in the reaction. It can be anticipated that the REE content of a metamorphite will change with the mineralogy, provided that the metamorphic fluid is capable of complexing the REE and removing them from the system.

Table 2 is a partial listing of investigators who have determined the REE contents of metamorphic rocks. The results of some of those determinations are depicted in Figures 27 through 39. The figures are arranged to show REE patterns of a general compositional group (i.e., mafic rocks) as a function of metamorphic grade. It is obvious that in such a broad lumping of compositions, ages and metamorphic grade that subtile changes in REE patterns and abundances are obscured. For example, the coherent mobility of REE that Hellman et al. (1977) suggested was the result of prehnite-pumpellyite facies metamorphism is totally obscured by the differences in scales between Figure 26 and Figure 30 which include those data.

However, two general types of REE patterns for metabasic rocks are seemingly preserved throughout the various metamorphic grades. A LREE-enriched trend and a flat trend are obvious in the data selected to represent the different metamorphic grades. This suggests that if the gross REE patterns reflect those of the protolith, they may be preserved throughout the spectrum of metamorphic grades. Most of the samples however, were carefully selected (some using the criteria of Gelinas et al., 1977) to avoid altered samples. Those samples which were avoided may be the ones that record systematic variations in whole-rock REE contents.

Table 2. Authors addressing whole-rock REE distributions.

Mafic and Ultramafic Metamorphites

Zeolite Facies (and low-temperature processes)

Church, 1987	Masuda & Nakamura, 1971
Dostal & Strong, 1983	Michard et al., 1983
Frey et al., 1968	Montigny et al., 1973
Frey et al., 1974	Nystrom, 1984
Garmann et al., 1975	Philpotts et al., 1969
Hellman & Henderson, 1977	Seifert et al., 1985
Hellman et al., 1977	Seifert et al., 1987
Humphris et al., 1978	Smewing & Potts, 1976
Kay & Senechal, 1976	Staudigel & Hart, 1983
Ludden & Thompson, 1978	Williams & Floyd, 1981
Ludden & Thompson, 1979	Wood et al., 1976

Greenschist Facies

Bhaskar Rao & Drury, 1982	Jahn et al., 1974
Bodinier et al., 1984	Jahn et al., 1980
Brooks et al., 1982	Jahn et al., 1982
Capdevila et al., 1982	Kite & Stoddard, 1984
Condie et al., 1977	Ludden et al., 1982
Drury, 1983	Menzies et al., 1977
Gelinas et al., 1984	Nelson et al., 1984
Hellman et al., 1979	Potocka, 1987
Herrmann et al., 1974	Sun & Nesbitt, 1978
Jahn, 1986	Thurston & Fryer, 1983
Jahn & Sun, 1979	Ueng et al., 1988

Amphibolite Facies

Collerson et al., 1984	Menzies & Seyfried, 1979
Dostal et al., 1985	Muecke et al., 1979
Evans et al., 1981	Sevigny, 1988
Garrison, 1981	Stille & Tatsumoto, 1985
Hajash, 1984	van de Kamp, 1969
Knoper & Condie, 1988	Weaver et al., 1982
MacLean et al., 1982	

Granulite Facies

Bernard-Griffiths et al., 1985
Compton, 1978
Hanson, 1975
Jahn et al., 1987

Felsic Metamorphites

Zeolite Facies (and low-temperature processes)

Chikhaoui et al., 1978

Greenschist Facies

Dickin, 1988	Sylvester et al., 1987
Jolly, 1987	Vocke et al., 1987
Norman et al., 1987	

Amphibolite Facies

Chamberlain et al., 1988	Nutman & Bridgewater, 1986
Helvaci & Griffin, 1983	Robertson & Condie, 1989
Lindahl & Grauch, 1988	

Granulite Facies

Condie et al., 1982	O'Nions & Pankhurst, 1974
Fowler, 1986	Pride & Muecke, 1980
Green et al., 1972	Sighinolfi et al., 1981
Jahn & Zong-qing, 1984	Stahle et al., 1987

Metasediments

Greenschist Facies

Jenner et al., 1981	Wronkiewicz & Condie, 1987
Lajoie & Ludden, 1984	

Amphibolite Facies

Bavinton & Taylor, 1980	Kerrich & Fryer, 1979
Condie & Martell, 1983	Lausch et al., 1974
Cullers, et al., 1974	McLennan et al., 1984
Day & Weiblen, 1986	Roaldset, 1975
Dostal, 1975	Ujike, 1984
Jarvis et al., 1975	Walker et al., 1986

Granulite Facies

Dostal & Capedri, 1979	Taylor et al., 1986

Metamorphite-hosted Mineral Deposits

Pride & Muecke, 1981	Barrett & Jarvis, 1988	Lindahl & Grauch, 1988
Sivell, 1986	Barrett et al., 1988	Lottermoser, 1989
Weaver, 1980	Bence & Taylor, 1985	MacLean, 1988
	Campbell et al., 1982	Strong, 1984
	Finlow-Bates & Stumpfl, 1981	Taylor & Fryer, 1983
	Fowler & Doig, 1983	Uiterdijk Appel &
	Kerrich & Fryer, 1979	Mahabaleswar, 1988
	Kerrich et al., 1981	Whitford et al., 1988

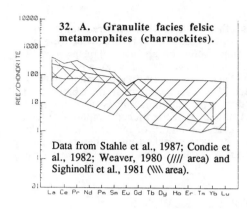

32. A. Granulite facies felsic metamorphites (charnockites).

Data from Stahle et al., 1987; Condie et al., 1982; Weaver, 1980 (//// area) and Sighinolfi et al., 1981 (\\\\ area).

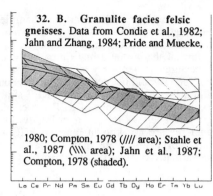

32. B. Granulite facies felsic gneisses. Data from Condie et al., 1982; Jahn and Zhang, 1984; Pride and Muecke,

1980; Compton, 1978 (//// area); Stahle et al., 1987 (\\\\ area); Jahn et al., 1987; Compton, 1978 (shaded).

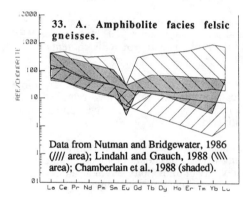

33. A. Amphibolite facies felsic gneisses.

Data from Nutman and Bridgewater, 1986 (//// area); Lindahl and Grauch, 1988 (\\\\ area); Chamberlain et al., 1988 (shaded).

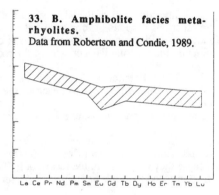

33. B. Amphibolite facies metarhyolites.
Data from Robertson and Condie, 1989.

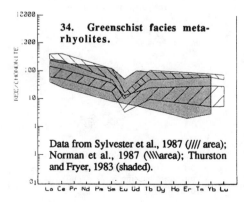

34. Greenschist facies metarhyolites.

Data from Sylvester et al., 1987 (//// area); Norman et al., 1987 (\\\\area); Thurston and Fryer, 1983 (shaded).

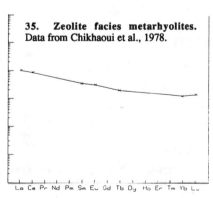

35. Zeolite facies metarhyolites.
Data from Chikhaoui et al., 1978.

Possible indications that samples other than the least altered may provide information on REE mobility are shown in Figures 37 and 33. In Figure 37 the pattern of a monazite-bearing biotite schist is shown in comparison to metapelites and metawackes. The nature of the biotite schist's protolith is unknown. It was probably not a heavy mineral concentrate (other heavy minerals are missing) and even a marine phosphorite (Goldberg et al., 1963) would not normally contain enough REE to account for that in the biotite schist. Did a P-rich sediment act as a REE sink during metamorphism or is there some other explanation? The range of REE values of a suite of HREE-enriched, amphibolite facies, granitic gneisses is shown in Figure 33. Again, the nature of the protoliths for these gneisses are not known, but the REE and Zr were enriched either before or during metamorphism. Were they introduced prior to metamorphism or were they simply redistributed during metamorphism?

SUGGESTIONS FOR FUTURE WORK

1.   Experimental and empirical data describing REE partitioning between coexisting minerals and between minerals and fluid phases are needed. Fluid phase compositions require attention, especially in regard to pH and carbonate and halogen content.

2.   Investigations of the REE content of metamorphites should involve an evaluation of the reactions responsible for the final mineral assemblage and a REE budget. The REE budget would include a modal analyses of the rock, a listing of REE contents of the minerals and the total REE content of the rock.

3.   Mass balance analyses are usually underdetermined because there are more unknowns than equations to describe the rock system. Therefore, new criteria for the constraint of mass balance analyses need to be developed and alternative approaches to classic mass balance analyses should be investigated.

4.   Considerably more work on atypical, lithologically and chemically distinct types of sample suites may help in our understanding of REE contents of metamorphic rocks because they may contain the record of REE mobility or immobility and indications of what REE-enriched minerals participate in continuous and discontinuous reactions.

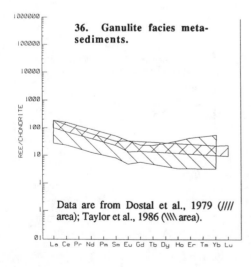

36. Granulite facies metasediments.

Data are from Dostal et al., 1979 (//// area); Taylor et al., 1986 (\\\\ area).

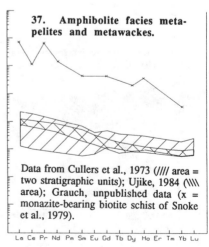

37. Amphibolite facies metapelites and metawackes.

Data from Cullers et al., 1973 (//// area = two stratigraphic units); Ujike, 1984 (\\\\ area); Grauch, unpublished data (x = monazite-bearing biotite schist of Snoke et al., 1979).

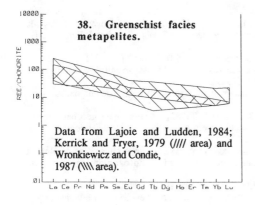

38. **Greenschist facies metapelites.**

Data from Lajoie and Ludden, 1984; Kerrick and Fryer, 1979 (//// area) and Wronkiewicz and Condie, 1987 (\\\\ area).

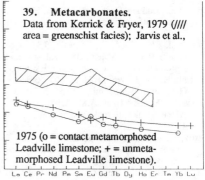

39. **Metacarbonates.**
Data from Kerrick & Fryer, 1979 (//// area = greenschist facies); Jarvis et al.,

1975 (o = contact metamorphosed Leadville limestone; + = unmetamorphosed Leadville limestone).

ACKNOWLEDGMENTS

If in my "zeal" to meet manuscript deadlines, I overlooked your favorite data set, please accept my apology and send me a reprint. I thank my family (especially Tien and Andy), friends and colleagues (especially Bohdan Kribek) for their support and indulgence while I prepared this review. A.R. Grauch and R.R. Grauch were a great help in last minute data entry which permitted plotting REE patterns at consistent scales. I also thank Donald Burt and Gordon McKay for very constructive reviews of this manuscript.

REFERENCES

Adams, J.W. (1968) Distribution of the lanthanides in minerals. 7th Rare Earth Res. Conf., San Diego, Calif., October, 1968, 2, 639-650.

Alderton, D.H.M., Pearce, J.A., and Potts, P.J. (1980) Rare earth element mobility during granite alteration: evidence from southwest England. Earth Planetary Science Letters, 49, 149-165.

Amli, R. (1975) Mineralogy and rare earth geochemistry of apatite and xenotime from the Gloserheia granite pegmatite, Froland, southern Norway. Amer. Mineral. 60, 607-620.

Barrett, T.J. and Jarvis, I. (1988) Rare-earth element geochemistry of metalliferous sediments from DSDP Leg 92: the East Pacific Rise transect. Chem. Geol. 67, 243-259.

Barrett, T.J., Fralick, P.W., and Jarvis, I. (1988) Rare-earth-element geochemistry of some Archean iron formations north of Lake Superior, Ontario. Canadian J. Earth Sci. 25, 570-580 .

Bartley, J.M. (1986) Evaluation of REE mobility in low-grade metabasalts using mass-balance calculations. Norsk Geol. Tidssk. 66, 145-152.

Bavinton, O.A. and Taylor, S.R. (1980) Rare earth element geochemistry of Archean metasedimentary rocks from Kambalda, Western Australia. Geochim. Cosmochim. Acta 44, 639-648.

Bence, A.E. and Taylor, B.E. (1985) Rare earth element systematics of West Shasta metavolvanic rocks: petrogenesis and hydrothermal alteration. Econ. Geology, 80, 2164-2176.

Bernard-Griffiths, J., Peucat, J.-J., Iglesias Ponce de Leon, M., and Gil Ibarguchi, J.I. (1985) U-Pb, Nd isotope and REE geochemistry in eclogites from the Cabo Ortegal Complex, Galicia, Spain: an example of REE immobility conserving MORB-like patterns during high-grade metamorphism. Chem. Geol. 52, 217-225.

Beswick, A.E. and Soucie, G. (1978) A correction procedure for metasomatism in an Archean greenstone belt. Precambrian Res. 6, 235-248.

Bhaskar Rao, Y.J. and Drury, S.A. (1982) Incompatible trace element geochemistry of Archean metavolcanic rocks from the Bababudan volcanic-sedimentary belt, Karnataka. J. Geol. Soc. India 23, 1-12.

Bodinier, J.L., Dupuy, C., and Dostal, J. (1984) Geochemistry of Precambrian ophiolites from Bou Azzer, Morocco. Contrib. Mineral. Petrol. 87, 43-50.

Bowles, J.F.W. and Morgan, D.J. (1984) The composition of rhabdophane. Mineral. Mag.48, 146-148.

Boynton, W.V. (1984) Cosmochemistry of the rare earth elements: meteorite studies: in P. Henderson [Ed.], Rare earth element geochemistry. Elsevier, Amsterdam, 63-114.

162

Brooks, C., Ludden, J., Pigeon, Y., and Hubregtse, J.J.M.W. (1982) Volcanism of shoshonite to high-K andesite affinity in an Archean arc environment, Oxford Lake, Manitoba. Canadian J. Earth Sci. 19, 55-67.

Campbell, I.H., Coad, P., Franklin, J.M., Gorton, M.P., Scott, S.D., Sowa, J., and Thurston, P.C. (1982) Rare earth elements in volcanic rocks associated with Cu-Zn massive sulfide mineralization: a preliminary report. Canadian J. Earth Sci. 19, 619-623.

Cantrell, K.J. and Byrne, R.H. (1987) Rare earth element complexation by carbonate and oxalate ions. Geochim. Cosmochim. Acta 51, 597-605.

Capdevila, R., Goodwin, A.M., Ujike, O., and Gorton, M.P. (1982) Trace-element geochemistry of Archean volcanic rocks and crustal growth in southwestern Abitibi belt, Canada. Geology 10, 418-422.

Cerny, P., Fryer, B.J., Longstaffe, F.J., and Tammemagi, H,Y. (1987) The Archean Lac du Bonnet batholith, Manitoba: igneous history, metamorphic effects, and fluid overprinting. Geochim. Cosmochim. Acta 51, 421-438.

Chamberlain, V.E., Lambert, R.St.J., Duke, M.J.M., and Holland, J.G. (1988) Geochemistry of the gneissic basement complex near Valemount, British Columbia: further evidence for a varied origin. Canadian J. Earth Sci. 25, 1725-1739.

Chauvel, C., Dupre, B., and Jenner, G.A. (1985) The Sm-Nd age of Kambalda volcanics is 500 Ma too old! Earth Planetary Science Letters 74, 315-324.

Chikhaoui, M., Dupuy, C., and Dostal, J. (1978) Geochemistry of Late Proterozoic volcanic rocks from Tassendjanet area (N.W. Hoggar, Algeria). Contrib. Mineral. Petrol. 66, 157-164.

Church, W.R. (1987) REE mobility due to alteration of Indian Ocean basalt: discussion. Canadian J. Earth Sci. 24, 192.

Collerson, K.D. and Fryer, B.J. (1978) The role of fluids in the formation and subsequent development of early continental crust. Contrib. Mineral. Petrol. 67, 151-167.

Collerson, K.D., McCulloch, M.T., and Bridgwater, D. (1984) Nd and Sr isotopic crustal contamination patterns in an Archean meta-basic dyke from northern Labrador: Geochim. Cosomchim. Acta, 48, 71-83.

Compton, P. (1978) Rare earth evidence for the origin of the Nuk gneisses Buksefjorden region, southern Norway. Contrib. Mineral. Petrol. 66, 283-293.

Condie, K.C. and Martell, C. (1983) Early Proterozoic metasediments from north-central Colorado: metamorphism, provenance, and tectonic setting. Geol. Soc. Amer. Bull. 94, 1215-1224.

Condie, K.C., Allen, P., and Narayana, B.L. (1982) Geochemistry of the Archean low- to high-grade transition zone, southern India. Contrib. Mineral. Petrol. 81, 157-167.

Condie, K.C., Viljoen, M.J., and Kable, E.J.D. (1977) Effects of alteration on element distributions in Archean tholeiites from the Barberton greenstone belt, South Africa. Contrib. Mineral. Petrol. 64, 75-89.

Copeland, R.A., Frey, F.A., and Wones, D.R. (1971) Origin of clay minerals in a Mid-Atlantic Ridge sediment. Earth Planetary Science Letters, 10, 186-192.

Cullers, R.L., Chaudhuri, Sambhudas, Arnold, Bill, Lee, Moon, and Wolf, C.W., Jr. (1975) Rare earth distributions in clay minerals and in the clay-sized fraction of the Lower Permian Havensville and Eskridge shales of Kansas and Oklahoma. Geochim. Cosmochim. Acta 39, 1691-1703.

Cullers, R.L., Medaris, L.G., and Haskin, L.A. (1973) Experimental studies of the distribution of rare earths as trace elements amoung silicate minerals and liquids and water. Geochim. Cosmochim. Acta 37, 1499-1512.

Cullers, R.L., Yeh, Long-Tsu, Chaudhuri, S., and Guidotti, C.V. (1974) Rare earth elements in Silurian pelitic schists from N.W. Maine. Geochim. Cosmochim. Acta 38, 389-400.

Day, W.C. and Weiblen, P.W. (1986) Origin of late Archean granite: geochemical evidence from the Vermilion Granitic Complex of northern Minnesota. Contrib. Mineral. Petrol. 93, 283-296.

Dickin, A.P. (1988) Evidence for limited REE leaching from the Roffna Gneiss, Switzerland - a discussion of the paper by Vocke et al. (1987) (CMP95:145-154). Contrib. Mineral. Petrol. 99, 273-275.

Dostal, J. and Capedri, S. (1979) Rare earth elements in high-grade metamorphic rocks from the western Alps. Lithos 12, 41-49.

Dostal, J. and Strong, D.F. (1983) Trace-element mobility during low-grade metamorphism and silicification of basaltic rocks from Saint John, New Brunswick. Canadian J. Earth Sci. 20, 431-435.

Dostal, J., Dupuy, C., and Poidevin, J.L. (1985) Geochemistry of Precambrian basaltic rocks from the Central African Republic (Equatorial Africa). Canadian J. Earth Sci. 22, 653-652.

Dostal, Jaroslav (1975) The origin of garnet-cordierite-sillimanite bearing rocks from Chandos Township, Ontario. Contrib. Mineral. Petrol. 49, 163-175.

Drury, S.A. (1983) The petrogenesis and setting of Archean metavolcanics from Karnataka State, south India. Geochim. Cosmochim. Acta 47, 317-329.

Dypvik, H. and Brunfelt, A.O. (1976) Rare-earth elements in Lower Paleozoic epicontinental and eugeosynclinal sediments from the Oslo and Trondheim regions: Sedimentology 23, 363-378.

Evans, B.W., Trommsdorff, V. and Goles, G.G. (1981) Geochemistry of high-grade eclogites and metarodingites from the central Alps. Contrib. Mineral. Petrol. 76, 301-311.

Ferry, J.M. [Ed.] (1982) Characterization of metamorphism through mineral equilibria: Rev. Mineral. 10, 397 p.

Finlow-Bates, T. and Stumpfl, E.F. (1981) The behaviour of so-called immobile elements in hydrothermally altered rocks associated with volcanogenic submarine-exhalative ore deposits. Mineral. Dep. 16, 319-328.

Fleischer, M. (1967) Some aspects of the geochemistry of yttrrium and the lanthanides. Geochim. Cosmo-chim. Acta 29, 755-772.

Fleischer, M. and Aultschuler, Z.S. (1969) The relationship of the rare-earth composition of minerals to geologic environment. Geochim. Cosmochim. Acta 33, 725-732.

Flynn, R.T. and Burnham, C.W. (1978) An experimental determination of rare earth partition coefficients between chloride containing vapor phase and silicate melts. Geochim. Cosmochim. Acta 42, 685-701.

Fowler, A.D. and Doig, R. (1983) The significance of europium anomalies in the REE spectra of granites and pegmatites, Mont Laurier, Quebec. Geochim. Cosmochim. Acta 47, 1131-1137.

Fowler, M.B. (1986) Large-ion lithophile element characteristics of an amphibolite facies to granulite facies transition at Gruinard Bay, north-west Scotland. J. Metamorphic Geol. 4, 345-359.

Frey, F.A., Haskin, M.A., Poetz, J.A., and Haskin, L.A. (1968) Rare earth abundances in some basic rocks. J. Geophys. Res. 73, 6085-6098.

Frey, F.A., Bryan, W.B., and Thompson, G. (1974) Atlantic Ocean floor: geochemistry and petrology of basalts from Legs 2 and 3 of the Deep Sea Drilling Project. J. Geophys. Res. 79, 5507-5527.

Fujimki, H. (1986) Partition coefficients of Hf, Zr, and REE between zircon, apatite, and liquid. Contrib. Mineral. Petrol. 94, 42-45.

Ganguly, J. and Saxena, S.K. (1987) Mixtures and mineral reactions. Springer-Verlag, Berlin, 291.

Garmann, L.B., Brunfelt, A.O., Finstad, K.G., and Heier, K.S. (1975) Rare-earth element distribution in basic and ultrabasic rocks from west Norway. Chem. Geol. 15, 103-116.

Garrison, J.R., Jr. (1981) Metabasalts and metagabbros from the Llano uplift, Texas: petrologic and geochemical characterization with emphasis on tectonic setting. Contrib. Mineral. Petrol. 78, 459-475.

Gelinas, L., Trudel, P., and Hubert, C. (1984) Chemostratigraphic division of the Blake River Group, Rouyn-Noranda area, Abitibi, Quebec. Canadian J. Earth Sci. 21, 220-231.

Gelinas. L., Brooks, C., Perrault, G., Carignan, J., Trudel, P., and Grasso, F. (1977) Chemostratigraphic divisions within the Abitibi volcanic belt, Rouyn Noranda district, Quebec. In: W.R.A. Baragar, L.C. Coleman, and J.M. Hall [Ed.], Volcanic regimes in Canada: Geol. Assoc. Canada Spec. Paper 16, Waterloo, Ontario, 265-295.

Giere, Reto (1986) Zirconolite, allanite and hoegbomite in a marble skarn from the Bergell contact aureole: implications for mobility of Ti, Zr and REE. Contrib. Mineral. Petrol. 93, 459-470.

Goldberg, E.D., Koide, M., Schmitt, R.A., and Smith, R.H. (1963) Rare earth distributions in the marine environment. J. Geophys. Res. 68, 4209-4217.

Grant, J. A. (1986) The isocon diagram - a simple solution to Gresens' equation for metasomatic alteration. Econ. Geol. 81, 1976-1982.

Green, T.H., Brunfelt, A.O., and Heier, K.S. (1972) Rare-earth element distribution and K/Rb ratios in granulites, mangerites and anorthosites, Lofoten-Vesteraalen, Norway. Geochim. Cosmochim. Acta 36, 241-257.

Gresens, R. L. (1976) Composition-volume relationships of metasomatism. Chem. Geol. 2, 47-65.

Gromet, L.P. and Silver, L.T. (1983) Rare earth element distributions among minerals in a granodiorite and their petrogenetic implications. Geochim. Cosmochim. Acta 47, 925-940.

Hajash, A., Jr. (1984) Rare earth element abundances and distribution patterns in hydrothermally altered basalts: experimental results. Contrib. Mineral. Petrol. 85, 409-412.

Hanson, G.N. (1975) REE analyses of the Morton and Montevideo gneisses from the Minnesota River Valley. Geol. Soc. America Abstr. Programs, 7, 7, 1099.

Harrison, W.J. (1981) Partition coefficients for REE between garnets and liquids: implications for non-Henry's Law behaviour for models of basalt origin and evolution. Geochim. Cosmochim. Acta 45, 1529-1544.

Hellman, P.L. and Henderson, P. (1977) Are rare earth elements mobile during spilitisation? Nature 267, 38-40.

Hellman, P.L., Smith, R.E., and Henderson, P. (1977) Rare earth element investigation of the Cliefden outcrop, N.S.W., Australia. Contrib. Mineral. Petrol. 65, 155-164.

164

Hellman, P.L., Smith, R.E., and Henderson, P. (1979) The mobility of rare earth elements: evidence and implications from selected terrains affected by burial metamorphism. Contrib. Mineral. Petrol. 71, 23-44.

Helvaci, C. and Griffin, W.L. (1983) Metamorphic feldspathization of metavolcanics and granitoids, Avnik area, Turkey. Contrib. Mineral. Petrol. 83, 309-319.

Henderson, P. (1980) Rare earth element partition between sphene, apatite and other coexisting minerals of the Kangerdlugssuaq Intrusion, E. Greenland. Contrib. Mineral. Petrol. 72, 81-85.

Henderson, P. (1984) General geochemical properties and abundances of the rare earth elements in Henderson, Paul [Ed.], Rare Earth Element Geochemistry. Elsevier, Amsterdam, Netherlands, 1-32.

Herrmann, A.G., Potts, M.J., and Knake, D. (1974) Geochemistry of the rare earth elements in spilites from oceanic and continental crust. Contrib. Mineral. Petrol. 44, 1-16.

Hickmott, D.D. (1988) Trace element zoning in garnets: implications for metmorphic processes. Unpub. Ph.D. Dissertation, Mass. Inst. Technology, 449 p..

Hickmott, D.D., Shimizu, N., Spear, F.S., and Selverstone, J. (1987) Trace-element zoning in a meta-morphic garnet. Geology 15, 573-576.

Humphris, S.E., Morrison, M.A., and Thompson, R.N. (1978) Influence of rock crystallisation history upon subsequent lanthinide mobility during hydrothermal alteration of basalts. Chem. Geol. 23 125-137.

Jahn, B.M., Auvray, B., Cornichet, J., Bai, Y.L., Shen, Q.H., and Liu, D.Y. (1987) 3.5 Ga amphibolites from eastern Hebei Province, China: field occurrence, petrography, Sm-Nd isochron age and REE geochemistry. Precambrian Res. 34, 311-346.

Jahn, Bor-ming and Sun, Shen-Su (1979) Trace element distribution and isotopic composition of Archean greenstones. In: Ahrens, L.H. [Ed.], Origin and distribution of the elements. Physics and Chemistry of the Earth 11. Pergamon Press, Oxford, England, 597-618.

Jahn, Bor-ming and Zong-qing Zhang (1984) Archean granulite gneisses from eastern Hebei Province, China: rare earth geochemistry and tectonic implications. Contrib. Mineral. Petrol. 85, 224-243.

Jahn, Bor-ming (1986) Mid-ocean ridge or marginal basin origin of the East Taiwan Ophiolite: chemical and isotopic evidence. Contrib. Mineral. Petrol. 92, 194-206.

Jahn, Bor-ming, Auvray, B., Blais, S., Capdevila, R., Cornichet, J., Vidal, F., and Hameurt, J. (1980) Trace element geochemistry and petrogenesis of Finnish greenstone belts. J. Petrology 21, 2, 201-244.

Jahn, Bor-ming, Gruau, G., and Glikson, A.Y. (1982) Komatiites of the Onverwacht Group, S. Africa: REE geochemistry, Sm/Nd age and mantle evolution. Contrib. Mineral. Petrol. 80, 25-40.

Jahn, Bor-ming, Shih, Chi-Yu, and Rama Murthy, V. (1974) Trace element geochemistry of Archean volcanic rocks. Geochim. Cosmochim. Acta 38, 611-627.

Jarvis, J.C., Wildeman, T.R., and Banks, N.G. (1975) Rare earths in the Leadville Limestone and its marble derivatives. Chem. Geol. 16, 27-37.

Jenner, G.A., Fryer, B.J., and McLennan, S.M. (1981) Geochemistry of the Archean Yellowknife Supergroup. Geochim. Cosmochim. Acta 45, 1111-1129.

Jensen, B.B. (1973) Patterns of trace element partitioning. Geochim. Cosmochim. Acta 37, 2227-2242.

Jolliff, B.L., Papike, J.J., and Laul J.C. (1987) Mineral recorders of pegmatite internal evolution: REE contents of tourmaline from the Bob Ingersoll pegmatite, South Dakota. Geochim. Cosmochim. Acta 51, 2225-2232.

Jolly, W.T. (1987) Geology and geochemistry of Huronian rhyolites and low-Ti continental tholeiites from the Thessalon region, central Ontario. Canadian J. Earth Sci. 24, 1360-1385.

Kay, R., Hubbard, N.J., and Gast, P.W. (1970) Chemical characteristics and origin of oceanic ridge volcanic rocks. J. Geophys. Res. 75, 1585-1613.

Kay, R.W. and Senechal, R.G. (1976) The rare earth geochemistry of the Troodos ophiolite complex. J. Geophys. Res. 81, 964-970.

Kerrich, R. and Fryer, B.J. (1979) Archean precious-metal hydrothermal systems, Dome Mine, Abitibi greenstone belt. II. REE and oxygen isotope relations. Canadian J. Earth Sci. 16, 440-458.

Kerrich, R., Fryer, B.J., Milner, K.J., and Peirce, M.G. (1981) The geochemistry of gold-bearing chemical sediments, Dickenson Mine, Red Lake, Ontario: a reconnaissance study. Canadian J. Earth Sci. 18, 624-637.

Khomyakov, A.P. (1967) Chemical and crystallochemical factors in the distribution of rare earths. Geochem. Int'l. 4, 127-135.

King, R.W., Kerrich, R.W., and Daddar, R. (1988) REE distributions in tourmaline: an INAA technique involving pretreatment by B volatilization. Amer. Mineral. 73, 424-431.

Kite, L.E. and Stoddard, E.F. (1984) The Halifax County complex: oceanic lithosphere in the eastern North Carolina Piedmont. Geol. Soc. Amer. Bull. 95, 422-432.

Knoper, M.W. and Condie, K.C. (1988) Geochemistry and petrogenesis of Early Proterozoic amphibolites, west-central Colorado, U.S.A. Chem. Geol. 67, 209-225.

165

Kosterin, A.V. (1959) The possible modes of transport of the rare earths by hydrothermal solutions. Geochem. 381-387.

Lajoie, J. and Ludden, J. (1984) Petrology of the Archean Pontiac and Kewagama sediments and implications for the stratigraphy of the southern Abitibi belt. Canadian J. Earth Sci. 21, 1305-1314.

Lausch, J., Moller, P., and Morteani, G. (1974) Die Verteilung der Seltenen Erden in den Karbonaten und penninischen Gneisen der Zillertaler Alpen (Tirol, Österreich). N. Jahrb. Mineral. Monat. 11, 490-507.

Lee, D.E. and Bastron, H. (1967) Fractionation of rare-earth elements in allanite and monazite as related to geology of the Mt. Wheeler mine area, Nevada. Geochim. Cosmochim. Acta 31, 339-356.

Lemarchand, F., Villemant, B., and Calas, G. (1987) Trace element distribution coeffients in alkaline series. Geochim. Cosmochim. Acta 51, 1071-1081.

Leroy, J.L. and Turpin, L. (1988) REE, Th and U behaviour during hydrothermal and supergene processes in a granitic environment. Chem. Geol. 68, 239-251.

Lindahl, I. and Grauch, R.I. (1988) Be-REE-U-Sn mineralization in Precambrian granitic gneisses, Nordland County, Norway. In: Zachrisson, E. [Ed.], Proc. 7th Quadrennial IAGOD Symp., E. Scheizerbart'sche Verlagsbuchhandlung, Stuttgart, 583-594.

Lottermoser, B.G. (1989) Rare earth element study of exhalites within the Willyama Supergroup, Broken Hill Block, Australia. Mineral. Dep. 24, 92-99.

Ludden, J.N. and Thompson, G. (1978) Behaviour of rare earth elements during submarine weathering of tholeiitic basalt. Nature 274, 147-149.

Ludden, J.N. and Thompson, G. (1979) An evaluation of the behaviour of the rare earth elements during the weathering of sea-floor basalt. Earth Planetary Science Letters, 85-92.

Ludden, J., Gelinas, L., and Trudel, P. (1982) Archean metavolcanics from the Rouyn Noranda district, Abitibi greenstone belt, Quebec. 2. Mobility of trace elements and petrogenetic constraints. Canadian J. Earth Sci. 19, 2276-2287.

MacLean, W.H. (1988) Rare earth element mobility at constant inter-REE ratios in the alteration zone at the Phelps Dodge massive sulphide deposit, Matagami, Quebec. Mineral. Dep. 23, 231-238.

MacLean, W.H., St. Seymour, K., and Prabhu, M.K. (1982) Sr, Y, Zr, Nb, Ti, and REE in Grenville amphibolites at Montauban-les-Minews, Quebec. Canadian J. Earth Sci. 19, 633-644.

Martin, R.F., Whitley, J.E., and Woolley, A.R. (1978) An investigation of rare-earth mobility: fenitized quartzites, Borralan complex, N.W. Scotland. Contrib. Mineral. Petrol. 66, 69-73.

Masuda, A., Nakamura, N., and Tanaka, T. (1971) Rare earth elements in metagabbros from the Mid-Atlantic Ridge and their possible implications for the genesis of alkali olivine basalts as well as the Lizard peridotite. Contrib. Mineral. Petrol. 32, 295-306.

McLennan, S.M., Taylor, S.R., and McGregor, V.R. (1984) Geochemistry of Archean metasedimentary rocks from West Greenland. Geochim. Cosmochim. Acta 48, 1-13.

Menzies, M. and Seyfried, Jr., W. (1979) Experimental evidence of rare earth element immobility in greenstones. Nature 282, 398-399.

Menzies, M., Blanchard, D., and Jacobs, J. (1977) Rare earth and trace element geochemistry of metabasalts from the Point Sal ophiolite, California. Earth Planetary Science Letters 37, 203-215.

Michard, A., Albarede, F., Michard, G., Minster, J.F., and Charlou, J.L. (1983) Rare-earth elements and uranium in high temperature solutions from East Pacific Rise hydrothermal vent field (13°N). Nature 303, 795-797.

Michard, Annie (1989) Rare earth element systematics in hydrothermal fluids. Geochim. Cosmochim. Acta 53, 745-750.

Miyashiro, A. (1973) Metamorphism and metamorphic belts: John Wiley & Sons, New York, 492 p.

Montigny, R., Bougault, H., Bottinga, Y., and Allegre, C.J. (1973) Trace element geochemistry and genesis of the Pindos ophiolite suite. Geochim. Cosmochim. Acta 37, 2135-2147.

Muecke, G.K., Pride, C., and Sarkar, P. (1979) Rare-earth element geochemistry of regional metamorphic rocks. In: L.H. Ahrens [Ed.], Origin and distribution of the elements 2. Pergamon, London, 449-464.

Nabelek, P.I. (1987), General equations for modeling fluid/rock interaction using trace elements and isotopes. Geochim. Cosmochim. Acta 51, 1765-1769.

Nelson, D.R., Crawford, A.J., and McCulloch, M.T. (1984) Nd-Sm isotopic and geochemical systematics in Cambrian boninites and tholeiites from Victoria, Australia. Contrib. Mineral. Petrol. 88, 164-172.

Norman, D.I., Condie, K.C., Smith, R.W., and Thomann, W.F. (1987) Geochemical and Sr and Nd isotopic constraints on the origin of late Proterozoic volcanics and associated tin-bearing granites from the Franklin Mountains, west Texas. Canadian J. Earth Sci. 24, 830-839.

166

Nutman, A.P. and Bridgwater, D. (1986) Early Archean Amitsoq tonolites and granites of the Isukasia area, southern West Greenland: development of the oldest-known sial. Contrib. Mineral. Petrol. 94, 137-148.

Nystrom, J.O. (1984) Rare eare element mobility in vesicular lava during low-grade metamorphism. Contrib. Mineral. Petrol. 88, 328-331.

O'Nions, R.K. and Pankhurst, R.J. (1974) Rare-earth element distribution in Archean gneisses and anorthosites, Godthab area, west Greenland. Earth Planetary Science Letters 22, 328-338.

Olivarez, A.M. and Owen, R.M. (1989) REE/Fe variations in hydrothermal sediments: Implications for the REE content of seawater. Geochim. Cosmochim. Acta 53, 757-762.

Ottonello, G., Ernst, W.G., and Joron, J.L. (1984) Rare earth and 3d transition element geochemistry of peridotitic rocks: I. peridotites from the western Alps. J. Petrol. 25, 343-372.

Patocka, F. (1987) The geochemistry of mafic volcanics: implications for the origin of the Devonian massive sulfide deposits at Zlate Hory, Czechoslovakia. Mineral. Dep. 22, 144-150.

Philpotts, J.A. and Schnetzler, C.C. (1969) Submarine basalts: some K, Rb, Sr, Ba, Rare-earth, H_2O, and CO_2 data bearing on their alteration, modification by plagioclase, and possible source materials. Earth Planetary Science Letters 7, 293-299.

Pride, C. and Muecke, G.K. (1980) Rare earth element geochemistry of the Scourin Complex, N.W. Scotland - evidence for the granite-granulite link. Contrib. Mineral. Petrol. 73, 403-412.

Pride, C. and Muecke, G.K. (1981) Rare earth element distributions amoung coexisting granulite facies minerals, Scourian Complex, NW Scotland. Contrib. Mineral. Petrol. 76, 463-471.

Puchelt, H. and Emmermann, R. (1976) Bearing of rare earth patterns of apatites from igneous and metamorphic rocks. Earth Planetary Science Letters 31, 279-286.

Reiitan, P.H., Roelandts, I., and Brunfelt, A.O. (1980) Optimum ionic size for substitution in the M(2)-site in metamorphic diopside: N. Jahrb. Mineral. Monat. 4, 181-191.

Reitan, P.H. and Roelandts, I. (1973) Rare earth partitioning: coexisting metamorphic pyroxenes. Geol. Soc. Amer. Abstr. Programs 5, 7, 778-779.

Roaldset, Elen (1975) Rare earth element distributions in some Precambrian rocks and their phyllosilicates, Numedal, Norway. Geochim. Cosmochim. Acta 39, 455-469.

Robertson, J.M. and Condie, K.C. (1989) Geology and geochemistry of early Proterozoic volcanic and subvolcanic rocks of the Pecos greenstone belt, Sangre de Cristo Mountains, New Mexico. In: Grambling, J.A. and Tewksbury, B.J. [Ed.], Proterozoic geology of the southern Rocky Mountains. Geol. Soc. Amer. Spec. Paper 235, Boulder, Colorado, 119-146.

Rollinson, H.R. and Windley, B.F. (1980) Selective elemental depletion during metamorphism of Archean granulites, Scourie, NW Scotland. Contrib. Mineral. Petrol. 72, 257-263.

Seifert, K.E., Cole, M.R.W., and Brunotte, D.A. (1985) REE mobility due to alteration of Indian Ocean basalt. Canadian J. Earth Sci. 22, 1884-1887.

Seifert, K.E., Cole, M.R.W., and Brunotte, D.A. (1987) REE mobility due to alteration of Indian Ocean basalt: reply. Canadian J. Earth Sci. 24, 193.

Sevigny, J.H. (1988) Geochemistry of Late Proterozoic amphibolites and ultramafic rocks, southeastern Canadian Cordillera. Canadian J. Earth Sci. 25, 1323-1337.

Sighinolfi, G.P., Figueredo, M.C.H., Fyfe, W.S., Kronberg, B.I., and Tanner Oliveira, M.A.F. (1981) Geochemistry and petrology of the Jequie granulitic complex (Brasil): an Archean basement complex. Contrib. Mineral. Petrol. 78, 263-271.

Sivell, W.J. (1986) A basaltic-ferrobasaltic granulite association, Oonagalabi gneiss complex, central Australia: magmatic variation in an early Proterozoic rift. Contrib. Mineral. Petrol. 93, 381-394.

Smewing, J.D. and Potts, P.J. (1976) Rare-earth abundances in basalts and metabasalts from the Troodos Massif, Cyprus. Contrib. Mineral. Petrol. 57, 245-258.

Snoke, A.W., McKee, E.H., and Stern, T.W. (1979) Plutonic, metamorphic, and structural chronology in the northern Ruby Mountains, Nevada: a preliminary report. Geol. Soc. Amer. Abstr. Programs 11, 520-521.

Stahle, H.J., Raith, M., Hoernes, S., and Delfs, A. (1987) Element mobility during incipient granulite formation at Kabbaldurga, southern India. J. Petrology 28, 5, 803-834.

Staudigel, Hubert and Hart, S.R. (1983) Alteration of basaltic glass: Mechanisms and significance for the oceanic crust-seawater budget. Geochim. Cosmochim. Acta 47, 337-350.

Stille, P. and Tatsumoto, M. (1985) Precambrian tholeiitic-dacitic rock-suites and Cambrian ultramafic rocks in the Pennine nappe system of the Alps: evidence from Sm-Nd isotopes and rare earth elements. Contrib. Mineral. Petrol. 89, 184-192.

Strong, D. F. (1984) Rare earth elements in volcanic rocks of the Buchans area, Newfoundland. Canadian J. Earth Sci. 21, 775-780.

Sun, Shen-Su and Nesbitt, R.W. (1978) Petrogenesis of Archean ultrabasic and basic volcanics: evidence from rare earth elements. Contrib. Mineral. Petrol. 65, 301-325.

167

Sylvester, P.J., Attoh, Kodjo, and Schulz, K.J. (1987) Tectonic setting of late Archean bimodal volcanism in the Michipicoten (Wawa) greenstone belt, Ontario. Canadian J. Earth Sci. 24, 1120-1134.

Taylor, R.P. and Fryer, B.J. (1983) Rare earth element lithogeochemistry of granitoid mineral deposits. CIM Bulletin 76, 74-84.

Taylor, S.R., Rudnick, R.L., McClennan, S.M., and Eriksson, K.A. (1986) Rare earth element patterns in Archean high-grade metasediments and their tectonic significance. Geochim. Cosmochim. Acta 50, 2267-2279.

Thurston, P.C. and Fryer, B.J. (1983) The geochemistry of repetitive cyclical volcanism from basalt through rhyolite in the Uchi-Confederation greenstone belt, Canada. Contrib. Mineral. Petrol. 83, 204-226.

Ueng, W.C., Fox, T.P., Larue, D.K., and Wilband, J.T. (1988) Geochemistry and petrogenesis of the early Proterozoic Hemlock volcanic rocks and Kiernan sills, southern Lake Superior region. Canadian J. Earth Sci. 25, 528-546.

Uitterdijk Appel, P.W. and Mahabaleswar, B. (1988) Secular trends in rare earth element patterns of Precambrian iron-formations from India and Greenland. J. Geol. Soc. India, 32, 3, 214-226.

Ujike, Osamu (1984) Chemical composition of Archean Pontiac metasediments, southwestern Abitibi belt, Superior Province. Canadian J. Earth Sci. 21, 727-731.

van Breemen, O. and Hawkesworth, C.J. (1980) Sm-Nd isotopic study of garnets and their metamorphic host rocks: Royal Soc. Edinburg, Trans. 71, 2, 97-102.

van de Kamp, P.C. (1969) Origin of amphibolites in the Beartooth Mountains, Wyoming and Montana: new data and interpretation. Geol. Soc. Amer. Bull. 80, 1127-1136.

Vocke, R.D., Jr., Hanson, G.N., and Grunenfelder, M. (1987) Rare earth element mobility in the Roffna Gneiss, Switzerland. Contrib. Mineral. Petrol. 95, 145-154.

Walker, R.J., Hanson, G.N., Papike, J.J., and O'Neil, J.R. (1986) Nd, O and Sr isotopic constraints on the origin of Precambrian rocks, southern Black Hills, South Dakota. Geochim. Cosmochim. Acta 50, 2833-2846.

Watson, E.B. (1985) Henry's law behavior in simple systems and magmas: Criteria for discerning concentration-dependent partition coefficients in nature. Geochim. Cosmochim. Acta 49, 917-923.

Weaver, B.L., Tarney, John, Windley, B.F., and Leake, B.E. (1982) Geochemistry and petrogenesis of Archean metavolcanic amphibolites from Fiskenaesset, S.W. Greenland. Geochim. Cosmochim. Acta 46, 2203-2215.

Weaver, B.L. (1980) Rare-earth element geochemistry of Madras granulites. Contrib. Mineral. Petrol. 71, 271-279.

Wendlandt, R.F. and Harrison, W.J. (1979) Rare earth partitioning between immiscible carbonate and silicate liquids and CO_2 vapor: results and implications for the formation of light rare earth-enriched rocks. Contrib. Mineral. Petrol. 69, 409-419.

Whitford, D.J., Korsch, M.J., Porritt, P.M., and Craven, S.J. (1988) Rare-earth element mobility around the volcanogenic polymetallic massive sulfide deposit at Que River, Tasmania, Australia. Chem. Geol. 68, 105-119.

Willams, C.T. and Floyd, P.A. (1981) The localised distribution of U and other incompatible elements in spilitic pillow basalts. Contrib. Mineral. Petrol. 78, 111-117.

Wood, D.A., Gibson, I.L., and Thompson, R.N. (1976) Element mobility during zeolite facies metamorphism of the Tertiary basalts of eastern Iceland. Contrib. Mineral. Petrol. 55, 241-254.

Wronkiewicz, D.J. and Condie, K.C. (1987) Geochemistry of Archean shales from the Witwatersrand Supergroup, South Africa: source-area weathering and provenance. Geochim. Cosmochim. Acta 51, 2401-2416.

Zayats, A.P. and Kuts, V.P. (1964) Rare earth elements in the accessory minerals of gneisses in the Ukrainian crystalline shield. Geochem. Int'l. 1, 1126-1128.

Zielinski, R.A. and Frey, F.A. (1974) An experimental study of the partitioning of a rare earth element (Gd) in the system diopside-aqueous vapor. Geochim. Cosmochim. Acta 38, 545-565.

RARE EARTH ELEMENTS IN SEDIMENTARY ROCKS:
INFLUENCE OF PROVENANCE AND SEDIMENTARY PROCESSES

INTRODUCTION

The geochemistry of sedimentary rocks has considerable consequence for our understanding of the the earth and the processes that shape it. Within their composition, sediments preserve a record of their sources (or provenance) and, consequently, allow us to examine the relationships between the composition of upper crustal sources and the nature and distribution of sediments. The processes of sedimentation, including weathering, erosion, sedimentary sorting and diagenesis, essentially involve water/rock interaction and result in many fundamental chemical changes. In turn, the composition of sedimentary rocks may provide useful insights into the chemistry and nature of these interactions, including fluid compositions, fluid/rock ratios, and the mechanisms of element mobility in crustal environments.

Within this framework, rare earth element (REE) distributions in sedimentary rocks have played a central role. The subtle, but well understood, variations in the properties of the REE make them sensitive to mineral/melt equilibria, leading to their special utility in geochemical studies of igneous systems. For sedimentary rocks, these characteristics are equally important, but in a less direct manner. There are general relationships between processes of formation (and bulk composition) and REE distributions in igneous systems. Because REE are not easily fractionated during sedimentation, sedimentary REE patterns may provide an index to average provenance compositions. REE studies in sedimentary rocks have basically followed four lines of investigation, including: (1) crustal evolution; (2) relationships between sedimentation and plate tectonics; (3) sedimentary processes (weathering, diagenesis, etc.); (4) paleoceanography. In this review, the importance of REE distributions in terrigenous sediments and sedimentary rocks will be emphasized. This is due both to the research interests of the author and the wealth of data from which to draw. Accordingly, questions of relating REE distributions to sedimentary provenance will be considered in terms of crustal evolution models and plate tectonic associations. Other topics will be addressed where they bear directly on the major questions under consideration. Over the past several years, there have been several reviews of REE geochemistry that emphasize, or at least include, significant discussion of the problems related to sedimentary systems (Henderson, 1984; Taylor and McLennan, 1985, 1988; Elderfield, 1988).

RARE EARTH ELEMENT PROPERTIES AND SEDIMENTARY ROCKS

Rare earth elements comprise the lanthanide elements, La – Lu, as well as Y and Sc (e.g., Puddephatt, 1972). In terms of geochemical behavior, Y mirrors the heavy lanthanides Dy-Ho, and typically is included with them for discussion. On the other hand, Sc is a substantially smaller cation (Table 1) with a geochemical behavior that differs from other rare earths (but similar to elements such as Cr, Ni, V, Co), normally entering early crystallizing minerals during igneous processes (see below). Accordingly, some inconsistency in terminology has arisen in the geochemical literature. In most geochemical discussion, the rare earths include only La–Lu and Y. Scandium generally is included with smaller ferromagnesian cations that enter six-fold coordinated mineral sites. On the other hand, in low temperature aqueous fluids (i.e., seawater, river water), Sc behaves similarly to other REE in having exceptionally low concentrations. This makes Sc of interest for sedimentary provenance studies and is thus useful to consider within the context of this review. The standard geochemical usage will be adhered to; where relevant, Sc will be discussed separately

170

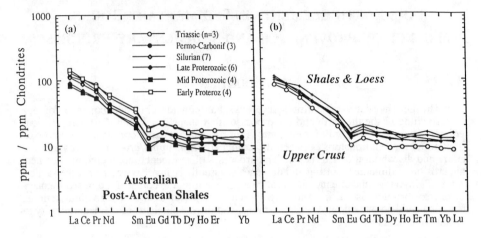

Figure 1.   Chondrite-normalized REE diagrams for sediments and sedimentary rocks of various age. (a) Average post-Archean Australian shales of various age. Early Proterozoic data from Ewers and Higgins (1985; Nourlangie Schist); other data from Nance and Taylor (1976). (b) Various averages and composites of shales and loess (data from Table 2). Note the similarity in the patterns with LREE enrichment, flat HREE distributions and the ubiquitous negative Eu-anomaly. The average sedimentary REE pattern is taken to be parallel to the average REE pattern of the upper continental crust. Also plotted is an estimate of upper crustal abundances which is based on PAAS (Taylor and McLennan, 1985).

Table 1. Some basic REE data.

	Z	Ionic Radius (Å)[1] CN6	Ionic Radius (Å)[1] CN8	Atomic Weight [2]	Ground State Configuration	50%Oxide Condensation(K)[3] 10^{-3} bar	50%Oxide Condensation(K)[3] 10^{-6} bar	
La	57	1.032	1.160	138.9055	$[Xe]5d^16s^2$	1621	1410	Lanthanum
Ce	58	1.01	1.143	140.115	$[Xe]4f^15d^16s^2$	1532	1286	Cerium
Pr	59	0.99	1.126	140.9077	$[Xe]4f^36s^2$	1636	1420	Praseodymium
Nd	60	0.983	1.109	144.24	$[Xe]4f^46s^2$	1640	1430	Neodymium
Sm	62	0.958	1.079	150.36	$[Xe]4f^66s^2$	1633	1433	Samarium
Eu	63	0.947	1.066	151.96	$[Xe]4f^76s^2$	1398	1232	Europium
Gd	64	0.938	1.053	157.25	$[Xe]4f^75d^16s^2$	1674	1462	Gadolinium
Tb	65	0.923	1.040	158.9254	$[Xe]4f^96s^2$	1676	1463	Terbium
Dy	66	0.912	1.027	162.50	$[Xe]4f^{10}6s^2$	1675	1463	Dysprosium
Ho	67	0.901	1.015	164.3033	$[Xe]4f^{11}6s^2$	1676	1463	Holmium
Er	68	0.890	1.004	167.26	$[Xe]4f^{12}6s^2$	1676	1463	Erbium
Tm	69	0.880	0.994	168.9342	$[Xe]4f^{13}6s^2$	1676	1463	Thulium
Yb	70	0.868	0.985	173.04	$[Xe]4f^{14}6s^2$	1549	1392	Ytterbium
Lu	71	0.861	0.977	174.967	$[Xe]4f^{14}5d^16s^2$	1676	1463	Lutetium
Sc	21	0.745	0.870	44.95591	$[Ar]3d^14s^2$	1724	1524	Scandium
Y	39	0.900	1.019	88.9059	$[Kr]4d^15s^2$	1692	1499	Yttrium
Eu^{2+}		1.17	1.25					
Ce^{4+}		0.87	0.97					

[1] - Trivalent Ionic Radii unless otherwise stated; from Shannon, 1976
[2] - From De Bievre et al., 1984
[3] - Calculated 50% oxide condensation temperatures for ideal solid solution in perovskite, for oxides in form $(REE)_2O_3$; From Kornacki and Fegley, 1986

High quality REE analyses date from the early 1960's with the advent of precise neutron activation techniques. From the earliest developments, sedimentary studies have been at the fore. There are at least two reasons for this. First, the earliest optical emission spectrographic analyses of REE for meteorites (Noddack, 1935) and sediment composites (Minami, 1935) were found to differ. This observation was the source of considerable discussion since it was generally held that the REE were not fractionated by anything other than the most extreme geological processes (e.g., Suess and Urey, 1956). Some priority was thus given to sedimentary analyses when rapid and precise analytical techniques were developed. The second reason was the early recognition of the remarkable uniformity of sedimentary REE patterns in spite of the considerable diversity in igneous rocks (Fig. 1). This observation led to the conclusion that sedimentation was providing an efficient sampling of the REE in the exposed continental crust.

The application of radiogenic isotopes involving REE, including the Sm-Nd and Lu-Hf isotopic sytems, to problems of sedimentary geochemistry has advanced considerably since the pioneering work of McCulloch and Wasserburg (1978). Most effort has been directed towards Nd-model age studies. Since the REE are generally considered immobile, the Nd-model age of a terrigenous sedimentary rock is interpreted as an estimate of the average age of mantle extraction of the various provenance components. This age information provides considerable constraints on the detailed provenance of sedimentary rocks and is useful in developing models of crustal evolution. Detailed consideration of this topic is outside the scope of this paper and the reader is referred elsewhere for a more thorough discussion (e.g., Goldstein et al., 1984; Patchett et al., 1984; Miller et al., 1986; Nelson and DePaolo, 1988; McLennan et al., 1990a,b).

Cosmochemical considerations

Rare earth elements possess a number of physical and chemical properties that make them especially useful for geochemical studies in general and sedimentary studies in particular. Although the various REE display significant differences in volatility, they are all refractory elements (Table 1) with oxide condensation temperatures comparable to Sr, U and Th (Kornacki and Fegley, 1986). Temperatures achieved during the formation of the earth were insufficient to cause significant large scale fractionation among the REE on the basis of volatility. The general similarity of REE patterns for chondritic meteorites and the solar photosphere further reinforces this conclusion (Table 2). In contrast, severe REE fractionations, on the basis of volatility, are associated with formation of refractory calcium - aluminum inclusions (CAI) found in some meteorites; this provides evidence of high temperature processing in the early nebula or during pre-nebular history (see review in Taylor and McLennan, 1988). There is general agreement that the bulk earth has a REE pattern essentially identical to CI meteorites, but with abundances about 1.5 times higher due to loss of volatiles (Evensen et al., 1978; Anders and Ebihara, 1982; Taylor and McLennan, 1985; Table 2). Assuming constant mantle ratios among refractory elements, upper mantle major element compositions (i.e., Ca, Al, Ti abundances) suggest that the REE content of the primitive mantle (i.e., present mantle plus crust) is enriched further by a factor of about 1.5 (Table 2).

Geochemical considerations

Under the pressure-temperature conditions found within the earth, the REE mostly exist in a trivalent state. There are two exceptions to this:

(1) Under reducing conditions, such as those found within the mantle or lower crust, europium may exist in the divalent state (Eu^{2+}). This results in an increase in ionic radius of about 17% (Table 1) making it essentially identical to Sr^{2+} (IR = 1.26 Å for CN = 8). The consequence of this is that Eu substitutes freely in place of Sr in

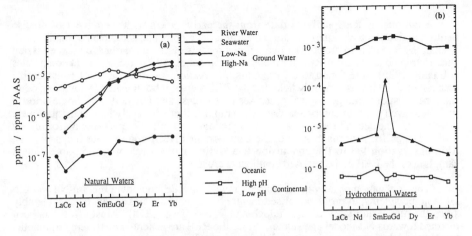

Figure 2. REE diagram, normalized to PAAS, for (a) selected natural waters and (b) selected hydrothermal waters. Note the extremely low abundances of REE in natural waters, especially for seawater. Except for the depletion of Ce, the general shape of most REE patterns for seawaters and river waters are approximately parallel to PAAS. In detail, seawater is typically severely depleted in Ce and is enriched in HREE relative to PAAS. Ground waters differ significantly from other natural waters in that significant fractionation of REE, relative to PAAS may be encountered, likely the result of differing stabilities of the various REE carbonate complexes (Michard et al., 1987). Hydrothermal solutions are enriched in REE compared to seawater but only acidic varieties have REE concentrations exceeding the ppb range. Oceanic hydrothermal solutions typically are greatly enriched in Eu, a consequence of plagioclase breakdown during fluid/rock interaction. Data from Table 3.

Table 2. REE abundances in various cosmochemical and geochemical reservoirs (ppm except where noted).

	Atom Abundance(Si=10[6]) Solar[1]	CI Chondrite[2]	Chondrite[3]	Sediments PAAS[4]	NASC[5]	ES[6]	Loess[7]	Continental Crust[8] Upper	Total	Igneous Rocks[9] Andesite	MORB
La	0.31	0.46	0.367	38.2	32	41.1	35.4	30	16	19	3.7
Ce	0.81	1.2	0.957	79.6	73	81.3	78.6	64	33	38	11.5
Pr	0.12	0.18	0.137	8.83	7.9	10.4	8.46	7.1	3.9	4.3	1.8
Nd	0.39	0.85	0.711	33.9	33	40.1	33.9	26	16	16	10.0
Sm	0.14	0.27	0.231	5.55	5.7	7.3	6.38	4.5	3.5	3.7	3.3
Eu	0.1	0.099	0.087	1.08	1.24	1.52	1.18	0.88	1.1	1.1	1.3
Gd	0.30	0.34	0.306	4.66	5.2	6.03	4.61	3.8	3.3	3.6	4.6
Tb	-	0.060	0.058	0.774	0.85	1.05	0.81	0.64	0.60	0.64	0.87
Dy	0.26	0.40	0.381	4.68	5.8	-	4.82	3.5	3.7	3.7	5.7
Ho	-	0.089	0.0851	0.991	1.04	1.20	1.01	0.80	0.78	0.82	1.3
Er	0.19	0.26	0.249	2.85	3.4	3.55	2.85	2.3	2.2	2.3	3.7
Tm	0.042	0.039	0.0356	0.405	0.50	0.56	-	0.33	0.32	0.32	0.54
Yb	0.2	0.25	0.248	2.82	3.1	3.29	2.71	2.2	2.2	2.2	3.7
Lu	0.13	0.038	0.0381	0.433	0.48	0.58	-	0.32	0.30	0.30	0.56
ΣREE			3.89	184.8	173	204	181	146	87	96	52.6
La_N/Yb_N			1.00	9.15	6.98	8.44	8.83	9.21	4.91	5.84	0.68
La_N/Sm_N			1.00	4.33	3.53	3.54	3.49	4.20	2.88	3.23	0.71
Eu/Eu^*			1.00	0.65	0.70	0.70	0.66	0.65	0.99	0.92	1.02
Sc	28	34.5	8.64	16	14.9	-	8.4	11	30	30	38
Y	4.0	4.33	2.25	27	27	-	25	22	20	22	32

1 - From Ross & Aller (1976); Aller (1987)
2 - From Evensen et al. (1978); Schmitt et al. (1964)
3 - From Taylor & McLennan (1985). Note that Primitive Earth Mantle abundances (ie.- present mantle plus crust) are taken as 1.5 times these values.
4 - Average 23 post-Archean shales from Australia (adapted from Taylor & McLennan, 1985); see text for details

5 - Composite 40 shales, mainly N.American (Haskin et al., 1968; Gromet et al., 1984)
6 - Composite of numerous European shales (Haskin & Haskin, 1966)
7 - Carbonate-free loess (Taylor et al., 1983)
8 - From Taylor and McLennan (1985)
9 - Averages (Taylor & McLennan, 1985)

feldspars, notably Ca-plagioclase, leading to distinctive geochemical behavior compared to the other REE.

(2) Under oxidizing conditions, Ce^{3+} may be oxidized to Ce^{4+} leading to a decrease in ionic radius of about 15% (Table 1). The only place on earth where this reaction occurs on a large scale is in the marine environment, associated with the formation of manganese nodules. The distinctive Ce depletion found in ocean waters and phases precipitated in equilibrium with seawater is the immediate consequence of this reaction.

Apart from these anomalies, the REE behave as an unusually coherent group of elements. There is a regular decrease in ionic radius (Table 1) from La^{3+} to Lu^{3+} (termed the lanthanide contraction) which is due to an increase in effective nuclear charge as the shielded, non-valence $4f$ or $5f$ electron shell is filled (and contracts) during the development of the lanthanide series. This property also affects the stability of various complexes of REE cations, which is also of geochemical significance (e.g., Woyski and Harris, 1963).

In most rock-forming minerals, the REE fill lattice sites in eight-fold coordination. For the trivalent cations, there is no common mineral-forming element that is geochemically similar. The closest analogues are Ca^{2+} (IR = 1.12 Å for CN = 8) and Na^+ (IR = 1.18 Å for CN = 8), but charge imbalances exist. Thus, a combination of charge and size imbalances restricts the entry of REE into common rock-forming minerals. For major rock-forming minerals in which REE have significant concentrations, they primarily substitute for Ca.

Although the behavior of the REE during melting and fractional crystallization is complex in detail, for common processes the trivalent cations can be considered as incompatible elements with the level of incompatibility related to size (e.g., Hanson, 1980, 1989). Thus, partial melting of mantle or crustal rocks tends to cause an enrichment of the light rare earth elements (LREE; La-Sm) over the heavy rare earth elements (HREE; Gd-Lu). In detail, HREE patterns are most strongly fractionated when garnet (and to a lesser extent, amphibole and pyroxene) is an equilibrium phase during melting or crystal fractionation. Of considerable interest is the observation that significant variations in sedimentary HREE patterns appears to be restricted to Archean occurrences (see below). Although a number of minerals may affect the distribution of Eu during igneous processes, by far the most important is feldspar, particularly plagioclase. In plagioclase, substantial Eu^{2+} may substitute for Ca^{2+}, in place of Sr^{2+}, leading to considerable enrichments compared to associated trivalent REE. Thus, liquids that have formed where plagioclase is a stable residual phase, or from which plagioclase is crystallized and lost, will tend to be significantly depleted in Eu. Plagioclase stability on the earth is restricted to about 40km depth and for crustal rocks, evidence of plagioclase fractionation is generally indicative of intracrustal melting or fractionation processes. The REE pattern of average sediments is interpreted to reflect the average upper continental crust and thus the negative Eu-anomaly found in most sedimentary rocks (Fig. 1) similarly is interpreted to be a feature of the upper continental crust. This indicates that shallow, intracrustal differentiation involving plagioclase fractionation (either through partial melting or fractional crystallization) must be a fundamental process in controlling the composition and element distribution within the continental crust (Taylor and McLennan, 1981, 1985; see below).

<u>Aqueous geochemistry</u>

It is not possible to fully understand REE distributions in sedimentary rocks without a basic appreciation of the aqueous geochemistry of the rare earths. A first order observation is that REE abundances in all natural waters, both low temperature surface waters and hydrothermal waters, are exceedingly low, typically 10^{-7}–10^{-2} of the levels found in most rocks. Several typical compositions are given in Table 3 and plotted in Figure 2. Note that only high temperature acidic hydrothermal solutions have REE concentrations exceeding

Table 3.   REE in selected natural waters (parts per trillion).

	River[1]	Seawater[2]	Ground Water[3] Low-Na	High-Na	Hydrothermal Waters Oceanic[4]	Continental high pH[5]	low pH[6]
La	192	4.08					
Ce	451	3.68	75.7	33.1	325	52	46000
Nd	264	3.61	60.0	34.8	175	21.4	32000
Sm	58.9	0.71	20.0	14.7	39	5.3	8400
Eu	14.7	0.13	6.99	6.38	158	0.6	1700
Gd	57.7	1.13	37.1	33.0	32	3.2	8170
Dy	43.4	0.991	51.0	63.1	22	2.8	6500
Er	23.6	0.851	39.0	53.0	8	1.7	2570
Yb	19.7	0.829	45.9	56.1	6	1.3	2700
Eu/Eu*	0.77	0.46	0.78	0.89	13.7	0.45	0.63

[1] - Luce River, filtered at <12μm (Hoyle et al., 1984)
[2] - Atlantic Ocean, >2500m (Elderfield and Greaves, 1982)
[3] - Vals-les Bains CO_2-rich (Michard et al., 1987)
[4] - Mid Atlantic Ridge (Michard, 1989)
[5] - Bulgaria (Michard and Alberede, 1986)
[6] - Valles Caldera (Michard, 1989)

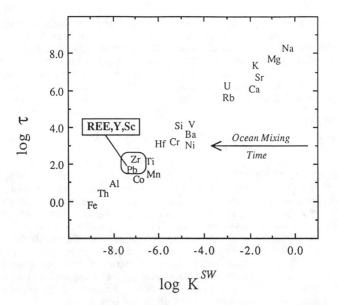

Figure 3.   Plot of log τ (residence time in years) against log K^{SW} (concentration in seawater / concentration in upper continental crust) for selected elements. Also shown is the approximate mixing time for the oceanic mass. Elements that are strongly partitioned into seawater, such as Na, Mg, Ca, U, and Rb, are readily mobilized during sedimentary processes, such as weathering and diagenesis, and are of limited use in discerning provenance of sedimentary rocks. On the other hand, elements with very low solubilities in natural waters, such as REE, Sc and Th, are likely to be transferred near quantitatively into terrigenous sediments and therefore preserve a more accurate record of the nature of the provenance. Because the residence times of these elements are less than the mixing time of the ocean, they tend to have heterogeneous distributions throughout seawater. Additional complications exist and are discussed in the text (cf. Taylor and McLennan, 1985, 1988).

the ppb range. As a result, water/rock interaction, especially at low temperatures, is unlikely to cause substantial changes in REE distributions of most rocks. One exception that has been noted is possible effects on Eu distributions from high temperature, acidic, chloride-rich hydrothermal solutions, such as those found at mid-ocean ridges or certain continental geothermal fields such as the Salton Sea (Michard, 1989). This has been interpreted to be related to reduction of Eu^{3+} to Eu^{2+}, associated with high temperature conditions (Michard and Alberede, 1986; also see Sverjensky, 1984).

The major assumption behind equating the REE pattern of siliciclastic sedimentary rocks with the average REE pattern of the source rocks is that the REE are transferred nearly quantitatively in the terrigenous component during erosion and sedimentation. The fraction of REE that may be carried in solution (and thus complicate the provenance/sediment REE relationship) is deemed to be trivial. The strength of this assumption can be evaluated by examining the partitioning of elements between the exposed continental crust and the major aqueous reservoir, the oceans. The residence time (τ) of various elements is plotted against the ratio of the concentration of the element in seawater to that in the upper continental crust (K^{SW}) in Figure 3. For elements that are known to be mobile in near surface, low temperature conditions, such as weathering and diagenesis (e.g., Na, Ca, Sr, Rb, U), there is a strong partitioning into the ocean relative to the exposed crust. In contrast, the REE, Sc and several other lithophile elements (e.g., Th, Zr, Hf, Ti, Al) have extremely low concentrations in seawater (and most other natural waters), relative to the exposed crust. Because the residence times of these elements are less than the mixing time of the oceans, these elements are heterogeneously distributed throughout seawater. It is concluded that these elements are primarily carried by the solid load rather than dissolved load during erosion and sedimentation. Several additional complications exist for some elements. Iron and manganese solubility is severely affected by changing redox conditions and Fe, along with Pb, may exhibit chalcophile characteristics. A consequence is that solubility in low temperature aqueous solutions may increase substantially under common diagenetic conditions, resulting in secondary mobility. Elements such as Zr, Hf and Sn tend to be most closely associated with heavy minerals and the distribution of these elements may be dominantly controlled by heavy mineral concentrations.

<u>Normalizing and notation</u>

For this review, two REE normalizing schemes are mostly used: average chondritic meteorites and average shale (PAAS). The average chondritic REE pattern is likely parallel to primordial abundances in the solar nebula and is also parallel to bulk earth abundances. The values adopted here are for primitive CI carbonaceous chondrites on a 'volatile-free' basis (Evensen et al., 1978). The values are about 17-19% higher than those commonly adopted but are preferred on cosmochemical grounds (see Taylor and McLennan, 1988 for further discussion). PAAS (Post-Archean Australian Average Shale) is convenient for normalizing since the REE pattern of average shale is thought to be parallel to the average upper continental crust (e.g. Taylor and McLennan, 1985). PAAS is considered preferable to shale composites, such as NASC (North American Shale Composite; Gromet et al., 1984), because there is always the potential for inclusion of aberrant material in a composite sample. Values are based on an average of 23 shales of post-Archean age (Nance and Taylor, 1976) but the exact values adopted here differ slightly from those reported previously (e.g.- Taylor and McLennan, 1981, 1985, 1988). A re-evaluation of PAAS was considered necessary to eliminate slight Dy anomalies that consistently appeared when REE data analysed by isotope dilution mass spectrometric techniques were normalized to PAAS. The procedure was as follows. The original data from Nance and Taylor (1976) were recompiled and plotted on a chondrite-normalized graph against ionic radii of the trivalent REE. The LREE data (La-Sm) and HREE data (Gd-Yb) were then fit separately by least square regression to a quadratic equation, in order to produce a smooth chondrite-normalized REE pattern, and the individual values were corrected, accordingly.

Tm and Lu are estimated respectively by interpolation and extrapolation of chondrite-normalized plots. This correction procedure resulted in changes to previously published PAAS values (e.g., Taylor and McLennan, 1981, 1985) of less than about 2% except for Nd and Dy which changed by about 6%.

In the following discussion, a subscript 'N' denotes chondrite-normalized values. Europium is commonly enriched (positive Eu-anomaly) or depleted (negative Eu-anomaly) relative to other REE on chondrite-normalized diagrams and this can be quantified by the term Eu/Eu*, where Eu* is the expected Eu value for a smooth chondrite-normalized REE pattern (against atomic number), such that:

$$Eu/Eu^* = Eu_N / (Sm_N * Gd_N)^{0.5}$$

An arithmetic mean (i.e., $[Sm_N + Gd_N] / 2$) is commonly used to calculate Eu* values, however, this is incorrect and may lead to serious errors, especially for steep chondrite-normalized REE patterns.

SEDIMENTARY PROCESSES

Although there is a considerable literature devoted to the question of REE behavior during various sedimentary processes, much of it is rather confused and contradictory. There appear to be several reasons for this. (1) A number of studies rely on incomplete REE analyses or data of uncertain quality. (2) In many cases, there is a failure to clearly address the distinction between small scale REE mobility associated with mineral reactions and the larger scale transport of REE into or out of the system under consideration. (3) REE aqueous geochemistry is not well suited for experimental study at low temperatures and few relevant data are available (e.g., Humphris, 1984). In spite of these difficulties, a number of important observations have been made.

Weathering

The nature of REE redistribution during mineralogical reactions associated with weathering is very poorly understood. This is an important avenue of research both because more complete understanding of REE chemistry during weathering would provide important additional constraints on the understanding of the process of weathering and because documentation of significant REE mobility could affect interpretations of sedimentary REE patterns. Since most natural waters have extremely low REE abundances, it is generally assumed that substantial REE mobility and fractionation is unlikely for expected fluid/rock ratios (e.g., Taylor and McLennan, 1985). On the other hand, weathering solutions commonly have extreme chemistries with regard to fO_2, pH and concentrations of potential complexing agents (e.g., organic acids, carbonate ions (CO_3^{2-}), chlorine ions (Cl^-), etc.), and suggestions of a potential for significant transport during weathering continue to appear.

Several field-based studies have suggested REE mobility during weathering but there is little agreement regarding the overall magnitude or the potential for fractionation among the REE (e.g., Ronov et al., 1967; Nesbitt, 1979; Duddy, 1980; Topp et al., 1984; Cullers et al., 1987, 1988; Middelburg et al., 1988). Because of significant volume changes associated with weathering, it is especially difficult to interpret differences in absolute abundances. Attempts have been made to account for volume changes (e.g., Duddy, 1980) or assumptions made about immobility of elements other than REE (e.g., Nesbitt, 1979).

The best documented study to date is that of weathering on the Torrongo grano-diorite, southeastern Australia (Nesbitt, 1979). Fresh granodiorite, variably weathered granodiorite and residual clay material were analysed, accounting for most major components (other than aqueous weathering solutions) resulting from the weathering process. Assuming that titanium is immobile and using it as a normalizing element, REE enrichments, relative to parent rock, of up to a factor of two were noted in samples representing moderate to later stages of weathering. HREE are preferentially enriched over LREE. Highly weathered residual clays, mainly kaolinite and illite, found in fractures and thought to have been derived from overlying soils by mechanical transport (cf. Nesbitt et al., 1980), display complementary REE patterns to early stage weathering products, being depleted in REE relative to the parent rock and preferentially depleted in HREE (Fig. 4).

The model proposed by Nesbitt (1979) is as follows (see Fig. 5):

(1) aggressive CO_2 and organic acid-charged rainwater (i.e., pH=4-5) percolates through soil horizons where some REE are removed into solution, most likely as complexes (e.g., carbonate (CO_3^{2-}) or hydroxide (OH^-) complexes) or free ions.

(2) waters continue to migrate down and react with less altered granodiorite, converting plagioclase, biotite and other phases to clay. During these weathering reactions, pH is increased and the rare earth elements come out of solution, either precipitated as compounds, exchanged for H^+ on suitable clays or adsorbed on mineral surfaces.

Thus, the REE are mobilized during weathering but are primarily recycled within the weathering profile rather than transported significant distances in solution. In this scheme, fractionation of the REE patterns is ascribed to mineralogical factors (Nesbitt, 1979), although differences in solubility of LREE versus HREE complexes could also play a role.

For sedimentary studies, the importance of such work is that it supports the suggestion that for the most common weathering conditions, REE are dominantly transported from the weathering profile by mechanical processes (Nesbitt, 1979; also see Middleburg et al., 1988).

Diagenesis

REE behavior during diagenesis is even less well understood than during weathering, at least for siliciclastic sediments, and represents an important subject for future research. The only significant study for siliciclastic sediments is from Chaudhuri and Cullers (1979), who examined sediments in cores from the Gulf Coast through the mixed-layer/illite transformation interval. They suggested that essentially no changes in REE patterns could be attributed to diagenetic reactions. However, the results are difficult to evaluate because changes in provenance also likely occurred during the same interval.

Because substantial changes in redox conditions may occur during diagenesis, Sverjensky (1984) considered the theoretical redox equilibria of Eu in aqueous solutions. In detail, the Eu^{3+}/Eu^{2+} redox equilibria is quite strongly temperature dependent. However, for most diagenetic conditions (<100°C), the aqueous chemistry of Eu is dominated by the trivalent state and thus preferential fractionation of Eu from the other trivalent REE is unlikely (Fig. 6). The only possible exception where significant Eu^{2+} may be encountered is in highly reducing, alkaline pore waters of anoxic marine sediments. Complexing of Eu with ligands, such as SO_4^{2-}, CO_3^{2-} and Cl^-, is also common under aqueous conditions but at temperatures less than about 100°C does not appear to have a large effect on Eu^{3+}/Eu^{2+} equilibria.

Some insight into possible effects from diagenesis can be obtained from examining carbonate systems, where mineralogy is simple and diagenetic reactions can commonly be reasonably well documented. In addition, identifying REE mobility is more likely in carbonate sediments because typically they have substantially lower total REE abundances than do siliciclastic sediments. Recently, Banner et al. (1988) have examined the REE and Nd-isotope geochemistry of regionally dolomitized Mississippean carbonates. Their data

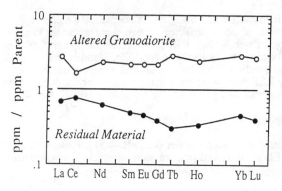

Figure 4. REE diagram, normalized to unweathered parent granodiorite, of weathered material from the Torrongo Granodiorite of southeastern Australia (Nesbitt, 1979). The most altered granodiorite and residual soil material taken from fractures are shown. The patterns are essentially complementary, suggesting that in this example, REE mobility takes place mainly within the weathering profile with little gain or loss of REE from the overall sequence.

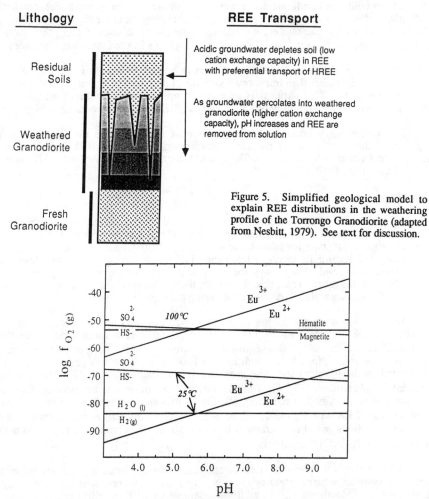

Lithology

REE Transport

Residual Soils

Acidic groundwater depletes soil (low cation exchange capacity) in REE with preferential transport of HREE

Weathered Granodiorite

As groundwater percolates into weathered granodiorite (higher cation exchange capacity), pH increases and REE are removed from solution

Fresh Granodiorite

Figure 5. Simplified geological model to explain REE distributions in the weathering profile of the Torrongo Granodiorite (adapted from Nesbitt, 1979). See text for discussion.

Figure 6. Plot of fO_2 versus pH showing equal activities of Eu^{3+} and Eu^{2+} at 25°C and 100°C. Unit activity of water is assumed. Also displayed are the redox equilibria of SO_4^{2-}/HS^- and Hematite/Magnetite. Note that Eu^{3+} dominates under most conditions relevant to sedimentation and diagenesis. Eu^{2+} is favored only under highly reducing and alkaline conditions, at elevated temperatures. From Sverjensky (1984).

indicated that REE patterns and Nd-isotopic signatures were not affected during development of dolomites. Quantitative modelling indicates that extremely large water/rock ratios ($>10^3$) are required to alter REE patterns in diagenetic carbonates (Fig. 7).

Sedimentary sorting

Three major factors associated with sedimentary sorting may affect REE patterns in sediments: grain size contrasts, general mineralogy and heavy mineral fractionation. REE abundances in various grain sizes have been examined by Cullers et al. (1979; also see Cullers et al., 1987, 1988). They found that the bulk of REE reside in the silt and clay size fraction, although there is no direct correlation with clay mineralogy, perhaps suggesting that trivalent REE may be readily accomodated in most clay minerals. Silt fractions tend to have lower overall abundances in comparison to clay fractions although the shape of the REE pattern is similar, suggesting the major control is a dilution effect from quartz, a mineral with very low REE abundances. Sand fractions tend to have the lowest REE abundances and commonly display lower La/Yb ratios than finer grain sizes. The low abundances can also be attributed to quartz dilution and the enrichment in HREE is likely a result of the heavy mineral content (see below). Heavy mineral suites extracted from both silt and sand fractions tend to be HREE enriched. The presence and magnitude of Eu-anomalies is similar for all grain sizes.

Other mineralogical factors are difficult to evaluate. Many mineralogical changes take place during the formation of sedimentary rocks and their diagenesis and there are no systematic studies to evaluate the relationship between mineralogy (especially clay mineralogy) and REE content. For framework grains in sandstones and siltstones, the REE chemistry of the various mineral grains will be dependent on the igneous/metamorphic history of the provenance (e.g., Hanson, 1980, 1989). Some mineralogical controls are very straightforward (see Taylor and McLennan, 1985, 1988 for review). Quartz is known to contain very low abundances of REE (with much of the content likely to be within inclusions) and enrichments of quartz in sands and silts (leading to a high quartz/clay ratio) results in a simple dilution effect. Calcite and dolomite also typically have fairly low REE abundances (although commonly higher than that of quartz) similarly resulting in a dilution effect. If carbonate minerals (or other minerals) are precipitated in equilibrium with seawater, they typically possess negative Ce-anomalies which may also be reflected in the bulk REE pattern (e.g., Piper, 1974; Palmer, 1985). On the other hand, clay minerals typically have much higher total REE and concentration of clay minerals in muds and clays results in high REE abundances.

In principle, other non-heavy minerals may be concentrated during sedimentary sorting and significantly affect REE patterns. Nance and Taylor (1977) observed evidence for enrichments of plagioclase in volcanogenic greywackes from the Devonian Baldwin Formation of Australia. Two samples with unusually high plagioclase contents had Eu/Eu* values of about 1.5 whereas most samples had Eu/Eu* of about 1.0 (Fig. 8; also see Bhatia, 1985). This was interpreted to be the result of local plagioclase enrichments during sedimentary sorting. Mixing calculations with typical island arc plagioclase compositions (e.g., Schnetzler and Philpotts, 1970) indicate that as much as 50% or more additional plagioclase would be required. It is likely that some Archean sedimentary rocks with positive Eu anomalies (relative to chondrite) are also affected by this process (Nance and Taylor, 1977). To date, this phenomenon has only been recognized in volcanogenic sediments where weathering effects are minor and plagioclase is an abundant constituent. In sediments that are chemically more mature, with little plagioclase, such enrichments have not been documented.

There is good evidence that a very large proportion of the complement of REE in some granitic rocks may be found within minor phases, such as sphene, zircon, allanite and monazite (e.g., Gromet and Silver, 1983). Many of these minerals, due to their high

180

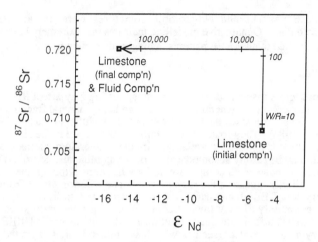

Figure 7. Plot of $^{87}Sr/^{86}Sr$ versus ε_{Nd}, illustrating the effects of fluid/rock interaction in the form of open system recrystallization of a limestone. Model adapted from Banner (1986) and Banner et al. (1988). Initially, limestone contains 1000 ppm Sr and 10 ppm Nd; fluid contains 100 ppm Sr, 0.0001 ppm Nd and 1000 ppm Ca; $K_D{}^{Sr-Ca}$ = 0.05; $K_D{}^{Nd-Ca}$ = 100; T = 50°C; porosity is 50%. Molar fluid/rock ratios (W/R) shown by tick marks on rock evolution path. Modelling illustrates that, in contrast to an element such as Sr, extremely high fluid/rock ratios (in this example, $>10^3$) are required to significantly affect Nd, and presumably other REE.

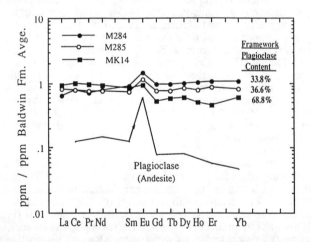

Figure 8. REE diagram for selected Eu-enriched Baldwin Formation greywackes, normalized to the (non-Eu-enriched) Baldwin Formation average (data from Nance and Taylor, 1977; Bhatia, 1985). The amount of plagioclase for each of these samples is shown on the right and compares with a formation average of about 16% plagioclase (Chappell, 1968; Bhatia, 1981). Also shown is a typical REE pattern for plagioclase from an andesite (Schnetzler and Philpotts, 1967). It is likely that the Eu enrichment found is some Baldwin samples results from enrichment of plagioclase during sedimentary sorting; petrographic data also indicate high plagioclase content for these samples.

density and resistance to weathering, may be concentrated during sedimentary sorting processes. Where the REE pattern of such heavy minerals differs significantly from the average source rock composition, a serious effect on sedimentary REE patterns may result. For sedimentary rocks, the most important heavy minerals to consider include zircon, monazite and allanite (Fig. 9) because of their high REE abundances, REE patterns which may differ from typical igneous rocks and ubiquitous occurrence as heavy minerals (Brenninkmeyer, 1978; Morton, 1985). In considering the importance of these minerals in sedimentary REE patterns, it is the enrichment above normally expected levels that is especially relevant. For example, the average Zr content of the exposed upper continental crust is about 190 ppm (Taylor and McLennan, 1985), mostly contained in zircon. Typical shales have about 200 ± 100 ppm Zr (Taylor and McLennan, 1985) and thus do not normally display any substantial enrichments or depletions of zircon. In contrast, sandstone may have higher Zr contents, on the order of 400 ± 200 ppm, likely reflecting heavy mineral (i.e., zircon) enrichments.

Many of the major heavy minerals have high concentrations of elements, other than the REE, that are found at trace levels in most sedimentary rocks. Accordingly, in some cases it is possible to evaluate the influence of heavy mineral fractionations on REE patterns by examining other geochemical data. For example, zircon typically contains nearly 50% Zr with up to several per cent Hf and monazite typically contains in excess of 6% Th. In Figures 10, 11 and 12, the effects on REE patterns and other trace element abundances are shown for adding various amounts of heavy minerals to typical sandstones and shales. In each case, the initial REE patterns are assumed to be parallel for sandstone and shale but with absolute abundances lower in the sandstone by a factor of five. For simplicity, the initial concentration of Zr (for zircon mixing) and Th (for monazite and allanite mixing) are also assumed to be the same for sandstone and shale. Several features are notable. First, because sandstones tend to have substantially lower REE abundances than do shales, their REE patterns are considerably more prone to being dominated by heavy minerals. In shales, even substantial amounts of heavy minerals do not necessarily result in substantial changes to the REE patterns. For example, adding 600ppm Zr (i.e., sediments in Fig. 10 have 800ppm Zr) in the form of zircon (approximately 0.1%) increases Yb abundances by only about 1X chondritic levels and decreases the Gd/Yb ratio by less than 10%. Addition of even small amounts of monazite (and to a lesser extent allanite) would significantly affect the sedimentary REE patterns (Figs. 11 and 12), but in the case of monazite, additions of as little as 0.02% are associated with anomalously high Th abundances.

The possibility of a dominating influence of heavy minerals in most sedimentary rocks can be excluded in many cases by the examination of sedimentary REE patterns themselves. Most post-Archean sedimentary rocks have fairly uniform REE patterns with $La_N/Yb_N<15$ (with very few >20). HREE patterns are flat with Gd_N/Yb_N rarely outside the range of 1.0-2.0. For low REE abundance sediments (such as many sandstones), even modest amounts of zircon result in $Gd_N/Yb_N<1.0$, a feature that would be recognized as anomalous (Fig. 10). Allanite addition of about 0.02% results in anomalously high La_N/Yb_N (>15) for low REE abundance sediments (Fig. 12). The influence of monazite is even more dramatic (Fig. 11); addition of as little as 0.005% to sandstones and 0.02% to shales results in anomalously high Gd_N/Yb_N ratios (>2.0). Addition of mixtures of these various heavy minerals could, of course, mask any effect on sedimentary REE patterns, but in such cases elevated (and possibly correlated) Zr, Hf and Th abundances would result.

182

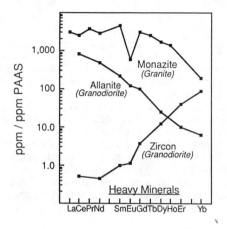

Figure 9. REE diagram, normalized to PAAS, of selected minerals that are common in heavy mineral suites of sedimentary rocks. Data from Gromet and Silver (1983) and Lee and Bastron (1967). Note the very high total REE abundances and that the patterns are substantially fractionated relative to PAAS.

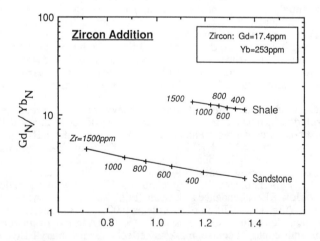

Figure 10. Plot of Yb_N versus Gd_N/Yb_N to illustrate the effects of concentrating zircon on sedimentary REE patterns. Sandstone is taken to have a REE pattern parallel to shale (PAAS) but with lower overall abundances by a factor of five. Initial Zr abundances are taken at 200 ppm for both sandstone and shale. It can be seen that the addition of zircon can influence the HREE pattern of sediments with sandstones being particularly affected. However, zircon concentration sufficient to cause substantial changes in Gd_N/Yb_N ratios also result in abnormally high Zr content and thus would be readily identified. Zircon REE data from Gromet and Silver (1983).

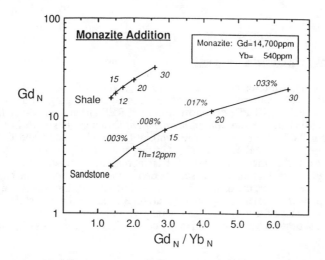

Figure 11. Plot of Gd_N versus Gd_N/Yb_N to illustrate the effects of concentrating monazite on sedimentary REE patterns. Sediment REE patterns as in Figure 10. Initial Th abundances are taken at 10 ppm for both sandstone and shale. Monazite addition of more than about 0.015% above normally expected background levels results in both significantly higher REE abundances, especially for sandstones, and in unusually high Th abundances. Concentration of monazite also results in highly fractionated HREE patterns which are virtually absent in post-Archean sediments. Post-Archean sediments have Gd_N/Yb_N ratios almost universally <2.0. Monazite REE data from Lee and Bastron (1967).

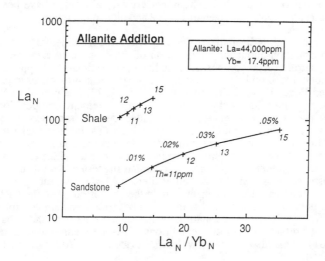

Figure 12. Plot of La_N versus La_N/Yb_N to illustrate the effects of concentrating allanite on sedimentary REE patterns. Sediment REE patterns and initial Th abundances as in Figures 10 and 11. Allanite concentrations result in strongly fractionated REE patterns and somewhat higher overall REE abundances. For post-Archean sediments, La_N/Yb_N typically is less than about 15 with very few ratios greater than 20. Allanite is also strongly fractionated in HREE but lower overall abundances result in considerably less HREE depletion than would addition of comparable amounts of monazite (see Figs. 9 and 11). Allanite REE data taken from Gromet and Silver (1983).

184

Sedimentary rocks and crustal abundances

Perhaps the most interesting feature of sedimentary REE data is the remarkable uniformity of the REE patterns, with absolute abundances ranging from about 100X chondritic for La to about 10-15X chondritic for Yb, LREE enrichment, fairly flat HREE patterns (Gd_N/Yb_N=1.0-2.0) and ubiquitous negative Eu-anomalies with Eu/Eu* fairly constant at about 0.60-0.70 (e.g., Taylor, 1960, 1962, 1964; Haskin and Gehl, 1962; Haskin et al., 1966; Nance and Taylor, 1976; Gromet et al., 1984; Taylor and McLennan, 1981, 1985). As the number of analyses has increased, more variability has been recognized due to several factors such as secular trends and variable provenance on a local scale (see below). However, the observation that sedimentary REE patterns are generally uniform in comparison to potential igneous/metamorphic source rocks is still valid and of fundamental significance for terrestrial geochemistry. In Figure 1, chondrite-normalized REE patterns are shown for various sedimentary composites and averages of post-Archean age.

This uniformity is generally interpreted as resulting from efficient mixing of various provenance components from the upper crust. Since the REE are likely transported primarily in particulate matter rather than in solution and since secondary mobilization of the REE is not normally a significant process, the average sedimentary REE pattern is equated to the average upper crust REE pattern (see Taylor and McLennan, 1985 for further discussion). REE abundances are considerably greater in shales in comparison to sandstones or carbonates and fine grained sediments constitute about 70% of the sedimentary record. Accordingly, shales dominate the mass balance and have received most attention. To a first order approximation, the REE patterns of sandstones and carbonates are parallel to those of shales and mass balance calculations indicate that the average shale REE pattern should be parallel to the upper crustal pattern but lower in absolute abundances by about 20%. The upper crustal values that have been adopted (Taylor and McLennan, 1985) are given in Table 2 and plotted in Figure 1.

The upper crustal REE pattern derived from the sedimentary data conveys considerable information about the composition of the continental crust and about the processes that control crustal composition. Independent estimates of upper continental crustal composition, based on large scale sampling programs, approximates to granodiorite (e.g., Fahrig and Eade, 1968; Eade and Fahrig, 1971, 1973; Shaw et al., 1967, 1976). REE abundances cannot be used to constrain major element compositions in any meaningful way, however, the upper crustal REE pattern derived from the sedimentary data is broadly consistent with a granodioritic composition (Taylor and McLennan, 1981, 1985). The LREE enrichment seen in the upper continental crust indicates that there is an overall enrichment in large ion lithophile elements compared to the ultimate mantle sources. The level of LREE enrichment is also greater than that suggested by most models of bulk continental crustal composition (Taylor and McLennan, 1985) suggesting that there is a further upward enrichment of LIL elements within the crust itself. The flat HREE pattern inferred for the upper crust suggests that there has been no dominant control on crustal compositions from a HREE-fractionating phase such as garnet, and thus provides constraints on the pressure/temperature conditions of mantle melting to form continental crust (Gill, 1981). This also contrasts with REE evidence for Archean crustal compositions (see below).

The occurrence of a substantial negative Eu-anomaly for the average upper continental crust is of special interest. Most models of bulk crustal composition suggest that there is no significant depletion or enrichment of Eu (Taylor and McLennan, 1985); few primary mantle-derived igneous rocks display anomalous distributions of Eu (Basaltic Volcanism Study Project, 1981; Gill, 1981). Consequently, the presence of Eu-depletion may be interpreted as reflecting shallow, intracrustal differentiation resulting in Eu-depletion in the upper continental crust, associated with the production of granitic rocks, and the complementary Eu-enrichment in the lower continental crust. These intracrustal processes, which may include both partial melting where plagioclase is a stable residual phase or fractional crystallization where plagioclase is a fractionating phase, can be expected to have a strong influence on crustal geochemistry (Hanson, 1980, 1989). Although both of these processes undoubtedly play a role, intracrustal partial melting is generally deemed to be dominant since most upper crustal granitic rocks are thought to have been derived through partial melting (Wyllie, 1977). In summary, the negative Eu-anomaly in the upper continental crust indicates that intracrustal differentiation is a dominant process in controlling element distributions within the continental crust.

Since Sc appears to behave similarly to the REE during sedimentary processes, it is also concluded that average Sc abundances in sedimentary rocks can be simply related to the abundances in the upper continental crust. In addition, it has been noted that Th displays coherent behavior with the REE such that La/Th and Th/Sc ratios are fairly constant in shales at about 2.8 and 1.0, respectively (McLennan et al., 1980; Taylor and McLennan, 1985; Fig. 13). The considerably greater scatter on the Th vs. Sc diagram is to be expected since the Th/Sc ratio is a much more sensitive index of average provenance composition than is the La/Th ratio (Taylor and McLennan, 1985). Since Th also has very low abundances in natural waters (see Fig. 3) it is concluded that this element too is providing an approximation of upper continental crustal abundances in sedimentary rocks. Using well established elemental ratios for crustal rocks, such as Th/U = 3.8, K/U = 10,000 and K/Rb = 250 (McLennan et al., 1980), it is also possible to estimate the upper crustal abundances of K, U and Rb directly from the sedimentary data. These values are given in Table 4 where they are compared to other independent estimates of upper crustal abundances. Values for heat producing elements (K, U, Th) are important since heat flow data provide one of the few independent geophysical constraints on any estimates of crustal abundances. Crustal abundance estimates derived directly from sedimentary data are particularly useful because, in principle, it is possible to examine variations in crustal abundances back through time.

What fraction of the continental crust is represented by these values? In principle, sediments provide information only on the crust that is exposed to weathering and erosion. Heat flow data indicate that the heat-producing elements (K, Th, U) are strongly enriched in the upper part of the continental crust (e.g., Morgan, 1984). The exact thickness is model dependent but a value of about 10 km is generally accepted. This upward enrichment of the heat producing elements, which are also large ion lithophile (LIL) elements, is generally taken as evidence that upper crustal abundances are not representative of the entire continental crust. A major complexity in this approach is the possible preferential upward transport of heat-producing elements by fluid phases during granulite grade metamorphism (Heier, 1973; Rudnick et al., 1985). Accordingly, we assume that the upper crustal abundances, including those derived from sedimentary data, are also valid for a comparable thickness, representing the upper 25-35% of the continental crust. It is important to realize that this is a simplified model to explain gross heat distributions, and by inference general geochemical distributions, within the crust and does not necessarily correspond to any simple lithological boundary.

Sedimentation and plate tectonics

A fundamental conclusion that may be drawn from the preceeding discussion is that for terrigenous sedimentary rocks, the REE patterns generally reflect the average compositions of the provenance. In detail, complications may arise, mainly for coarser sediments such as sandstones, primarily related to heavy mineral concentrations. The uniformity of REE distributions in terrigenous sedimentary rocks has been emphasized and attributed to the efficient mixing of source lithologies during sedimentary processes. Most sedimentary rocks that are preserved in the geological record have had a significant and extended

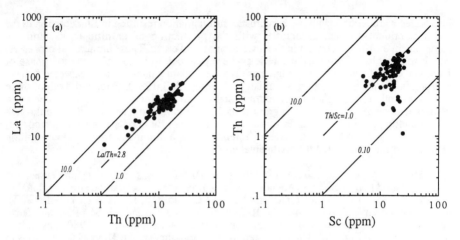

Figure 13. Plots of (a) La versus Th and (b) Th versus Sc for post-Archean fine grained sedimentary rocks. Note that the ratios La/Th and Th/Sc are mostly within a restricted range and average ~2.8 and ~1.0 respectively. It is concluded that these are also the La/Th and Th/Sc ratios of the upper continental crust.

Table 4. Composition of the upper continental crust determined from sedimentary REE abundances and from direct measurement.

	From Sediments[1]	Basis of Estimate[1]	Direct Measure[2]
La (ppm)	30	0.8 x PAAS	30
Th (ppm)	10.7	La/Th=2.8	10.5
U (ppm)	2.8	Th/U=3.8	2.5
K_2O (wt.%)	3.4	$K/U=10^4$	3.30
Rb (ppm)	112	K/Rb=250	110
Sc (ppm)	11	Th/Sc=1.0	10

[1]- From Taylor and McLennan, 1985
[2]- Compiled from Shaw (1967); Shaw et al. (1967, 1976); Fahrig and Eade (1968); Eade and Fahrig (1971, 1973)

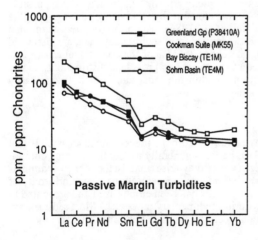

Figure 14. Chondrite-normalized REE diagram for selected Recent and Phanerozoic turbidites deposited at passive margin tectonic settings. Note that the patterns are essentially identical to typical post-Archean shales (Fig. 1) with LREE-enrichment, flat HREE patterns and the ubiquitous negative Eu-anomaly. Data from Nathan (1976), Bhatia (1985), and McLennan et al. (1990a).

sedimentary recycling history (Veizer and Jansen, 1979, 1985) which facilitates mixing and homogenization of REE and other geochemical signatures. However, where sedimentary recycling processes are less efficient and where a certain lithogy may dominate the provenance, REE patterns that differ substantially from PAAS may be encountered. This occurs mostly at volcanically active tectonic settings.

Studies which have examined the relationship between sedimentary REE patterns and plate tectonic setting are not abundant (Nathan, 1976; Dypvik and Brunfelt, 1976; Nance and Taylor, 1977; Bhatia, 1985; White et al., 1985; Andre et al., 1986; Floyd and Leveridge, 1987; also see Wildeman and Haskin, 1965; Liu and Schmitt, 1984; Hu et al., 1988; Sholkovitz, 1988). Bhatia (1985) attempted to establish a scheme for the discrimination of plate tectonic setting of sediments on the basis of REE geochemistry. Such an approach is considered premature for a number of reasons (Taylor and McLennan, 1988), including:

(1)   The data base for REE in sedimentary rocks is small and studies examining the relationship between tectonics and sedimentary REE geochemistry are few.

(2)   There are few studies that document systematic relationships between modern sediments of known tectonic setting and REE geochemistry, although such work is underway (e.g., McLennan et al., 1989, 1990a,b). Most studies have dealt with ancient sediments where tectonic conditions must be interpreted on the basis of other evidence.

(3)   Sediment may be transported across tectonic boundaries and thus may not necessarily be indicative of the tectonic setting of deposition (e.g., Mack, 1984; Velbel, 1985). In a trace element - Sr-Nd-Pb isotope study of sediments deposited in front of the Lesser Antilles island arc, White et al. (1985) demonstrated that Eu-depleted sediments were mainly derived from the Guyana Shield (i.e., passive margin) rather than the exposed island arc.

(4)   Attempts to discriminate the plate tectonic setting of igneous rocks on the basis of trace element signatures (e.g., Pearce and Cann, 1973) have shown that any such relationship is complex and that many exceptions are to be expected.

Although considerable variability in REE (and other) geochemistry of sediments deposited at active continental margins is encountered, such variability is still very much less than that seen in potential igneous sources. This suggests that even where more localized provenance is dominant and sedimentary recycling is less severe, the processes of sedimentation still accomplish some degree of homogenization of source rocks.

Continental margins may be classified into passive and active types. Passive margin settings are the primary sites of major river systems (Potter, 1978a) and typically have a provenance consisting of recycled sedimentary debris and/or older plutonic/metamorphic material, with a relatively small volcanic component (Potter, 1978b). This provenance is clearly exhibited in the REE chemistry of sediments deposited at passive margins, with patterns being fairly uniform and similar to PAAS (e.g., Nathan, 1976; Bhatia, 1985; McLennan et al., 1990a) (Fig. 14).

It is a common assertion that sedimentary rocks deposited at active continental margins, such as island arcs or continental arcs, will have REE characteristics similar to undifferentiated volcanic rocks of the arc itself (i.e., similar to andesites), with lower REE abundances, La/Sm and La/Yb ratios and without negative Eu-anomalies (e.g., Gromet et al., 1984; Gibbs et al., 1986; Condie and Wronkiewicz, 1989). In a few cases, such as with the volcanogenic greywackes of the Devonian Baldwin Formation or turbidites deposited in fore arc regions of the Marianas Trench and Soloman islands, this is indeed the case (Fig. 15a). However, sediments deposited at active continental margins generally show REE patterns intermediate between a 'typical andesite pattern' and PAAS or in some cases, indistinguishable from PAAS itself (Figs. 15b, 16). Thus, most continental margin sediments display intermediate REE abundances, variable LREE enrichment and variable negative Eu-anomalies, with Eu/Eu* in the range 0.60-1.00. This latter feature is important when REE data for Archean sedimentary rocks from greenstone belts are considered (see below).

There are probably several major source components for sediments deposited at active continental margins, which can be inferred on the basis of petrographic, geochemical and isotopic data (e.g., McLennan et al., 1989, 1990a,b). One component is old upper continental crust which may be either old igneous/metamorphic terranes or old, recycled sedimentary rocks. Such sources are likely to have REE patterns fairly similar to average, or typical upper continental crust. A second major source is a young component derived from the island arc or continental arc itself. Igneous material which has not undergone significant crystal fractionation at shallow depths (i.e., within the field of plagioclase stability) will display less fractionated REE patterns with no Eu-anomalies. On the other hand, fractionated igneous rocks, in the form of felsic volcanics and plutonic rocks, commonly have had plagioclase removed during their igneous history and in terms of REE geochemistry, are effectively young upper continental crust. This third source, which may be incorporated into sediments during explosive volcanism or, more likely, when the arc is matured and dissected to expose plutonic root zones, is likely to display variable REE patterns but in general will exhibit the signature of intracrustal differentiation in the form of negative Eu-anomalies. Locally, it is also possible to incorporate MORB-like components, such as found in the Marianas fore arc region (McLennan et al., 1990a,b).

Archean sedimentary rocks and the Archean crust

The earliest studies of REE distributions in Archean sedimentary rocks raised the possibility that patterns may be different from those in younger sediments, suggesting that the composition of the exposed continental crust would also be different (e.g., Haskin and Gehl, 1962; Haskin et al., 1966; Schnetzler and Philpotts, 1967). Studies of Wildeman and Haskin (1973) and Wildeman and Condie (1973) confirmed that Archean sedimentary rocks were significantly enriched in Eu relative to post-Archean shales. Jakes and Taylor (1974) first provided an explanation for differing sedimentary REE patterns, within a framework of crustal evolution. They pointed out that REE patterns of Archean sedimentary rocks were similar to modern island arc volcanics and suggested that ancient island arcs were the source of Archean sediments. In addition, they suggested that the ubiquitous chondrite-normalized negative Eu-anomaly in post-Archean shales resulted from intracrustal partial melting processes to form a granodioritic upper continental crust and a plagioclase-rich (i.e., Eu-enriched) residual lower crust.

Taylor (1977, 1979) pointed out that an 'island arc' model for Archean sedimentary rocks may be inconsistent with observations of high Cr and Ni abundances in many Archean sedimentary rocks, and other geochemical data. An alternative model is that these Archean sedimentary data may represent a mixture of a bimodal suite of mafic volcanics and felsic volcanics/plutonics. Such a bimodal suite, with suitable REE patterns, appears common in many Archean terranes. In this model, incorporation of igneous rocks of intermediate composition (e.g., calc-alkaline volcanics) cannot be excluded since the presence of such rocks cannot be distinguished from geochemical evidence (Taylor and McLennan, 1981, 1985). Considerable data have been collected in order to evaluate these models (e.g., Nance and Taylor, 1977; Bavinton and Taylor, 1980; Jenner et al., 1981; McLennan et al., 1983a,b, 1984; Ujike, 1984; Sawyer, 1986) and it has been generally concluded that the bimodal mixing model was likely to be an important, and probably dominant, process. Much of this work has been reviewed in much greater detail by McLennan and Taylor (1984, 1990) and Taylor and McLennan (1985) and the reader is referred to these publications for further discussion.

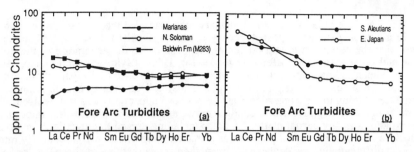

Figure 15. Chondrite-normalized REE diagrams for selected Recent and Phanerozoic turbidites deposited in island-arc fore arc settings. (a) Samples that do not display negative Eu-anomalies. Note that these samples also tend to be less enriched in LREE. (b) Samples displaying significant negative Eu-anomalies. Note that these samples are derived from arcs with older basement (Japan) or more mature arcs with exposed plutonic roots (Aleutians). In most cases, the magnitude of Eu-depletion is less than that seen for PAAS. Data from Nance and Taylor (1977) and McLennan et al. (1990a).

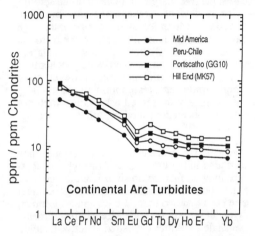

Figure 16. Chondrite-normalized REE diagram for selected Recent and Phanerozoic turbidites deposited in continental arc settings. Samples are typical and commonly display REE patterns comparable to typical post-Archean shales. In detail, a wide range of REE patterns is found in active tectonic settings, but the signature of intracrustal differentiation, in the form of negative Eu-anomalies, is present to varying degrees in most samples. Data from Bhatia (1985), Floyd and Leveridge (1987), McLennan et al. (1990a).

Figure 17 (right). Chondrite-normalized REE diagrams for selected greywacke-shale turbidites from Archean greenstone belts. Patterns are divided into three types on the basis of La/Sm and Gd/Yb ratios; Type 1 is most common. Occurrence of HREE-depleted samples with $Gd_N/Yb>2.0$ appears essentially restricted to the Archean. Such HREE depletion is probably the signature of either garnet fractionation sometime in the igneous history of the provenance or a reflection of the mantle source REE pattern. Data from Wildeman and Condie (1973), Nance and Taylor (1977), Bavinton and Taylor (1980), Jenner et al. (1981), McLennan et al. (1983b).

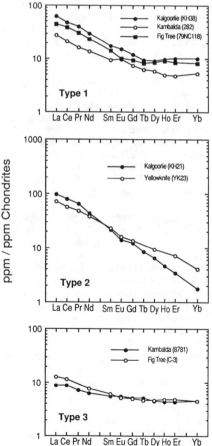

If it is assumed that Archean sedimentary rocks sampled the contemporary exposed crust in a similar manner to post-Archean sediments, then the differing REE patterns suggest that the composition and processes responsible for the generation of the upper continental crust may have differed (e.g., Taylor, 1979; Taylor and McLennan, 1981, 1985). Archean sedimentary rocks tend to have less fractionated LREE patterns and other geochemical characteristics (e.g., lower Th/Sc ratios) indicative of an overall more mafic provenance, leading to the suggestion that the Archean upper crust was less evolved. The general paucity of negative Eu-anomalies may be taken as evidence that intracrustal differentiation (mostly partial melting but also including crystal fractionation), involving plagioclase as a stable residual phase or as a fractionating phase, was a relatively minor process in differentiating the Archean continental crust whereas this has been the dominant process during the post-Archean (also see McLennan, 1988).

Virtually all of the Archean sedimentary REE data, prior to about 1985, was from sedimentary rocks preserved in greenstone belts. The bulk of sediments preserved in greenstone belts are greywacke - shale turbidites, generally thought to have been deposited in active tectonic settings (although in some cases, passive margin and intracratonic settings have been proposed; see Windley (1984) for review of Archean tectonic models). Sedimentary rocks from greenstone belts received more attention than those from the high metamorphic grade terranes for three reasons: (1) greenstone belts appear to be volumetrically more important than high grade terranes; (2) more geological data are available for low grade terranes; (3) greenstone belts are less prone to possible geochemical fractionations due to high grades of metamorphism and/or partial melting. However, more recently it has been recognized that some high grade terranes contain fundamentally different sedimentary facies associations. A commonly preserved association includes quartzite - arkose - shale - carbonate, thought to have been deposited in a platform setting, presumably associated with a relatively stable craton.

The observation that sedimentary rocks preserved in Archean high grade terranes may possess REE and other geochemical characteristics indistinguishable from typical post-Archean shales has led to suggestions that there may, in fact, be no significant differences in sedimentary REE patterns (and therefore the upper continental crust) over geological time. In this interpretation, differences that have been claimed (e.g., Taylor and McLennan, 1985) would be the result of biassed sampling or preservation. Greenstone belt sediments would be equivalent to Phanerozoic greywacke-shale turbidites deposited at active tectonic settings and platform sequences in high grade terranes would be equivalent to post-Archean shales with PAAS-like REE patterns. Recent discussions of this problem may be found in Gibbs et al. (1986, 1988) and McLennan and Taylor (1988, 1990).

Archean greenstone belts. Representative REE patterns for Archean turbidites from greenstone belts are plotted in Figure 17. These have been divided into three major types (McLennan and Taylor, 1984, 1990):

Type 1. The most common varieties display LREE enrichment, flat HREE (Gd_N/Yb_N = 1.0-2.0) and no or only slight negative Eu anomalies. These REE patterns are thought to have been derived mainly from mixing of a bimodal igneous suite with REE patterns similar to Types 2 and 3 described below. Provenance components characterized by intermediate REE patterns, such as those found at modern island arcs (i.e., andesites) could also play a role.

Type 2. Steep REE patterns with substantial HREE depletion (Gd_N/Yb_N > 2.0) are common but less abundant than Type 1. These patterns are typical of many Archean Na-rich granitic rocks (tonalites, trondhjemites, granodiorites) and felsic volcanics found within greenstone belts.

Type 3. REE patterns from a few Archean greenstone belt sediments are essentially flat, with or without slight LREE enrichment or depletion. Comparable REE patterns are found in greenstone belt mafic-ultramafic volcanic suites which likely represent the second component in the mixing model described above.

How do these REE characteristics compare to those of post-Archean sedimentary rocks deposited in broadly comparable tectonic settings? In Figure 18 Archean turbidites and post-Archean active margin turbidites are compared in terms of Eu/Eu* and Gd/Yb ratios. Although there is considerable overlap, the two data sets are clearly distinctive. Post-Archean active margin turbidites have variable Eu/Eu* with samples mostly within the range of Eu/Eu* = 0.60-1.00; a large proportion of the samples have substantial negative Eu-anomalies (Eu/Eu* < 0.85). In contrast, Archean turbidites tend to have considerably less depletion of Eu, suggesting that the processes of intracrustal differentiation were less important in forming the sources of these sediments. Also, Archean turbidites commonly display significant HREE depletion with $Gd_N/Yb_N > 2.0$ whereas such HREE depletion is essentially absent throughout the post-Archean. HREE depleted components of the Archean crust include Na-rich granitic and felsic volcanic rocks. The origin of such depletion is best explained by garnet fractionation sometime during their igneous history (see Taylor and McLennan, 1985 for review) or by a mantle source with similar HREE depletion (e.g., Stern et al., 1989). In either case, such characteristics appear restricted to the Archean crust.

High grade terranes. At least two varieties of high grade terranes are preserved, that are relevant to the present discussion. The first is simply a highly metamorphosed equivalent to the greenstone belts. The Kapuskasing Structural Zone is an example of this (Fig. 19). In these cases, the REE characteristics of metasedimentary rocks are essentially the same as those found in low grade terranes (Taylor et al., 1986). One exception is that where local melting has occurred (not always apparent), REE patterns may be affected.

In contrast, many high grade regions contain sedimentary associations typical of cratonic environments and these metasedimentary rocks display quite different REE (and other geochemical) characteristics (e.g., Gibbs et al., 1986; Taylor et al., 1986; also see McLennan and Taylor, 1990). Notably, REE patterns that are indistinguishable from PAAS are common (Fig. 20). In contrast to the post-Archean, in a number of locations studied, these sedimentary rocks appear to to be in close physical association with others that have essentially 'Archean-like' characteristics, suggesting that homogenization of the provenance components has not been particularly efficient (Fig. 21). In any case, the presence of cratonic sedimentary rocks with significant negative Eu-anomalies suggests that intracrustal differentiation occurred on a local scale in the Archean crust.

Archean/Proterozoic transition. The change in sedimentary REE characteristics is correlated with the profound unconformity associated with the Archean/Proterozoic transition. In the ca. 2.5-2.2 Ga Huronian Supergroup, a gradual change in REE patterns from 'Archean-like' at the base to indistinguishable from PAAS at the top is observed (McLennan et al., 1979). In the Hamersley Basin of Western Australia, the change is fairly abrupt (McLennan, 1981). Wronkiewicz and Condie (1987, 1989) have suggested that for southern Africa, the change in patterns took place over an extended period of time with several cycles in the REE characteristics. Although in any given area, the change in REE characteristics appear relatively abrupt, on a global scale it takes place at different times in different locations, and correlates with the transition from unstable crustal conditions associated with widespread greenstone belts to relatively stable cratonic conditions (e.g., Cloud, 1976). Thus, in South Africa, the transition took place between about 3.2-2.8 Ga, in the Pilbara Block at about 2.8-2.6 Ga and in most other Precambrian regions at about 2.7-2.5 Ga. (see Taylor and McLennan (1985) for review). Excavation of the late Archean upper crust, which itself changed composition during an especially important period of crustal growth/evolution at the end of the Archean, is considered to be responsible for the changes in sedimentary REE patterns. This event also marks the time when intracrustal processes began to dominate the composition of the upper continental crust.

192

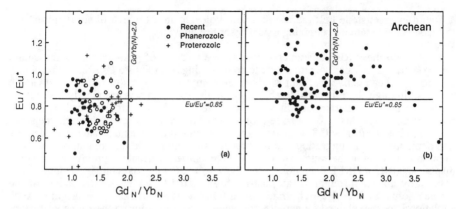

Figure 18. Plot of Eu/Eu* versus Gd_N/Yb_N for turbidites of (a) Post-Archean age and (b) Archean age. Post-Archean samples include Recent deep-sea turbidites from various active tectonic settings, Phanerozoic and Proterozoic greywacke- shale turbidites interpreted to have been deposited in active tectonic settings. Archean samples are greywacke - shale turbidites from greenstone belts. Note that turbidites deposited in active tectonic settings are virtually identical throughout the post-Archean, however, Archean turbidites are distinctive in that samples tend to be considerably less depleted in Eu (Eu/Eu* generally >0.85) and samples with HREE-depletion (i.e., $Gd_N/Yb_N > 2.0$) are essentially restricted to the Archean. Data sources include (a) Nance and Taylor (1977), Bhatia (1985), Gibbs et al. (1986), Floyd and Leveridge (1987), McLennan et al. (1990a), unpublished data; (b) Wildeman and Condie (1973), Nance and Taylor (1977), Bavinton and Taylor (1980), Jenner et al. (1981), McLennan (1981), McLennan et al. (1983a,b), Lajoie and Ludden (1984), Ujike (1984), Sawyer (1986).

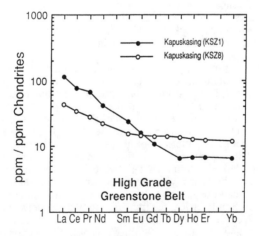

Figure 19. Chondrite-normalized REE diagram for metasedimentary samples from the Archean Kapuskasing Structural Zone in the Canadian Shield, considered to be the highly metamorphosed equivalent to a typical low grade Archean greenstone belt. Note that the patterns are essentially identical to those found in greenstone belts (Fig. 17). Data from Taylor et al. (1986).

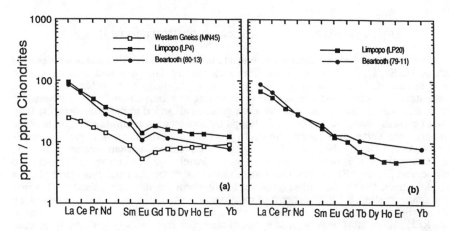

Figure 20. Chondrite-normalized REE diagrams for selected metasedimentary samples from Archean high grade terranes considered to have been deposited in shallow water conditions in stable platformal environments. Two types of patterns are found including: (a) those with REE characteristics essentially identical to post-Archean shales and (b) those with REE characteristics comparable to samples found in greenstone belts. Samples with these differing patterns may be found in fairly close physical association. Data from Gibbs et al. (1986) and Taylor et al. (1986).

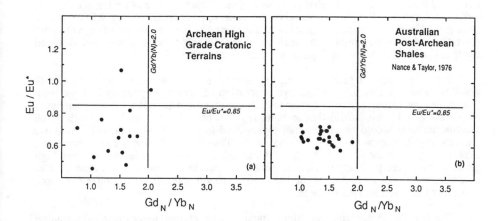

Figure 21. Plot of Eu/Eu* versus Gd_N/Yb_N for (a) metasedimentary rocks from Archean high grade terranes considered to have been deposited in stable platformal conditions and (b) typical post Archean shales from Australia. Note that most of the high grade samples, though somewhat scattered, have REE characteristics essentially identical to typical post-Archean shales. This is interpreted to suggest that intra-crustal differentiation processes, leading to the formation of small relatively stable cratons, took place throughout the Archean. Also note that a few of the high grade samples tend to lack Eu-depletion. The close association of samples from a cratonic setting with such diverse patterns may be distinctive of the Archean but the significance is not well understood. Data sources include: (a) Gibbs et al. (1986); Taylor et al. (1986); (b) Nance and Taylor (1976); McLennan (1981).

REE IN SEDIMENTARY ROCKS AND CRUSTAL EVOLUTION

In order to precisely estimate upper crustal abundances of the REE and other elements such as Th, Sc, U, K, Rb (see above) from sedimentary data, it would be necessary to derive reliable estimates for each major sedimentary lithology (e.g., shales, sandstones, carbonates, evaporites) and average these according to relative mass. To study secular trends, this would have to be done for each geological period under consideration (and would be greatly complicated by the liklihood of biassed preservation of certain tectonic domains – see Veizer and Jansen, 1985). In practice, however, the task is greatly simplified by nature. During the post-Archean, shales are the dominant sedimentary lithology, constituting about 70% of the sedimentary mass (Garrels and Mackenzie, 1971) and have considerably higher REE content than most sandstones, carbonates and evaporites (Taylor and McLennan, 1985). Accordingly, shales totally dominate the mass balance. To a first order approximation, the REE patterns of most post-Archean sandstones and carbonates are near parallel to those found in shales (Taylor and McLennan, 1985) and so the average shale REE pattern must be very close to that of the upper crust, with only slight differences in absolute abundances (see Fig. 1).

Is there any evidence for secular trends in sedimentary REE abundances that can be related to changes in upper crustal abundances? The available evidence suggests that there is no discernable change in sedimentary REE patterns throughout the post-Archean (Taylor and McLennan, 1985; McLennan and Taylor, 1990; see Fig. 1), but some caution is warranted in interpreting this observation in terms of crustal evolution. Terrigenous sedimentation is fundamentally a cannibalistic process and, although sedimentary recycling rates are highly variable and related to tectonic setting, most terrigenous sediment is derived from pre-existing sedimentary rocks (Veizer and Jansen, 1979, 1985). An obvious consequence is that sedimentary compositions are strongly buffered by this recycling process and only major, fundamental changes in upper crustal composition would be translated into an discernable change to sedimentary compositions (Veizer and Jansen, 1979; McLennan, 1988). Accordingly, the constancy of sedimentary REE patterns throughout the post-Archean suggests only that there has been no major change in upper crustal compositions during that time; minor changes, short term variations and localized changes in composition may take place without being recognized.

In the Archean both turbidites from greenstone belts and cratonic sediments from high grade terranes are found. The relative proportion of present exposures has not been determined but greenstone belts appear to be somewhat more abundant. It has been noted by Veizer and Jansen (1985) that tectonic regimes broadly comparable to high grade cratonic terranes would be preferentially preserved in the geological record over those regimes broadly comparable to greenstone belts. Thus it is likely that the greenstone belt terranes dominated the Archean crust. Cratonic regions, presently preserved in some high grade terranes, are probably of limited dimension but do indicate that the fundamental processes of crustal evolution, involving crustal growth and intracrustal differentiation, took place back well into the Archean.

The Archean/Proterozoic transition remains as a benchmark in the chemical evolution of sedimentary rocks and, in turn, the continental crust. The REE distributions in sedimentary rocks provide especially persuasive evidence. In terms of simple averages of dominant sedimentary patterns, there is a remarkable contrast in Archean and post-Archean REE patterns. The post-Archean is particularly depleted in Eu, has somewhat higher LREE-enrichment, has higher total REE abundances and lacks evidence for any provenance components with HREE depletion. When Archean greenstone belt turbidites are compared with Phanerozoic turbidites of broadly comparable tectonic affinity, the distinction in Eu/Eu* remains (although not as strong as for the comparison with post-Archean cratonic shales) and the contrast in HREE distributions is also apparent.

A change in sedimentary REE patterns (and other geochemical features) at about the Archean/Proterozoic transition provides important insights into the overall evolution of the continental crust (Taylor and McLennan, 1985; McLennan and Taylor, 1990). During the Archean, a wide variety of sedimentary REE patterns are found but with relatively few displaying Eu-depletion. In greenstone belts, the patterns are best explained through derivation from mixtures of the two dominant igneous lithologies comprised of Na-rich granitic and volcanic rocks (tonalites, tronhjemites, dacites, etc.) and mafic/ultramafic volcanics. The common occurrence of HREE depletion is in total contrast with the post-Archean and suggests differences in the pressure/temperature conditions of mantle melting to form continental crust, or differences in the composition of mantle sources of continental crust. In many Archean high-grade terrains, where preserved sedimentary environments indicate shallow cratonic conditions, sedimentary REE patterns commonly are indistinguishable from typical post-Archen shales, with significant Eu-depletion. It is likely that these cratonic terrains have been preferentially preserved in the geological record (Veizer and Jansen, 1985; McLennan and Taylor, 1988; McLennan, 1988) and thus were of minor importance during much of the Archean. Nevertheless, the sedimentary REE patterns, with Eu-depletion, are of great significance because they suggest that, for both Archean and post-Archean, intracrustal differentiation is a fundamental process involved with the maturing of continental crust into stable cratons. During an extended period of time from about 3.2 Ga to 2.6 Ga (with precise age differing for individual cratonic regions) an especially important episode of crustal growth took place. This was followed by intracrustal differentiation resulting in widespread production of K-rich granitic rocks with Eu-depletion (granites, granodiorites), stabilization of these new crustal additions into cratonic terranes, and an effective change in the composition of the upper continental crust. As these new additions were incorporated into the sedimentary mass, the Archean REE signature was overwhelmed and the average composition of sedimentary rocks changed to that seen in PAAS. The abrupt change in sedimentary compostions, in spite of the buffering effects induced by sedimentary recycling processes, attests to the magnitude of this event (e.g., Veizer and Jansen, 1979; McLennan, 1988).

A common view is that crustal growth and evolution proceeds in an episodic fashion (e.g., Moorbath, 1976). Although long considered puzzling, given the slow but gradual cooling of the earth, episodic crustal evolution may be understood within a framework of cycles of supercontinent assembly and fragmentation, which appears to have begun during the latest Archean (Worsley et al., 1986; Hoffman, 1989). During the post-Archean, such cycles were not accompanied by any long term changes in sedimentary REE patterns. What is different about the Archean/Proterozoic transition? An obvious difference is that the earth was much hotter during the Archean and the potential for both crustal addition and crustal recycling back into the mantle was also greater (Gurnis and Davies, 1985, 1986). Differing pressure/temperature conditions of mantle melting (or differing mantle source compositions) could give rise to crustal compositions (e.g., HREE-depleted compostions) that differed from the post-Archean (McLennan, 1989). By the end of the Archean the earth had cooled considerably and new additions to the crust may have had compositions similar to that seen today, in arc settings. During the post-Archean, a much cooler earth may have resulted in smaller volumes of new crust being added during each supercontinental cycle and any short term changes to upper crustal compositions would be minor and not easily recognized in the highly buffered sedimentary record. If this general secular evolutionary scheme is superimposed on an episodic pattern of crustal growth/evolution, associated with supercontinental cycles, then the unique and distinct features associated with the Archean/Proterozoic transition begin to make sense.

ACKNOWLEDGMENTS

I am grateful to Sidney Hemming, Gary Hemming, Gordon McKay and Ross Taylor for commenting on the manuscript. Support was provided by National Science Foundation Grants EAR-8816386 and EAR-8957784.

REFERENCES

Aller, L.H. (1987) Chemical abundances. In: A. Dalgarno and D. Layzer (Eds.), Spectroscopy of Astrophysical Plasmas. Cambridge Univ. Press, Cambridge, p. 89-124.

Anders, E. and Ebihara, M. (1982) Solar-system abundances of the elements. Geochim. Cosmochim. Acta 46, 2363-2380.

Andre, L., Deutsch, S. and Hertogen, J. (1986) Trace-element and Nd isotopes in shales as indexes of provenance and crustal growth: the Early Paleozoic from the Brabant Massif (Belgium). Chem. Geol. 57, 101-115.

Banner, J.L. (1986) Petrologic and geochemical constraints on the origin of regionally extensive dolomites of the Burlington-Keokuk Fms., Iowa and Missouri. Ph.D. Dissertation, State University of New York at Stony Brook, 368 p.

Banner, J.L., Hanson, G.N. and Meyers, W.J. (1988) Rare earth element and Nd isotopic variations in regionally extensive dolomites from the Burlington-Keokuk Formation (Mississippian): Implications for REE mobility during carbonate diagenesis. J. Sed. Pet., 58, 415-432.

Basaltic Volcanism Study Project (1981) Basaltic Volcanism on the Terrestrial Planets (Pergamon, New York) 1286pp.

Bavinton, O.A. and Taylor, S.R. (1980) Rare earth element geochemistry of Archean metasedimentary rocks from Kambalda, Western Australia. Geochim. Cosmochim. Acta 44, 639-648.

Brenninkmeyer, B.M. (1978) Heavy Minerals. In: R.W. Fairbridge and J. Bourgeois (Eds.) The Encyclopedia of Sedimentology. Dowden, Hutchinson and Ross, Stroudsburg, p. 400-402.

Bhatia, M.R. (1981) Petrology, geochemistry and tectonic setting of some flysch deposits. Ph.D. Dissertation, The Australian National University, Canberra, 382 p.

Bhatia, M.R. (1985) Rare earth element geochemistry of Australian Paleozoic graywackes and mudrocks: provenance and tectonic control. Sediment. Geol. 45, 97-113.

Chappell, B.W. (1968) Volcanic greywackes from the Upper Devonian Baldwin Formation, Tamworth - Barraba district, New South Wales. J. Geol. Soc. Australia 15, 87-102.

Chaudhuri, S. and Cullers, R.L. (1979) The distribution of rare-earth elements in deeply buried Gulf Coast sediments. Chem. Geol. 24, 327-338.

Cloud, P. (1976) Major features of crustal evolution. Trans. Geol. Soc. South Afr. vol. 79 Annex., 32pp.

Condie, K.C. and Wronkiewicz, D.J. (1989) A new look at the Archean - Proterozoic boundary: sediments and the tectonic setting constraint. In: S.M. Naqvi (ed.) Precambrian Continental Crust and its Economic Resources. Elsevier (in press).

Cullers, R.L., Chaudhuri, S., Kilbane, N. and Koch, R. (1979) Rare earths in size fractions and sedimentary rocks of Pennsylvanian-Permian age from the mid-continent of the U.S.A. Geochim. Cosmochim. Acta 43, 1285-1302.

Cullers, R.L., Barrett, T., Carlson, R. and Robinson, B. (1987) Rare-earth element and mineralogic changes in Holocene soil and stream sediment: a case study in the Wet Mountains, Colorado, U.S.A. Chem. Geol. 63, 275-297.

Cullers, R.L., Basu, A. and Suttner, L.J. (1988) Geochemical signature of provenance in sand-size material in soils and stream sediments near the Tobacco Root Batholith, Montana, U.S.A. Chem. Geol. 70, 335-348.

De Bievre, P., Gallet, M., Holden, N.E. and Barnes, I.L. (1984) Isotopic abundances and atomic weights of the elements. J. Phys. Chem. Ref. Data 13, 809-891.

Duddy, I.R. (1980) Redistribution and fractionation of rare-earth and other elements in a weathering profile. Chem. Geol. 30, 363-381.

Dypvik, H. and Brunfelt, A.O. (1976) Rare-earth elements in Lower Paleozoic epicontinental and eugeosynclinal sediments from the Oslo and Trondheim regions. Sedimentology 23, 363-378.

Eade, K.E. and Fahrig, W.F. (1971) Chemical evolutionary trends of continental plates - a preliminary study of the Canadian Shield. Bull. Geol. Surv. Can. 179, 51pp.

Eade, K.E. and Fahrig, W.F. (1973) Regional, lithological and temporal variation in the abundances of some trace elements in the Canadian Shield. Geol. Surv. Can. Paper 72-46, 46pp.

Elderfield, H. (1988) The oceanic chemistry of the rare-earth elements. Phil. Trans. Roy. Soc. London A325, 105-126.

Elderfield, H. and Greaves, M.J. (1982) The rare earth elements in seawater. Nature 296, 214-219.

Evensen, N.M., Hamilton, P.J. and O'Nions, R.K. (1978) Rare-earth abundances in chondritic meteorites. Geochim. Cosmochim. Acta 42, 1199-1212.

Ewers, G.R. and Higgins, N.C. (1985) Geochemistry of the Early Proterozoic metasedimentary rocks of the Alligator Rivers Region, Northern Territory, Australia. Precambrian Res. 29, 331-357.

Fahrig, W.F. and Eade, K.E. (1968) The chemical evolution of the Canadian Shield. Can. J. Earth Sci., 5, 1247-1252.

Floyd, P.A. and Leveridge, B.E. (1987) Tectonic environment of the Devonian Gramscatho basin, south Cornwall: framework mode and geochemical evidence from turbiditic sandstones. J. Geol. Soc. London 144, 531-542.

Garrels, R.M. and Mackenzie, F.T. (1971) Evolution of Sedimentary Rocks. Norton, New York, 397 p.

Gibbs, A.K., Montgomery, C.W., O'Day, P.A. and Erslev, E.A. (1986) The Archean - Proterozoic transition: evidence from the geochemistry of metasedimentary rocks of Guyana and Montana. Geochim. Cosmochim. Acta 50, 2125-2141.

Gibbs, A.K., Montgomery, C.W., O'Day, P.A. and Erslev, E.A. (1988) Crustal evolution revisited: Reply to Comments by S.M. McLennan and S.R. Taylor and J. Veizer, on "The Archean - Proterozoic transition: evidence from the geochemistry of metasedimentary rocks of Guyana and Montana". Geochim. Cosmochim. Acta 52, 793-795.

Gill, J. (1981) Orogenic Andesites and Plate Tectonics. Springer-Verlag, Berlin, 390 p.

Goldstein, S.L., O'Nions, R.K. and Hamilton, P.J. (1984) A Sm-Nd isotopic study of atmospheric dusts and particulates from major river systems. Earth Planet. Sci. Lett. 70, 221-236.

Gromet, L.P. and Silver, L.T. (1983) Rare earth element distributions among minerals in a granodiorite and their petrogenetic implications. Geochim. Cosmochim. Acta 47, 925-939.

Gromet, L.P., Dymek, R.F., Haskin, L.A. and Korotev, R.L. (1984) The "North American shale composite": Its compilation, major and trace element characteristics. Geochim. Cosmochim. Acta 48, 2469-2482.

Gurnis, M. and Davies, G.F. (1985) Simple parametric models of crustal growth. J. Geodynamics 3, 105-135.

Gurnis, M. and Davies, G.F. (1986) Apparent episodic crustal growth arising from a smoothly evolving mantle. Geology 14, 396-399.

Hanson, G.N. (1980) Rare earth elements in petrogenetic studies of igneous systems. Ann. Rev. Earth Planet. Sci. 8, 371-406.

Hanson, G.N. (1989) An approach to trace element modeling using a simple igneous system as an example. Rev. Mineral. 21 (this volume).

Haskin, L.A. and Gehl, M.A. (1962) The rare-earth distribution in sediments. J. Geophys. Res. 67, 2537-2541.

Haskin, L.A., Wildman, R.T., Frey, F.A., Collins, K.A., Keedy, C.R. and Haskin, M.A. (1966) Rare earths in sediments. J. Geophys. Res. 71, 6091-6105.

Haskin, L.A., Haskin, M.A., Frey, F.A. and Wildman, T.R. (1968) Relative and absolute terrestrial abundances of the rare earths. In: L.H. Ahrens (Ed.) Origin and Distribution of the Elements, Pergamon, New York, p. 889-912.

Haskin, M.A. and Haskin, L.A. (1966) Rare earths in European shales: a redetermination. Science 154, 507-509.

Heier, K.S. (1973) Geochemistry of granulite facies rocks and problems of their origin. Phil. Trans. Roy. Soc. Lond. A273, 429-442.

Henderson, P. (ed.) (1984) Rare Earth Element Geochemistry. Elsevier, Amsterdam, 510 p.

Hoffman, P.F. (1989) Speculations on Laurentia's first gigayear (2.0 to 1.0 Ga). Geology 17, 135-138.

Hoyle, J., Elderfield, H., Glendhill, A. and Greaves, M. (1984) The behavior of the rare earth elements during mixing of river and sea waters. Geochim. Cosmochim. Acta 48, 143-149.

Hu, X., Wang, Y.L. and Schmitt, R.A. (1988) Geochemistry of sediments on the Rio Grande Rise and the redox evolution of the South Atlantic Ocean. Geochim. Cosmochim. Acta 52, 201-207.

Humphris, S.E. (1984) The mobility of the rare earth elements in the crust. In: P. Henderson (Ed.) Rare Earth Element Geochemistry, Elsevier, Amsterdam, p. 317-342.

Jakes, P. and Taylor, S.R. (1974) Excess europium content in Precambrian sedimentary rocks and continental evolution. Geochim. Cosmochim. Acta 38, 739-745.

Jenner, G.A., Fryer, B.J. and McLennan, S.M. (1981) Geochemistry of the Yellowknife Supergroup. Geochim. Cosmochim. Acta 45, 1111-1129.

Kornacki, A.S. and Fegley, B., Jr. (1986) The abundance and relative volatility of refractory trace elements in Allende Ca, Al-rich inclusions: implications for chemical and physical processes in the solar nebula. Earth Planet. Sci. Lett. 79, 217-234.

Lajoie, J. and Ludden, J. (1984) Petrology of the Archean Pontiac and Kewagama sediments and implications for the stratigraphy of the southern Abitibi belt. Can. J. Earth Sci. 21, 1305-1314.

Lee, D.E. and Bastron, H. (1967) Fractionation of rare-earth elements in allanite and monazite as related to geology of the Mt. Wheeler mine area, Nevada. Geochim. Cosmochim. Acta 31, 339-356.

Liu, Y.-G. and Schmitt, R.A. (1984) Chemical profiles in sediment and basalt samples from Deep Sea Drilling Project Leg 74, Hole 525A, Walvis Ridge. Init. Repts. Deep Sea Drilling Proj. LXXIV, 713-730.

Mack, G.H. (1984) Exceptions to the relationship between plate tectonics and sandstone composition. J. Sed. Pet. 54, 212-220.

McCulloch, M.T. and Wasserburg, G.J. (1978) Sm-Nd and Rb-Sr chronology of continental crust formation. Science 200, 1003-1011.

McLennan, S.M. (1981) Trace element geochemistry of sedimentary rocks: Implications for the composition and evolution of the continental crust. Ph.D. Dissertation, The Australian National University, Canberra, 609 p.

McLennan, S. M. (1988) Recycling of the continental crust. Pure Appl. Geophys. (PAGEOPH) 128, 683-724.

McLennan, S.M. (1989) Archean sedimentary rocks and the Archean mantle. In: L.D. Ashwal (Ed.) Workshop on the Archean Mantle, LPI Tech. Rept. 89-05, p. 57-59, Lunar and Planetary Institute, Houston, TX.

McLennan, S.M. and Taylor, S.R. (1984) Archaean sedimentary rocks and their relation to the composition of the Archaean continental crust. In: (A. Kroner, G.N. Hanson and A.M. Goodwin (Eds.) Archaean Geochemistry, Springer-Verlag, Berlin, p. 47-72.

McLennan, S.M. and Taylor, S.R. (1988) Crustal evolution: Comments on "The Archean - Proterozoic transition: evidence from the geochemistry of metasedimentary rocks of Guyana and Montana" by A.K. Gibbs, C.W. Montgomery, P.A. O'Day and E.A. Erslev. Geochim. Cosmochim. Acta 52, 785-787.

McLennan, S.M. and Taylor, S.R. (1990) Sedimentary rocks and crustal evolution revisited: Tectonic setting and secular trends. J. Geol. (submitted).

McLennan, S.M., Fryer, B.J. and Young, G.M. (1979) Rare earth elements in Huronian (Lower Proterozoic) sedimentary rocks: Composition and evolution of the post-Kenoran upper crust. Geochim. Cosmochim. Acta 43, 375-388.

McLennan, S.M., Nance, W.B. and Taylor, S.R. (1980) Rare earth element - thorium correlations in sedimentary rocks, and the composition of the continental crust. Geochim. Cosmochim. Acta 44, 1833-1839.

McLennan, S.M., Taylor, S.R. and Eriksson, K.A. (1983a) Geochemistry of Archean shales from the Pilbara Supergroup, Western Australia. Geochim. Cosmochim. Acta 47, 1211-1222.

McLennan, S.M., Taylor, S.R. and Kroner, A. (1983b) Geochemical evolution of Archean shales from South Africa. I. The Swaziland and Pongola Supergroups. Precambrian Res. 22, 93-124.

McLennan, S.M., Taylor, S.R. and McGregor, V.R. (1984) Geochemistry of Archean metasedimentary rocks from West Greenland. Geochim. Cosmochim. Acta 48, 1-13.

McLennan, S.M., McCulloch, M.T., Taylor, S.R. and Maynard, J.B. (1989) Effects of sedimentary sorting on neodymium isotopes in deep-sea turbidites. Nature 337, 547-549.

McLennan, S.M., Taylor, S.R., McCulloch, M.T. and Maynard, J.B. (1990a) Geochemical and Nd-Sr isotopic composition of deep sea turbidites. Part I: Implications for crustal evolution. Geochim. Cosmochim. Acta (submitted).

McLennan, S.M., Taylor, S.R., McCulloch, M.T. and Maynard, J.B. (1990b) Geochemical and Nd-Sr isotopic composition of deep sea turbidites. Part II: Sedimentological and plate tectonic controls. Geochim. Cosmochim. Acta (submitted).

Michard, A. (1989) Rare earth element systematics in hydrothermal fluids. Geochim. Cosmochim. Acta 53, 745-750.

Michard, A. and Alberede, F. (1986) The REE content of some hydrothermal fluids. Chem. Geol. 55, 51-60.

Michard, A., Beaucaire, C. and Michard, G. (1987) Uranium and rare earth elements in CO_2-rich waters from Vals-les-Bains (France). Geochim. Cosmochim. Acta 51, 901-909.

Middelburg, J.J., van der Weijden, C.H. and Woittiez, J.R.W. (1988) Chemical processes affecting the mobility of major, minor and trace elements during weathering of granitic rocks. Chem. Geol. 68, 253-273.

Miller, R.G., O'Nions, R.K., Hamilton, P.J. and Welin, E. (1986) Crustal residence ages of clastic sediments, orogeny and continental evolution. Chem. Geol. 57, 87-99.

Minami, E. (1935) Gehelte an seltenen Erden in europaishen und japaneschen Tonschiefern. Nach. Gess. Wiss. Goettingen, 2, Math-Physik KL.IV, 1, 155-170.

Moorbath, S. (1976) Age and isotope constraints for the evolution of Archaean crust. In: B.F. Windley (Ed.) The Early History of the Earth, Wiley, London, p. 351-360.

Morgan, P. (1984) The thermal structure and thermal evolution of the continental lithosphere. Phys. Chem. Earth 15, 107-193.

Morton, A.C. (1985) Heavy minerals in provenance studies. In: G.G. Zuffa (Ed.) Provenance of Arenites. D. Reidel, Dordrecht, The Netherlands, p. 249-277.

Nance, W.B. and Taylor, S.R. (1976) Rare earth patterns and crustal evolution - I. Australian post-Archean sedimentary rocks. Geochim. Cosmochim. Acta 40, 1539-1551.

Nance, W.B. and Taylor, S.R. (1977) Rare earth element patterns and crustal evolution - II. Archean sedimentary rocks from Kalgoorlie, Australia. Geochim. Cosmochim. Acta 41, 225-231.

Nathan, S. (1976) Geochemistry of the Greenland Group (Early Ordovician), New Zealand. New Zealand J. Geol. Geophys. 19, 683-706.

Nelson, B.K. and DePaolo, D.J. (1988) Comparison of isotopic and petrographic provenance indicators in sediments from Tertiary continental basins of New Mexico. J. Sed. Pet. 58, 348-357.

Nesbitt, H.W. (1979) Mobility and fractionation of rare earth elements during weathering of a granodiorite. Nature 279, 206-210.

Nesbitt, H. W., Markovics, G. and Price, R.C. (1980) Chemical processes affecting alkalis and alkaline earths during continental weathering. Geochim. Cosmochim. Acta 44, 1659-1666.

Noddack, I. (1935) Die Haufrgdeiten du seltenen Erden in meteoriten. Z. Anorg. Allg. Chem. 225, 337-364.

Palmer, M.R. (1985) Rare earth elements in foraminifera tests. Earth Planet. Sci. Lett. 73, 285-298.

Patchett, P.J., White, W.M., Feldmann, H., Kielinczuk, S. and Hofmann, A.W. (1984) Hafnium/rare earth element fractionation in the sedimentary system and crustal recycling into the Earth's mantle. Earth Planer. Sci. Lett. 69, 365-378.

Pearce, J.A. and Cann, J.R. (1973) Tectonic setting of basic volcanic rocks determined using trace element analyses. Earth Planet. Sci. Lett. 19, 290-300.

Piper, D.Z. (1974) Rare earth elements in the sedimentary cycle: a summary. Chem. Geol. 14, 285-304.

Potter, P.E. (1978a) Significance and origin of big rivers. J. Geol. 86, 13-33.

Potter, P.E. (1978b) Petrology and chemistry of modern big rivers. J. Geol. 86, 423-449.

Puddephatt, R.J. (1972) The Periodic Table of the Elements, Oxford Univ. Press, Oxford, 84 p.

Ronov, A.B., Balashov, Yu.A. and Migdisov, A.A. (1967) Geochemistry of the rare earths in the sedimentary cycle. Geochem. Int. 4, 1-17.

Ross, J.E. and Aller, L.H. (1976) The chemical composition of the Sun. Science 191, 1223-1229.

Rudnick, R.L., McLennan, S.M. and Taylor, S.R. (1985) Large ion lithophile elements in rocks from high-pressure granulite facies terrains. Geochim. Cosmochim. Acta 49, 1645-1655.

Sawyer, E.W. (1986) The influence of source rock type, chemical weathering and sorting on the geochemistry of clastic sediments from the Quetico Metasedimentary Belt, Superior Province, Canada. Chem. Geol. 55, 77-95.

Schmitt, R.A., Smith, R.H. and Olehy, D.A. (1964) Rare-earth, yttrium and scandium abundances in meteoritic and terrestrial matter - II. Geochim. Cosmochim. Acta 28, 67-86.

Schnetzler, C.C. and Philpotts, J.A. (1967) Has the earth's crust changed with time? Rare-earth abundances in ancient sediments. Goddard Space Flight Ctr. Rept. X-641-67-237. Greenbelt, MD. 12 p.

Schnetzler, C.C. and Philpotts, J.A. (1970) Partition coefficients of rare-earth elements between igneous matrix material and rock-forming mineral phenocrysts- II. Geochim. Cosmochim. Acta 34, 331-340.

Shannon, R.D. (1976) Revised effective ionic radii and systematic studies of interatomic distances in halides and chalcogenides. Acta Cryst. A32, 751-767.

Shaw, D.M. (1967) U, Th and K in the Canadian Precambrian shield and possible mantle compositions. Geochim. Cosmochim. Acta 31, 1111-1113.

Shaw, D.M., Reilly, G.A., Muysson, J.R., Pattenden, G.E. and Campbell, F.E. (1967) An estimate of the chemical composition of the Canadian Precambrian shield. Can. J. Earth Sci. 4, 829-853.

Shaw, D.M., Dostal, J. and Reays, R.R. (1976) Additional estimates of continental surface Precambrian shield composition in Canada. Geochim. Cosmochim. Acta 40, 73-83.

Sholkovitz, E.R. (1988) Rare earth elements in the sediments of the North Atlantic Ocean, Amazon Delta, and East China Sea: reinterpretation of terrigenous input patterns to the oceans. Am. J. Sci. 288, 236-281.

Stern, R.A., Hanson, G.N. and Shirey, S.B. (1989) Petrogenesis of mantle-derived, LILE-enriched Archean monzodiorites and trachyandesites (sanukitoids) in southwestern Superior Province. Can. J. Earth Sci. (in press).

Suess, H.E. and Urey, H.C. (1956) Abundances of the elements. Rev. Mod. Phys. 28, 53-74.

Sverjensky, D.M. (1984) Europium redox equilibrium in aqueous solutions. Earth Planet. Sci. Lett. 67, 70-78.

Taylor, S.R. (1960) The abundance of the rare earth elements in relation to their origin. Geochim. Cosmochim. Acta 19, 100-112.

Taylor, S.R. (1962) Meteoritic and terrestrial rare earth abundance patterns. Geochim. Cosmochim. Acta 26, 81-88.

Taylor, S.R. (1964) Abundance of chemical elements in the continental crust: a new table. Geochim. Cosmochim. Acta 28, 1273-1285.

Taylor, S.R. (1977) Island arc models and the composition of the continental crust. Am. Geophys. Union Maurice Ewing Series 1, 325-335.

200

Taylor, S.R. (1979) Chemical composition and evolution of the continental crust: the rare earth element evidence. In: M.W. McElhinny (Ed.) The Earth: Its Origin, Structure and Evolution. Academic Press, New York, p. 353-376.

Taylor, S.R. and McLennan, S.M. (1981) The composition and evolution of the continental crust: rare earth element evidence from sedimentary rocks. Phil. Trans. Roy. Soc. London A301, 381-399.

Taylor, S.R. and McLennan, S.M. (1985) The Continental Crust: Its Composition and Evolution. Blackwell, Oxford, 312 p.

Taylor, S.R. and McLennan, S.M. (1988) The significance of the rare earths in geochemistry and cosmochemistry. In: K.A. Gschneidner, Jr. and L. Eyring (Eds.), Handbook on the Physics and Chemistry of Rare Earths, Vol. 11, 485-578, Elsevier, Amsterdam.

Taylor, S.R., McLennan, S.M. and McCulloch, M.T. (1983) Geochemistry of loess, continental crustal composition and crustal model ages. Geochim. Cosmochim. Acta 47, 1897-1905.

Taylor, S.R., Rudnick, R.L., McLennan, S.M. and Eriksson, K.A. (1986) Rare earth element patterns in Archean high-grade metasediments and their tectonic significance. Geochim. Cosmochim. Acta 50, 2267-2279.

Topp, S.E., Salbu, B., Roaldset, E. and Jorgensen, P. (1984) Vertical distribution of trace elements in laterite soil (Suriname). Chem. Geol. 47, 159-174.

Ujike, O. (1984) Chemical composition of Archean Pontiac metasediments, southwestern Abitibi Belt, Superior Province. Can. J. Earth Sci. 21, 727-731.

Veizer, J. and Jansen, S.L. (1979) Basement and sedimentary recycling and continental evolution. J. Geol. 87, 341-370.

Veizer, J. and Jansen, S.L. (1985) Basement and sedimentary recycling - 2: time dimension to global tectonics. J. Geol. 93, 625-643.

Velbel, M.A. (1985) Mineralogically mature sandstones in accretionary prisms. J. Sed. Pet. 55, 685-690.

White, W.M., Dupre, B. and Vidal, P. (1985) Isotope and trace element geochemistry of sediments from the Barbados Ridge-Demerara Plain region, Atlantic Ocean. Geochim. Cosmochim. Acta 49, 1875-1886.

Wildeman, T.R. and Condie, K.C. (1973) Rare earths in Archean graywackes from Wyoming and from the Fig Tree Group, South Africa. Geochim. Cosmochim. Acta 37, 439-453.

Wildeman, T.R. and Haskin, L. (1965) Rare-earth elements in ocean sediments. J. Geophys. Res. 70, 2905-2910.

Wildeman, T.R. and Haskin, L.A. (1973) Rare earths in Precambrian sediments. Geochim. Cosmochim. Acta 37, 419-439

Windley, B.F. (1984) The Evolving Continents, 2nd Ed. Wiley, London, 399 p.

Worsley, T.R., Nance, R.D. and Moody, J.B. (1986) Tectonic cycles and the history of the Earth's biogeochemical and paleoceanographic record. Paleoceanography 1, 233-263.

Woyski, M.M. and Harris, R.E. (1963) The rare earths and rare-earth compounds. In: I.M. Kolthoff and P.J. Elving (Eds.) Tritise on Analytical Chemistry. Part II - Analytical Chemistry of the Elements. Vol. 8, 1-14. Interscience, New York.

Wronkiewicz, D.J. and Condie, K.C. (1987) Geochemistry of Archean shales from the Witwatersrand Supergroup, South Africa: Source-area weathering and provenance. Geochim. Cosmochim. Acta 51, 2401-2416.

Wronkiewicz, D.J. and Condie, K.C. (1989) Geochemistry and provenance of sediments from the Pongola Supergroup, South Africa: Evidence for a 3.0 Ga-old continental craton. Geochim. Cosmochim. Acta 53, 1537-1549.

Wyllie, P.J. (1977) Crustal anatexis: an experimental review. Tectonophys. 43, 41-71.

Chapter 8   D. G. Brookins

AQUEOUS GEOCHEMISTRY OF RARE EARTH ELEMENTS

INTRODUCTION

Rare earth elements, and especially the lanthanides (La-Lu), have received a great deal of attention over the last few decades because their unique chemical behavior allows them to be used for tracers of a wide variety of geochemical processes. While the use of lanthanide distribution and abundance is well known for petrogenesis theory and application--there are a number of papers in this volume that address various aspects of petrogenesis--the aqueous behavior of the lanthanides has not received much attention. At the same time, use of $^{143}Nd/^{144}Nd$ isotopy, along with Sm-Nd dating, has allowed many problems of rock genesis to be addressed. Relatively few efforts have addressed the problems of aqueous behavior of lanthanides.

In this brief report I shall address the chemical characteristics and behavior of the lanthanides, including the trivalent species and Ce^{4+} and Eu^{2+}, and other chemical elements, especially Y and Sc. I shall review the lanthanide aqueous geochemistry and emphasize trivalent species with additional commentary on the others. This section will include discussion of species in solution, "anomalous" behavior of some lanthanides, and Eh-pH diagrams as appropriate. I shall then briefly discuss the behavior of the lanthanides in sea waters and other waters. In other parts of this volume the behavior of lanthanides in sediments and sedimentary rocks is discussed by McLennan, and Patchett writes about isotopes of Nd. My effort is to focus on how bottom sediments, continental input, and other sources affect the lanthanides in sea water with some discussion on Nd isotopics as well.

To benefit the reader, this review provides an up to date compilation of critical thermochemical data for the lanthanides as well as a moderate overview of appropriate literature. A comprehensive treatment of thermochemical data for the rare earths is given by Wood (in press). Finally, this review focuses on the inorganic aqueous geochemistry of the rare earth elements primarily because there is sparse information available on the abundance of organic rare earth complexes. Much of the behavior of the rare earths in common water systems can be approached by use of the available inorganic data and models.

THE TRIVALENT LANTHANIDES (Ln III)

With the exception of Ce(IV) and Eu(II), all the lanthanides occur in nature as Ln(III) species. These Ln(III) ions are characterized by large ionic radii which, with their high valence, tends to segregate them from other M(III) ions (Fig. 1). In the series Y-La-Ac, there is the expected increase in ionic radii from 0.9 A (Y^{3+}) to 1.03 A (La^{3+}) to 1.12 A (Ac^{3+}), but due to the lanthanide contraction, the systematic filling of f orbitals occurs while leaving 5d, 6s, and 6p orbitals empty (Vander Sluis and Nugent, 1974; Baes and Mesmer, 1976; Jorgeson, 1988). Hence the ionic radii of the lanthanides decrease from La^{3+} (1.032 A) to Lu^{3+} (0.861 A) (CN 6). The ionic radii of the lanthanides as Ln(III) species are shown in Table 1 along with values for other relevant ions (From Shannon, 1976).

As Thompson (1979) has pointed out, the Ln(III) ions are essentially spherical and form complexes, very much like alkaline and alkaline earth ions. The partially filled f orbitals are effectively shielded from most chemical bonding, hence the bonding characteristics of the Ln(III) ions are largely ionic. Thompson (1979) notes that this shielding is so effective that crystal field effects in the lanthanides are on the order of 100 cm^{-1} as opposed to 30,000 cm^{-1} for many d-transition elements. Yttrium (III), however, with an inert gas

202

Table 1a. Ionic radii for geochemically
important species of the rare earths

Ion	Ionic radius (A)	
	CN6	CN8
Sc^{3+}	0.745	0.870
Y^{3+}	0.900	1.019
La^{3+}	1.032	1.160
Ce^{3+}	1.011	1.143
Ce^{4+}	0.870	0.970
Pr^{3+}	0.990	1.126
Nd^{3+}	0.983	1.109
Sm^{3+}	0.958	1.079
Eu^{3+}	0.947	1.066
Eu^{2+}	1.170	1.250
Gd^{3+}	0.938	1.053
Tb^{3+}	0.923	1.040
Dy^{3+}	0.912	1.027
Ho^{3+}	0.901	1.015
Er^{3+}	0.890	1.004
Tm^{3+}	0.880	0.994
Yb^{3+}	0.868	0.985
Lu^{3+}	0.861	0.977

Table 1b. Ionic radii for other geochemically
important species

Ion	Ionic radius (A)	CN	Ion	Ionic radius (A)	CN
Cs^+	1.88	12	Nb^{3+}	0.72	6
Rb^+	1.72	12	Nb^{5+}	0.64	6
K^+	1.64	12	Cr^{3+}	0.62	6
Ba^{2+}	1.61	12	V^{3+}	0.64	6
			Fe^{3+}	0.65	6
Pb^{2+}	1.29	8	Ti^{4+}	0.61	6
Sr^{2+}	1.26	8	Ni^{2+}	0.69	6
Eu^{2+}	1.25	8	Co^{2+}	0.75	6
Na^+	1.18	8	Cu^{2+}	0.73	6
Ca^{2+}	1.12	8	Fe^{2+}	0.78	6
Th^{4+}	1.05	8	Mn^{2+}	0.83	6
U^{4+}	1.00	8	Mg^{2+}	0.72	6
Ce^{4+}	0.97	8			
Zr^{4+}	0.84	8			
Hf^{4+}	0.83	8			

Notes: Ionic radii are from Shannon (1976) in angstroms
CN = coordination number

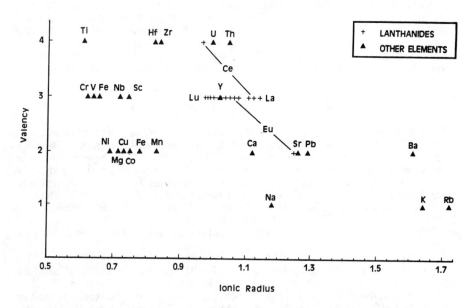

Figure 1. Relationship between valency and ionic radius for the rare earth elments (lanthanides and Sc and
Y) and other elements of interest. The trivalent lanthanides and Y are isolated from most other elements
shown. The solid lines connect Ce(III) with Ce(IV) (upper line) and Eu(III) with Eu(II) (lower line). See
text for details. This figure is reproduced with the kind permission of Elsevier Science Publishers.

structure, is much more prone to covalency. The Ln(III) ions possess low polarizability due to their high z/r ratio. Further, the Ln(III) ions, and Y(III) are normally found in solids in CN (coordination number) 8 whereas the smaller Sc(III) is found in CN 6. The lanthanides and Y do, however, tend to bind the water molecules, and the most common that forms in water is $Ln(H_2O)_6^{3+}$ although $Sc(H_2O)_5(OH)^{2+}$ and other polymers are also known (Leskela and Niinisto, 1986; Komissarova, 1980; Biedermann and Ferri, 1981).

Types of complexes in solution

Moeller et al. (1965) and Cantrell and Byrne (1987) have tabulated the dominant types of trivalent rare earth element complexes in solution. These are given in Table 2.

There are subtle differences, however, even within the lanthanide complexes. Habenschuss and Spedding (1974) established that the coordination of water molecules about the inner lanthanide sphere is nine from La^{3+} through Nd^{3+}, and this number is eight from Tb^{3+} through Lu^{3+}. In the Eu^{3+} to Gd^{3+} range, this coordination of water molecules about the inner sphere is a mixture of 8 and 9, with Eu more 9-prevalent and Gd more 8-prevalent. This change in water molecule coordination may have other implications for lanthanide behavior in natural waters (see below; gadolinium-terbium anomalies?).

The lanthanides are hard acids in the terminology of Pearson (1963), which means that they will bond preferentially with hard bases which contain oxygen as donor atoms. They will, as pointed out by Thompson (1979), bond more weakly with soft bases such as those with donor S or P. Thus the dominant lanthanide complexes contain ligands with at least one donor oxygen atom. In this the lanthanides are quite different from the M(III) ions of the d-transition series elements. Kinetically, the rare earth elements react quickly to form complexes (Reidler and Silber, 1973; Thompson, 1979).

In response to more vigorous research, better equipment, and refined theory, tremendous progress has made in studies of the Ln(III) species. Below, I shall discuss the available data for Ln complexes with water, dissolved carbon dioxide, halides, sulfate, and phosphate. The important data for many of the complexes are presented in Tables 3 and 4.

Table 2. Lanthanide Complexes in Solution*

Inorganic Ligands	Organic Ligands
LnX^{2+} (X = F, Cl, Br, I)	$Ln(C_2H_3O_2)_n^{(3-n)+}$ (n = 1-3)
LnX_2^+ (X = Cl, Br)	$Ln(HOCH_2COO)_n^{(3-n)+}$ (n = 1-4)
$LnClO_4^{2+}$	$Ln(EDTA)^-$
$LnNO_3^{2+}$	$Ln(NTA)_n^{(3-n)+}$ (n = 1,2)
$Ln(P_2O_7)_n^{(3-4n)+}$ (n = 1,2)	$Ln(HEDTA)(IMDA)^{2-}$
$Ln(SO_4)^+$	$Ln(HEDTA)(OH)^-$
$Ln(SO_4)_2^-$	$Ln(b-diketone)_n^{(3-n)+}$ (n = 1-3)
$Ln(CO_3)^+$	$Ln(PDC)_n^{(3-2n)+}$ (n = 1-3)
$Ln(CO_3)_2^-$	$Ln(NO_3)_3 \cdot 3TBP$

* Modified from Moeller and others (1965), Thompson (1979, Turner and others (1981), and Cantrell and Byrne (1987).

Table 3 – Thermochemical Data for the Lanthanides, Sc, and Y

Species	log Q°	log Q*(I)
La (III)		
La(OH)²⁺	-8.50	—
La(OH)₂⁺	-17.40	-17.40 (0.0)
La(OH)₃	-27.40	-27.40 (0.0)
La(OH)₄⁻	-38.80	-38.80 (0.0)
LaF²⁺	+3.60	—
LaCl²⁺	+0.80	-0.10 (1.0)
LaCl₂⁺	-0.29	-0.60 (4.0)
La(SO₄)⁺	+3.64	—
La(SO₄)₂⁻	+5.29	+2.46 (1.0)
La(CO₃)⁺	+6.16	+5.31 (0.7)
La(CO₃)₂⁻	—	+9.12 (0.7)
Ce (III)		
Ce(OH)²⁺	-8.30	-8.30 (0.0)
Ce(OH)₂⁺	-17.10	-17.10 (0.0)
Ce(OH)₃	-26.80	-26.80 (0.0)
Ce(OH)₄⁻	-37.60	-37.60 (0.0)
CeF²⁺	+4.00	—
CeCl²⁺	+0.80	-0.10 (1.0)
CeCl₂⁺	+1.19	-0.50 (1.0)
Ce(SO₄)⁺	+3.59	—
Ce(SO₄)₂⁻	+5.27	+2.90 (0.5)
Ce(CO₃)⁺	+6.78	+5.42 (0.7)
Ce(CO₃)₂⁻	—	+9.29 (0.7)
Pr (III)		
Pr(OH)²⁺	-8.10	-8.10 (0.0)
Pr(OH)₂⁺	-17.00	-17.00 (0.0)
Pr(OH)₃	-26.60	-26.60 (0.0)
Pr(OH)₄⁻	-37.20	-37.20 (0.0)
PrF²⁺	+3.91	+3.01 (1.0)
PrCl²⁺	+0.80	-0.10 (1.0)
PrCl₂⁺	-0.29	-0.60 (4.0)
Pr(SO₄)⁺	+3.62	+3.62 (0.0)
Pr(SO₄)₂⁻	+4.92	+4.92 (0.0)
Pr(CO₃)⁺	+6.62	+6.62 (0.0)

Species	log Q°	log Q*(I)
Nd (III)		
Nd(OH)²⁺	-8.00	-8.00 (0.0)
Nd(OH)₂⁺	-16.90	-16.90 (0.0)
Nd(OH)₃	-26.50	-26.50 (0.0)
Nd(OH)₄⁻	-37.10	-37.10 (0.0)
NdF²⁺	+3.99	+3.09 (1.0)
NdCl²⁺	+0.80	-0.10 (1.0)
NdCl₂⁺	-0.29	-0.60 (4.0)
Nd(SO₄)⁺	+3.64	+3.64 (0.0)
Nd(SO₄)₂⁻	+5.10	+5.10 (0.0)
Nd(CO₃)⁺	+6.72	+6.72 (0.0)
Sm (III)		
Sm(OH)²⁺	-7.90	-7.90 (0.0)
Sm(OH)₂⁺	-16.60	-16.60 (0.0)
Sm(OH)₃	-25.80	-25.80 (0.0)
Sm(OH)₄⁻	-35.70	-35.70 (0.0)
SmF²⁺	+4.02	+3.12 (1.0)
SmCl²⁺	+0.80	-0.10 (1.0)
SmCl₂⁺	-0.29	-0.60 (4.0)
Sm(SO₄)⁺	+3.67	+3.67 (0.0)
Sm(SO₄)₂⁻	+5.20	+5.20 (0.0)
Sm(CO₃)⁺	+6.86	+6.86 (0.0)
Eu (III)		
Eu(OH)²⁺	-7.80	-7.80 (0.0)
Eu(OH)₂⁺	-16.60	-16.60 (0.0)
Eu(OH)₃	-25.60	-25.60 (0.0)
Eu(OH)₄⁻	-35.30	-35.30 (0.0)
EuF²⁺	+4.09	+3.19 (1.0)
EuCl²⁺	+0.80	-0.10 (1.0)
EuCl₂⁺	+0.99	-0.70 (1.0)
Eu(SO₄)⁺	+3.59	—
Eu(SO₄)₂⁻	+5.41	+5.86
Eu(CO3)+	+6.83	+10.1

Species	log Q°	log Q*(I)
Gd (III)		
Gd(OH)²⁺	-8.00	-8.00 (0.0)
Gd(OH)₂⁺	-16.40	-16.40 (0.0)
Gd(OH)₃	-25.20	-25.20 (0.0)
Gd(OH)₄⁻	-34.40	-34.40 (0.0)
GdF²⁺	+4.30	—
GdCl²⁺	+0.80	-0.10 (1.0)
GdCl₂⁺	-0.29	-0.60 (4.0)
Gd(SO₄)⁺	+3.66	—
Gd(SO₄)₂⁻	+5.21	—
Gd(CO₃)⁺	+7.29	+7.29 (0.0)
Tb (III)		
Tb(OH)²⁺	-7.90	-7.90 (0.0)
Tb(OH)₂⁺	-16.30	-16.30 (0.0)
Tb(OH)₃	-25.10	-25.10 (0.0)
Tb(OH)₄⁻	-34.30	-34.30 (0.0)
TbF²⁺	+4.32	+3.42 (1.0)
TbCl²⁺	+0.80	-0.10 (1.0)
TbCl₂⁺	-0.29	-0.60 (4.0)
Tb(SO₄)⁺	+3.64	+3.64 (0.0)
Tb(SO₄)₂⁻	-5.15	+5.15 (0.0)
Tb(CO₃)⁺	+6.93	+6.93 (0.0)
Dy (III)		
Dy(OH)²⁺	-8.00	-8.00 (0.0)
Dy(OH)₂⁺	-16.20	-16.20 (0.0)
Dy(OH)₃	-24.70	-24.70 (0.0)
Dy(OH)₄⁻	-33.50	-33.50 (0.0)
DyF²⁺	+4.36	+3.46 (1.0)
DyCl²⁺	+0.80	-0.10 (1.0)
DyCl₂⁺	-0.29	-0.60 (4.0)
Dy(SO₄)⁺	+3.62	+3.62 (0.0)
Dy(SO₄)₂⁻	+4.80	+4.80 (0.0)
Dy(CO₃)⁺	+7.17	+7.17 (0.0)

Table 3 continued

Species	log Q^0	log $Q'(I)$
Ho (III)		
$Ho(OH)^{2+}$	-8.00	-8.00 (0.0)
$Ho(OH)_2^+$	-16.10	-16.10 (0.0)
$Ho(OH)_3$	-24.60	-24.60 (0.0)
$Ho(OH)_4^-$	-33.40	-33.40 (0.0)
HoF^{2+}	+4.42	+3.52 (1.0)
$HoCl^{2+}$	+0.80	-0.10 (1.0)
$HoCl_2^+$	-0.29	-0.60 (4.0)
$Ho(SO_4)^+$	+3.59	+3.59 (0.0)
$Ho(SO_4)_2^-$	+4.90	+4.90 (0.0)
$Ho(CO_3)^+$	+7.23	+7.23 (0.0)
Er (III)		
$Er(OH)^{2+}$	-7.90	-7.90 (0.0)
$Er(OH)_2^+$	-15.90	-15.90 (0.0)
$Er(OH)_3$	-24.20	-24.20 (0.0)
$Er(OH)_4^-$	-32.60	-32.60 (0.0)
ErF^{2+}	+4.44	+3.54 (1.0)
$ErCl^{2+}$	+0.80	-0.10 (1.0)
$ErCl_2^+$	-0.29	-0.60 (4.0)
$Er(SO_4)^+$	+3.59	+3.59 (0.0)
$Er(SO_4)_2^-$	+5.10	+5.10 (0.0)
$Er(CO_3)^+$	+7.32	+7.32 (0.0)
Tm (III)		
$Tm(OH)^{2+}$	-7.70	-7.70 (0.0)
$Tm(OH)_2^+$	-15.90	-15.90 (0.0)
$Tm(OH)_3$	-24.10	-24.40 (0.0)
$Tm(OH)_4^-$	-32.60	-32.60 (0.0)
TmF^{2+}	+4.46	+3.56 (1.0)
$TmCl^{2+}$	+0.80	-0.10 (1.0)
$TmCl_2^+$	-0.29	-0.60 (4.0)
$Tm(SO_4)^+$	+3.59	+3.59 (0.0)
$Tm(SO_4)_2^-$	+5.14	+5.14 (0.0)
$Tm(CO_3)^+$	+7.08	+7.08 (0.0)

Species	log Q^0	log $Q'(I)$
Yb (III)		
$Yb(OH)^{2+}$	-7.70	-7.70 (0.0)
$Yb(OH)_2^+$	-15.80	-15.80 (0.0)
$Yb(OH)_3$	-24.10	-24.10 (0.0)
$Yb(OH)_4^-$	-32.70	-32.70 (0.0)
YbF^{2+}	+4.48	+3.58 (1.0)
$YbCl^{2+}$	+0.70	-0.20 (1.0)
$YbCl_2^+$	-0.29	-0.60 (4.0)
$Yb(SO_4)^+$	+3.58	+3.58 (0.0)
$Yb(SO_4)_2^-$	+5.20	+5.20 (0.0)
$Yb(CO_3)^+$	+7.60	+7.60 (0.0)
Lu (III)		
$Lu(OH)^{2+}$	-7.60	—
$Lu(OH)_2^+$	-15.70	-15.70 (0.0)
$Lu(OH)_3$	-23.70	-23.70 (0.0)
$Lu(OH)_4^-$	-31.80	-31.80 (0.0)
LuF^{2+}	+4.51	+3.61 (1.0)
$LuCl^{2+}$	+0.50	-0.40 (1.0)
$LuCl_2^+$	-0.29	-0.60 (4.0)
$Lu(SO_4)^+$	+3.52	—
$Lu(SO_4)_2^-$	+5.30	—
$Lu(CO_3)^+$	+7.57	+7.57 (0.0)
Sc (III)		
$Sc(OH)^{2+}$	-4.30	—
$Sc(OH)_2^+$	-9.70	—
$Sc(OH)_3$	-16.10	—
$Sc(OH)_4^-$	-26.00	—
ScF^{2+}	+7.03	+6.18 (0.5)
$ScCl^{2+}$	+0.92	+0.04 (0.7)
$ScCl_2^+$	+1.57	-0.10 (0.7)
$Sc(SO_4)^+$	+4.40	+2.59 (0.5)
$Sc(SO_4)_2^-$	+6.33	+3.96 (0.5)
$Sc(CO_3)^+$	+10.10	+10.10 (0.0)

Species	log Q^0	log $Q'(I)$
Y (III)		
$Y(OH)^{2+}$	-7.70	—
$Y(OH)_2^+$	-16.40	—
$Y(OH)_3$	-26.00	—
$Y(OH)_4^-$	-36.50	—
YF^{2+}	+4.80	—
YCl^{2+}	+0.80	-0.10 (1.0)
$Y(SO_4)^+$	+3.47	+3.47 (0.0)
$Y(SO_4)_2^-$	+5.30	+5.30 (0.0)
$Y(CO_3)^+$	+6.94	+5.98 (0.7)
$Y(CO_3)_2^-$	+ ____	+10.3 (0.7)

Q^0 = total stability constant for I = 0
$Q'(I)$ = total stability constant for I > 0

Sources: Turner and others (1981), Baes and Mesmer (1976), Smith and Martell (1976), Cantrell and Byrne (1987), deBaar and others (1988).

Hydrolysis products

The Ln(III) species and Y^{3+} do not readily hydrolyze in contrast to Sc^{3+} which, because of its smaller ionic radius and larger polarizability easily forms hydrolysis products. The values for formation of $Ln(OH)^{2+}$ range from log $Q^0 = -8.50$ (La) to -7.80 (Eu), then from -8.00 (Gd) to -7.60 (Lu) (Q^0 = stability constant at Ionic Strength, I, = 0). All of these values are significantly less than the log Q^0 value for $Sc(OH)^{2+}$ of -4.30. Yttrium ($Y[OH]^{2+}$) yields a log Q^0 value of -7.70, identical to values for other heavy Ln (Ho, Er) species.

The stepwise formational constants for $Ln(OH)_2^+$, $Ln(OH)_3(aq)$, and $Ln(OH)_4^-$ decrease significantly. Baes and Mesmer (1976) point out that at acidic to neutral pH media the dominant form of Ln(III) ions in water may be simple M^{3+} ions. Wood (in press) adds that the stability constants for all Ln hydrolysis products are uncertain.

Data are also given in Table 3 for total formational constants at ionic strengths greater than zero. The right column of Table 3 is taken from the excellent compilation of Turner et al. (1981) and from Cantrell and Byrne (1987). The data support the systematic behavior of all the Ln(III) ions. Due to the lack of pronounced hydrolysis, simple Ln^{3+} ions will be assumed for construction of the Eh-pH diagrams to follow (see section later in this paper). Not shown are poorly obtained data for formation of dimers and some other polymers.

Phosphate complexes

Phosphate complexes of the lanthanides have not received the attention they deserve. According to Krauskopf (1977), total dissolved P in sea water is on the order of 0.6 ppm and Berner and Berner (1987)) state river waters average 0.025 ppm. These are large values when compared to the average total lanthanide concentrations reported for sea water and river waters (Elderfield and Greaves, 1982; Klinkhammer et al., 1983; Martin et al., 1976; Taylor and McLennan, 1988). Typically, Ln values in ocean water range from about 1 ppt (parts per trillion) for surface ocean waters to 4-6.5 ppt for 2500 m deep ocean waters; Ln values for some river waters are about 200 ppt.

Under calculated surface water conditions, phosphate complexes of the type $Ln(P_2O_7)^{2+}$ appear to be quite stable. For example, Herman and Langmuir (1978) report that phosphate complexes may, under certain pH and other conditions, transport Th and perhaps other actinides. This may also may apply to the lanthanides. Alfonin and Pechurova (1987) report that at acidic pH the stable form of phosphate complexes is $LnH_2PO_4^+$, and Wood (in press) cites evidence for a $LnHPO_4^+$ complex above pH = 4. I have calculated (unpub.) that in the absence of other dissolved ligands the $Ln(P_2O_7)^{2+}$ complex will be more important than Ln-hydrolysis products over the neutral pH range in natural waters. Thermodynamic data for $Ln(P_2O_7)^{2+}$ species are given in Table 4.

Carbonate complexes

Cantrell and Byrne (1987) have conducted an extensive study of the importance of Ln-carbonate complexes for sea water conditions and have determined stability constants for carbonate and oxalate of trivalent Ce, Eu, and Yb; these are incorporated into Table 3. They present convincing evidence and data for the importance of a $Ln(CO_3)^+$ species and $Ln(CO_3)_2^-$ species. The former is important in the pH range of the oceans. From the results and equations obtained, the stability constants for the entire lanthanide series were estimated. Cantrell and Byrne's (1987) study reconfirm the usefulness of the oxalate-carbonate similarities in parallel linear arrays of stability constants for oxalates and carbonates as advocated by the geochemical work by Langmuir (1979). Figure 2 shows the calculation by Cantrell and Byrne (1987) of total carbonate complexes of the lanthanides and the role of $Ln(CO_3)^+$ and $Ln(CO_3)_2^-$ where carbonate complexes constitute 86% of total metal in solution for La to 98% for Lu. Their work also shows that the $Ln(CO_3)+$ is important for the light lanthanides and the $Ln(CO_3)_2^-$ complex for the heavier lanthanides

Table 4. Thermodynamic data[a] [G$_f^0$(kcal/gfw)] for the lanthanides and yttrium A

a) M(III) species

Element	Ln³⁺ (aq)	Ln(OH)₃ (c)[b]	Ln₂O₃ (c)	Ln₂(CO₃)₃ (c)	Ln(OH)₄⁻ (aq)	Ln₂(P₂O₇)⁺²
La	-163.4	(-305.8)	-407.7	-750.9	-	-
Ce	-160.6	(-303.6)	-407.8	(-744.1	-	-
Pr	-162.3	-307.1	-----	(-747.1)	-	-
Nd	-160.5	(-305.2)	-411.3	-741.4	-336.7	-
Sm	-159.3	(-306.9)	-414.6	-741.4	-	-804.8
Eu	-137.2	-285.5	-372.2	(-697.3)	-	-760.7
Gd	-158.0	(-306.9	-----	(-738.9)	-338.0	-----
Tb	-155.8	(-303.4)	-----	(-734.5)	-	-798.3
Dy	-159.0	(-307.2)	-423.4	-739.8	-339.9	-804.8
Ho	-161.0	(-310.1)	-428.1	(-744.9)	-	-809.2
Er	-159.9	(-309.5)	-432.3	(-742.7	-342.2	-
Tm	-158.2	(-307.8)	-428.9	(-739.3)	-	-
Yb	-153.9	(-303.9)	-412.6	-729.1	-336.0	-
Lu	-150.0	(-300.3)	-427.6	(-722.9	_	-789.0
Y	-165.8	-308.6	-434.2	-752.4	-341.2	-

(b) Other rare earth element data

Species (state)	G$_f^0$(kcal/gfw)	Reference
Eu²⁺ (aq)	-129.10	Wagman et al. (1982)
CeO₂ (c)	-244.40	Wagman et al. (1982)
Ce⁴⁺ (aq)	-120.44	Wagman et al. (1982)

[a]See Brookins (1983 , 1988) for discussion of errors. Data in parentheses ( )
calculated by Brookins (1983), the remainder of the data are from Schumm et al.
(1973) and Smith and Martell (1976).
[b](c) = crystalline
[c](aq) = aqueous

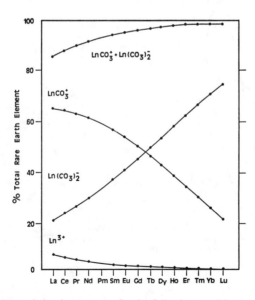

Figure 2. Lanthanide speciation in sea water for the following conditions: salinity = 35.1 o/oo, total
dissolved CO₃ = 1.4 x 10⁻⁴, 25°C, 1 atm total pressure. Only uncomplexed Ln³⁺ and carbonate-Ln
complexes are shown. This figure is reproduced in modified form from Canfield and Byrne (1987) by the
kind permission of Pergamon Press.

208

(Fig. 2). Other species in sea water are less important. For La, the uncomplexed ion La^{3+} makes up about 7% of the total dissolved metal, and the remainder is made of La complexes with hydroxide, sulfate, and halides (Cantrell and Byrne, 1987; deBaar et al., 1988). For Lu, the remaining 2% is comprised of 0.5% Lu^{3+} and the balance is complexes of hydroxide, sulfate, and halides (Cantrell and Byrne, 1987).

More recently, deBaar et al. (1988) studied the rare earth distribution in the Cariaco Trench anoxic waters. Their results for Ce(III) (Fig. 3) show that $CeCO_3^+$ is the dominant species along with smaller amounts of $Ce(CO_3)_2^-$ and Ce^{3+}. Also shown is that the sulfate and halide complexes are of minor importance even at very low pH relative to Ln^{3+}. But at very high pH, $Ln(OH)_3^0$ and $Ce(OH)_2^+$ are important (deBaar et al., 1988). This work shows that at sea water pH over 80% of dissolved Ce(III) is present as a carbonate complex.

Of further interest, Cantrell and Byrne (1987) found that the stability constants for the lanthanide carbonate complexes were not temperature sensitive and remained approximately constant over the range $-2°C$ to $+35°C$. This is important as it shows that the 25°C data (from many sources) may apply to the entire water column in the oceans.

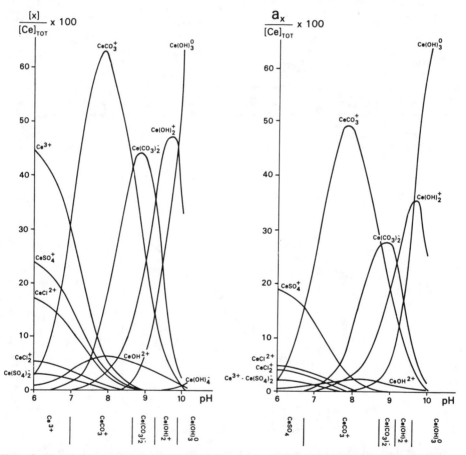

Figure 3. The relative concentration (left) and relative activity (right) of aqueous Ce species in sea water (salinity = 35 o/oo, 25°C, 1 atm total pressure) as a function of pH disregarding all Ce(IV) species. See text for details. This figure is reproduced from deBaar and others (1988) with the kind permission of Pergamon Press.

Turner et al. (1981) have summarized the available total stability constant (Q^0) data for rare earth element carbonate complexes of the type $Ln(CO_3)^+$. Certainly, it is reasonable to postulate carbonate complexes for Ln transport. It is well established that carbonate complexes are extremely important for transport of uranium in the formation of uranium ore deposits (Brookins, 1979, 1982; Langmuir, 1978; Taylor and McLennan, 1988; others). Since the lanthanides are commonly enriched and highly fractionated in most uranium deposits (ibid.; Della Valle, 1980), then carbonate complexes are attractive for Ln transport. One problem, of course, is that reduction of U(VI) to U(IV) is responsible for the U fixation, and these conditions of redox are fairly well established (Langmuir, 1978; Brookins, 1979, 1982, 1984), yet the lanthanides (except for Eu) are unaffected by redox in the same Eh-pH range. Further, the lanthanides are fixed not only in uranium minerals but also in clay minerals shown to be penecontemporaneous with the uranium mineralization. This is a complex problem. The generation of calcite often encountered with sandstone uranium ores in the United States contains isotopic C derived from organic sources (Brookins, 1982; Leventhal and Threlkeld, 1978). Were this calcite to contain C derived only from extrinsic marine sources, then its isotopic carbon should be reflected, but it is not.

Kosterin (1959) has argued for the existence of the $Ln(CO_3)_3^{3-}$ complex at very high total dissolved carbonates to explain lanthanide transport and this species has been presented as possible-to-probable by Fryer (1977) and Taylor and McLennan (1979). Yet there is uncertainty about the importance of this species.

In sea water and in river water, however, the lanthanide carbonate complexes are likely to be $Ln(CO_3)^+$ and $Ln(CO_3)_2^-$ especially at the higher pH ranges. At lower pH, Ln^{3+} will dominate. Under high total dissolved P conditions, the $Ln(P_2O_7)^{2+}$ ion may compete with the $Ln(CO_3)^+$ and $Ln(CO_3)_2^-$ species.

Halide complexes

Halide complexes of the lanthanides in solution are well known (see Table 3). The fluoro-complexes are important in the formation of Ln-rich fluorite deposits (Bilal et al., 1979; Bilal and Becker, 1979; Bilal and Kob, 1980; McLennan and Taylor, 1979; Ekambaram et al., 1986).

Log Q^0 values are given in Table 3 for halide complexes of the lanthanides (from Turner et al., 1981; Smith and Martell, 1976). Anhydrous halides of the lanthanides are easily prepared in the laboratory (Haschke, 1979). The non-hydroscopic LnF_3 compounds yield low stability constants in the range of 10^{-19} (La) to 10^{-15} (Lu). The $Ln(Cl,Br,I)$ halides, on the other hand, are all rapidly hydrolyzed or hydrated.

Aqueous complexes of the Ln halides are dominated by LnF^{2+}, $LnCl^{2+}$, and $LnCl_2^+$ species. The stability constants for LnF^{2+} species systematically increase from log $Q^0 = +3.60$ (La) to +4.51 (Lu), based on the data compiled by Turner et al. (1981). Bilal and Becker (1979) suggest a maximum at Tb for some LnF^{2+} complexes, but this is not supported by the data of Turner et al. (1981). The chloride complexes, $LnCl^{2+}$ and $LnCl_2^+$, show smaller stability constants of formation than the fluoride complexes. Even for chlorine in sea water, however, the lanthanide chlorides are not significant in the pH range 5 and above.

Under acidic, hydrothermal conditions, it is likely that some lanthanide transport by halide complexes will occur. A controlling factor may well be the competing effect of sulfate-lanthanide complexes that may be present in such systems (Jorgenson, 1979).

Complexes with total dissolved sulfur

Sulfides, selenides, and tellurides of the lanthanides are known in the laboratory (Flahout, 1979), and in the absence of other ligands, fields of Ln_2S_3 are observed.

However, under most natural conditions the lanthanides will not complex with S(-II) species (H_2S, HS^-, S_x^{2x-}, etc.) due to the strong affinity of the hard acid lanthanides for oxygen-donors from hard bases. Under very alkaline, sulfur-rich, carbonate-poor conditions, some heavy lanthanides may be incorporated into metal sulfides. It would be interesting to analyze natural pyrrhotites and study this reaction.

Dissolved S as SO_4^{2-} ion, however, will readily complex with the lanthanides as well as with Y and Sc. The data are given in Table 3. The data for log Q^0 for the lanthanides is essentially unchanged from La through Lu for both $Ln(SO_4)^+$ and $Ln(SO_4)_2^-$ species. These complexes are of possible significance under conditions of acidic, oxidizing environments such as various metal mill tailings, mine wastes, and similar settings. It is established (Langmuir, 1978) that the actinides readily complex with sulfate under similar conditions, hence it is probable that the lanthanides do so as well. The various studies carried out on the near field environment for fission-produced lanthanides from radioactive wastes demonstrate the importance of the $Ln(SO_4)_x^{2-}$ species (see Brookins, 1984).

Under the mid-pH range of natural waters, however, the sulfate complexes of the lanthanides are not as important as Ln^{3+}, $Ln(CO_3)^+$ and, in some instances, $Ln(P_2O_7)^{2+}$ species.

GADOLINIUM - TERBIUM FRACTIONATION ?

Several authors have argued for fractionation of the lanthanides by means other than the traditionally evoked Ce and Eu anomalies which are caused by redox mechanisms. Recently, deBaar et al. (1985) proposed that the behavior of Gd^{3+}, due to its special electronic configuration in water, may be slightly different from neighboring Eu^{3+} and Tb^{3+}. In Gd, the 4f shell is exactly half-filled; in a plot of ionization potential, deBaar et al. (1985) note that a minimum number of electrons occur in the inner shell at Gd and maxima at Eu and Yb. They suggest that the positive Gd anomaly reported for measurements of lanthanides in sea water is due to this electronic configuration subtlety; i.e., in some way Gd is enriched in sea water--relative to Eu^{3+} and Tb^{3+}. Presumably, the $Gd(H_2O)_x^{3+}$ species or possibly the $Gd(OH)^{2+}$(?) are slightly more stable than neighboring Eu and Tb species. As pointed out earlier, and shown in Table 3, the values of log Q^0 shift noticeably between Eu and Gd (see $Ln(OH)^{2+}$ values), but it remains to be demonstrated that this factor accounts for the Gd anomaly noted by deBaar et al. (1985) and other investigators (Roaldset and Rosenqvist, 1971; Masuda and Ikeuchi, 1979).

Recently, Taylor and McLennan (1988) have questioned the Gd anomaly proposed by others (above). They correctly point out that there is a genuine concern about the Gd data used by the earlier workers: there is a problem with the neutron activation method for Gd and Tb (see also Brookins, 1979; Della Valle, 1980). In addition, old normalizing factors were used to determine the anomalies relative to a shale composite. With newer values, some of the purported anomalies disappear. Further, Taylor and McLennan (1988) note that Elderfield and Greaves (1982), who analyzed sea water by refined isotope dilution techniques, do not observe any Gd anomaly. Finally, since there is a natural fractionation of heavy and light lanthanides in sea water, then any inflections could give rise to apparent anomalies.

Cantrell and Byrne (1987) have also presented arguments based on their carbonate work that no unusually large stability complex exist for Gd relative to Eu and/or Tb. Thus the proposed Gd anomaly of deBaar et al. (1985), if real, must be due to other causes.

Yet this problem is still unresolved. The shift in coordination of water molecules about the inner Ln sphere at about Eu (see earlier discussion and Thompson, 1979) could affect the hydrated ion radius at Gd rendering it slightly more stable in sea water. This would not necessarily show up in the data base of Baes and Mesmer (1976), Turner et al. (1981) and others since, for the most part, the data were taken with ionic strength considerably greater than zero. Our knowledge of exact speciation of lanthanides in sea water is improving rapidly, and any such effect as proposed by deBaar et al. (1985) must be consistent with the new data. It is suggested that where possible, the critical samples analyzed by deBaar et al. (1985) be cross checked by isotope dilution techniques.

SCANDIUM AND YTTRIUM

Although scandium and yttrium are both rare earth elements, they differ from the lanthanides in several important ways. First, Sc^{3+} hydrolizes readily. The stability constants for formation of $ScOH^{2+}$ and $Sc(OH)^{2+}$ are both significantly larger than those for the equivalent Ln(III) species (Table 3). Yttrium, however, is more like the lanthanides in that the ionic radius of Y^{3+} (0.900 Å; CN 6) is identical to that of Ho^{3+} (0.901 Å; CN 6), and consequently, like the lanthanides, it does not readily hydrolize in the low to intermediate pH range of natural waters.

Second, both Sc and Y lack f electrons which characterize the lanthanides, and thus Sc and Y are more like the d-type transition metals (see discussion in Leskela and Niinisto, 1986). Further, as shown in Figure 1, while Y falls with the lanthanides in a plot of ionic radius vs. charge, Sc falls with Cr, Fe, V and Nb. The well established Sc-Fe(III) diadochy is testimony to this fact.

EUROPIUM (II)

Europium is the only lanthanide that, under natural conditions, has a (II) valence, and this readily explains the segregation of Eu from the other lanthanides under reducing conditions. Under such conditions, Eu^{2+} is almost identical to Sr^{2+} in its geochemical characteristics (see Table 1). Camouflage of Eu^{2+} and Sr^{2+} in Ca^{2+}-sites during rock genesis is well documented and is considered in several other papers in this short-course symposium.

Like the alkaline earths, then, Eu^{2+} does not hydrolyze readily (Baes and Mesmer, 1976; Thompson, 1979). Under concentrated salt conditions, Eu(II) will form ion pairs, etc. like the alkaline earths, but for most terrestrial waters, the uncomplexed ion Eu^{2+} will predominate.

In nature, very reducing waters will contain Eu^{2+}, or Eu(II)-organo complexes (Jorgeson, 1979). In Eh-pH space, for conditions of low total dissolved Eu and moderate total dissolved CO_2, Eu^{2+} occupies a small part of the diagram (see Fig. 5 below). This is discussed later in this paper (Eh-pH diagrams). Europium (II) also forms a simple carbonate, $EuCO_3$, but there are no reported thermodynamic data for this species.

CERIUM (IV)

Cerium (IV) is best known for its segregation from the M^{3+} lanthanides due to oxidation of Ce(III) to Ce(IV). The emf for the half cell:

$$Ce^{3+} = Ce^{4+} + e^-$$

is 1.74 volts, which places Ce^{4+} well above the upper stability limit of water (1.23 V at pH = 0). It is well known that under laboratory conditions Ce^{4+} exhibits a very complex chemistry and is readily aquated and hydrolized. The dominant species are given in Table 5. Thompson (1979), analyzes the Ce(IV) species in his discussion of rare earth element complexes.

In nature, Ce(IV) is important only in solids. In Eh-pH space (see Fig. 4), there exists a large stability field of CeO_2 (or $Ce(OH)_4$), which explains the segregation of Ce(IV) from the M(III) lanthanides in sea water and perhaps other media. This will be discussed in the next section.

212

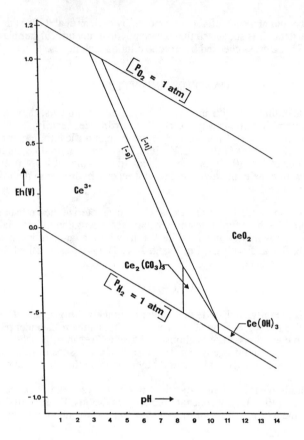

Figure 4. Eh-pH diagram for part of the system Ce-C-O-H (25°C, 1 atm total pressure). Activities: Total dissolved $CO_3 = 10^{-3}$, total dissolved $Ce = 10^{-9}$ and 10^{-11}. Fields for aqueous complexes not shown. See text for details.

Table 5. Ce(IV) Hydrolysis Products

Species		Log Q°
$Ce(OH)^{3+}$	+	1.1
$Ce(OH)_2^{2+}$	+	0.3
$Ce_2(OH)_2^{6+}$	+	3.6
$Ce_2(OH)_3^{5+}$	+	4.1
$Ce_2(OH)_4^{2+}$	+	3.5
$Ce_6(OH)_{12}^{12+}$	+	15.4

Data from Baes and Mesmer (1976); see also
deBaar and others (1988).

Eh-pH DIAGRAMS

Eh-pH diagrams for some of the lanthanides are shown in Figures. 4 through 7. Other published Eh-pH diagrams are given in Brookins (1983, 1988), and for Ce, see deBaar et al. (1988). The thermodynamic data used in the construction of the diagrams are given in Table 4, and the methods of construction, discussion of errors, and other constraints are given in Brookins (1988). For all the figures given here, total dissolved individual lanthanide activity is taken as 10^{-9} to 10^{-11} which very nearly brackets their concentrations in river and ocean waters (see Table 7). For dissolved carbonate, chosen as HCO_3^- for the ocean, a value of 10^{-3} is assumed.

Cerium

Cerium in waters shows a different abundance than the other lanthanides due to the formation of Ce(IV) from Ce (III). The stable form of Ce(IV) is CeO_2, while $Ce(OH)_4$ is metastable (see Brookins, 1988; deBaar et al., 1988). For Ce(III) species, Ce^{3+} is the dominant form at low pH although $Ce(CO_3)^+$ becomes dominant at pH close to that for sea water. Ce(III) is also present in the carbonate phase $Ce_2(CO_3)_3$ which occupies a small stability field in Eh-pH space (Fig. 4). Figure 4 also shows a restricted field for $Ce(OH)_3$, although the thermodynamic data for this species are not well defined (Brookins, 1988; deBaar et al., 1988).

Figure 4 shows the relative ease of Ce(IV) removal as CeO_2 from the other trivalent lanthanides. Further, a small field of $Ce_2(CO_3)_3$ appears that may be important in scavenging any Ce(III) along with the other Ln(III) from sea water in sediments associated with manganese nodules.

Europium

Europium possesses a Eu(II) oxidation state as well as Eu(III). This Eu(III) is responsible, directly or indirectly, for the well known Eu anomalies used extensively in petrogenetic theory and application. The Eh-pH diagram for the system Eu-O-H is shown in Figure 5 (Brookins, 1988).

The stability field of Eu(II) as simple Eu^{2+} is very small in Eh-pH space, and its importance for ocean processes is probably slight. For example, in their studies of the Cariaco Trench, deBaar et al. (1988) do not find any evidence for the presence or impact of Eu(II). Most of the Eh-pH space is occupied by Eu^{3+} and $Eu(OH)_3$. The stability fields of carbonate complexes and $Eu(OH)_4^-$ are not shown due to lack of data. There is no apparent anomaly of Eu relative to Sm or Gd when compared to a North American Shale (NAS) normalized plot (Taylor and McLennan, 1988).

Other lanthanides

Earlier versions of Eh-pH diagrams for the lanthanides have used high activities of 10^{-6} to 10^{-8} for dissolved lanthanides (Brookins, 1983, 1988). While such activities for Ln species may be appropriate for some terrestrial waters and associated deposits of rare earths, they are much too great for sea water conditions.

The stability constants for $Ln(OH)_4^-$ species are largely uncertain (Table 4). For sea water activities of dissolved lanthanides, the field of $Ln(OH)_4^-$ is of only minor importance as the pH boundaries between $Ln(OH)_3^0$ and $Ln(OH)_4^-$ species are at 12 and higher for most lanthanides, and $Ln(OH)_3(C)$ is stable with respect to $Ln(OH)_3^0$. deBaar et al. (1988) also show the importance of $Ce(OH)_3^0$ at very high pH values (see Fig. 4); $Ce(OH)_4^-$ is only of importance at pH above 12.2 (calculated from data of deBaar et al., 1988).

214

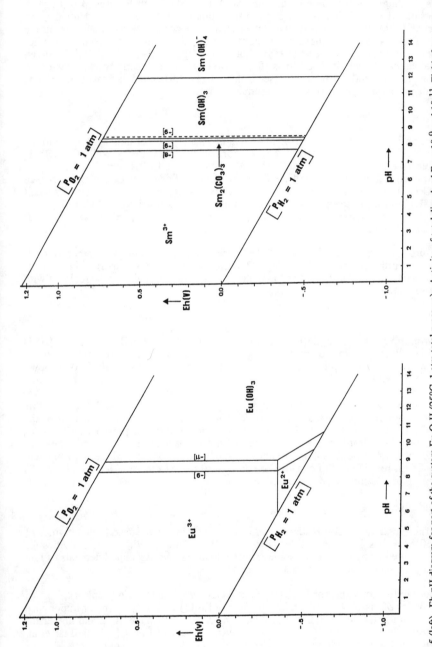

Figure 5 (left). Eh-pH diagram for part of the system Eu-O-H (25°C, 1 atm total pressure). Activity of total dissolved Eu = 10^{-9} and 10^{-11}. Fields for aqueous comple

Figure 6 (right). Eh-pH diagram for part of the system Sm-C-O-H (25°C, 1 atm total pressure). Activities: Total dissolved $CO_3 = 10^{-3}$, total dissolved Sm : complexes other than $Sm(OH)_4^-$ not shown. See text for details.

The Eh-pH diagrams for Sm and Nd are given in Figures 6 and 7. For both Sm and Nd, small fields of $Ln_2(CO_3)_3$ appear at pH just below the fields for $Ln(CO_3)^+$ species. Complexes of Sm and Nd with carbonate in solution are not shown, but from Figure 2 from Cantrell and Byrne (1987), the proportions of $Ln(CO_3)^+$ and $Ln(CO_3)_2^-$ change rapidly in the vicinity of Sm and Nd, thus some partitioning of Sm from Nd may occur. While unimportant for modern sea water due to the short residence times, etc., this may account for the spread in Sm/Nd ratios in authigenic marine minerals (cf. Della Valle, 1980; others). For both Sm and Nd species, the sea water conditions in Eh-pH show the probability of formation of solids, probably as $Ln(OH)_3$ but possibly in carbonate phases. An interesting question is whether the lanthanides found in calcitic phases on the ocean floors are due to co-precipitated Ln-Ca mix crystals or admixed Ln-carbonate with Ca-carbonate.

LANTHANIDES IN OCEAN WATERS

The abundance and distribution of the lanthanides in sea water has been the subject of extensive investigation over the last two decades, and the reader is referred to the works of Taylor and McLennan (1988), Piepgras and Jacobsen (1988), deBaar et al. (1985, 1988), Elderfield and Greaves (1982), and references contained therein.

The residence time of the lanthanides is given in Table 6 (from Elderfield and Greaves, 1982). There it is noted that values range from just over 100 years for Ce relative to aeolian and river output to just under 1000 years for the heavy lanthanides relative to sediment output. Since the mixing time of the oceans is about 1000 years, a large variability among the lanthanides is expected (Taylor and McLennan, 1988).

In surface and near-surface waters, the lanthanides are thought to be fixed into particulate matter, including organics. The surface waters are presumably dominated by aeolian input (Taylor and McLennan, 1988). In large part, lanthanide adsorption on organic matter is the dominant process in surface waters (Cantrell and Byrne, 1987). However, carbonate complexing is dominant for organic lanthanides. The ratio of lanthanide sorbed on particulates to $Ln(CO_3)^+$ will change only slightly with pH, but $Ln(CO_3)_2^-$ will change more rapidly (Cantrell and Byrne, 1987). These researchers show that the stability constant of $Ln(CO_3)^+$ increases only by a factor of eight from La to Lu whereas the stability constant for $Ln(CO_3)_2^-$ increases by a factor of 80.

Table 6. Residence times (in years) of lanthanides in seawater[a]

Element	(1)	(2)
La	690	240
Ce	400	110
Pr	(640)	(360)
Nd	620	450
Sm	600	220
Eu	460	470
Gd	720	300
Tb	(760)	(320)
Dy	810	340
Ho	(890)	(380)
Er	980	420
Tm	(980)	(420)
Yb	970	410
Lu	(970)	(410)

[a] From Elderfield and Greaves (1982) and Taylor and McLennan (1988).
(1) Relative to sediment output.
(2) Relative to aeolian and river input (probably minimum values).

With depth, other processes are important. deBaar et al. (1988) argue that at 280 m in the Cariaco Trench, as manganese oxide particulates form at the oxic-anoxic water interface, the particulates again scavenge the lanthanides. The patterns shown by deBaar et al. (1988) are for a decrease in abundance of all lanthanides until the oxic-anoxic interface. Then a rapid increase in concentration occurs to about 800 m. Below this depth, the abundance remains fairly constant.

The open ocean lanthanides values from Elderfield and Greaves (1982), Klinkhammer et al. (1983) and Martin et al. (1976) are shown in Table 7. The chondrite-normalized REE patterns for these data (from Taylor and McLennan, 1988) are given as

216

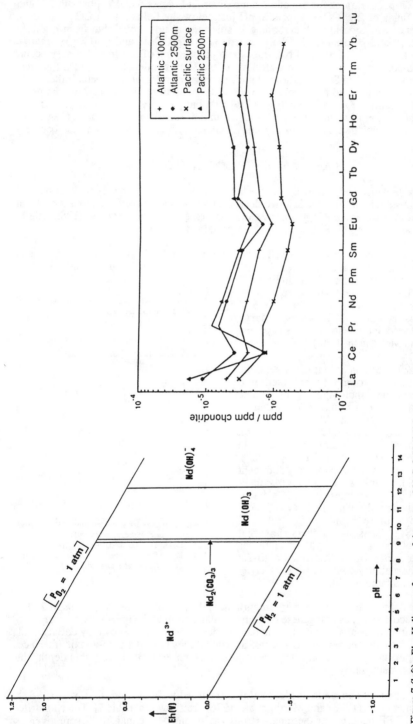

Figure 7 (left). Eh-pH diagram for part of the system Nd-C-O-H (25°C, 1 atm total pressure). Activities: Total dissolved $CO_3 = 10^{-3}$, total dissolved $Nd = 10^{-9}$ and other than $Nd(OH)_4^-$ not shown. See text for details.

Figure 8 (right). Lanthanide abundance patterns for selected sea water samples (from Taylor and McLennan, 1988). Figure 8 is reproduced from kind permission of Elsevier Science Publishers.

Figure 8. A greater abundance of lanthanides are found at 2500 m than at 100 m, and the quantity of lanthanides are different between the two oceans. Table 7 also gives data from the Garonne and Dardogne Rivers (from Martin et al., 1976), and it is noted that the total riverine lanthanide concentration is at least an order of magnitude higher than the ocean values. Taylor and McLennan (1988; cf. Elderfield and Greaves, 1982) point out that the surface waters must be dominated by aeolian and riverine input, with REE patterns like those of the upper continental crust, as well as some bottom source that involves recycling and perhaps new input.

Table 7. Lanthanide content (Parts per trillion) of seawater and river water[a]

Element	Atlantic Ocean		N.W. Pacific Ocean		Garonne and Dordogne Rivers
	100 m	2500 m	surface	2500 m	
La	1.81	4.08	1.2	6.5	47.5
Ce	2.35	3.68	1.4	1.3	79.0
Pr	-	-	-	-	7.3
Nd	1.85	3.61	0.74	4.3	37.9
Sm	0.401	0.714	0.15	0.80	7.8
Eu	0.0979	0.136	0.050	0.21	1.48
Gd	0.536	1.13	0.25	1.3	8.5
Tb-	-	-	-	-	1.24
Dy	0.777	0.991	0.33	1.6	-
Ho	-	-	-	-	-1.44
Er	0.681	0.851	0.28	1.6	4.2
Tm	-	-	-	-	0.61
Yb	0.614	0.829	0.19	1.4	3.64
Lu	-	-	-	-	0.64
Total	13	16	5	19	202
La/Yb	8.31	4.92	6.32	4.64	13.0
La_N/Yb_N	5.61	3.33	4.27	3.14	8.82
Eu/Eu^*	0.65	0.46	0.79	0.63	0.56

[a]From Taylor and McLennan (1988).
N = chondrite normalized values
Eu^* = hypothetical Eu value for no anomaly between Sm and Gd.

The presence of a negative Ce anomaly in sea water is well known (cf. Piper, 1974), as is the fact that overall lanthanide abundances in sea water are very low, for the most part they range over a few ppt. This Ce anomaly is readily explained since Ce(IV) as CeO_2 forms from Ce(III) (see Fig. 4), and CeO_2 has a large Eh-pH stability field. As shown by deBaar et al. (1988), when anoxic conditions are encountered, there is a tendency for some reduction of Ce(IV) to Ce(III) (as Ce^{3+} or the appropriate complex; see Fig. 4) which, in turn, can be correlated with HS- build-up and other evidences for increased reducing conditions. These workers also calculated scavenging times for the lanthanides in sea waters which show the light lanthanides more easily scavenged than the heavy lanthanides, consistent with the observation that the heavy lanthanides are enriched relative to the light lanthanides in sea water.

To fully explain lanthanide behavior in waters, we must recognize the effect of weathering on lanthanide release and subsequent migration; however, the behavior of the lanthanides during weathering has not been fully investigated (see summary articles by Humphris, 1984; Fleet, 1984; Taylor and McLennan, 1988). The lanthanides are mobile under acidic conditions, as attested to by their fixation in kaolinitic clays, where there has been lanthanide removal. Similarly, under higher pH conditions, precipitation or some other removal mechanism occurs so that the lanthanides are fixed. Yet we do not have a quantitative understanding of the nature of the migration-fixation nor the mechanisms.

This is also true of most waters. In the oceans, the shelf waters possess normal La/Ce ratios (see Humphris, 1984; Fleet, 1984), but as pointed out earlier, the La/Ce are high in the open ocean due to Ce removal preferential to the other lanthanides. The Ce removal by formation of CeO_2 is well established (deBaar et al., 1988; references therein), and its negative anomaly relative to the shale normalized REE is well known. Yet, all of the lanthanides are undersaturated in the ocean. Fleet (1984) has proposed and summarized the following removal mechanisms: (1) inorganic precipitation, (2) lanthanide incorporation into biogenic and/or hydrogenous material, (3) halmyrolytic reactions between lithogenous material and sea water and (4) hydrothermal processes in ocean bottom-vent areas.

A simple inorganic precipitation of the lanthanides is not consistent with the observation that values of individual lanthanides predicted from insoluble salts are higher than the values measured in sea water. Hence, inorganic removal mechanisms must be only a minor contributing factor.

The fixation of lanthanides on biogenic and/or hydrogenous material is attractive. Turner and Whitfield (1979) argue for adsorption of the lanthanides onto biogenic and/or hydrogenous particle surfaces in response to their polarizing power, large valence, and affinity for calcitic matter. By their model, the light lanthanides are more readily fixed than the heavy lanthanides due to differences in polarizability (i.e., Lu has a greater polarizing power than La), thus, the heavy lanthanides are more enriched in sea water relative to the light lanthanides. This is entirely consistent with the observations made by numerous workers (deBaar et al., 1985a,b, 1988; Elderfield and Greaves, 1982; others). A major problem is encountered when reviewing the literature on this complex subject; the exact material housing the lanthanides is not precisely known in most cases. What is known, however, is that authigenic minerals such as glauconite, phillipsite, and especially manganese nodules are enriched in the lanthanides; manganese nodules, especially, are enriched in Ce. Spirn's (1965) data argue convincingly for lanthanide removal by growth of calcitic material, although contamination by lithogenous material may play a major role (see Fleet, 1984).

Although there is a question concerning the efficiency of biogenic material such as plankton for lanthanide scavenging, it is established that fish residue and debris are excellent scavengers of the lanthanides (Bernat, 1975). It is proposed that apatite in the residue is an effective getter for lanthanides (cf. Arrhenius et al., 1957; Bernat, 1975; Fleet, 1984). That the residue in sea water is a source of the lanthanides is proven by the negative anomaly of Ce (relative to NAS). Such a negative anomaly is also found in foraminifera and other authigenic material such as biogenic ooze (Spirn, 1965; Elderfield et al., 1981).

More information is available for volcanic detritus. Clay minerals, zeolites and other minerals form readily from submarine volcanics by halmyrolytic processes. Courois and Hoffert (1977) have shown that smectites from South Pacific ocean bottom samples are enriched in the heavy lanthanides and contain a negative Ce anomaly, thus attesting to the interaction of sea water with the volcanics. Studies of marine phillipsites are less clear, often because there is a possibility of lithogenous and ocean sources for the lanthanides (see Piper, 1974; Bernat, 1975; Fleet, 1984).

Fleet (1984) has summarized the scant information on the interaction of lithogenous material with sea water. Halmyrolysis may be effective in the estuarine environment, but exactly how important a role this plays in the open seas is questionable. Fleet (1984) points out, using data from Martin et al. (1976), that it is not certain if the lanthanides are released or taken up by the presence of continental-derived sediments. However, Brewer et al. (1980) point out that there is a correlation of Al with La. Since Al is conventionally assumed to be highly insoluble in ocean waters, then it is postulated that La is also not affected by halmyrolysis during deposition of sediments.

Surface adsorption of the lanthanides is apparently an important process. The work of Piper (1974), Spirn (1965), and others (see Fleet, 1984) has established the high content of lanthanide associated with clay minerals and the surfaces of lithogenous material. Della Valle (1980) has noted very large lanthanide build up in clay rims associated with saline waters in marine to non-marine rocks.

Material that is presumed to be truly authigenic includes ocean floor or manganese nodules, phosphorites, barites, hydrothermal vent-related sediments and others. These materials all effectively scavenge the lanthanides and their lanthanide/NAS patterns show a clear influence of sea water. The manganese nodules, however, are very different from the others. For the heavy lanthanides, the marine barite, phosphorites, and vent-related sediments all show a clear negative Ce anomaly and overall lanthanide enrichment relative to shale. Diagrams depicting this are to be found in Piper and Graef (1974), Goldberg et al. (1963), Altschuler (1980), Guichard et al. (1979), and others. Robertson and Fleet (1976, 1977) and Elderfield (1976) have shown that many of the vent-related sediment crusts are enriched in the lanthanides, and that sulfide mineralization is unimportant as a direct scavenging means. When the sulfides are weathered to form oxyhydroxide crusts, however, these become efficient scavengers for sea water lanthanides (see discussion Fleet, 1984; Michard, 1989; Olivarez and Owen, 1989).

Manganese nodules have been the subject of extensive research, and good summaries of this work are given by Fleet (1984) and Taylor and McLennan (1988). If, as proposed by deBaar et al. (1988; and referenced work therein), CeO_2 precipitation is correlated with $Mn(IV)$ oxide formation with resultant precipitation at relatively shallow depths (est. 250-300 m), then the cerium-bearing manganese precipitate settles to the ocean floor where it accretes to nucleated precursors. All the lanthanides generally accumulate in these nodules. Models proposed by Elderfield et al. (1981) include direct precipitation of $Ln(III)$ into nodules, co-precipitation of the $Ln(III)$ with $Fe(III)$, or surface fixation by some other (or mixed) processes, reworking of sea water-derived lanthanides from fish remains, etc., with resultant coprecipitation onto the nodules, or derivation of the lanthanides from underlying sediments with fixation into the nodules by ion exchange processes. Fleet (1984) and Taylor and McLennan (1988) have commented on these models in some detail.

References to the Eh-pH diagram presented here (Fig. 4) and also to that of deBaar et al. (1988) may help explain some of the lanthanide occurrences in manganese nodules. If the $Ce(IV)$ is brought to the nodules with precipitated $Mn(IV)$ oxyhydroxides, then this material must accrete into the nodules so efficiently that, with decreasing Eh with depth, reduction of $Ce(IV)$ and $Mn(IV)$ is kinetically prevented. Thus when the stability field of $Ce(III)$ is encountered (Fig. 4) there is insufficient time to reduce $Ce(IV)$ to soluble Ce^{3+} or $CeCO_3^+$. As also noted in Figure 4, a small field of $Ce_2(CO_3)_3$ appears that may help explain how the trivalent lanthanides are fixed in these nodules. Deep sea pH is close to the stability field acidic (lower) limit for trivalent $Ln(OH)_3$, and in most cases small fields of $Ln_2(CO_3)_3$ appear on the Eh-pH diagrams (Brookins, 1988). Data for $Ln-CO_3$ complexes are not reliable enough to warrant plotting their species in Eh-pH space. While the existence of these carbonates is debatable, they do indicate a mechanism for lanthanide scavenging by calcitic ooze, etc., in proximity to the nodules from which they can be exchanged with or in other ways fixed on the nodules. This model is similar to those proposed by Piper (1974) and Elderfield et al. (1981).

Studies of Nd isotopes in manganese nodules show that the nodules and associated sediments often possess $^{143}Nd/^{144}Nd$ ratios of a sea water source, thus some transfer of Nd from sediments to nodules during diagenesis is proposed (see Elderfield et al., 1981). Faure (1987) has summarized much of the important Nd isotopy of the oceans, as have Taylor and McLennan (1988) to a lesser extent. The Nd isotopes of sediments and associated materials are covered in more detail in these proceedings by Patchett and by McLennan, and only a brief overview will be given here. The $^{143}Nd/^{144}Nd$ ratios will vary from ocean to ocean and even within oceans since different average continental crustal material surrounds the oceans and because of the short residence time of Nd in the oceans. Overall, ocean $^{143}Nd/^{144}Nd$ ratios are much less than those for mid-ocean ridge basalts (MORB) and testify to the derivation of the content of sea water lanthanide from continental sources. For example, based on Nd studies, Piepgras and Jacobsen (1988) show that 90% of the lanthanides in the Atlantic Ocean have been derived from continental sources. Faure (1987; Fig. 14.3) shows that the Nd isotopy of manganese nodules clearly reflects the ocean Nd isotopy in which they occur. Thus, Atlantic Ocean $^{143}Nd/^{144}Nd$ ratios for both water and nodules are 0.5121, whereas Pacific materials fall close to 0.51225, and both values are quite different than the MORB values of 0.5128 to 0.5134. The use of $^{143}Nd/^{144}Nd$ ratios as tracers for mixing of oceanic and other water masses is not treated here, and the reader is referred to Faure (1987) for summary and key references. The sea water studies by Nd isotopic methods are a powerful approach to the distribution of the lanthanides in the oceans in general. Faure (1987) also notes that unlike Sr isotopic systematics, which require a water/rock ratio of about 10^5 for resetting, a water/rock ratio of perhaps 10^8 is required for Nd isotopic system resetting. This is important because if the Sr data indicate incomplete resetting, then any anomalous Nd isotopic behavior is probably not a result of resetting and instead can be investigated in terms of mixed ocean versus lithogenous or other source.

LANTHANIDES AND ACTINIDES

The actinides are often compared to the lanthanides and vice versa because each group is characterized by filling an inner shell of electrons (i.e., 4f in lanthanides, 5f in actinides) until 14 electrons have been so placed. Just as the lanthanides possess very similar chemistry throughout their series, the actinides also exhibit many similarities. Yet the structure of the actinide series is much more complex than lanthanides (Baes and Mesner, 1976). Unlike the lanthanides, of course, all the actinides contain only radioactive isotopes. Those actinides with long half lives (i.e., ^{232}Th, ^{235}U, ^{238}U) occur in nature, and isotopes of Ac and Pa occur as intermediate radioactive daughter products of the Th, U chains. The elements from Np and higher are referred to as transuranics.

Among the actinides, the 5f orbitals are not as well shielded as are the 4f orbitals in the lanthanides, hence the former electrons more readily participate in bonding. The actinides from Ac to Pu are especially complex and contain valences ranging from (III) to (VII). For actinides heavier than Pu, however, valences are from (II) to (IV) and most show a wide stability range of (III) species. The trans-plutonium actinides are thus very much like their lanthanide homologs. The actinides Ac, Th, and Pa posses an important (IV) valence. However, for these actinides the species MO_2 occupies wide extents of Eh-pH space and thus, even the stability fields of the M^{4+} ions are scarce, usually existing at a pH of < 2 to 3. The (V) valent species of Ac and Pa also hydrolyze readily to $MO_2(OH)$ (Aq) (see discussion in Baes and Mesmer, 1976). Dissolved sulfate will readily affect the Th budget, such as in mill tailings, with formation of $Th(SO_4)^{2+}$, $Th(SO_4)_3^{2-}$, and $Th(SO_4)_4^{4-}$. Data are lacking for Ac and Pa species involving dissolved S. Actinium, Pa, and Th also bond readily with F, forming MF^{3+}, MF_2^{2+}, and MF_3^+ complexes (Wagman et al., 1983).

Uranium has a complex chemistry (Langmuir, 1978; Brookins, 1984). Uranium (IV) is often present as UO_2 which occupies a wide stability field of Eh and pH in waters; hence, only a small area for M^{4+} of $M(OH)^{3+}$ occurs. Uranium (VI) occurs as UO_2^{2+} which may hydrolyze to $UO_2(OH)^+$ and subsequently to $(UO_2)_2(OH)_2^{2+}$ and higher polymers (Baes and Mesmer, 1976; Langmuir, 1978). For acidic conditions, and for appreciable dissolved sulfate, U complexes readily with sulfate to form UO_2SO_4. Uranium also complexes with F- to form species from UF^{3+} through UF_6^{2-} (Langmuir, 1978). Uranium is well known for its ability to complex with carbonate, forming the important species UO_2CO_3, $UO_2(CO_3)_2^{2-}$ and $UO_2(CO_3)_3^{4-}$ (Langmuir, 1978; Brookins, 1984).

Of the transuranics, neptunium has valences of IV, V, and VI. The simple Np^{4+} hydrolyzes to $Np(OH)^{3+}$ at about pH = 2, but like the lower-series actinides, the field of NpO_2 commonly masks higher hydrolysis polymers (Baes and Mesmer, 1976). Plutonium occurs in III, IV, V, and VI valences. Of these, Pu^{4+} disproportionates to Pu^{3+} and

PuO_2^+, but PuO_2^+ is otherwise unimportant (Baes and Mesmer, 1976). The Pu^{3+} is an important species (Brookins, 1984, 1988), but it does not hydrolyze readily, thus behaving similarly to the lanthanides. Trans-plutonium actinides have important III valences analogous to the lanthanides, at least through Bk (Baes and Mesmer, 1976). The importance of Am^{3+}, for example, has been noted by Brookins (1988) in the Eh-pH diagram for the system Am-O-H. The only important Am(IV) aqueous species may be $Am(OH)_5^-$. Carbonate complexes are also important, as shown by the work of Lundqvist (1982) and Cantrell and Byrne (1987). Using the oxalate-carbonate method of determining stability constants (Langmuir, 1979; Cantrell and Byrne, 1987), stability constants for carbonate complexes of trivalent Am, Cm, Bk and Cf are given (see Fig. 9 and Table 8). The data of Lundqvist (1982) and Cantrell and Byrne (1987) for $Am(CO_3)^+$ and $Am(CO_3)_2^{2-}$ compare favorably.

There are many similarities and significant dissimilarities between the lanthanides and actinides. Both series contain large stability fields of insoluble oxides and/or hydroxides, with M(IV) dominant in the actinides and M(III) dominant in the lanthanides. Both An^{4+} and Ln^{3+} do not hydrolyze, but the Ln^{3+} are more stable than the An^{4+} species. The Ln^{3+} and An^{3+} species are quite similar (Brookins, 1988).

CONCLUDING REMARKS

In the foregoing discussion, attention was directed at the inorganic complexes and behavior of the lanthanides since data in support of inorganic processes dominating lanthanide behavior under many conditions was available. Meanwhile, the role of organics must be quantitatively assessed and will require more rigorous study in the future.

In ocean water, residence times of the lanthanides are short relative to the mixing times of the oceans, hence, as reflected in Nd-isotopic studies, the abundance and distribution of lanthanides varies not only from ocean to ocean but within oceans as well. Most of the lanthanides come from continental weathering as opposed to deep sea sources. The lanthanides are highly undersaturated in sea water, especially with respect to the solubilities of probable solid phases such as $Ln(OH)_3$. Scavenging of the Ln(III) species with $Fe(OH)_3$ is an attractive mechanism as well as the documented scavenging by phosphate-bearing and carbonate-bearing phases. Cerium is more depleted than the other lanthanides due to its removal as CeO_2, probably co-precipitated with, or scavenged by, precipitating MnO_2 species. Europium behaves similarly to the other Ln(III) species, and the role of Eu(II) is apparently of minor importance in sea water.

Several complexes of the lanthanides are important and fairly pH-specific. At very low pH conditions, Ln^{3+} and $LnSO_4^+$ are important. The halide complexes of the lanthanides are only of minor importance, even at low pH. In the neutral to slightly basic pH range, the carbonate complexes, $Ln(CO_3)^+$ and $Ln(CO_3)_2^-$ are dominant. As shown by Cantrell and Byrne (1987), the light lanthanides prefer $Ln(CO_3)^+$ and the heavy lanthanides $Ln(CO_3)_2^-$. At pH values approaching 10, hydrolysis products become important for the trivalent lanthanides as first $LnOH^{2+}$ and then as $Ln(OH)_3^0$. In the pH range of sea water, both hydrolysis products, however, are metastable with respect to crystalline $Ln(OH)_3$ species.

By use of the oxalate-carbonate method of estimating thermochemical data (cf. Langmuir, 1979), some of the trivalent actinides and confirming data for the lanthanides have been estimated by Cantrell and Byrne (1987). Finally, data are presented here for the lanthanides and for Y and Sc which include the research of Baes and Mesmer (1976), Turner et al. (1981), Cantrell and Byrne (1987), and deBaar et al. (1988). While these data are useful, it must be remembered that such information is always subject to revision.

222

Table 8. Preliminary carbonate stability constants for
actinides (25°C, 1 bar)

M^3	I	log Q^*_1	log Q^*_2	Reference
Am	0.7	5.73	9.88	Canfield and Byrne (1987)
Am	1.0	5.81	9.72	Lundqvist (1982)
Cm	0.7	5.73	9.88	Canfield and Byrne (1987)
Bk	0.7	5.95	10.20	Canfield and Byrne (1987)
Cf	0.7	6.00	10.40	Canfield and Byrne (1987)

I = Ionic strength
Q^*_1 = total stability constant fo formation of $An(CO_3)'$
Q^*_2 = total stability constant for formation of $An(CO_3)_2^-$

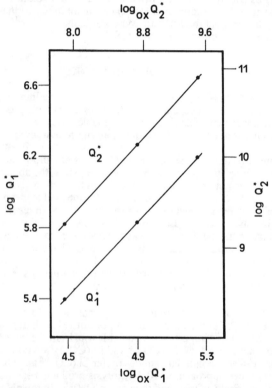

Figure 9. Stability constants for lanthanide carbonates (Q_1 for $LnCO_3^+$; Q_2 for $Ln(CO_3)_2^-$) and oxalates ($_{ox}Q_1$, $_{ox}Q_2$) at 0.68 ionic strength and 25°C. From Cantrell and Byrne (1987) and reproduced here in modified form with the kind permission of Pergamon Press.

ACKNOWLEDGMENTS

I thank Bruce Lipin not only for inviting me to prepare this report but also for his advice and encouragement. He and Gordon McKay reviewed the manuscript. Ms. Judith Binder prepared the manuscript, and Mr. Aaron Johnson drafted Figures 4-7. The kind co-operation of and permission to publish materials from Elsevier Science Publishers, North Holland Physics Publishing Division, and Pergamon Press, Inc. is gratefully recognized.

REFERENCES

Alfonin, E.G., and Pechurova, N.I., 1987, Complex formation of Nd(III) with phosphoric acid. Zh. Neorg. Kh., 33, 2141-2142.

Alstschuler, Z.S., 1980, The geochemistry of trace elements in marine phosphorites, 1. characteristic abundances and enrichment. Soc. Econ. Paleon. Min. Spec. Pub. 29, 19-30.

Arrhenius, G., Bramlette, M.N., and Piciolto, E., 1957, Localization of radioactive and stable heavy nuclides. Nature, 180, 85-86.

Baes, C.F., and Mesmer, R.E., 1976, The Hydrolysis of Cations. Wiley-Interscience, New York, 489 p.

Bernat, M., 1975, Les isotopes de l'uranium et due thorium et les terres rares dans l'environment marin. Cah. ORSTOM Sr. Geol., 7, 65-83.

Berner, M.K., and Berner, R.A., 1987, The Global Water Cycle. Prentice-Hall Pubs., Englewood Cliffs, NJ, 397 p.

Biedermann, G, and Ferri, D., 1982, On the polynuclear hydrolysis of the indium ion; In^{3+}. Acta Chem. Scand., A 36, 611-618.

Bilal, B.A., and Becker, P., 1979, Complex formation of trace elements in geochemical systems--II. Stability of rare earths fluoro complexes in fluorite bearing model system at various ionic strengths. J. Inorg. Nuc. Chem., 41, 1607-1608.

Bilal, B.A., and Kob, V., 1980, Complex formation of trace elements in geochemical systems--3. Studies on the distribution of fluoro complexes of rare earth elements in fluorite bearing model systems. J. Inorg. Nuc. Chem., 42, 629-630.

Bilal, B.A., Hermann, F., and Fleischer, W., 1979, Complex formation of trace elements in geochemical systems--I. Potentiometric study of fluoro complexes of rare earth elements in fluorite bearing model systems. J. Inorg. Nuc. Chem., 41, 347-350.

Brewer, P.G., Nozaki, Y., Spencer, D.W., and Fleet, A.P., 1980, Sediment trap experiments in the deep North Atlantic: isotopic and elemental fluxes. J. Mar. Res., 38, 703-728.

Brookins, D.G., 1979, Geochemical Studies of the Grants, NM Uranium Deposits. DOE Report 79-029E, 459 p.

Brookins, D.G., 1982, Geochemistry of clay minerals for uranium exploration in the Grants Mineral Belt, New Mexico. Mineral. Deposita, 17, 37-53.

Brookins, D.G., 1983, Eh-pH diagrams for the rare earth elements at 25°C and one bar pressure. Geochem. J., 17, 223-229.

Brookins, D.G., 1984, Geochemical Aspects of Radioactive Waste Disposal. Springer-Verlag, Pubs., New York, 347 p.

Brookins, D.G., 1988, Eh-pH Diagrams for Geochemists. Springer-Verlag Pubs., New York, 176 p.

Cantrell, K.J., and Byrne, R.H., 1987, Rare earth element complexation by carbonate and oxalate ions. Geochim. Cosmochim. Acta, 51, 597-606.

Courtois, C, and Hoffert, M., 1977, Distribution des terres rares dans les sdiments superficiel du Pacifique sud-est. Bull. Soc. Gol. Fr., 19, 1245-1251.

deBaar, H.J.W., German, C.R., Eldergield, H., and Van Gaans, P., 1988, Rare earth element distributions in anoxic waters of the Cariaco Trench. Geochim. Cosmochim. Acta, 52, 1203-1220.

deBaar, H.J.W., Bacon, M.P., Brewer, P.G., and Bruland, K.W., 1985a, Rare earth elements in the Pacific and Atlantic Oceans. Geochim. Cosmochim. Acta, 49, 1943-1954.

deBaar, H.J.W., Brewer, P.G., and Bacon, M.P., 1985b, Anomalies in rare earth distribution in seawater. Geochim. Cosmochim. Acta., 49, 1955-1963.

Della Valle, R.S., 1980, Geochemical studies of the Grant Mineral Belt, New Mexico. Unpub. Ph.D. dissertation, Univ. New Mexico, Albuquerque, NM, 658 p.

Ekambaram, V., Brookins, D.G., Rosenberg, P.H., and Emanuel, K.M., 1986, Rare-earth element geochemistry of fluorite-carbonate deposits in western Montana, USA. Chem. Geol., 54, 319-331.

Elderfield, H., 1976, Hydrogenous material in marine sediments; excluding manganese nodules. In: J.P. Riley and R. Chester, Eds., Chemical Oceanography 5. Academic Press, New York, p. 137-215.

Elderfield, H., and Greaves, M.J., 1982, The rare earth elements in sea water. Nature, 296, 214-219.

Elderfield, H., Hawkesworth, C.J., Greaves, M.J., and Calvert, S.E., 1981, Rare earth element geochemistry of oceanic gerromanganese nodules and associated sediments. Geochim. Cosmochim. Acta, 45, 513-528.

Faure, G., 1987, Principles of Isotope Geology, 2nd edition. Wiley-Interscience, New York, 589 p.

Flahaut, J., 1979, Sulfides, selenides and tellurides. In: K.A. Gschneider and L. Eyring, Eds., Handbook on the Physics and Chemistry of the Rare Earths. Elsevier Sci. Pubs., North Holland Physics Pub. Div., Amsterdam, Vol. 4, p. 1-88.

Fleet, A.J., 1984, Aqueous and sedimentary geochemistry of the rare earth elements. In: P. Henderson, Ed., Rare Earth Element Geochemistry. Elsevier Science Pubs., Amsterdam, p. 343-373.

Fryer, B.J., 1977, Trace element geochemistry of the Sukoman Iron Formation. Can. J. Earth Sci., 14, 1598.-1612.

224

Goldberg, E.D., Koide, M., Schmitt, R.A., and Smith, R.H., 1963, Rare earth distribution in the marine environment. J. Geophs. Res., 68, 4209-4217.

Guichard, F., Church, T.M., Treuil, M., and Jaffrezic, H., 1979, Rare earths in barites: distribution and effects on aqueous partitioning. Geochim. Cosmochim. Acta, 43, 983-997.

Habenschuss, A., and Spedding, F.H., 1974, A survey of some properties of aqueous rare earth salt solutions, I. volume, thermal expansion, Raman spectra and x-ray diffraction. In: J.M. Haschke and H.A. Eick, Eds., Proc. 11th Rare Earth Conference. U.S. Atomic Energy Commission, Oak Ridge, TN, Vol. 2, p. 9.

Haschke, J.M., 1979, Halides. In: K.A. Gschneider and L. Eyring, Eds., Handbook on the Physics and Chemistry of Rare Earths. Elsevier Sci. Pubs., North Holland Physics Pub. Div., Amsterdam, Vol. 4, p. 89-150.

Hawkesworth, C.J., and Van Calsteren, P.W.C., 1984, Radiogenic isotopes--some geological applications. In: P. Henderson, Ed., Rare Earth Element Geochemistry. Elsevier Science Pubs., Amsterdam, p. 375-421.

Herman, J.S., and Langmuir, D., 1978, Thorium complexes in natural waters. Geol. Soc. Amer.Abstr. Program, 10, 419-420.

Humphris, S.E., 1984, The mobility of the rare earth elements in the crust. In: P. Henderson, Ed., Rare Earth Element Geochemistry. Elsevier Science Pubs., Amsterdam, p. 317-342.

Jorgenson, C.K., 1979, Theoretical chemistry of rare earths. In: K.A. Gschneider and L. Eyring, Eds., Handbook on the Physics and Chemistry of the Rare Earths. Elsevier Sci. Pubs., North Holland Physics Pub. Div., Amsterdam, Vol. 4, p. 111-172.

Jorgenson, C.K., 1988, Influence of rare earths on chemical understanding and classification. In: K.A. Gschneider and L. Eyring, Eds., Handbook on the Physics and Chemistry of the Rare Earths. Elsevier Sci. Pubs., North Holland Physics Pub. Div., Amsterdam, Vol. 11, p. 197-292.

Klinkhammer, G., Elderfield, H., and Hudson, A., 1983, Rare earth elements in seawater near hydrothermal vents. Nature, 305, 185-188.

Komissarova, L.N., 1980, Zh. Neorg. Khim., 25, 143 (Russian J. Inorg. Chem., 25, , 75, Engl. Trans.).

Kosterin, A.V., 1959, The possible modes of transport of the rare earths by hydrothermal solutions. Geochemistry, p. 381-387.

Krauskopf, K.B., 1977, Introduction to Geochemistry, 2nd edition. McGraw-Hill Pubs., New York, 617 p.

Langmuir, D., 1978, Uranium solution-mineral equilibria at low temperatures with applications to sedimentary ore genesis. Geochim. Cosmochim. Acta, 42, 547-570.

Langmuir, D., 1979,Techniques of estimating thermodynamic properties for some aqueous complexes of geochemical interest. In: E. Jenne, Ed., Chemical Modelling of Aqueous Systems. c. 18, Amer. Chem. Soc., 353-387.

Leskel, A., and Niinist, L., 1986, Inorganic complex compounds I. In: K.A. Gschneider and L. Eyring, Eds., Handbook on the Physics and Chemistry of the Rare Earths Elsevier Sci. Pubs., North Holland Physics Pub. Div., Amsterdam, Vol. 8, p. 203-327.

Leventhal, J.S., and Threlkeld, C.N., 1978, Carbon-13/carbon-12 fractionation of organic matter associated with uranium ores induced byalpha irradiation. Science, 202, 430-432.

Lundqvist, R., 1982, Hydrophilic complexes of the actinides. I. carbonates of trivalent americium and europium. Acta Chem. Scand, A36, 471-450.

Martin, J.M., Hogdahl, O., and Phillpott, 1976, Rare earth element supply for the ocean. J. Geophys. Res., 81, 3119-3124.

Masuda, A., and Ikeuchi, Y., 1979, Lanthanide tetrad effect observed in marine environment. Geochem. J., 13, 19-22.

McLennan, S.M., and Taylor, S.R., 1979, Rare earth element mobility associated with uranium mineralization. Nature, 282, 247-250.

Michard, A., 1989, Rare earth element systematics in hydrothermal fluids. Geochim. Cosmochim. Acta, 53, 745-750.

Moeller, T., and Vicentini, G., 1965, Observations on the rare earths-LXXVII. J. Inorg. Nuc. Chem., 27, 1477-1482.

Olivarez, A.M., and Owen, R.M., 1989, REE/Fe variations in hydrothermal fluids. Geochim. Cosmochim. Acta, 53, 757-762.

Pearson, R.G., 1963, Hard and soft acids and bases: anhydrous N,N-dimethylacetamide adducts of the tripositive perchlorates. J. Amer. Chem. Soc., 85, 3533-3539.

Piepgras, D.J., and Jacobsen, S.B., 1988, The isotopic composition of neodymium in the North Pacific. Geochim. Cosmochim. Acta, 52, 1373-1382.

Piper, D.Z., 1974, Rare earth elements in ferromanganese nodules and other marine phases. Geochim. Cosmochim. Acta, 38, 1007-1022.

Piper, D.Z, and Graef, P.A., 1974, Gold and rare-earth elements in sediments from the East Pacific Rise. Mar. Geol., 17, 287-297.

Reidler, J., and Silber, H., 1973, Deuterium isotope effects in complexzation kinetics II. Lanthanide (III) sulfate systems. J. Phys. Chem., 77, 1275-1280.

Roaldset, E., and Rosenqvist, I.T., 1971, Unusual lanthanide distribition. Nature, 231, 153-154.

Robertson, A.H.F., and Fleet, A.J., 1976, The origins of rare earths in metalliferous sediments of the Troodos Massif, Cyprus. Earth Planet. Sci. Lett., 28, 285-294.

Robertson, A.H.F., and Fleet, A.J., 1977, REE evidence for the genesis of the metalliferous sediments of Troodos, Cyprus. Spec. Pub. Geol. Soc. London, 7, 78-79.

Schumm, R.H., Wagman, D.D., Bailey, S., Evans, W.H., and Parker, V.B., 1973, Selected values of chemical thermodynamic properties. Tables for the lanthanide (rare earth) elements. Nat'l Bur. Stand. Tech. Note 270-7, 75.

Shannon, R.D., 1976, Revised effective ionic radii and systematic studies of interatomic distances in halides and chalcogenides. Acta Cryst., A32, 751-767.

Smith, R.M., and Martell, A.E., 1976, Critical Stability Constants, Vol. 4. Inorganic Complexes. Plenum Press, New York, 176 p.

Spirn, R.V., 1965, Rare earth distributions in the marine environment. Unpub. Ph.D. dissertation, Mass. Inst. Tech.,Cambridge, MA 165 p.

Taylor, S.R., and McLennan, S.M., 1988, The significance of the rare earths in geochemistry and cosmochemistry. In: K.A. Gschneider and L. Eyring, Eds., Handbook on the Physics and Chemistry of Rare Earths. Elsevier Sci. Pubs., North Holland Physics Pub. Div., Amsterdam, Vol. 11, p. 485-580.

Thompson, L.C., 1979, Complexes. In: K.A. Gschneider and L. Eyring, Eds., Handbook on the Physics and Chemistry of Rare Earths. Elsevier Sci. Pubs., North Holland Physics Pub. Div., Amsterdam, Vol. 4, p. 209-295.

Turner, D.R., and Whitfield, M., 1979, Control of seawater composition. Nature, 281, 468-469.

Turner, D.R., Whitfield, M., and Dickson, A.G., 1981, The equilibrium speciation of dissolved components in freshwater and seawater at 25°C and 1 atm, pressure. Geochim. Cosmochim. Acta, 45, 855-881.

Vander Sluis, K.L., and Nugent, L.J., 1972, Relative energies of the lowest levels of the $f^q ps^2$, $f^q ds^2$, and $f^{q+1}s^2$ electron configurations of the lanthanide and actinide neutral atoms. Phys. Rev., A6, 86-94.

Wagman, D.D., Evans, W.H., Parker, V.B., Schumm, R.H., Halow, I., Bailey, S.M., Churney, K.L., and Buttall, R.L., 1982, The NBS tables of chemical thermodynamic properties. Selected values for inorganic and C1 and C2 organic substances in SI units. J. Phys. Chem. Ref. Data 11, suppl. 2, 392

Wood, S.A., in press, The aqueous geochemistry of the rare earth elements and yttrium. Part 1. Review of available low temperature data for inorganic complexes and the inorganic REE speciation of natural waters. Chem. Geol.

RARE EARTH ELEMENTS IN LUNAR MATERIALS

INTRODUCTION

Before the Surveyor and Apollo missions, scientists could still believe that the Moon had accreted cold from the solar nebula and, like the chondritic meteorites, had never undergone igneous differentiation. This general proposition was set forth in detail by Urey (1952). If that had been so, we might have expected the relative abundances of the rare-earth elements (REE) for samples of materials from the Moon (as well as for the Moon as a whole) to match those in the chondritic meteorites. In contrast, it was also possible to believe that the Moon had undergone extensive internal melting and that the lunar seas were dark in albedo from volcanic lavas (e.g., Baldwin, 1949; Kuiper, 1954). Processes of internal melting leading to formation of lavas would surely produce fractionation among the REE similar to that found in terrestrial basalts. Today, with some 382 kg of rocks and "soils" collected from the Moon, plus newly discovered lunar meteorites, we know that the Moon underwent extensive chemical differentiation and the relative REE abundances of its crustal materials differ substantially from the solar nebula distribution as we infer it from chondrites and the solar atmosphere.

It is the purpose of this chapter to provide an introduction to lunar REE distributions and how they may be related to the processes that produced the materials sampled at the lunar surface. Despite the universality of chemical separation processes and many similarities in rock types, the Moon lacks the plate tectonic framework, the active mantle, the continued recycling of crustal materials, and the surface weathering processes that characterize terrestrial geology. Frozen into its crust are features that may resemble those of the Earth's earliest crust, which Earth's active surface processes have long since obscured. Thus, we begin with an overview of some major features of the Moon and a model for its early history.

Our sampling of the lunar surface is meager, and most samples were collected from the surface where they had been thrown by impacts into the crust, so their specific provenances are not known. This situation lends itself more readily to speculation about whole-planet geochemical properties and to development of global models for the formation and differentiation of the Moon than to the systematic understanding of individual rock types or formations. REE patterns observed for common lunar materials have provided both clues to and constraints on the prevalent model for lunar igneous differentiation, that of the global magma ocean, set forth shortly after the Apollo 11 samples had been acquired (e.g., Anderson et al., 1970; Wood et al., 1970). This model accounts for several major features of the Moon and provides an outline of the principal aspects of the Moon's early chemical differentiation. In its simplest form, it does not account for the entire variety of known rock types or all of the relationships among the REE and the other elements. Nor should we expect a global model still under development to explain every separation of materials on anything as vast and complex as even a small planet. The magma ocean model in its simplest form is nevertheless a convenient means of introducing some rather general features of the Moon and how the REE patterns of common lunar materials relate to lunar geochemical evolution, so we use it as a framework here.

THE NATURE OF PLANET MOON

We begin by listing some observations about the Moon (for a more detailed review, see Taylor, 1982). Its density is 3.34 g/cm^3, nearly identical to that of ordinary chondritic meteorites. Its density and moment of inertia (nearly that of a homogeneous sphere) preclude the presence of a large metallic or sulfide core; seismic evidence for a small core is ambiguous. The Moon's crust is compositionally evolved compared with any plausible starting material for planets in the solar system. Surface highland materials average about ~26 wt % Al_2O_3, which corresponds to ~70 wt % calcic plagioclase. Most lunar plagioclase is very calcic (average An >90%) because the Moon, presumably ever since it accreted, is deficient compared to the Earth in Na and other elements of equal or greater volatility. The parent regions of the mare basalts are so depleted in water, for example, that whatever hydrogen the rising magmas may have contained has been overwhelmed by the trace amounts of that element produced in the sampled lavas by cosmic ray spallation reactions during their residence on the lunar surface. The extensive depletion in volatile substances is also reflected in the highly reducing environment of the lunar interior. Lunar crustal rocks contain no Fe^{3+}; their equilibrium is between Fe^{2+} and FeO.

The depletion of the Moon in volatile substances is matched by a depletion in siderophile elements. Concentrations of Au, Ir, etc. in lunar lavas are as low or lower than in their terrestrial counterparts, despite the absence of any substantial lunar core. Concentrations of Ni in mare basalts are typically 10 to 100 parts-per-million (ppm). This seems consistent with the currently popular theory for the formation of the Moon, namely, that it condensed from an Earth-encircling ring of debris that formed when, late in the era of planetary accretion, a Mars-sized object collided with the proto-Earth (e.g., Hartmann and Vail, 1986; Stevenson, 1987). In this model, both the Mars-sized object and the Earth had cores, which merged as a result of the collision. The ring of silicate would thus have been deficient in siderophile elements.

The feldspathic crust has the average composition of anorthositic norite or anorthositic gabbro, based on analyses of bulk lunar soils and interpretation of the ~20% of the surface that was remotely sensed by the gamma-ray experiments of Apollo 15 and 16. The center-of-gravity of the Moon is offset in the direction of Earth by about 2 km from its center-of-figure. Accompanying the observation of a greater distance of the farside surface from the center-of-gravity is the virtual absence of mare fill in farside basins. If we attribute the offset to a greater depth of plagioclase on the farside, we calculate that the thickness of the farside crust may exceed that of the nearside crust by some 60 km, a result not yet tested by seismic observation. Available seismic evidence indicates that the crust beneath the Oceanus Procellarum region of the nearside is some 20 km thick, and the exterior of the Moon is frozen to a depth of at least 900 km, below which some partially molten material appears to be present. The Moon is seismically quiet, releasing only about 10^{-10} as much seismic energy per year as the Earth.

Surviving fragments of several plutonic highland rocks give crystallization ages >4.4 Ga, evidence that igneous differentiation occurred very early in solar-system history to form a stable crust. The crust was thick enough by the time lavas flooded the mare basins to support the lavas and maintain substantial gravity anomalies ("mascons"). The principal tectonic style of the Moon, whose effects are frozen into its ancient crust, was that of impacts of planetoids, the last large group of which hit an already thick crust some 3.9-4 Ga ago.

These last major impacts produced the large, near-side basins that filled with dark lavas to form the features the ancients called the maria, or lunar seas. The oldest sampled mare basalts have ages about 4 Ga and the youngest have ages of about 3.1 Ga; smaller lava fields with low densities of impact craters may be substantially younger. The meteoroid impacts that produced the extensively cratered surface and subsequent meteorite bombardment produced an extensive regolith of dust ("soil"), breccias, new and apparently undifferentiated igneous material ("melt rocks"), and glassy products that range from microscopic glass mixed with mineral and lithic fragments ("agglutinates") found in the soils, to breccias with glassy matrices, to continuous glassy coatings splashed onto rocks and soils. (Igneous crusts of the majority of the solid bodies of the solar system are presumably hidden beneath analogous regoliths of debris from impact cratering; exceptions are planets with extensive surface processes, e.g., Earth, Io, Venus, and to an extent, Mars).

Unaltered samples of ancient mare basalts are common in the Apollo sample collection because the mare lavas postdate the era of heaviest impact bombardment, but the proportion of Apollo samples that consists of recognizable highland igneous rocks with diameters exceeding a centimeter is very small. However, impact brecciation and mixing did not homogenize the highland regolith to such an extent that it has the same composition everywhere. We obtain clues to the nature of the original igneous rocks of the lunar highlands by detailed studies of small fragments of minerals and rocks found as particles in the soils and clasts within the breccias and of compositional trends among bulk soils and breccias. However, direct evidence in sampled materials for the presence of some inferred rock types and compositions has been obscured. Nearside provenances several kilometers in diameter with different mafic mineral composition are being identified by Earth-based remote sensing (e.g., Pieters, 1986).

Impact processes mixed previously differentiated materials together, but are not yet known not to have fostered new chemical separation of those mixed materials (except for volatile trace elements). It is an important but often conveniently overlooked testimony to the inadequacy of our sampling of highland igneous rocks that the compositions of the highland soils and of many breccias cannot be accounted for as simple mixtures of the identified igneous rocks. This striking deficiency of our knowledge should be borne in mind as the literature on lunar differentiation is perused.

THE MAGMA OCEAN HYPOTHESIS AND ITS PRESUMED PRODUCTS

The elegant idea of a global ocean of magma was conceived to account for the major features of early lunar igneous differentiation and crust formation (see Warren, 1985, for a review). In oversimplified form, but one that provides a convenient framework for introducing lunar igneous rock types, the magma ocean concept of the main features of the Moon's early igneous history amounts to this: At the time it formed or very shortly thereafter, the Moon underwent total melting to at least a depth of 300-400 km (roughly half of its volume) to produce a global ocean of magma of bulk Moon composition. As this convecting ocean cooled by radiating its heat away to space, it crystallized. Its initial precipitates were cumulates mainly of olivine and orthopyroxene that came to rest on its floor. After more than half of the ocean had crystallized and its composition had evolved sufficiently, it also precipitated clinopyroxene, ilmenite, and plagioclase. Less dense than its parent magma, the plagioclase floated to form a thick anorthositic crust. The final fraction of the magma, a trace-element-rich liquid given the name "urKREEP" (Warren and Wasson, 1979a), was trapped between the thick pile of mafic cumulates and the crust. (Certain early observed glassy fragments were called "KREEP" by Hubbard et al., 1971a, because of their high concentrations of K, REE, and P; they contain a common set of relative abundances for incompatible trace elements—ITE—in the lunar highlands. The term "KREEP" has since been used in conjunction with any samples that show this set of ITE relative abundances.) Later partial remelting of the mafic cumulates gave rise to the mare lavas.

The principal rock types we might expect to result from crystallization of this type of magma ocean would be sinking dunite, pyroxenite, and peridotite, and floating anorthosite. From the residual urKREEP liquid and from remelting of the mafic cumulates we could expect two general types of basalt; the urKREEP layer would produce a trace-element-rich variety of highland volcanic rock, and the remelting of mafic cumulates would produce a more mid-ocean-ridge (MORB)-like basalt depleted in the most incompatible trace elements. The observed average regolith composition of noritic anorthosite would then correspond to mixtures of anorthositic rocks, volcanic rocks, and cumulates exhumed from beneath the anorthositic crust (if meteoroid impacts produced sufficiently deep craters, which is questionable), plus a small amount of meteoritic debris. The observed anorthositic norite composition need not represent the entire crust, however, but might only be that of a veneer over the bulk crust.

How do these predictions compare with what we find in the sample collections? Among the several investigator groups who continue to discover new rock types, Warren and coworkers (Warren et al., 1978, 1979b, 1980, 1981, 1982, 1983, 1986, 1987) have made particularly extensive and systematic searches for large and small fragments of plutonic rocks. They label as "pristine" those fragments whose mineralogical, textural, and compositional characteristics appear to stem from the original igneous process that produced the parent rock, without extensive metamorphism or chemical contamination from impact brecciation. The bulk of these "pristine" plutonic rocks, including the largest samples of plutonic rocks in the Apollo collection, are anorthosites (most of which contain of 95-98 vol % plagioclase), norites, and troctolites. Rarer dunite, anorthositic norite, noritic anorthosite, gabbro, quartz monzodiorite, and felsite have been observed as fragments of plutonic rocks, and a KREEPy form of basalt has also been identified. Most of the "pristine" rocks occur as small clasts in breccias. Modern analytical techniques (mainly neutron activation analysis and mass spectrometric isotope dilution) have enabled us to obtain extensive compositional data from very small samples (<1-100 mg), but extrapolation of such observations to types and compositions of rock formations leaves room for substantial uncertainty. Even the largest samples of anorthosite collected contain a relatively small number of grains or crystals.

Such a substantial proportion of norite and troctolite as found among the samples is not predicted by the magma ocean concept. Their presence implies failure of plagioclase and mafic minerals to separate from each other if we take them to be products of the magma ocean. For this and other reasons discussed below, most investigators regard them as later additions to the crust (e.g., Warren, 1986). Were it not for a disparity in Mg# between the most common type of anorthosite and the norite and troctolites, we could adjust the model slightly to allow that the separation of plagioclase from residual magma was not a very clean one, owing to the small difference in density between those two materials. Then we might expect to find a substantial proportion of noritic anorthosite and anorthositic norite among the crystallization products of the magma ocean. The average surface composition even suggests that this may have happened; "pristine" samples of these rock types are rare in the sample collections, however.

Crystallization of a real magma ocean would be a complicated affair (e.g., Longhi and Boudreau, 1979). The reader may wish to peruse the lunar literature for the evolution of fixes to bring the magma ocean model into harmony with the properties of the lunar rocks and soils. Whether there actually was a magma ocean with the general properties and evolutionary path introduced above remains an open question. Either there was, and we are learning how it functioned and what changes occurred in the crust subsequent to its crystallization, or there was not, and some other model will eventually supplant it.

LUNAR REE PATTERNS

Despite the paucity of information about the Moon in comparison to that about the Earth, we cannot do justice to the extensive information about REE in lunar rocks or to the investigators who have provided the data and the many ideas for their interpretation. We limit our discussions mainly to the general REE characteristics of materials, emphasizing igneous rocks and processes, and more or less to the mainstream of their interpretation. More specifically, we provide an overview of REE concentrations and relate them to lunar crustal evolution in the framework of the magma ocean model. We make no attempt to reference all pertinent works, but for more detail and alternate interpretation, we refer the reader to the two- and three-page abstracts of the twenty Lunar Science Conferences (initially, the Apollo 11 Lunar Science Conference, and later the Lunar and Planetary Science Conferences), to the proceedings of those conferences, to the forthcoming Lunar Sourcebook (Heiken et al., 1990), and to the references therein.

If we take the mean relative REE abundances for the whole Moon to be that of chondrites, we can predict the REE patterns we would expect for lunar rocks. We can compare these predictions with observed REE patterns, and use the results to support or modify whatever model for their origin we may propose. We cannot emphasize too much that, useful as REE are as indicators of the nature and extent of igneous processes, they provide only a single set of clues and constraints. Data for other trace elements, for major elements, and for isotopes, plus petrographic and field observations and the context of experimental studies are all required to provide a convincing demonstration of the origin of any rock formation.

Highland plutonic rocks

Anorthosites. Grains of plagioclase interpreted as fragments of anorthosite were first observed when the Apollo 11 soil was investigated, and the concept of a thick, floated anorthositic crust soon followed (e.g., Anderson et al., 1970; Wood et al., 1970). The discovery of the "genesis rock," a 600-g sample collected during the Apollo 15 mission, confirmed the existence of lunar anorthosite. The "genesis rock" belongs to the principal class of lunar anorthosite, called "ferroan" anorthosite because the Mg# (mols of MgO divided by sum of mols of MgO and FeO) of its sparse olivine and pyroxene is low relative to the trend of Mg# of mafic minerals versus An content of plagioclase observed in most other lunar plutonic rocks. Its proportion of plagioclase is high (>95%). Its concentrations of ITE are low, implying, along with its low proportion of mafic minerals, near absence of a trapped liquid component. Ferroan anorthosite has been taken by some investigators to contain the prototypical plagioclase accumulated from the magma ocean.

We would expect an almost exclusively plagioclase rock crystallized from a magma ocean to have low REE concentrations, a negative slope, and a substantial positive Eu anomaly; in fact, this is observed. REE concentrations for two whole-rock fragments of the "genesis rock," analyzed simultaneously, are shown in Figure 1a. Note the Eu anomaly, whose size exceeds that normally found among terrestrial rocks and stems from the highly reducing lunar environment. (Note also the apparent difference in relative Lu abundances between the two samples. Such differences for plagioclase-rich rocks appear in data obtained by both neutron activation and mass spectrometric isotope dilution, and still escape explanation; Haskin et al., 1981.) The proportion of plagioclase in the ferroan anorthosites is so high and the proportion of cocrystallized mafic minerals or frozen trapped liquid is so low that the REE pattern is essentially that of pure plagioclase. The mechanism for producing anorthosite with such a high proportion of plagioclase of nearly constant An content remains a puzzle, and an especially important one if ferroan anorthosite as represented by this and several Apollo 16 specimens is abundant on the Moon. We note, however, that the suite of rocks called ferroan anorthosites was initially defined in a manner that excluded small, rare anorthositic samples containing higher percentages of mafic minerals (see Haskin and Lindstrom, 1988, for a discussion of this problem).

Large fragments of anorthosite were also found at the Apollo 16 site, and small fragments are present in the Luna 20 core. Only rare, tiny fragments have been found at the Apollos 14 and 17 highland sites. Although ferroan anorthosite was once regarded as the definitive product of the magma ocean, the crust probably never consisted of a shell of nearly pure ferroan anorthosite; if it did, the part of it we have sampled was contaminated with later plutons and extrusive materials. No site has yielded the preponderance of anorthosite expected from the extreme hypothesis of flotation of a nearly pure plagioclase crust.

Figures 1b and 1c show the ranges and means of REE concentrations in large and small ferroan anorthosites from the Apollo 15 and 16 sites. The small fragments available for analysis from the Apollo 17 and Luna 20 missions have somewhat higher REE concentrations than the "purest" samples from Apollos 15 and 16, and nearly horizontal slopes for the HREE that reflect higher proportions of mafic minerals. They also have correspondingly higher concentrations of Fe and Sc.

232

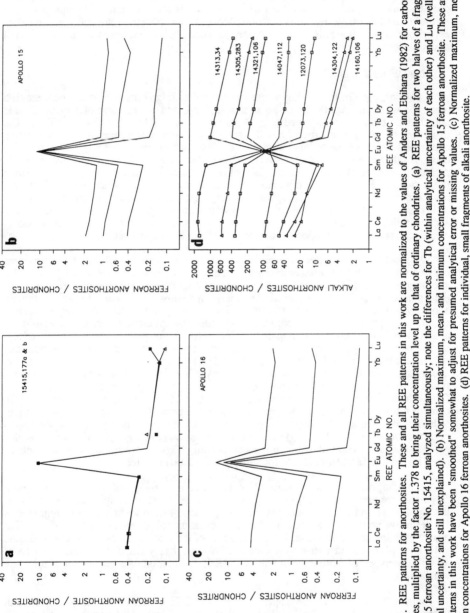

Figure 1. REE patterns for anorthosites. These and all REE patterns in this work are normalized to the values of Anders and Ebihara (1982) for carbonaceous chondrites, multiplied by the factor 1.378 to bring their concentration level up to that of ordinary chondrites. (a) REE patterns for two halves of a fragment of Apollo 15 ferroan anorthosite No. 15415, analyzed simultaneously; note the differences for Tb (within analytical uncertainty of each other) and Lu (well outside analytical uncertainty, and still unexplained). (b) Normalized maximum, mean, and minimum concentrations for Apollo 15 ferroan anorthosite. These and most REE patterns in this work have been "smoothed" somewhat to adjust for presumed analytical error or missing values. (c) Normalized maximum, mean, and minimum concentrations for Apollo 16 ferroan anorthosites. (d) REE patterns for individual, small fragments of alkali anorthosite.

(The REE patterns in most of the figures in this paper have been "smoothed" somewhat to compensate for suspected analytical errors and for missing data for some elements in some samples. This has been done mainly for the elements Nd, Gd, Tb, and Dy. The reader is warned that reported values produce significantly bumpier patterns than those shown in some cases, and that such differences, while highly suspect, could be real. Treating the actual patterns as real, however, would not aid this introduction to the overall nature of lunar REE patterns and would require forms of explanation that the author is unprepared to justify. Any use of the data to support a particular hypothesis for the origin and nature of a particular rock, however, should not so readily ignore the possibility of errors introduced by smoothing. The carbonaceous chondrite values of Anders and Ebihara, 1982, multiplied by 1.378 to account for the difference between ordinary chondrites, were used for the normalization in the figures in this paper. Values in parts-per-million used for normalization are as follows: La, 0.325; Ce, 0.849; Nd, 0.630; Sm, 0.205; Eu, 0.0772; Gd, 0.272; Tb, 0.049; Dy, 0.338; Yb, 0.219; Lu, 0.0338. References from which the data for the figures were gathered are given in Table 1.)

Recent studies (James et al., 1989a) suggest that the ferroan anorthosites do not all derive from a single, homogeneous source, but represent several different sampled groups, and the implication is that each group represents a specific environment or pluton. Anorthosites from a single landing site do not all belong to a single group. We note that, as more studies of lunar samples are done, the tendency is toward recognizing that compositional groups within rock types seem to represent products of localized differentiation. Whether that contradicts the notion of a relatively homogeneous magma ocean or indicates important detail about how a magma ocean works awaits further investigation and argument.

There is a compositional gap between the ferroan anorthosites (members of a presumed "ferroan suite" of lunar plutonic rocks) and other plutonic rock types ("magnesian suite") that models for cogenetic differentiation find difficult or impossible to bridge. Two non-ferroan types of anorthosite have been found as small particles or clasts in the lunar samples. Slightly more alkaline anorthosite (An_{75-87} versus An_{96} for ferroan anorthosite) occurs as small fragments in the Apollo 12 and 14 samples (e.g., Warren et al., 1981). In fact, the breccia clast with the highest observed REE concentrations found so far for any lunar rock is a fragment of alkali anorthosite (Fig. 1d). These rocks contain whitlockite (a calcium phosphate) as the host phase of the high REE concentrations. How to produce such REE-rich, plagioclase-phosphate rocks remains an interesting question. Another type is called "magnesian" anorthosite because its Mg# is that expected for anorthosite from the magnesian suite of plutonic rocks (Lindstrom et al., 1984). REE concentrations in one sample of magnesian anorthosite are quite high, and the REE pattern shows no obvious effect of the presence of plagioclase.

Dunites, troctolites, norites, and gabbros. If plagioclase floated and mafic minerals sank, as predicted by our oversimplified magma ocean model, mafic minerals should be found in dunitic and pyroxenitic rocks, unassociated with plagioclase (except for small proportions formed from trapped liquid). Rocks such as troctolite and norite (or gabbro), which consist of mixtures of plagioclase and mafic minerals, are not predicted to be abundant. Troctolites and norites are nevertheless, next to anorthosite, the most abundant "pristine" rock types. Also, the majority of the troctolites and norites belong to the magnesian suite of lunar plutonic rocks, and thus appear not to be genetically related to the anorthosites. If we accept the hypothesis of an early crust of ferroan affinity derived from a magma ocean, we must ascribe the presence of the majority of the norites and troctolites to later intruding plutons.

We might predict the occurrence of abundant troctolite and norite (but not the ones we have found) from a slightly modified magma ocean model in which the lunar crust did not consist mainly of accumulated plagioclase, but of frozen, residual, cotectic magma in which

234

Table 1. References to data used in figures [*]

References	Figures	References	Figures
Blanchard and Budahn (1979)	2e,3,9	Ma et al. (1981)	5a
Blanchard et al. (1975)	2e,9	Marvin and Lindstrom (1983)	1c,2d,9
Blanchard et al. (1977)	2c,3	Marvin and Warren (1980)	2d,9
Blanchard et al. (1978)	5e,6a,6b,9	McKay et al. (1986)	6a,6b,9
Boynton and Hill (1983)	8a	Nava and Philpotts (1973)	6b
Dickinson et al. (1985)	5f	Neal et al. (1988a)	5f
Fukouka et al. (1986)	8a	Neal et al. (1988b)	5f
Gast et al. (1970)	5b	Neal et al. (1989)	5f
Gillum et al. (1972)	6a,6b,9	Nyquist et al. (1979)	5c
Goles et al. (1970)	5b	Ostertag et al. (1986)	8a
Goodrich et al. (1984)	8a	Palme et al. (1978)	9
Goodrich et al. (1986)	2a,5f,9	Palme et al. (1983)	8a,9
Gromet et al. (1984)	6b	Philpotts et al. (1972)	5e
Haskin et al. (1970)	5b	Quick et al. (1977)	3,9
Haskin et al. (1973)	7a,7b,9	Rhodes and Blanchard (1980)	5b,9
Haskin et al. (1974)	2b,9	Rhodes and Blanchard (1981)	5c,6a,6b,9
Haskin et al. (1981)	1a,1b,1c,1e,9	Rhodes and Hubbard (1973)	5d
Helmke and Haskin (1972)	5e,6a	Rhodes et al. (1974)	5d
Helmke et al. (1972)	5f,6a	Rhodes et al. (1976)	5a,9
Helmke et al. (1973a)	5d	Rhodes et al. (1977)	5c,9
Helmke et al. (1973b)	6a	Shervais et al. (1983)	5f
Hubbard et al. (1971b)	1d	Shervais et al. (1984)	1d,9
Hubbard et al. (1972a)	7b	Shervais et al. (1985a)	5f,9
Hubbard et al. (1972b)	5e	Shervais et al. (1985b)	2a,5f,9
Hubbard et al. (1973)	4,9	Shih et al. (1975)	5a
Hubbard et al. (1974)	1c	Shih et al. (1985)	4
Hughes et al. (1989)	5a	Simon et al. (1988)	1b,4,9
James et al. (1984)	1c	Tatsumoto et al. (1976)	4
James et al. (1989a)	1e,9	Taylor et al. (1974)	1c,9
James et al. (1989b)	2d	Wänke et al. (1970)	5b,6a,6b,9
Jerome et al. (1973)	6a,6b,9	Wänke et al. (1971)	5c,6a,6b,7b,9
Kallemeyn and Warren (1983)	8a,9	Wänke et al. (1972)	6a,6b,7b,9
Korotev et al. (unpubl.)	5e,8a,9	Wänke et al. (1973)	6a,6b,7b,9
Korotev (1981)	6a,6b,9	Wänke et al. (1974)	1c,5a,6a,6b,7b,9
Korotev et al. (1983)	8a,9	Wänke et al. (1975)	1b,1e,5a,5d,6a,6b,7b,9
Laul and Schmitt (1975)	2b,9	Wänke et al. (1976)	9
Laul and Schmitt (1973a)	6a,6b,7b,9	Wänke et al. (1977)	6a,6b,7b,9
Laul and Schmitt (1973b)	1b,1c,9	Warner et al. (1979)	5a
Laul et al. (1977)	5e,9	Warren and Kallemeyn (1986)	8a,9
Laul et al (1983a)	9	Warren and Wasson (1978)	1c,2b,2e,4
Laul et al. (1983b)	8a,9	Warren and Wasson (1979b)	1c,1e,2b,2d,2e,9
Laul et al. (1989)	8b,9	Warren and Wasson (1980)	1b,1c,1d,1e,2a,2c,2d,9
Lindstrom (1984)	1c,2a,7b,9	Warren et al. (1978)	1b,1e,4,5a,6a,6b,7b,9
Lindstrom et al. (1984)	9	Warren et al. (1981)	2a,2b,2c,9
Lindstrom et al. (1986)	8a	Warren et al. (1982)	1d,2a,2e,9
Lindstrom et al. (1987)	8a	Warren et al. (1983)	1b,1c,1d,9
Lindstrom et al. (1988)	1b,2c,4,9	Warren et al. (1986)	9
Lindstrom et al. (1989)	4,9	Warren et al. (1987)	1d,3,9
Lindstrom & Lindstrom (1986)	8a	Wentworth et al. (1979)	5a,9
Lindstrom and Salpas (1983)	1c,2d,8b,9	Winzer et al. (1974)	2e,9
Ma et al. (1977)	5e,6a,6b,9	Winzer et al. (1975)	2e

[*] Data for figures were taken from these references. Note that 1e refers to Apollos 14, 16, and 17 troctolitic anorthosites (not shown), 2e refers to Apollo 17 norites (not shown), and that most of the data used to derive the figures are not individually shown.

earlier crystallized plagioclase accumulated but from which cocrystallizing mafic minerals sank. This would provide a crust enriched in plagioclase relative to a cotectic liquid and would be in line with the anorthositic norite overall composition of the lunar surface crust (about 26% Al_2O_3). The straightforward implication of that composition, if it is extrapolated to the entire highland crust, is that the crust, while enriched in plagioclase, consists mainly of anorthositic varieties of troctolite and norite. The majority of the sampled troctolites and norites, however, in addition to having Mg#'s too high to be related to the ferroan anorthosites, tend toward cotectic compositions without any excess of accumulated plagioclase crystals. Anorthositic varieties of either magnesian or ferroan affinity are rare, although several small clasts have been analyzed.

Our initial thought might be that most norites and troctolites from a magma ocean would carry an excess of plagioclase relative to a cotectic liquid and thus have positive Eu anomalies. However, as long as Eu-enriched plagioclase remained mixed in the proper proportions with its Eu-depleted residual parent liquid, there would be no Eu anomaly except where fluctuations about these proportions might occur on a local scale. The same considerations would hold on a smaller scale for single plutons that might float plagioclase (e.g., Korotev and Haskin, 1988).

The kinds of fluctuation in REE pattern we would predict from this modified magma ocean model are observed, at least qualitatively, in the REE patterns for the actual lunar norites and troctolites as shown in Figure 2. REE patterns for norites and troctolites tend to be similar, and comments made for one rock type hold for the other. The average Apollo 14 troctolite shows a slight negative Eu anomaly and the average Apollo 17 troctolite a small positive one. Individual samples show positive or negative Eu anomalies. These samples may reflect accurately the general characteristics of their parent formations. In that case, regions with an excess of accumulated plagioclase relative to the amount of associated residual liquid would have relatively low, positive Eu anomalies. Magmas with an excess of residual liquid relative to plagioclase would have higher REE concentrations and negative Eu anomalies (this could include, for example, magmas that assimilated urKREEP during ascent, e.g., Warren, 1988). Qualitatively, these trends in REE concentration and Eu anomaly are observed; quantitatively, it is difficult to evaluate this idea because without field evidence there is no reason to believe that the different rocks derived from a common parental magma.

Alternatively, the Eu anomalies and REE concentration levels might be artifacts of the small samples collected or of the even smaller samples analyzed from these relatively coarse-grained rocks. Overall, the parental magmas or rock formations might not deviate significantly from cotectic proportions of plagioclase and pyroxene or olivine. Individual samples with a slight excess of the final dregs of crystallizing liquid relative to the average proportion would have negative Eu anomalies and relatively high REE concentrations, and samples deficient in such ITE-rich liquid would have positive Eu anomalies and relatively low REE concentrations, much as is observed.

Parent liquid REE concentrations can be estimated for these rocks. This is sometimes done by analyzing separated minerals and dividing their REE concentrations by values of the distribution coefficients for that mineral. This is an incorrect procedure if the entire rock consists of crystallized initial liquid or if the rock contains a mixture of cumulus minerals and trapped parental liquid (Haskin and Korotev, 1977). In the first case, the whole rock represents the liquid. In the second case, it is necessary to correct the composition of the mineral separates for their orthocumulus portions. This is usually a difficult task, because the rocks tend to be metamorphically equilibrated so their minerals are homogeneous in composition and the orthocumulus contribution for REE is apt to exceed the cumulus contribution by an order of magnitude. In some cases, the composition of the trapped, intercumulus liquid, which may represent the parental liquid, can be estimated (Haskin et al., 1974). Most careful efforts to estimate parent liquid REE concentrations yield La concentrations in the approximate range of 10 to 40 times the chondritic value, with negative slopes such as characterize most highland samples. Many predict Eu anomalies, but this is difficult to establish accurately.

Noritic rocks (mainly plagioclase and orthopyroxene) are more abundant than gabbros (mainly plagioclase and clinopyroxene) among the highland samples, and the rare gabbros are found mainly as small clasts. A REE-rich form called sodic ferrogabbro has been reported; REE patterns for two samples are included in Figure 2d. The relatively shallow LREE slope of the ferrogabbros matches that of the Apollo 16 norites, suggesting that both the ferrogabbros and the norites may belong to a single geochemical province.

Samples of pure mafic cumulates are rare. One 32-gram sample of magnesian lunar dunite, a few much smaller probable dunites, and some tiny fragments of possible pyroxenites have been identified in the lunar collection. These materials are unlikely to be early cumulates from a simple magma ocean because such mafic, magnesian materials should be too deep within the lunar mantle to have been exhumed by impact cratering. Most investigators regard them as members of the magnesian suite of plutonic rocks, products of differentiation of plutons emplaced within the initial crust.

The REE pattern for the dunite is shown in Figure 2b. Laul and Schmitt (1975) analyzed several small samples of the rock and modeled the compositional variations they observed as mixtures of cumulus olivine and products of trapped liquid. The REE concentrations inferred for the parent liquid have the same characteristics as those for the troctolites and norites. It has a positive Eu anomaly similar to that in the whole rock. The small, positive Eu anomaly suggests that the dunite may contain an excess of plagioclase over that expected from crystallization of a trapped residual liquid, however. If so, the parent liquid might not have had a positive Eu anomaly.

Lunar felsite (granite). Most highland materials have REE patterns that show an almost linear negative slope, except for the Eu anomaly. A major exception is lunar felsite (granite), whose REE pattern shows a negative slope for LREE and a positive slope for HREE (Fig. 3). Lunar felsite has remarkably high K concentrations (for a lunar material) of some 5 wt %, which leads to the presence of alkali feldspar; the other and more abundant main component is silica, another rare lunar material. Lunar felsite has been found in pure form only as small fragments in breccias or soils, and as a major mixing component in an 82-gram complex breccia from Apollo 12.

Lunar felsite is the sort of highly differentiated material, apparently minor in volume, whose existence and nature would be difficult to predict a priori from a simple magma ocean model. Based on its high REE concentrations and peculiar REE pattern, it would seem to be related to late-stage processes of magma ocean crystallization, after the main evolution of crustal differentiation was complete, the very conditions it is most difficult for a general model to accomodate. A more proper concern is whether it can be produced from a straightforward, major product of magma ocean differentiation. The logical material to consider in this regard is urKREEP.

An origin of lunar felsite by a combination of phosphate fractionation and liquid silicate immiscibility of phosphate-rich KREEPy differentiates has been proposed (e.g., Shih et al., 1985; Rutherford et al., 1976; Neal and Taylor, 1989). In fact, the quantities of felsite in the crust are suggested to be greater than the paucity among the samples so far. Based on suspected assimilation of granite during formation of some mare basalts, Neal and Taylor (1989) have argued that substantial bodies of lunar granite be present beneath the surface.

Highland volcanic rocks: KREEP. Only one common type of material is known that seems to be unambiguously a volcanic product of the highlands, and that is KREEP. KREEP was first identified as glassy particles in Apollo 12 soils (Hubbard et al., 1971a), and given the name KREEP because it was noticed to have high concentrations of K, REE, and P (and also other ITE). The same relative abundances of REE and other ITE have

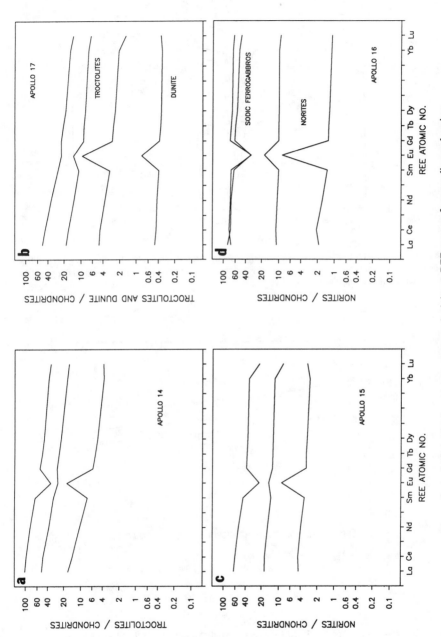

Figure 2. (a-c) Normalized maximum, mean, and minimum REE patterns for troctolites and norites. (d) REE patterns for two Apollo 16 norites and two ferrogabbros.

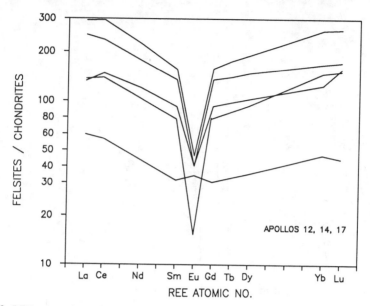

Figure 3. REE patterns for lunar felsites (granites). Note the deep Eu anomalies for most samples and the steep negative LREE slope and steep positive HREE slope, unusual among lunar materials, and a possible result of silicate liquid immiscibility.

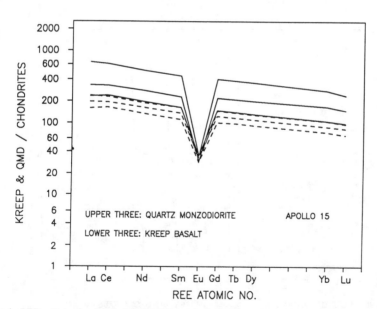

Figure 4. REE patterns for Apollo 15 KREEP basalt, the only known form of highlands igneous extrusive rock, and of three small fragments with the major-element composition and mineralogy of quartz monzo-diorite (QMD). The QMD samples have the same relative abundances of REE and other incompatible trace elements as the KREEP basalt. Many nearside highlands samples have this set of relative abundances and are referred to as "KREEPy."

since been found in significant amounts of material from most sampled sites, and such materials are commonly referred to as KREEP, or as KREEPy, so the definition of KREEP as a specific rock type is to be avoided, and any discussion of KREEPy material requires definition. Here, we will use "KREEP" and "KREEPy" only as adjectives to indicate that a material has fairly high concentrations of ITE with the KREEPy relative abundance pattern.

In our oversimplified version of magma ocean evolution, toward the end of crystallization, the bulk of the original ocean would consist of thick cumulus layers of mafic minerals, and a significant fraction of the last half would consist of plagioclase floating in residual liquid. In fact, some early modeling presumed that the plagioclase floated on the final liquid as a nearly continuous shell rather than in it. The final liquid would thus form a thin layer beneath the shell of (plagioclase-rich) anorthosite. Because of fractional crystallization, the final liquid would be highly enriched in REE and other ITE. This layer of liquid, the initial source of the ITE now present in KREEPy materials, has been dubbed "urKREEP" (Warren and Wasson, 1979a).

The highest frequency of encounter of KREEPy materials is in the samples collected at Apollo 14. The element Th, which can be observed by remote sensing, is well correlated with the REE other ITE in most highland materials. This correlation itself is mainly a consequence of the presence of KREEPy material in most sampled highland breccias and all soils. We infer that the entire ITE suite of elements is enriched in proportion to Th. Remote sensing for Th by gamma-ray spectrometry during the Apollos 15 and 16 missions indicates that the Mare Imbrium region has the highest surface concentrations of that element, which presumably means the highest exposures of KREEPy material. If there is a globe-encircling shell of urKREEP, it is not evident at the surface on the lunar farside. It can be argued that the greater distance from the center-of-mass to the farside surface may have precluded its ascension as an extrusive material onto the surface.

If we assume a value of 0.01 for the overall distribution coefficient of La during the process, we find that the average of 110 ppm for Apollo 14 high-K KREEP would require that the process of fractional crystallization of a chondritic ocean have proceeded to the extent of 99.7%; even beginning with a magma ocean with a La concentrations three times that of chondrites would require crystallization to the extent of 99.1%. These are minimum values; most KREEPy materials are regarded as dilutions of urKREEP by less REE-rich materials during breccia formation and, indeed, KREEPy materials with La concentrations as much as three times as high as that of typical Apollo 14 KREEP are present in the sample collection. The final liquid would also presumably be rich in FeO and TiO_2, and too dense to extrude onto the surface, whose plagioclase density would be ~2.8 g/cm^3. Also, since the Mg# of KREEPy materials is ~65, rather typical of lunar soils and far higher than expected for >99% fractional crystallization of mafic magma, the production of such material from urKREEP requires some extensive chemical changes. Warren (1987) proposed that some of the second-stage magnesian magmas (the same type that produced the magnesian suite of norites and troctolites) assimilated urKREEP or at least extracted ITE from it during passage through the urKREEP layer into the crust. The resulting mixtures would be extruded as KREEPy volcanic rocks.

The only confirmed igneous occurrences of KREEP basalt are small samples found at the Apollos 15 and 17 sites (Fig. 4). REE concentrations of these materials are only about two-thirds as high as typical Apollo 14 KREEPy melt rocks, some of which have basaltic texture, but are clearly products of impact processes (e.g., McKay et al., 1978, 1979). The isotopic ages are ~3.9 Ga, and overall isotopic characteristics can be interpreted as indicating separation of the source region (i.e., urKREEP) some 4.3-4.4 Ga ago, the same as indicated for the mare basalt source regions (see Warren and Wasson, 1979, for a review). Whether all original occurrences of KREEP were volcanic remains unknown. Apollo 15 breccias also include clasts of rare quartz monzodiorite, which has a KREEPy REE pattern but higher concentrations and a more evolved major-element composition (Fig. 4).

Mare basalts

It is a treat to work with samples of mare basalt. Most have not been substantially shocked or partly melted by impact events and, despite their great ages (most of which are between 3.1 and 3.9 Ga), they are beautifully unaltered and unweathered. Their excellent state of preservation and their abundance in the Apollo collection owing to the necessity of landing in relatively smooth (therefore mare) terrain, has resulted in studies of mare basalts far out of proportion to the small volumetric fraction of those materials in the lunar crust. From mare basalts we obtain our best estimates of the nature of the lunar interior, its mineral composition, isotopic and chemical composition, and heterogeneity. (For a general overview, see the section on mare basalts, Basaltic Volcanic Study Project, 1981).

On the other hand, it is an inconvenience and a limitation to lack such field observations as how many flows were sampled and what their relative stratigraphic positions were. The number of flows and whether any in-flow differentiation occurred have to be surmised to an extent even less satisfactory than that associated with sampling of mid-ocean ridge basalts (MORB) from bore holes. Only at the Apollo 15 site was layering of mare basalt visible, and that was on the opposite side of Hadley Rille from where samples were taken, although at least one specimen from the Rille's edge can be defended as having been sampled in situ. Nevertheless, there are such large differences in compositions among mare basalts as to require equally large differences in compositions of lunar source regions, and there are means of determining the number of significantly different magmas of generally similar character at each landing site.

The principal samples of mare basalt were obtained from the mare landing sites, Apollos 11, 12, 15, and 17. One sizeable and one smaller specimen were obtained at the Apollo 14 site. Small fragments were found in the cores of Lunas 16 and 24. Less abundant types of mare basalt have been found in many soils and breccias, and some of these from Apollos 14 and 17 have been characterized. Relatively coarser-grained basalt samples are somewhat heterogeneous in composition, as evidenced from results of multiple analyses of 1-gram aliquants. Finest-grained varieties are regarded as more nearly homogeneous on that size scale. The compositions of the main basalt types are based on many analyses of several specimens of substantial size; compositions of other varieties depend on analyses of small to very small fragments from soils and breccias.

Without field data, what criteria are used to determine the number of separate basalt flows sampled at each site, information that is required to determine the number of distinct magma source regions and the extent of heterogeneity within regions? This must be done on petrographic and isotopic and chemical compositional grounds, and the answer is accordingly model dependent. Using terrestrial lava flows as analogs, we expect basalt samples from a single lunar lava flow to have mutually similar major-element compositions and identical mineral compositions, or to show textural evidence, modal variation, and compositional trends commensurate with in-flow differentiation. Gaps in chemical composition between groups of samples is a criterion for separate flows, and how large the gaps need to be to support separate flows we estimate by analogy with compositional differences between terrestrial flows and compositional ranges for multiple samples of single terrestrial flows (e.g., Lindstrom and Haskin, 1978). Compositional trends among sample groups belonging to individual flows can indicate near-surface magma fractionation.

Based mainly on criteria such as these, the principal types of mare basalt are taken to be the following: A-11 high-K basalts, one group; A-11 high Ti (low-K) basalts, 4 groups, and the similar A-17 high-Ti and VHT (very high-Ti) basalts, 3 groups; Apollo 12 low-Ti basalts, with an olivine-pigeonite group, an ilmenite group, and a feldspathic group, and similar Apollo 15 low-Ti basalts, with one or two olivine-normative groups, one of which is picritic, and a quartz-normative group; Apollo 14 aluminous basalts, consisting of two "large" samples plus many fragments from breccias which may constitute several groups, and fragments of a VHK (very high K) subvariety; a Luna-16 aluminous variety; an Apollo 17 VLT (very low Ti) group made up of small fragments, and a Luna 24 VLT group, also made up of small fragments. The number of flows sampled is at least as large as the number of basalt groups. Based on spectral reflectance data, which provide information on Fe and Ti concentrations, not more than a third or a half of the general mare basalt types have yet been sampled (Pieters, 1978).

Compositional differences among some of the basalt groups are reflected in their REE patterns and concentrations, which are shown in Figure 5. However, from REE concentrations and patterns alone, not all groups are mutually distinguishable. Thus, in Figure 5, the mean for several samples and the ranges for a population of samples are given by Apollo site, rather than by known basalt group. At some sites, the groups are reasonably distinct in REE patterns; e.g., the Apollo 17 high-Ti basalts and the Apollo 17 VLT basalts (Fig. 5a). Most REE patterns for Apollo 11 high-K, high-Ti basalts are distinguishable from those for low-K, high-Ti basalts (Fig. 5b). There are small but defendable differences, for example, among the Apollo 12 ilmenite group, (steeper positive slopes for the LREE), the Apollo 12 olivine-pigeonite group, and the Apollo 12 feldspathic group (higher relative LREE abundances) (Fig. 5c), and these differences are important to quantitative modeling of their separate origins, but we ignore them here.

Most mare basalt REE patterns have negative Eu anomalies, shallow negative slopes for HREE, and steep to shallow positive slopes for LREE. REE concentrations vary substantially, with LREE to MREE up to ~100 times chondritic values in Apollo 11 Hi-K basalts to <5 times chondritic values in the Apollo 17 and Luna 24 VLT basalts. An exception is the low-REE VLT (very low Ti) Luna 24 basalt from Mare Crisium, which has a slight positive Eu anomaly (or, within uncertainty, no Eu anomaly; Fig. 5f); the VLT basalt from Apollo 17 is not an exception. Another exception is the Apollo-14 high-alumina basalts, which have negative LREE slopes (Fig. 5e). Luna-16 mare basalt is interesting for its shallow Eu anomaly (Fig. 5f).

(It is somewhat amusing that the mare basalts, with their LREE-depleted patterns, are the lunar analog of the terrestrial MORB in the sense that both derive from mantles that are partly depleted in ITE and the major elements required for basalt formation. Perhaps the ancients who called the dark regions of the Moon the "maria" were more perspicacious than we realize, since basalts with such properties occur on Earth on the ocean floors! Except for the magnitude of their Eu anomalies, REE patterns for some mare basalts could pass for those of MORB.)

Mare basalt sources as magma ocean products. Much consideration has been given to modeling the origin of the mare basalts by partial melting of mafic cumulates precipitated from a lunar magma ocean. The negative Eu anomalies are taken as complementing the positive anomalies once expected for the highland crust owing to its accumulation of plagioclase. The positive LREE slopes suggest an origin from LREE-depleted mantle sources. These are characteristics consistent with a mantle source consisting of cumulus mafic minerals beneath the urKREEP layer. In addition, most mare basalts do not have the mineral plagioclase on their liquidi, suggesting a plagioclase-free or plagioclase-poor source. More aluminous varieties (Apollo 12 feldspathic basalts, most Apollo 14 basalts, and Lunas 16 and 24 basalts) require more substantial proportions of plagioclase in their source regions or assimilation of plagioclase-rich material during ascent to the surface. Remelting of clinopyroxene-bearing cumulates is the model that has received the most attention for the origin of mare basalts.

Papers involving modeling of REE for mare basalt genesis from cumulates are too numerous to acknowledge individually, and the reader is referred to the lunar literature. Some particularly careful trace-element based models of this type have been developed for Apollo 12 and Apollo 17 mare basalts by Nyquist and coworkers (1976, 1977, 1979), who used very precise isotopic data in conjunction with REE and other trace-element data for the same samples. Among the key parameters they used are REE concentrations, including particularly ratios of Sm to Nd, and isotopic ratios for $^{143}Nd/^{144}Nd$ as principal constraints. (Note that the use of Sm-Nd isotopic systematics was first developed for applica-

242

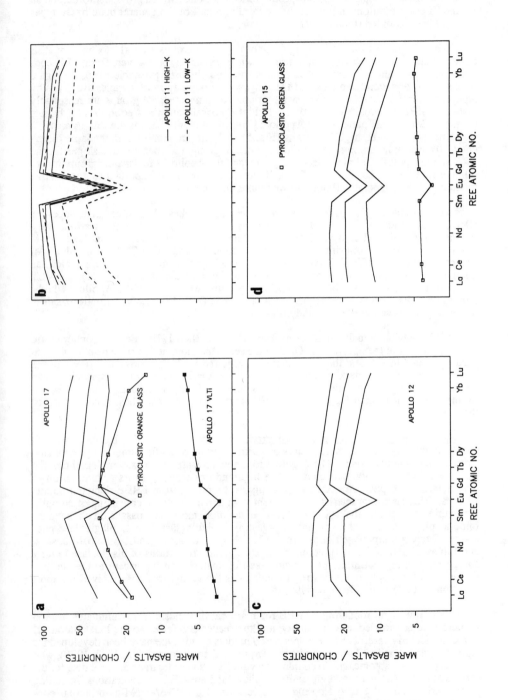

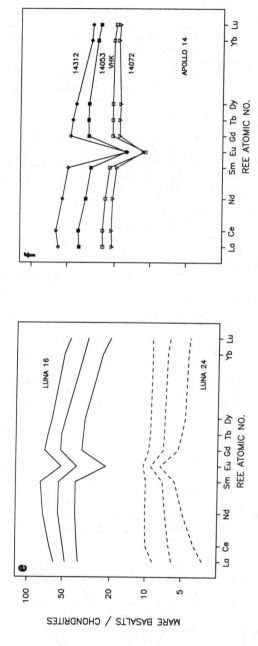

Figure 5. REE patterns for mare basalts and related pyroclastic glasses. (a) Note the striking differences in REE concentrations between the common Apollo 17 high-Ti basalts (normalized maximum, mean, and minimum concentrations) and the small fragments of very-low-Ti basalts. Note also the unusually steep slope of the HREE in the presumably pyroclastic orange glass. (b) The Apollo 11 high-K basalts (normal-ized maximum, mean, and minimum concentrations) have flatter patterns than the Apollo 11 low-K, high-Ti basalts (normalized maximum, mean, and minimum concentrations), which are similar to the Apollo 17 high-Ti basalts. (c, d) The Apollos 12 and 15 low-Ti basalts (normalized maximum, mean, and minimum concentrations) are similar in composition. The Apollo 15 green glass, which consists of spherules of probable pyroclastic origin, is more primitive in major element and REE characteristics but does not obviously represent the parental magma of the mare basalts. (e) Luna 16 and Luna 24 basalt compositions (normalized maximum, mean, and minimum concentrations shown) are based on analyses of many individual fragments taken from the cored soils collected on those two missions. Note the low concen-trations and the probable positive Eu anomaly for the Luna 24 (unique for mare basalts analyzed so far) and the unusually high Eu concentrations for the Luna 16 basalts. (f) REE patterns for individual Apollo 14 aluminous mare basalts. Two samples (14053 and 14072) were recognized early. The rest were extracted as clasts from breccias. Individual fragments of the aluminous basalts cover a range of REE concentrations from that of 14072 to the average for several fragments listed as 14321, the number of their parent breccia. The fragments of possibly related VHK basalts cover a similar range in REE concentrations; the average of several low in REE is shown to emphasize the absence of the coupling of K concentration and REE concentrations that is usually observed in highlands materials.

tion to lunar samples; Lugmair 1975). They obtained self-consistent results between these REE parameters and constraints from the Sr isotopic system and its associated trace elements for models based on cumulate remelting, where the cumulates consisted of olivine, orthopyroxene, clinopyroxene, and ilmenite.

Their scenario begins with a magma ocean some 4.6 Ga ago. Its solidification results in the formation of compositionally closed source regions some 4.4 Ga ago for the different basalt types, regions with Sm/Nd ratios that were higher than that of the bulk magma ocean. Separate regions are required to provide the different $^{143}Nd/^{144}Nd$ initial ratios for the different basalt groups. The source regions are taken to be portions of the cumulate pile, and the change in Sm/Nd is attributed to the preference of the cumulus mafic minerals for heavier REE over lighter REE. Then, some 3.8 Ga ago for Apollo 17 basalts and 3.2 Ga ago for Apollo 12 basalts (each of the three Apollo 12 magma types has a slightly different age), a decrease in Sm/Nd occurred to produce the values in the present-day basalts; this change would correspond to partial melting of the source regions to form the basaltic magmas. The magnitude of the change in Sm/Nd differs for the four basalt types, and is estimated to be ~11% for Apollo 12 feldspathic basalts, ~13% for Apollo 12 olivine-pigeonite basalts, ~24% for Apollo 12 ilmenite basalts, and 20% for Apollo 17 high-Ti basalts. Based on the general mineralogy anticipated for the parent cumulates and values for REE distribution coefficients, the mineral proportions of the source regions can be estimated. The principal host for most of the REE is clinopyroxene, whose proportion ranges from 7% for the parent of the Apollo 12 olivine-pigeonite basalts to 38% for the Apollo 12 ilmenite basalts. The mantle source for Apollo 12 basalts is thus somewhat heterogeneous in mineralogy and REE concentrations. It has a negative Eu anomaly and an even greater relative depletion in LREE than the basalts themselves, which are presumed to be liquids that formed in equilibrium with the HREE-preferring mafic residual minerals. A particularly interesting consequence of cumulate-remelting models is that such partitioning does not yield the shallow negative slope of the HREE observed in most basalt REE patterns if the cumulates contain HREE in chondritic relative abundances or have the positive HREE slope expected for mafic cumulus minerals precipitated from a magma with chondritic relative HREE abundances. Thus, Nyquist and coworkers conclude that the magma ocean, at least by the time the parent cumulates of most mare basalts had formed, had acquired a negative HREE slope. Their estimate gives the same HREE slope, in fact, that is derived for parental magmas of highland plutonic rocks. They regard this as possibly a feature of the bulk lunar REE pattern. Other investigators (e.g., Wentworth et al., 1979) reach this conclusion for the modelled cumulate source regions for Luna 24 VLT basalts and possibly Apollo 17 VLT basalts.

Considering other processes, Nyquist et al. rule out mixing of magma from the cumulate source region and primitive material from beneath the magma ocean as important for the Apollos 12 and 17 high-Ti basalts on isotopic grounds. Among other things, the three Apollo 12 mare basalt groups do not line up in the same way for the Nd-Sm isotopic system (feldspathic to olivine-pigeonite to ilmenite) as for the Rb-Sr isotopic system (feldspathic to ilmenite to olivine-pigeonite). Nor do major and trace-element data support mixing of two end members to produce these basalt types. Isotopic data from another REE, Lu, were used by Fujimaki and Tatsumoto (1984) to argue against assimilation to account for the variety of mare basalt types.

It is sometimes argued that more specific information on the genesis of mare basalts and the mare basalt source region depends on improved knowledge of distribution coefficients. Interest in the genesis of lunar rocks has spawned some of the best experimental measurements of REE D values (e.g., McKay, 1982; Colson et al., 1988; McKay et al., 1986; Nakamura et al., 1986). However, equally important to accurate D values is the correct choice and description of the separation processes that produced the magma and that modified its composition further to form the sampled rock and the accurate development of appropriate models to describe those processes. Accurate D values are just one parameter in the description and their uncertainty may not be the most significant source of error.

Assimilation of crustal material during basalt petrogenesis. Assimilation is considered likely in the formation of certain Apollo 14 aluminous mare basalts, however, which are known mainly as small fragments from Apollo 14 breccias. The discovery and development of information on Apollo 14 aluminous basalts and their high-K (VHK) variant is a good illustration of the further information being gained about the types and petrogenesis of lunar rocks by painstaking examination of complex breccias and analysis of tiny fragments of their components. Dickinson et al. (1985), who analyzed 36 fragments of Apollo 14 aluminous basalt, and Shervais et al. (1985), who analyzed 11 fragments, argued for assimilation of KREEPy material from the lower crust during the migration of mare basalt magma to the surface. REE patterns have negative slopes for many of these basalts. However, for a subset of Apollo 14 VHK basalts analyzed for isotopic compositions as well as trace elements, Shervais et al. (1985) and Shih et al. (1986) found that the ratio Sm/Nd scarcely changed at the time of melting to form the basalts, in contrast to a large change in Rb/Sr. They ruled out KREEP assimilation on isotopic grounds and attributed the combined REE and REE isotopic characteristics of Apollo 14 VHK basalts principally to melting events in a cumulate pile source (with nearly chondritic Sm/Nd) for aluminous mare basalts, and the unusually high ratios of K/La and the Rb/Sr isotopic characteristics to assimilation of small amounts of lunar granitic material (felsite), which is allowed by isotopic constraints. They suggested that assimilation of KREEPy material might have occurred during production of the Apollo 11 high-K basalts and certain Apollo 17 basalts (those from Station 4), for which there is evidence for LREE enhancement. By now, many fragments of the Apollo 14 aluminous and VHK basalts have been studied, and a continuum of compositions is observed, spanning a factor of ten or so in REE concentrations. A good case for assimilation of KREEP as well as felsite has been made (e.g., Neal et al., 1988a,b; 1989).

Glassy spherules. In addition to the mare basalts, there are beautiful microscopic glass beads of mare basalt affinity in many lunar soils. The most concentrated occurrences are in the Apollo 15 Apennine front soil (green glass) and in the Shorty crater soil in the Taurus-Littrow valley of Apollo 17 (orange glass). The concensus of many studies is that these are of pyroclastic origin. Delano has actively studied these in the context of the information they may provide about the lunar interior and the mare basalt source regions, and has summarized arguments for identifying "pristine" varieties (Delano, 1986). These glasses have more primitive compositions (higher concentrations of MgO and Ni, and higher values of Mg#) than the mare basalts themselves, and thus may represent primary lunar magmas. Delano argues for some 25 compositional groups, and the number grows as studies progress. He assigns these groups to two separate, general arrays that he takes to reflect broad types of mantle sources.

REE patterns for Apollo 15 green glass (an average of analyses of 55 individual beads, Ma et al., 1981) and for Apollo 17 orange glass (average of analyses of 16 beads, Hughes et al., 1989) are shown in Figures 5a and 5d. The compositions of the orange glasses, which were collected from a well defined layer at Shorty Crater, are virtually identical. Those for the Apollo 15 beads, which were collected as clods mixed with other regolith materials, show significant compositional variability. On the basis of subtle differences in major-element compositions of a large set of beads, Delano divided the Apollo 15 green glasses into five separate groups. Trace-element analyses of some 400 individual beads, with electron probe microanalysis of major elements of a subset, indicate at least five separate, distinguishable groups of beads (Steele, Haskin, and Korotev, unpublished). LREE concentrations vary substantially, but HREE vary little. REE and other ITE do not in general correlate with major-element concentrations among the groups. This provides an interesting puzzle in the petrogenetic origin of these differences, and yet another example of the use of REE and other compositional data to compensate for lack of field and stratigraphic information.

Soils and breccias

The distinction between Apollo 15 KREEP basalt, a "pristine" rock, and the plethora of

KREEPy melt rocks at the Apollo 14 site is the distinction between a surviving, well defined igneous rock and polymict breccias and soils produced by impact mixing and whose igneous precursors we must still determine. As this paper is concerned principally with igneous rocks and their relationship to lunar crustal differentiation, relatively little is included about the nature of impact processes and their products. This does not mean that they are less important. The processes are interesting, and the overwhelming majority of the lunar highland samples collected are polymict breccias and soils. These, not surviving igneous rocks, are the most typical lunar crustal materials accessible for present and future study, including remote sensing. Many of the igneous rocks identified and characterized so far have been extracted as small to medium sized particles and clasts from these polymict materials, and study of breccia clasts and soil particles continues to reveal further compositional variants of igneous rock types.

Soil compositions vary from site to site, and broad scale variations in regolith composition are observed by remote sensing. Typical compositions of soils from mare and highland sites are shown in Figures 6a and 6b. Some mare soils reflect the REE concentrations and patterns of the principal types of mare basalt collected at the sites, e.g., Apollos 11 and 17 (mare terrain) soils, and Lunas 16 and 24 soils. Others do not, e.g., the Apollos 12 and 15 (mare terrain) soils. The Apollos 12 and 15 soils reflect a considerable highland component. The source of highland material at the Apollo 15 site is self evident, as that site was at the interface between the Apennine Mountains and Palus Putredinis at the edge of Mare Imbrium. The Apollo 12 site in southern Oceanus Procellarum is not immediately adjacent to highlands, but not far distant. On the other hand, Apollo 11 is not appreciably farther from highlands than Apollo 12, and Apollo 17 is at a mare-highland interface just as Apollo 15. The relatively thin mare basalt formations are underlain by highland materials at depths easily reached by cratering impacts of modest size, which suggests that all mare regolith contains a significant highland component. The Apollos 11 and 17 soil REE patterns resemble those in the low-K, high Ti basalts because the REE concentrations in those basalts are high enough that they are not overshadowed by those from the KREEPy highland component; the Apollo 11 soil contains some 25% highland material.

The REE patterns of the highland soils (Fig. 6b) are not sufficiently diluted with mare material to have positive LREE slopes. The high proportion of KREEPy material in the Apollo 14 soil is evident in its unusually high REE concentrations. Significant proportions of KREEPy materials dominate the REE patterns at the Apollo 16 site and in the highland regions of the Apollos 15 and 17 sites. The proportion of KREEPy materials at the Luna 16 site is much lower.

The highland soils represent the averages of many materials of the lunar nearside highlands. As such, they provide a convenient comparison between a common lunar crustal REE pattern and a common terrestrial REE pattern, that of the so-called North American Shale Composite (NASC, Gromet et al, 1984; Fig. 6b). Note the close similarity between the slopes of the two patterns for the HREE, but the much steeper slope for the LREE of the NASC. While REE concentrations in lunar materials analyzed so far range over nearly four orders of magnitude, lunar REE patterns reflect less variability and less extensive REE fractionation than are common on Earth.

Materials making up lunar soils are believed to be primarily of local rather than exotic origin. Giant cratering events have certainly moved large quantities of lunar crust over substantial distances, but secondary excavation of local material by ballistic fragments falling from primary ejecta appears to dominate over the signature of the exotic fragments. Study of individual soil fragments >1 mm in diameter provides an indication of what types of materials of both local and exotic origins contribute to the substance of the soils. For example, variation in compositions among the <1 mm size fractions of the various soils collected at the Apollo 16 site is relatively small (except for material near the rim of North Ray crater). Compositional trends are observed, but appear to correspond mainly to mixtures of one or more local soils plus one or more rock types; e.g., the compositions of

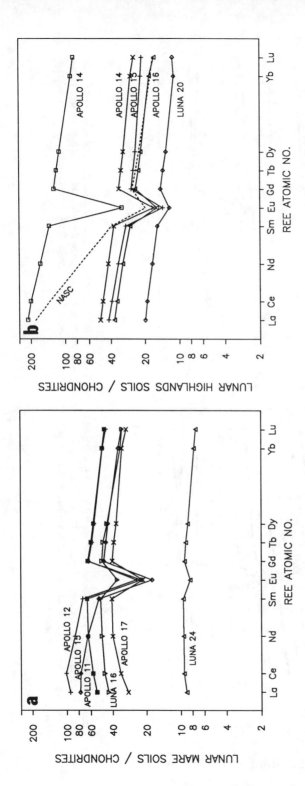

Figure 6. Normalized average concentrations for samples of the <1 mm size fraction of lunar soils (which is the overwhelming bulk of the soils) are shown. Few soils were collected by the Apollo 11-14 missions, and one each by the Luna missions. Many soils were collected by the Apollos 15-17 missions, but by no means all are included in the averages. (a) Mare soils. Note the positive LREE slopes for the Apollos 11 and 17 and Lunas 16 and 24 mare soils but the negative slopes for the Apollos 12 and 15 soils. The slopes reflect the interplay of REE concentrations in the local mare basalts and the proportion and KREEPiness of the highlands component. (b) Highlands soils. None of these highland soil averages reflects the presence of mare basalt because the combined proportions and REE concentrations of that material are overshadowed by the REE concentrations of the highlands components. The REE pattern of the North American Shale Composite is included to show the similarity of HREE slope but the greater steepness of the LREE slope.

248

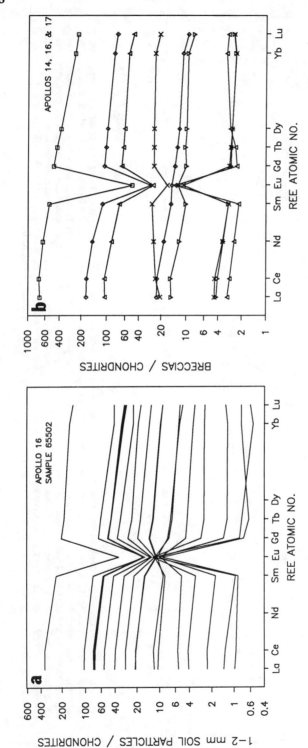

Figure 7. (a) Normalized REE concentrations for twenty 1-2 mm fragments from Apollo 16 soil No. 66502. Note the near constancy of slope for all REE patterns, except for the size and sign of the Eu anomaly. Groups of nearside highlands soils and breccias tend to show this relationship, and it stems in part from the admixing of KREEPy breccias into the soils, a condition that may not be present in the lunar farside highlands. (b) A relationship similar to that in (a) is seen for samples of breccias from three highlands sites.

Apollo 15 soils are mixtures of a well mixed highland soil, mare basalt, green glass, and a KREEPy basalt (Korotev, 1987). We have not observed any highland soil whose composition corresponds to that of a mixture of known igneous rock types.

Contrast that result for <1 mm soil fractions with the compositions of individual 1-2 mm fragments from a typical member of that soil group (Fig. 7a). La concentrations range from about one times to nearly 400 times the chondritic value. Most Eu concentrations fall in the narrow range of 10-20 times the chondritic value, so that the size and direction of the Eu anomalies varies systematically with the other REE concentrations. This is typical of highland materials in the Apollo and Luna collections overall. Figure 7b shows a similar result for breccias from several highland sites. Based on our oversimplified magma ocean model, we might expect just such a relationship for mixtures of ferroan anorthosite (low REE concentrations, high positive Eu anomaly) with KREEPy basalt or norite (high REE concentrations, large negative Eu anomaly). Tantalizing as this simple REE-based explanation is, it does not acceptably describe compositional data for other elements. Based on available samples as representative of lunar crustal materials as a whole, it can be shown that the bulk of the REE in the lunar crust is not contained in KREEPy materials with high REE concentrations, but is associated with the noritic-anorthositic materials that contain the bulk of the crustal Fe and Mg (e.g., Korotev et al, 1980; Korotev and Haskin, 1988).

Note, however, that the Apollo and Luna materials come from the KREEPy lunar nearside, and the bulk of the lunar highlands, with substantially lower Th concentrations, is on the lunar farside, distant from the Apollo 14 site. The spread of REE patterns shown in Figure 7 may not be typical of the farside highlands, where we infer that KREEPy materials with high REE concentrations are rare. In this regard, the Antarctic meteorites are particularly interesting.

In addition to the soils, there are soil-like breccias (regolith breccias). These breccias are not exactly equivalent to lunar soils, but all contain soil-like materials and consist of compacted, crushed, and mostly pulverized regolith (e.g., McKay et al., 1989). Four particularly intriguing examples of regolith breccias are the lunar meteorites (Fig. 8a) found during the past several years in Antarctica in regions of low precipitation and net sublimation of glacial ice. These highland regolith breccias have about the same Al concentration as the average for the lunar crust, and they have significantly lower REE concentrations than those in Apollo and Luna highland soils and positive Eu anomalies. They contain virtually no KREEP. Their Th concentrations, and presumably their REE and other ITE concentrations, are significantly lower than the average for the lunar highlands, as determined from the gamma-ray remote sensing experiments. Their bulk compositions are that of noritic anorthosite. They contain very few clasts of ferroan anorthosite. Their lunar provenance is unknown; they could have come from the nearside or the farside.

What materials were the precursors to the bulk soils and breccias? Although fragments of "pristine" ferroan troctolites, norites, and anorthositic norite are rare in the lunar sample collections, materials of this composition appear to be needed to account for the overall crustal composition. Some breccias appear to be nearly monomict, i.e., the bulk of their material may have come from a single rock type and formation (e.g., McGee, 1988, 1989). Certain lunar granulites of anorthositic norite composition, containing approximately the crustal average of 26% Al_2O_3, are particularly interesting in this regard (e.g., Lindstrom and Lindstrom, 1986; Figure 8b). Note the similarity between their REE patterns and those of the Antarctic lunar meteorites. Some are ferroan and others magnesian in character. This is especially important, because the soils are intermediate in Mg# and cannot be produced by mixing of nearly Fe- and Mg-free "pristine" ferroan anorthosite with "pristine" magnesian norites and troctolites. KREEP basalt has a range of Mg# spanning that of the soils, but its high REE concentrations limit its contribution of Fe and Mg to a small fraction of that contained in soils. Thus, we must turn to non-"pristine" materials to determine the principal types of igneous rocks that give the crust its overall anorthositic norite compositional character.

250

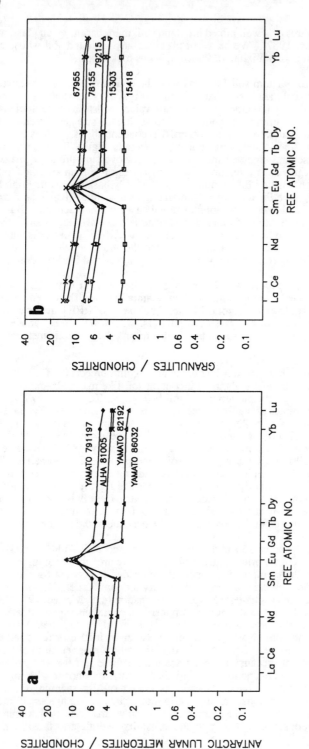

Figure 8. (a) Normalized REE concentrations for three highlands regolith breccias, materials similar to compacted soils, found in Antarctica as meteorites from the Moon. These KREEP-free materials have major element compositions similar to that estimated for the lunar highlands overall. The provenance of the breccias is unknown and could be farside. (b) Normalized REE concentrations for granulites with bulk compositions of anorthositic norite, similar to that estimated for the lunar crust. Representatives of both the "ferroan" and "magnesian" differentiation suites are shown. Although they are all polymict, some of the granulites appear to have derived mainly from a single rock type or formation.

There are numerous examples of compositional constraints that affect our estimates of the importance of various materials to the lunar crust. If we allow that the soils and breccias contain the most representative samples of the Moon's surface that we have at hand or can observe by remote sensing, then their compositions place constraints on the nature of the igneous rocks that were mixed to produce them.

Stemming from this assumption is the following example of REE constraints on the global nature of the lunar crust. Consider the proposition that, in accord with our simple version of the magma ocean model, most of the plagioclase in the lunar crust had the composition of the plagioclase in ferroan anorthosite. That plagioclase contributes 35.5% Al_2O_3 and 0.8 ppm Eu to lunar highland materials. If essentially all crustal Al and all Eu were in such plagioclase, a graph of Al_2O_3 concentration versus Eu concentration would yield a line with a slope of 0.8/35.5. Figure 9 is such a graph for characteristic lunar highland samples. We observe that, relative to their Al_2O_3 concentrations, most highland materials contain an excess of Eu over that estimated from this simple model. We may then

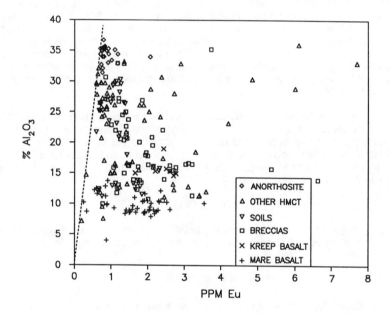

Figure 9. Percent Al_2O_3 is plotted against ppm Eu for a wide variety of lunar materials. Ferroan anorthosite, whose plagioclase has often been taken to represent that of the major crustal product of crystallization of the proposed magma ocean, is represented by diamonds; alkali and magnesian anorthosite are included with the other HMCT (highlands monomict rocks, presumed to represent igneous rocks of the early lunar crust). The dashed line represents the Al_2O_3/Eu concentration ratio for Apollo 15 ferroan anorthosite. All other lunar materials (including many ferroan anorthosites, owing to their trace non-plagioclase components) lie to the Eu-rich side of the line, indicating they have higher Al_2O_3/Eu concentration ratios than the plagioclase in ferroan anorthosites. This cannot result from admixture of small quantities of alkali anorthosites (the upright triangles at the upper right) or KREEP basalt (X's) because such admixtures would contain too much REE. This suggests that the plagioclase of ferroan anorthosite cannot represent the bulk of the plagioclase in the lunar crust.

ask whether there is a source of plagioclase highly concentrated in Eu that might account for the discrepancy. The only known sources are KREEP with up to 4 ppm Eu and alkali anorthosite with up to 8 ppm Eu. Thus, might we add from a few to several percent of KREEP or alkali anorthosite to account for this difference? The answer is no, because alkali anorthosite and KREEP also contain such high concentrations of Sm and other ITE that addition of the requisite amount to account for the "excess" Eu results in mixtures that contain substantially more Sm and ITE than found in the samples. Thus, a substantial fraction of the plagioclase in the highlands must have a higher Eu concentration than ferroan anorthosite. A limit can thus be placed on the fraction of crustal plagioclase that can consist of ferroan anorthosite. It is somewhat model dependent; an upper limit of some 40% of such plagioclase in the crust seems probable. (An extensive treatment of this is given by Korotev and Haskin, 1988.)

Caveat

This has been an introduction of REE in lunar materials from a particular point of view, that of a very simple version of a magma ocean model for lunar crustal differentiation. This was done to introduce the reader in a reasonably coherent way to lunar REE patterns and how they are perceived to fit into lunar geoscientific history. This treatment neither provides an adequate description or defense of magma ocean models. Nor does it provide a critique or a discussion of alternative models such as those of Binder (e.g., Binder, 1986, for an unusual magma-ocean based model of mare basalt petrogenesis), of Walker (1983) and Longhi and Ashwal (1985) for crustal formation and differentiation by serial magmatism rather than from magma ocean crystallization. It is left for the interested reader to peruse the literature to gain a more detailed and balanced sense of the results and problems of lunar research.

It is a fact that the Moon differentiated early and extensively. There is yet much to be done in the way of further exploration and sampling of the Moon and analysis of samples already on hand and samples acquired in the future before we know whether the path of differentiation actually involved a global ocean of magma and, if it did, exactly what products that ocean produced and how. Exploration and sampling of the Moon are, after all, barely underway.

ACKNOWLEDGMENTS

The author gratefully acknowledges the helpful suggestions of colleagues Randy L. Korotev and Bradley N. Jolliff during the preparation of this manuscript.

REFERENCES

Anders E. and Ebihara M. (1982) Solar-system abundances of the elements. Geochim. Cosmochim. Acta 46, 2363-2380.
Anderson A.T., Jr., Crewe A.V., Goldsmith J.R., Moore P.B., Newton J.C., Olsen E.J., Smith J.V., and Wyllie P.J. (1970) Petrologic history of the Moon suggested by petrography, mineralogy, and crystallography. Science 167, 587-590.
Baldwin R.B. (1949) The Face of the Moon, University of Chicago Press, Chicago, IL.
Basaltic Volcanism Study Project (1981) Basaltic Volcanism on the Terrestrial Planets. Pergamon, N.Y. Section 1.2.9, Lunar mare basalts, p. 214-233.
Binder A.B. (1986) Mare basalt genesis: Modeling trace elements and isotopic ratios. Proc. Lunar Planet. Sci. Conf. 16th, D19-D30.
Blanchard D.P., Brannon J.C., Aaboe E., and Budahn J.R. (1978) Major and trace element chemistry of Luna 24 samples from Mare Crisium. Proc. Conf. Luna 24, 613-630. Lunar and Planetary Institute, Houston.
_____ and Budahn J.R. (1979) Remnants from the ancient lunar crust: Clasts from consortium breccia 73255. Proc. Lunar Planet. Sci. Conf. 10th, 803-816.

Blanchard D.P., Haskin L.A., Jacobs J.W., Brannon J.C., and Korotev R. L. (1975) Major and trace element chemistry of boulder 1 at Station 2, Apollo 17. The Moon, 14, 359-371.

_____, Jacobs J.W., and Brannon J.C. (1977) Chemistry of ANT-suite and felsite clasts from consortium breccia 73215 and of gabbroic anorthosite 79215. Proc. Lunar Sci. Conf. 8th, 2507-2524.

Boynton W.V. and Hill D.H. (1983) Composition of bulk samples and a possible pristine clast from Allan Hills A81005. Geophys. Res. Lett. 10, 837-840.

Colson R.O., McKay G.A., and Taylor L.A. (1988) Temperature and composition dependencies of trace element partitioning: Olivine/melt and low-Ca pyroxene/melt. Geochim. Cosmochim. Acta 52, 539-553.

Delano, J.W. (1986) Pristine lunar glasses: Criteria, data, and implications. Proc. Lunar Planet. Sci. Conf. 16th, D201-D213.

Dickinson T., Taylor G.J., Keil K., Schmitt R.A., Hughes S.S., and Smith M.R. (1985) Apollo 14 aluminous mare basalts and their possible relationship to KREEP. Proc. Lunar Planet. Sci. Conf. 15th, C365-C374.

Fujimaka H. and Tatsumoto M. (1984) Lu-Hf constraints on the evolution of lunar basalts. Proc. Lunar Planet. Sci. Conf. 14th, B445-B458.

Fukuoka T., Laul J.C., Smith M.R., Hughes S.S., and Schmitt R.A. (1986) Chemistry of Yamato-791197 Antarctic meteorite: Evidence for its lunar highland origin. Proc. Tenth Symp. Antarctic Meteorites, Tokyo, p. 84-95.

Gast P.W., Hubbard N.J., and Wiesmann H. (1970) Chemical composition and petrogenesis of basalts from Tranquillity Base. Proc. Apollo 11 Lunar Sci. Conf., 1143-1163.

Gillum D.E., Ehmann W.D., Wakita H., and Schmitt R.A. (1972) Bulk and rare earth abundances in the Luna-16 soil levels A and D. Earth Planet. Sci. Lett. 13, 444-449.

Goles G.G., Randle K., Osawa, M., Schmitt R.A., Wakita H., Ehmann W.D., and Morgan J.W. (1970) Elemental abundances by instrumental activation analyses in chips from 27 lunar rocks. Proc. Apollo 11 Lunar Sci. Conf., 1165-1176.

Goodrich C.A., Taylor G.J., Keil K., Boynton W.V., and Hill D.H. (1984) Petrology and chemistry of hyperferroan anorthosites and other clasts from lunar meteorite ALHA81005. Proc. Lunar Planet. Sci. Conf. 15th, C87-C94.

Goodrich C.A., Taylor G.J., Keil K., Kallemeyn G.W., and Warren P.H. (1986) Alkali norite, troctolites, and VHK mare basalts from breccia 14304. Proc. Lunar Planet. Sci. Conf. 16th, D305-D318.

Gromet L.P., Dymek R.F., Haskin L.A., and Korotev R.L. (1984) The "North American Shale Composite": Its compilation, major and trace element characteristics. Geochim. Cosmochim. Acta 48, 2469-2482.

Hartmann W.K. and Vail S.M. (1986) Giant impactors: Plausible sizes and populations. In Origin of the Moon, W.K. Hartmann, R.J. Phillips, and G.J. Taylor, eds. Lunar and Planetary Institute, Houston, TX, p. 551-566.

Haskin L.A., Allen R.O., Helmke P.A., Paster, T.P., Anderson M.A., Korotev R.L., and Zweifel K.A. (1970) Rare earths and other trace elements in Apollo 11 lunar samples. Proc. Apollo 11 Lunar Sci. Conf., 1213-1231.

_____, Helmke P.A., Blanchard D.P., Jacobs J.W., and Telander K. (1973) Major and trace element abundances in samples from the lunar highlands. Proc. Lunar Sci. Conf. 4th, 1275-1296.

_____ and Korotev R.L. (1977) Test of a model for trace element partitioning during closed-system solidification of a silicate liquid. Geochim. Cosmochim. Acta 41, 921-939.

_____ and Lindstrom D.J. (1988) On identifying parent plutonic rocks from lunar breccia and soil fragments. Proc. Lunar Planet. Sci. Conf. 18th, 1-9.

_____, Lindstrom M.M., Salpas P.A., and Lindstrom D.J. (1981) On compositional variations among lunar anorthosites. Proc. Lunar Planet. Sci. Conf. 12th, 41-66.

_____, Shih C.-Y., Bansal B.M., Rhodes J.M., Wiesmann H., and Nyquist L.E. (1974) Chemical evidence for the origin of 76535 as a cumulate. Proc. Lunar Sci. Conf. 5th, 1213-1225.

Heiken G., Vaniman D., and French B. (In press) Lunar Sourcebook. Lunar and Planetary Science Institute, Houston.

Helmke P.A., Blanchard D.P., Haskin L.A., Telander K., Weiss C., and Jacobs J.W. (1973a) Major and trace elements in igneous rocks from Apollo 15. The Moon 8, 129-148.

_____, _____, Jacobs J.W. and Haskin, L.A. (1973b) Rare earths, other trace elements and iron in Luna 20 samples. Geochim. Cosmochim. Acta 37, 869-874.

Helmke P.A. and Haskin L.A. (1972) Rare earths and other trace elements in Luna 16 soil. Earth Planet. Sci. Lett. 13, 441-443.

254

_____, _____, Korotev R.L., and Ziege K.E. (1972) Rare earths and other trace elements in Apollo 14 samples. Proc. Lunar Sci. Conf. 3rd, 1275-1292.

Hubbard N.J., Meyer C., Jr., Gast P.W., and Wiesmann H. (1971a)The composition and derivation of Apollo 12 soils. Earth Planet. Sci. Lett. 10, 341-350.

_____, Gast, P.W., Meyer C., Jr., Nyquist, L.E., Shih, C., and Wiesmann, H. (1971b) Chemical composition of lunar anorthosites and their parent liquids. Earth Planet. Sci. Lett. 13, 71-75.

_____, _____, Rhodes J.M., Bansal B.M., Wiesmann H. and Church S.E. (1972a) Nonmare basalts: Part II. Proc. Lunar Sci. Conf. 3rd, 1161-1179.

_____, Nyquist L.E., Rhodes J.M., Bansal B.M., Wiesmann H., and Church, S.E. (1972b) Chemical features of the Luna 16 regolith sample. Earth Planet. Sci. Lett. 13, 423-428.

_____, Rhodes J.M., Gast P.W., Bansal B.M., Shih C.-Y., Wiesmann H., and Nyquist L.E. (1973) Proc. Lunar Sci. Conf. 4th, 1297-1312.

_____, _____, Wiesmann H., Shih C.-Y., and Bansal B.M. (1974) The chemical definition and interpretation of rock types returned from the non-mare regions of the moon. Proc. Lunar Sci. Conf. 5th, 1227-1246.

Hughes S.S., Delano J.W., and Schmitt R.A. (1989) Petrogenetic modeling of 74220 high-Ti orange volcanic glass and the Apollo 11 and 17 high-Ti mare basalts. Proc. Lunar Planet. Sci. Conf. 19th, 175-188.

James O.B., Flohr M.K., and Lindstrom M.M. (1984) Petrology and geochemistry of lunar dimict breccia 61015. Proc. Lunar Planet. Sci. Conf. 15th, C63-C86.

_____, Lindstrom M.M., and Flohr M.K. (1989a) Ferroan anorthosite from lunar breccia 66435: Implications for the origin and history of lunar ferroan anorthosites. Proc. Lunar Planet. Sci. Conf. 19th, 219-243.

_____, _____, and _____ (1989b) Petrology and geochemistry of alkali gabbronorites from lunar breccia 67975. Proc. Lunar Planet. Sci. Conf. 17th, E314-E330.

Jerome D.Y. and Phillipot J.-C. (1973) Chemical composition of Luna 20 soil and rock fragments. Geochim. Cosmochim. Acta 37, 909-914.

Kallemeyn G.W. and Warren P.H. (1983) Compositional implications regarding the lunar origin of the ALHA81005 meteorite. Geophys. Res. Lett. 10, 833-836.

Korotev, R.L. (1981) Compositional trends in Apollo 16 soils. Proc. Lunar Planet. Sci. Conf. 12th, 577-605.

_____ (1987) Mixing levels, the Apennine Front soil component, and compositional trends in Apollo 15 soils. Proc. Lunar Planet. Sci. Conf. 17th, E411-E431.

_____ and Haskin L.A. (1988) Europium mass balance in polymict samples and implications for plutonic rocks of the lunar crust. Geochim. Cosmochim. Acta 52, 1795-1813.

_____, _____, and Lindstrom M.M. (1980) A synthesis of lunar highlands data. Proc. Lunar Planet. Sci. Conf. 11th, 395-429.

_____, Lindstrom M.M., Lindstrom D.J., and Haskin L.A. (1983) Antarctic meteorite ALHA81005 - not just another lunar anorthositic breccia. Geophys. Res. Lett. 10, 829-832.

Kuiper G.P. (1954) On the origin of the lunar surface features. Proc. U.S. Nat. Acad. Sci. 40, 1096-1112.

Laul J.C., Gosselin D.C., Galbreath K.C., Simon S.B., and Papike J.J. (1989) Chemistry and petrology of Apollo 17 highland coarse fines: Plutonic and melt rocks. Proc. Lunar Planet. Sci. Conf. 19th, 85-97.

_____, Papike J.J., Simon S.B., and Shearer C.K. (1983a) Chemistry of the Apollo 11 highland component. Proc. Lunar Planet. Sci. Conf. 14th, B139-B149.

_____ and Schmitt R.A. (1973a) Chemical composition of Luna 12 rocks and soil and Apollo 16 soils. Geochim. Cosmochim. Acta 37, 927-942.

_____ and Schmitt R.A. (1973b) Chemical composition of Apollo 15, 16, and 17 samples. Proc. Lunar Sci. Conf. 4th, 1349-1367.

_____ and _____ (1975) Dunite 74217: A chemical study and interpretation. Proc. Lunar Sci. Conf. 6th, 1231-1254.

_____, _____, and Schmitt R.A. (1983b) ALHA 81005 meteorite: chemical evidence for a lunar highland origin. Geophys. Res. Lett. 10, 825-828.

_____, Vaniman D.T., and Papike, J.J. (1977) Chemistry, mineralogy and petrology of seven >1mm fragments from Mare Crisium. Proc. Conf. Luna 24, 537-568. Lunar and Planetary Institute, Houston.

Lindstrom M.M. (1984) Alkali gabbronorite, Ultra-KREEPy melt rock and the diverse suite of clasts in North Ray Crater feldspathic fragmental Breccia 67975. Proc. Lunar Planet. Conf. 15th, C50-C62.

_____ and Haskin (1978) Causes of compositional variations within mare basalt suites. Proc. Lunar Planet. Sci. Conf. 9th, 465-486.

_____, Knapp S.A., Shervais J.W., and Taylor L.A. (1984) Magnesian anorthosites and associated troctolites and dunite in Apollo 14 breccias. Proc. Lunar Planet. Sci. Conf. 15th, C41-C49

_____, Korotev R.L., Lindstrom D.J., and Haskin L.A. (1987) Lunar meteorites Y82192 and 82193: Geochemical and petrologic comparisons to other lunar breccias (abstr.). Proc. Twelfth Symp. Antarctic Meteorites. National Institute of Polar Research, Tokyo, p.19-21.

_____ and Lindstrom D.J. (1986) Lunar granulites and their precursor anorthositic norites of the early lunar crust. Proc. Lunar Planet. Sci. Conf. 16th, D263-D276.

_____, _____, Korotev R.L., and Haskin L.A. (1986) Lunar meteorite Yamato-791197: A polymict anorthositic norite breccia. Proc. Tenth Symp. Antarctic Meteorites, Tokyo, p. 58-75.

_____, Marvin U.B., and Mittlefehldt D.W. (1989) Apollo 15 Mg- and Fe-norites: A redefini-tion of the Mg-suite differentiation trend. Proc. Lunar Planet. Sci. Conf. 19th, 245-254.

_____, _____, Vetter S.K., and Shervais J.W. (1988) Apennine Front revisited: Diversity of Apollo 15 highland rock types. Proc. Lunar Planet. Sci. Conf. 18th, 169-185.

_____ and Salpas P.A. (1983) Geochemical studies of feldspathic fragmental breccias and the nature of the North Ray Crater ejecta. Proc. Lunar Planet. Sci. Conf. 13th, A671-A683.

Longhi J. and Ashwal L.D. (1985) Two-stage models for lunar and terrestrial anorthosites: Petrogenesis without a magma ocean. Proc. Lunar Planet. Sci. Conf. 15th, C571-C584.

_____ and Boudreau A.E. (1979) Complex igneous processes and the formation of the primitive lunar crustal rocks. Proc. Lunar Planet. Sci. Conf. 10th, 2085-2105.

Lugmair G. (1975) Sm-Nd systematics of some Apollo 17 basalts (abstr.). In Papers Presented to the Conference on Origins of Mare Basalts and Their Implications for Lunar Evolution. Lunar and Planetary Institute, Houston, TX, p. 107-109.

Ma M.-S., Liu Y.-G., and Schmitt R.A. (1981) A chemical study of individual green glasses and brown glasses from 15426: Implications for their petrogenesis. Proc. Lunar Planet. Sci. Conf. 12th, 915-933.

_____, Schmitt R.A., Taylor G.J., Warner R.D., Lange D.E., and Keil K. (1977) Chemistry and petrology of Luna 24 fragments and <250 *um soils: Constraints on the origin of VLT mare basalts. Proc. Conf. Luna 24, 569-592. Lunar and Planetary Institute, Houston.

Marvin U.B. and Lindstrom M.M. (1983) Rock 67015: A feldspathic fragmental breccias with KREEP-rich melt clasts. Proc. Lunar Planet. Sci. Conf. 13th, A659-A670.

_____ and Warren P.H. (1980) A pristine eucrite-like gabbro from Descartes and its exotic kindred. Proc. Lunar Planet. Sci. Conf. 11th, 507-521.

McGee J.J. (1988) Petrology of brecciated ferroan noritic anorthosite 67215. Proc. Lunar Planet. Sci. Conf. 18th, 21-31.

_____ (1989) Granulitic breccia clasts and feldspathic melt breccia clasts from North Ray Crater breccia 67975: Precursors and petrogenesis. Proc. Lunar Planet. Sci. Conf. 19th, 73-84.

McKay D.S., Bogard D.D., Morris R.V., Korotev R.L., Johnson P., and Wentworth S.J. (1986) Apollo 16 regolith breccias: Characterization and evidence for early formation in the mega-regolith. Proc. Lunar Planet. Sci. Conf. 16th, D277-D303.

_____, _____, _____, _____, Wentworth S.J., and Johnson P. (1989) Apollo 15 regolith breccias: Window to a KREEP regolith. Proc. Lunar Planet. Sci. Conf. 19th, 19-41.

McKay G.A. (1982) Partitioning of REE between olivine, plagioclase, and synthetic basaltic melts: Implications for the origin of lunar anorthosites (abstr.). Lunar Planet. Sci. XIII, Lunar and Planetary Institute, Houston, TX, p. 493-494.

_____, Wagstaff J., and Yang S.-R. (1986) Zirconium, hafnium, and rare earth partition coefficients for ilmenite and other minerals in high-Ti lunar mare basalts: An experimental study. Proc. Lunar Planet. Sci. Conf. 16th, D229-D237.

_____, Wiesmann H., Bansal B.M., and Shih C.-Y. (1979) Petrology, chemistry, and chronology of Apollo 14 KREEP basalts. Proc. Lunar Planet. Sci. Conf. 10th, 181-205.

_____, _____, Nyquist L.E., Wooden J.L., and Bansal B.M. (1978) Petrology, chemistry, and chronology of 14708: Chemical constraints on the origin of KREEP. Proc. Lunar Planet. Sci. Conf. 9th, 661-687.

Nakamura Y., Fujimaki H., Nakamura N., Tatsumoto M., McKay G.A., and Wagstaff J. (1986) Hf, Zr, and REE partition coefficients between ilmenite and liquid: Implications for lunar petrogenesis. Proc. Lunar Planet. Sci. Conf. 16th, D239-D250.

Nava D.F. and Philpotts J.A. (1973) A lunar differentiation model in light of new chemical data on Luna 20 and Apollo 16 soils. Geochim. Cosmochim. Acta 37, 963-973.

256

Neal C.R. and Taylor L.A. (1989) The nature of barium partitioning between immiscible melts: A comparison of experimental and natural systems with reference to lunar granite petrogenesis. Proc. Lunar Planet. Sci. Conf. 19th, 209-218.

_____, _____, and Lindstrom M.M. (1988a) Importance of lunar granite and KREEP in very high potassium (VHK) basalt petrogenesis. Proc. Lunar Planet. Sci. Conf. 18th, 121-137.

_____, _____, and _____ (1988b) Apollo 14 mare basalt petrogenesis: Assimilation of KREEP-like components by a fractionating magma. Proc. Lunar Planet. Sci. Conf. 18th, 139-153.

_____, _____, Schmitt R.A., Hughes S.S., and Lindstrom M.M. (1989) High alumina (HA) and very high potassium (VHK) basalt clasts from Apollo 14 breccias, part 2 - whole rock geochemistry: Further evidence for combined assimilation and fractional crystallization within the lunar crust. Proc. Lunar Planet. Sci. Conf. 19th, 147-161.

Nyquist L.E., Bansal B.M., and Wiesmann H. (1976) Sr isotope constraints on the petrogenesis of Apollo 17 mare basalts. Proc. Lunar Sci. Conf. 7th, 1507-1528.

_____, _____, Wooden J.L., and Wiesmann H. (1977) Sr-isotopic constraints on the petrogenesis of Apollo 12 mare basalts. Proc. Lunar Sci. Conf. 8th, 1383-1415.

_____, Shih C.-Y., Wooden J.L., Bansal B.M., and Wiesmann H. (1979) The Sr and Nd isotopic record of Apollo 12 basalts: Implications for lunar geochemical evolution. Proc. Lunar Planet. Sci. Conf. 10th, 77-114.

Ostertag R., Stöffler D., Bischoff A., Palme H., Schulz L., Spettel B., Weber H., Weckwerth G., and Wänke H. (1986) Lunar meteorite Yamato-791197: Petrography, shock history and chemical composition. Proc. Tenth Symp. Antarctic Meteorites, Tokyo, p. 17-44.

Palme H., Baddenhausen H., Blum K., Cendales M., Dreibus G., Hofmeister H., Kruse H., Palme C., Spettel B., Vilcsek E., and Wänke H. (1978) Proc. Lunar Planet. Sci. Conf.. 9th, 25-57.

_____, Spettel B., Weckwerth G. and Wänke H. (1983) Antarctic meteorite ALHA 81005, a piece from the ancient lunar crust. Geophys. Res. Lett. 10, 817-820.

Philpotts J.A., Schnetzler C.C., Bottino M.L., Schuhmann S., and Thomas H. H. (1972) Luna 16: Some Li, K, Rb, Sr, Ba, rare-earth, Zr, and Hf concentrations. Earth Planet. Sci. Lett. 13, 429-435.

Pieters C.M. (1978) Mare basalt types on the front side of the moon: A summary of spectral reflectance data. Proc. Lunar Planet. Sci. Conf. 9th, 2825-2849.

Quick J.E., Albee, A.L., Ma M.-S., Murali A.V., and Schmitt R.A. (1977) Chemical compositions and possible immiscibility of two silicate melts in 12013. Proc. Lunar Sci. Conf. 8th, 2153-2189.

Rhodes J.M. and Blanchard D.P. (1981) Apollo 11 breccias and soils: Aluminous mare basalts of multi-component mixtures? Proc. Lunar Planet. Sci. Conf. 12th, 607-620.

_____, _____, Dungan M.A., Brannon J.C., and Rodgers K.V. (1977) Chemistry of Apollo 12 mare basalts: Magma types and fractionation processes. Proc. Lunar Sci. Conf. 8th, 1305-1338.

_____ and Hubbard N.J. (1973) Chemistry, classification, and petrogenesis of Apollo 15 mare basalts. Proc. Lunar Sci. Conf. 4th, 1127-1148.

_____, _____, Wiesmann H., Rodgers K.V., Brannon J.C., and Bansal B.M. (1976) Chemistry, classification, and petrogenesis of Apollo 17 mare basalts. Proc. Lunar Sci. Conf. 7th, 1467-1489.

_____, Rodgers K.V., Shih C., Bansal B.M., Nyquist L.E., Wiesmann H., and Hubbard N.J. (1974) The relationships between geology and soil chemistry at the Apollo 17 landing site. Proc. Lunar Sci. Conf. 5th, 1097-1117.

Rutherford M.J., Hess P.C., Ryerson F.J., Campbell, H.W., and Dick P.A. (1976) The chemistry, origin and petrogenetic implications of lunar granite and monzonite. Proc. Lunar Planet. Sci. Conf. 7th, 1723-1740.

Shervais J.W., Taylor L.A., and Laul J.C. (1983) Ancient crustal components in the Fra Mauro breccias. Proc. Lunar Planet. Sci. Conf. 14th, B177-B192.

_____, _____, _____, Shih C.-Y., and Nyquist L.E. (1985a) Very high potassium (VHK) basalt: Complications in mare basalt petrogenesis. Proc. Lunar Planet. Sci. Conf. 16th, D3-D18.

_____, _____, _____, and Smith M.R. (1984) Pristine highland clasts in consortium breccia 14305: Petrology and geochemistry. Proc. Lunar Planet. Sci. Conf. 15th, C25-C40.

_____, _____, and Lindstrom M.M. (1985b) Apollo 14 mare basalts: Petrology and geochemistry of clasts from consortium breccia 14321. Proc. Lunar Planet. Sci. Conf. 15th, C375-C395.

Shih C.Y., Haskin L.A., Wiesmann H., Bansal B.M., and Brannon J.C. (1975) On the origin of high-Ti mare basalts. Proc. Lunar Sci. Conf. 6th, 1255-1285.

_____, Nyquist L.E., Bogard D.D., Bansal B.M., Wiesmann H., Johnson P., Shervais J.W., and Taylor L.A. (1986) Geochronology and petrogenesis of Apollo 14 very high potassium mare basalts. Proc. Lunar Planet. Sci. Conf. 16th, D214-D228.

257

_____, Nyquist L.E., Bogard D.D., Wooden J.L., Bansal B.M., and Wiesmann H. (1985) Chronology and petrogenesis of a 1.8 g lunar granitic clast: 14321,1062. Geochim. Cosmochim. Acta 49, 411-426.

Simon S.B., Papike J.J., and Laul J.C. (1988) Chemistry and petrology of the Apennine Front, Apollo 15, part I: KREEP basalts and plutonic rocks. Proc. Lunar Planet. Sci. Conf. 18th, 187-201.

Stevenson D.J. (1987) Origin of the Moon -- the collision hypothesis. Ann. Rev. Earth Planet. Sci. 15, 271-315.

Tatsumoto M. and Unruh D.M. (1976) KREEP basalt age: Grain by grain U-Th-Pb systematics study of the quartz monzodiorite clast 15405,88. Proc. Lunar Sci. Conf. 7th, 2107-2129.

Taylor S.R. (1982) Planetary Science: A Lunar Perspective. Lunar and Planetary Institute, Houston, TX, 481 p.

_____, Gorton M., Muir P., Nance W., Rudowski R., and Ware N. (1974) Geochemical zoning in the Moon (abstr.). Lunar Science V, Lunar and Planetary Institute, Houston, TX, p. 789-791.

Urey H.C. (1952) The Planets, Yale Univ. Press, New Haven, CT.

Walker D. (1983) Lunar and terrestrial crust formation. Proc. Lunar Planet. Sci. Conf. 14th, B17-B25.

Wänke H., Baddenhausen H., Balacescu A., Teschke F., Spettel B., Dreibus G., Palme H., Quijano-Rico M., Kruse H., Wlotzka F., and Begemann F. (1972) Multielement analyses of lunar samples and some implications of the results. Proc. Lunar Sci. Conf. 3rd, 1251-1268.

_____, _____, Blum K., Cendales M., Dreibus G., Hofmeister H., Kruse H., Jagoutz E., Palme C., Spettel B., Thacker R., and Vilsek E. (1977) On the chemistry of lunar samples and achon-drites. Primary matter in the lunar highlands: A re-evaluation. Proc. Lunar Sci. Conf. 8th, 2191-2213.

_____, _____, Dreibus G., Jagoutz E., Kruse H., Palme H., Spettel B., and Teschke F. (1973) Multielement analysis of Apollo 15, 16, and 17 samples and the bulk composition of the moon. Proc. Lunar Sci. Conf. 4th, 1461-1481.

_____, Palme H., Baddenhausen H., Dreibus G., Jagoutz E., Kruse H., Palme C., Spettel B., Teschke F., and Thacker R. (1975) New data on the chemistry of the lunar samples: Primary matter in the lunar highlands and the bulk composition of the moon. Proc. Lunar Sci. Conf. 6th, 1313-1340.

_____, _____, _____, Dreibus G., Jagoutz E., Kruse H., Spettel B., Teschke F., and Thacker (1974) Chemistry of Apollo 16 and 17 samples: Bulk composition, late stage accumulation and early differentiation of the moon. Proc. Lunar Sci. Conf. 5th, 1307-1335.

_____, _____, Kruse H., Baddenhausen H., Cendales M., Dreibus G., Hofmeister H., Jagoutz E., Palme C., Spettel B., and Thacker R. (1976) Chemistry of highland rocks: A refined evaluation of the composition of the primary matter. Proc. Lunar Sci. Conf. 7th, 3479-3499.

_____, Rieder R., Baddenhausen H., Spettel B., Teschke F., Quijano-Rico M., and Balacescu A. (1970) Major and trace elements in lunar materials. Proc. Apollo 11 Lunar Sci. Conf., 1719-1727.

_____, Wlotzka F., Baddenhausen H., Balacescu A., Spettel B., Teschke F., Jagoutz E., Kruse H., Quijano-Rico M., and Rieder R. (1971) Apollo 12 samples: Chemical composition and its relation to sample locations and exposure ages, the two component origin of the various soil samples and studies on lunar metallic particles. Proc. Lunar Sci. Conf. 2nd, 1187-1208.

Warner R.D., Taylor G.J., Conrad G. H., Northrop H.R., Barker S., Keil K., Ma M.-S., and Schmitt R.A. (1979) Apollo 17 high-Ti mare basalts: New bulk compositional data, magma types, and petrogenesis. Proc. Lunar Planet. Sci. Conf. 10th, 225-247.

Warren P.H. (1985) The magma ocean concept and lunar evolution. Ann. Rev. Earth Planet. Sci. 13, 201-240.

_____ (1986) Anorthosite assimilation and the origin of the Mg/Fe-related bimodality of pristine lunar rocks: Support for the magmasphere hypothesis. Proc. Lunar Planet. Sci. Conf. 16th, D331-D343.

_____ (1988) The origin of pristine KREEP: Effects of mixing between urKREEP and the magmas parental to the Mg-rich cumulates. Proc. Lunar Planet. Sci. Conf. 18th, 233-241.

_____, Afiattalab F., and Wasson J.T. (1978) Investigation of unusual KREEPy samples: Pristine rock 15386, Cone Crater soil fragments 14143, and 12023, a typical Apollo 12 soil. Proc. Lunar Planet. Sci. Conf. 9th, 653-660.

_____, Jerde E.A., and Kallemeyn G.W. (1987) Pristine Moon rocks: A "large" felsite and a metal-rich ferroan anorthosite. Proc. Lunar Planet. Sci. Conf. 17th, E303-E313.

_____ and Kallemeyn G. (1986) Geochemistry of lunar meteorite Yamato-791197: Comparison with ALHA81005 and other lunar samples. Proc. Tenth Symp. Antarctic Meteorites, Tokyo, p. 3-16.

_____, Shirley D.N., and Kallemeyn G.W. (1986) A potpourri of pristine Moon rocks, including a VHK mare basalt and a unique, augite-rich Apollo 17 anorthosite. Proc. Lunar Planet. Sci. Conf. 16th, D319-D330.

_____, Taylor G.J., Keil K., Kallemeyn G.W., Rosener P.S., and Wasson J.T. (1982) Sixth foray for pristine nonmare rocks and an assessment of the diversity of lunar anorthosites. Proc. Lunar Planet. Sci. Conf. 13th, A615-A630.

_____, _____, _____, _____, Shirley D.N., and Wasson J.T. (1983) Seventh foray: Whitlockite-rich lithologies, a diopside-bearing troctolitic anorthosite, ferroan anorthosites, and KREEP. Proc. Lunar Planet. Sci. Conf. 14th, B151-B164.

_____, _____, Keil K., Marshall C., and Wasson J.T. (1981) Foraging westward for pristine nonmare rocks: Complications for petrogenetic models. Proc. Lunar Planet. Sci. Conf. 12B, 21-40.

_____ and Wasson J.T. (1979a) The origin of KREEP. Rev. Geophys. Space Phys. 17, 73-88.

_____ and _____ (1979b) The compositional-petrographic search for pristine nonmare rocks: Third foray. Proc. Lunar Planet. Sci. Conf. 10th, 583-610.

_____ and _____ (1980) Further foraging for pristine nonmare rocks: Correlations between geochemistry and longitude. Proc. Lunar Planet. Sci. Conf. 11th, 431-470.

_____ and _____ (1978) Compositional-petrographic investigation of pristine nonmare rocks. Proc. Lunar Planet. Sci. Conf. 9th, 185-217.

Wentworth S., Taylor G.J., Warner R.D., Keil K., Ma M.-S., and Schmitt R.A. (1979) The unique nature of Apollo 17 VLT mare basalts. Proc. Lunar Planet. Sci. Conf. 10th, 207-223.

Winzer S.R., Nava D.F., Lum R.K.L., Schuhmann S., Schuhmann P., and Philpotts, J.A. (1975) Origin of 78235, a lunar cumulate norite. Proc. Lunar Sci. Conf. 6th, 1219-1229.

_____, _____, Schuhmann S., Kouns C.W., Lum R.K.L., and Philpotts J.A. (1974) Major, minor and trace element abundances in samples from the Apollo 17 Station 7 boulder: Implications for the origin of early lunar crustal rocks. Earth Planet. Sci. Lett. 23, 439-444.

Wood J.A., Dickey J.S., Jr, Marvin U.B., and Powell B.N. (1970) Lunar anorthosites. Science 167, 602-604.

COMPOSITIONAL AND PHASE RELATIONS AMONG
RARE EARTH ELEMENT MINERALS

INTRODUCTION

Rare earth elements (REE) occur in minerals both as essential constituents and as trace constituents. This review mainly treats minerals in which the REE are essential constituents (e.g., bastnaesite, monazite, xenotime, aeschynite, allanite), although the chemical mechanisms and limits of REE substitution in some rock-forming minerals (e.g., zircon, apatite, titanite, garnet) are also derived. Details on the crystallography and crystal chemistry of various rock-forming and accessory minerals are more appropriately given in other books in this series.

Recent comprehensive reviews of the mineralogy of REE have concentrated on geological occurrences (Clark, 1984) and on crystal structures (Ewing and Chakoumakos, 1982; Miyawaki and Nakai, 1987, 1988). Another mineralogical review is provided in Solodov et al. (1987); unlike an older comprehensive review (Semenov, 1964a; 1964b), it has not been translated from Russian.

This review concentrates on the vector representation of complex coupled substitutions in selected REE-bearing minerals, and ends with some comments on REE-partitioning between minerals as related to acid-base tendencies and mineral stabilities. Inasmuch as the same or analogous coupled substitutions involving the REE occur in a wide variety of mineral structures, they are appropriately discussed together.

Thinking of coupled substitutions as vector quantities facilitates the graphical representation of mineral composition spaces (e.g., Smith, 1959; Thompson, 1981, 1982; the vector section in Cerny and Burt, 1984; Burt, 1988, 1989a). This may be done on normal orthogonal graph paper or using computer x-y plotting routines. No barycentric triangles are needed, nor in many cases are they appropriate, because the accessible irregular composition spaces have up to five faces on a plane or up to six or more in space, as illustrated by the examples below.

GEOCHEMICAL BACKGROUND

Strictly speaking, REE (also called lanthanides) include lithophile elements with atomic numbers 57 through 71, although yttrium, with atomic number 39, is commonly included because of its similar chemical behavior. Scandium, of atomic number 21, is small enough to be chemically distinct and is generally not included. The actinides uranium (92) and especially thorium (90) commonly substitute for REE in nature, although they are not considered lanthanides.

REE of even atomic number (including Ce, Nd, Sm, Gd, Dy, Er, and Yb, as well as Y) are geochemically more abundant than those of odd atomic number (La, Pr, Pm, Eu, Tb, Ho, Tm, and Lu), which produces a "zig-zag" effect in plots of analyses. This effect is normally eliminated by plotting REE analyses normalized to (that is, relative to) average REE abundances in chondritic meteorites or shales.

Normally, all REE have an identical +3 valence and an ionic radius that gradually decreases with increasing atomic number (the "lanthanide contraction"). Exceptions in nature are mainly Ce which can have a valence of +4 under oxidizing conditions and Eu which can have a valence of +2 under reducing conditions. The +4 valence of cerium is important in weathering; the +2 valence of europium under reducing conditions can give rise to a "europium anomaly" (normally produced by fractional crystallization of plagioclase in igneous rocks). Lanthanum is commonly used in place of cerium in experimental studies because of the variable valence of Ce.

As discussed below, the light rare earths (called cerium-group or LREE) have relatively large ionic radii similar to that of Ca^{2+} and Th^{4+}, whereas the heavy rare earths (plus Y and therefore called the yttrium-group or HREE) have smaller ionic radii approaching that of Mn^{2+}. All of the REE commonly substitute for each other in minerals. The Ce-group or LREE tend to be concentrated in highly fractionated basic rocks such as carbonatites, whereas HREE and especally Y tend to be concentrated in fractionated acid rocks such as alkaline granites and pegmatites.

MINERALS

The nomenclature of the REE minerals is unique, in that simple end-member names no longer exist. Instead, since 1966 each mineral name is a combination of a structural formula name and the symbol of the dominant lanthanide element (Levinson, 1966; Nickel and Mandarino, 1987, Bayliss and Levinson, 1988). For example, "doverite" is now called synchysite-(Y). It was feared that end-members might be found for each of the 16 rare earth elements (incuding Y), or at least the nine most common ones (Y, La, Ce, Nd, Sm, Gd, Dy, Er, Yb). This has not yet happened; generally either Y or Ce is dominant. Bayliss and Levinson (1988) state that the predominant REE in mineral names are Y (59 examples), Ce (51), Nd (13), La (12), Yb (2), and Gd (1); the data base for this count (not stated) is presumably Fleischer (1987). Some pumpellyite-group minerals also share this type of notation, e.g., pumpellyite-(Fe^{2+}).

The formulas of approximately 190 minerals containing essential or significant REE are listed in alphabetical order in Table 1. The source of structural formulas in this listing is mainly Miyawaki and Nakai (1987, 1988), although Fleischer (1987, 1989) was also heavily consulted. These references should be examined for original literature citations. The crystal structures of many REE minerals are poorly-known, because the phases are metamict in nature (Th and U commonly substitute for REE in minerals, as mentioned above). Table 2 lists synonyms and currently discredited REE mineral names commonly seen in older literature. My impression is that this list seems to grow almost as fast as the list of new REE minerals does.

Comprehensive reviews of inorganic crystalline compounds of the REE are given by Felsche (1973, 1978), Kubach and Schubert (1984), Kubach and Töpper (1984), Leskela and Niinisto (1986), and Niinisto and Leskela (1987). Some of these compounds occur in nature as minerals; others could be sought as new minerals (e.g, $NaYSiO_4$, analogous to $CaMgSiO_4$, monticellite: Felsche, 1973), and at least one other (texasite, now discredited) has erroneously been reported as a mineral.

The classification of REE minerals can conventionally be done by anion groups (e.g., halides, carbonates, phosphates, silicates, etc.) or more structurally, according to the geometry of anionic groups (triangular, tetrahedral, octahedral, etc.). This approach (used by Miyawaki and Nakai, 1987, 1989) emphasizes the isostructural nature of phosphates such as xenotime-(Y) and silicates such as zircon, which otherwise might be discussed separately. (A more extreme example, disussed below, is provided by the apatite structure type, which can include silicates, sulfates, and perhaps even borates in addition to phosphates.) I avoid the problem in Table 1 by using an alphabetical listing; in what follows I shall discuss some mineral groups in the following order: halides, carbonates, fluorocarbonates, and phosphates first, niobates and titanates in the middle, and silicates last.

Table 1. Minerals containing rare earth elements (alphabetical listing). Formulas mostly from Miyawaki and Nakai ([M&N]: 1987; 1988) and Fleischer ([MF]: 1987; 1989).

Aeschynite-(Ce) $CeTiNbO_6$ (simplified w. Ti>Nb+Ta); or $(Ce,Y,Ca,Fe,Th)(Ti,Nb)_2(O,OH)_6$; compare formula of lucasite-(Ce)

Aeschynite-(Nd) $(Nd,Fe,Ca)(Ti,Nb)_2(O,OH)_6$

Aeschynite-(Y) $YTiNbO_6$ (simplified w. Ti>Nb+Ta); low T polymorph euxenite); or $(Y,Ca,Fe,Th)(Ti,Nb)_2(O,OH)_6$

Agardite-(La) $(La,Ca)Cu_6(AsO_4)_3(OH)_6 \cdot 3H_2O$; mixite group

Agardite-(Y) $(Y,Ca)Cu_6(AsO_4)_3(OH)_6 \cdot 3H_2O$

Agrellite $Na(Ca,RE)_2Si_4O_{10}F$

Allanite-(Ce) $CaCe(Fe^{2+}Al_2)(SiO_4)_3(OH)$ (simplified); or $Ca(Ce,Y,Ca)Al(Al,Fe^{3+})(Fe^{2+},Al)(SiO_4)_3(OH)$; epidote group

Allanite-(Y) $CaY(Fe^{2+}Al_2)(SiO_4)_3(OH)$ (simplified)

Ancylite-(Ce) $(Ce,Sr,Ca)(CO_3)(OH,H_2O)$; compare gysinite-(Nd)

Apatite (see below under "Fluorapatite")

Arsenoflorencite-(Ce) $CeAl_3(AsO_4)_2(OH)_6$; crandallite group

Ashcroftine-(Y) $K_5Na_5(Y,Ca)_{12}(Si_{28}O_{70})(OH)_2(CO_3)_8 \cdot 8H_2O$

Baiyuneboite-(Ce) $BaNaCe_2(CO_3)_4F$

Bastnaesite-(Ce) $(Ce,La)(CO_3)F$

Bastnaesite-(La) $(La,Ce)(CO_3)F$

Bastnaesite-(Y) $(Y,Ce)(CO_3)F$

Belovite $Sr_3NaCe(PO_4)_3(OH)$; apatite group

Bijvoetite-(Y) $(Y,Dy)_2(UO_2)_4(CO_3)_4(OH)_6 \cdot 11H_2O$

Braitschite-(Ce) $(Ca,Na_2)_7(Ce,La)_2B_{22}O_{43} \cdot 7H_2O$

Brannerite UTi_2O_6 (simplified) or $(U,Th,Ca,Y)(Ti,Fe)_2O_6$

Britholite-(Ce) $(Ce,Y,Ca)_5[(Si,P)O_4]_3(OH,F)$; related to apatite group

Britholite-(Y) $(Y,Ce,Ca)_5[(Si,P)O_4]_3(OH,F)$

Brockite $(Ca,Th,Ce)(PO_4) \cdot H_2O$; rhabdophane group

Burbankite $(Na,Ca)_3(Sr,Ba,Ca,Ce)_3(CO_3)_5$; compare khanneshite and remondite-(Ce)

Calcioancylite-(Ce) $(Ce,Ca,Sr)(CO_3)(OH,H_2O)$

Calciobetafite Ca_2NbTiO_6F (simplified w. 2Ti>Nb+Ta) or $(Ca,Na,Ce,Th)_2(Ti,Nb,Ta)_2O_6(O,F)$

Calkinsite-(Ce) $Ce_2(CO_3)_3 \cdot 4H_2O$

Cappelenite-(Y) $BaY_6Si_3B_6O_{24}F_2$

Carbocernaite $(Ca,Na)(Sr,Ce,Ba)(CO_3)_2$; compare ewaldite formula

Caysichite-(Y) $Ca_3Y_4GdSi_8O_{20}(CO_3)_6(OH) \cdot 2H_2O$ [MF] or $Y_2(Ca,Ln)_2Si_4O_{10}(CO_3)_3(H_2O,OH) \cdot 3H_2O$ [M&N]

Cebaite-(Ce) $Ba_3Ce_2(CO_3)_5F_2$

Cebaite-(Nd) $Ba_3(Nd,Ce)_2(CO_3)_5F_2$

Cerianite-(Ce) $(Ce^{4+},Th)O_2$

Ceriopyrochlore-(Ce) $(Ce,Ca,Y)_2(Nb,Ta)_2O_6(OH,F)$

Cerite-(Ce) $Ce_9([],Ca)(Fe^{3+},Mg)(SiO_4)_6[SiO_3(OH)](OH)_3$

Cerotungstite-(Ce) $(Ce,Ca)(W,Al)_2O_6(OH)_3$

Cervandonite-(Ce) $(Ce,Nd,La)(Fe^{3+},Fe^{2+},Ti,Al)_3SiAs(Si,As)O_{13}$ (Armbruster et al., 1988)

Cheralite $(Ca,Ce,Th)(P,Si)O_4$; monazite group

Chernovite-(Y) $YAsO_4$; compare xenotime-(Y), wakefieldite-(Y)

Chevkinite $Ce_4Mg(Ti_3Mg)Si_4O_{22}$ synthetic or $(Ca,Ce,Th)_4(Fe,Mg)(Ti,Mg,Fe)_4Si_4O_{22}$; dimorph with perrierite; compare strontiochevkinite

Chukhrovite-(Ce) $Ca_3(Ce,Y)Al_2(SO_4)F_{13} \cdot 10H_2O$

Chukhrovite-(Y) $Ca_3(Y,Ce)Al_2(SO_4)F_{13} \cdot 10H_2O$

Churchite-(Y) $YPO_4 \cdot 2H_2O$

Cordylite-(Ce) $BaCe_2(CO_3)_3F_2$

Crichtonite $(Sr,La,Pb,[])(Mn,Fe)Fe_2(Fe,Ti)_6Ti_{12}O_{38}$; crichtonite group

Daqingshanite $(Sr,Ca,Ba)_3Ce(PO_4)(CO_3)_3$

Davidite-(Ce) $(Ce,La)(Y,U,Fe^{2+})(Ti,Fe^{3+})_{20}(O<OH)_{38}$; crichtonite group

Davidite-(La) $(La,Ce,Ca,Th,[])(U,Y)(Fe,Mg,[])_2(Fe^{3+},Ti,[])_6Ti_{12}(O,OH)_{38}$

Dollaseite-(Ce) $CaCe^{3+}Mg_2AlSi_3O_{11}(OH)F$; epidote group (Peacor and Dunn, 1988)

Donnayite-(Y) $NaCaSr_3Y(CO_3)_6 \cdot 3H_2O$; compare mckelveyite-(Y)

Eudialyte, cerian $(Ca,Ce)_6Zr_3(Fe,[])_3Si_{21}(Si,[])_9([],K)_3([],Al)([],Mn)_3(Na,[])_{46}O_{69}(O,[])_6(OH,[])_6(Cl,[])_4$ [M&N]; cf. Harris et al (1982); Harris and Rickard (1987)

Euxenite-(Y) $YNbTiO_6$ (simplified w. Nb+Ta>Ti; high T polymorph of aeschynite-(Y)) or $(Y,Ca,Ce,U,Th)(Nb,Ta,Ti)_2O_6$

Ewaldite $Ba(Y,Na,K,Sr,U,[])(CO_3)_2$ [M&N]; compare carbocernatite formula

Fergusonite-(Y) $YNbO_4$; high T tetrag. dimorph of $YNBO_4$; isostr. w. scheelite; compare formanite-(Y)

Fergusonite-ß-(Ce) $(Ce,La,Nd)NbO_4$

Fergusonite-ß-(Nd) $(Nd,Ce)NbO_4$

Fergusonite-ß-(Y) $YNbO_4$; dimorph with fergusonite-(Y)

Fersmite $CaNb_2O_6$ (simplified; has columbite str. with octah. Ca) or $(Ca,RE)(Nb,Ti)_2(O,OH)_6$

Florencite-(Ce) $CeAl_3(PO_4)_2(OH)_6$; crandallite group

Florencite-(La) $(La,Ce)Al_3(PO_4)_2(OH)_6$
Florencite-(Nd) $(Nd,Ce)Al_3(PO_4)_2(OH)_6$
Fluocerite-(Ce) $(Ce,La)F_3$
Fluocerite-(La) $(La,Ce)F_3$
Fluorapatite, cerian or yttrian $(Ca,RE,Na)_5(PO_4)_3(F,OH)$
Fluorite, yttrian or cerian $(Ca,Y)F_2$ or, better, $Ca_{1-x}Y_xF_{2+x}$; distinct from tveitite-(Y)
Formanite-(Y) $YTaO_4$; compare fergusonite-(Y); formula of yttrotantalite-(Y)
Gadolinite-(Ce) $(Ce,La,Nd,Y)_2(Fe^{2+},[])Be_2Si_2O_8(O,OH)_2$; gadolinite group; compare datolite, $CaBSiO_4OH$
Gadolinite-(Y) $(Y,Ce)_2(Fe^{2+},[])Be_2Si_2O_8(O,OH)_2$
Gagarinite-(Y) $NaCaYF_6$ (simplified) or $(Y,Re,Ca)_2(Na,[])F_6$
Garnet group $(Ca,Fe,Mg,Mn,Y)_3(Al,Cr,Fe,Mn,Ti,V,Zr)_2(Si,Al)_3O_{12}$; synthetic YAl_5O_{12} (YAG) and YFe_5O_{12} (YIG), etc.
Gasparite-(Ce) $(Ce,La,Nd)AsO_4$; compare monazite-(Ce)
Goudeyite $(Al,Y,Ca)Cu_6(AsO_4)_3(OH)_6 \cdot 3H_2O$; mixite group
Gysinite-(Nd) $(Nd,Pb)(CO_3)(OH,H_2O)$; compare ancylite-(Ce)

Hellandite $(Ca,Y,[])_6(Al,Fe^{3+})B_4Si_4O_{18}(OH,O)_2(OH)_2$
Hibonite $(Ca,Ce)(Al,Ti,Mg)_{12}O_{19}$
Hingganite-(Ce) $Ce_2[]Be_2Si_2O_8(OH)$; gadolinite group
Hingganite-(Y) $(Y,Yb,Er)_2([],Fe^{2+})Be_2Si_2O_8(OH,O)_2$
Hingganite-(Yb) $(Yb,Y)_2([],Fe^{2+})Be_2Si_2O_8(OH,O)_2$
Huanghoite-(Ce) $BaCe(CO_3)_2F$
Hydroxylbastnaesite-(Ce) $(Ce,La)(CO_3)(OH,F)$
Hydroxylbastnaesite-(La) $(La,Ce)(CO_3)(OH,F)$
Hydroxylbastnaesite-(Nd) $(Nd,Be,La)(CO_3)(OH,F)$

Iimoriite-(Y) $Y_2(SiO_4)(CO_3)$
Ilimaussite-(Ce) $Ba_2Na_4CeFe^{3+}Nb_2Si_8O_{28} \cdot 5H_2O$
Ilmajokite $(Na,Ce,Ba)_{10}Ti_5(Si,Al)_{14}O_{22}(OH)_{44} \cdot nH_2O$
Iraqite $(K,[])(La,Th,U,Pb)(Ca,La,Na)_2(Si,Al)_8(O,F)_{20}$
Ishikawaite $(U,Fe,Y,Ca)(Nb,Ta)O_4$?

Joaquinite-(Ce) $NaFe^{2+}Ba_2Ce_2Ti_2Si_8O_{26}(OH) \cdot H_2O$; dimorph w. orthojoaquinite-(Ce); joaquinite group

Kainosite-(Y) $Ca_2(Y,Ce)_2[Si_4O_{12}](CO_3) \cdot H_2O$
Kamotoite-(Y) $U^{6+}_4(Y,Nd,Gd)_2O_{12}(CO_3)_3 \cdot 14.5H_2O$
Karnasurtite-(Ce) $(Ce,La,Th)(Ti,Nb)(Al,Fe^{3+})(Si,P)_2O_7(OH)_4 \cdot 3H_2O$
Keiviite-(Y) $(Y,Yb)_2Si_2O_7$; compare thorveitite
Keiviite-(Yb) $(Yb,Y)_2Si_2O_7$
Kemmlitzite $(Sr,Ce)Al_3(AsO_4)[(P,S)O_4](OH)_6$; beudantite group
Khanneshite $(Na,Ca)_3(Ba,Sr,Ce,Ca)_3(CO_3)_5$; compare burbankite and remondite-(Ce)
Kimuraite-(Y) $CaY_2(CO_3)_4 \cdot 6H_2O$
Kobeite-(Y) $(Y,U)(Ti,Nb)_2(O,OH)_6$?; compare euxenite-(Y)
Kuliokite-(Y) $(Y,Yb)_4Al(SiO_4)_2(OH)_2F_5$

Lanthanite-(Ce) $(Ce,La)_2(CO_3)_3 \cdot 8H_2O$
Lanthanite-(La) $(La,Ce)_2(CO_3)_3 \cdot 8H_2O$
Lanthanite-(Nd) $(Nd,La)_2(CO_3)_3 \cdot 8H_2O$
Laplandite-(Ce) $(Na,K,Ca)_4(Ce,Th)(Ti,Mg,Al,Nb)(P,Si)_8O_{22} \cdot 5H_2O$
Lepersonnite-(Gd) $CaGd_2(UO_2)_{24}(CO_3)_8Si_4O_{12} \cdot 60H_2O$
Lokkaite-(Y) $CaY_4(CO_3)_7 \cdot 9H_2O$
Loparite-(Ce) $(NaCe^{3+})Ti_2O_6$ (simplified) or $(Ce,Na,Ca)(Ti,Nb)O_3$; perovskite group
Loranskite-(Y) $(Y,Ce,Ca)ZrTaO_6$?; compare euxenite
Loveringite $(Ca,Ce,Th)(Zr,Mn,Ce)(Fe,Mg,[])_2(Cr,Fe,Ti,V,[])_6(Ti,Al)_{12}O_{38}$; crichtonite group
Lucasite-(Ce) $(Ce,La)Ti_2(O,OH)_6$; compare formula of aeschynite-(Ce)

Mckelveyite-(Y) $NaCaBa_3Y(CO_3)_6 \cdot 3H_2O$; compare donnayite-(Y)
Melanocerite-(Ce) $(Ce,Ca)_5(Si,B)_3O_{12}(OH<F) \cdot nH_2O$; compare fluorapatite
Minasgeraisite-(Y) $(Y,Bi)_2CaBe_2Si_2O_{10}$
Miserite $K(Ca,Ce)_6Si_8O_{22}(OH,F)_2$ [MF] or $Ca_{10}(Ca,K,[])K_2Y_2Si_{16}O_{44}(OH)_2F_2 \cdot nH_2O$ (n<1) [M&N]
Monazite-(Ce) $(Ce,La,Nd,Th)PO_4$
Monazite-(La) $(La,Ce,Nd)PO_4$
Monazite-(Nd) $(Nd,La,Ce)PO_4$
Monteregianite-(Y) $(Na,Ca[])_2([],Na)_2K_2(Y,Ce)_2Si_{16}O_{38} \cdot 10H_2O$
Moydite-(Y) $Y[B(OH)_4]CO_3$
Murataite $(Na,Y)_4(Zn,Fe^{2+})_3(Ti,Nb)_6O_{18}(F,OH)_4$

Nacareniobsite-(Ce) $Na_3Ca_3(Ce,La)(Si_2O_7)_2OF_3$ (Petersen et al., 1989); compare rinkite-(Ce)
Ningyoite $(U,Ca,Ce)(PO_4) \cdot nH_2O$; rhabdophane group
Niobo-aeschynite-(Ce) $CeNbTiO_6$ (simplified w. Nb+Ta>Ti) or $(Ce,Y,Ca,Th)(Nb,Ti)_2O_6$
Niobo-aeschynite-(Nd) $(Nd,Ce)(Nb,Ti)_2(O,OH)_6$
Nordite-(Ce) $(Ce,La)(Sr,Ca)Na_2(Na,Mn)(Zn,Mg)Si_6O_{17}$
Nordite-(La) $(La,Ce)(Sr,Ca)Na_2(Na,Mn)(Zn,Mg)Si_6O_{17}$

Okanoganite-(Y) $(Na,Ca)_3(Y,Ce)_{12}Si_6B_2O_{27}F_{14}$
Orthojoaquinite-(Ce) $NaFe^{2+}Ba_2Ce_2Ti_2Si_8O_{26}(OH)\cdot H_2O$; dimorph w. joaquinite-(Ce)

Parisite-(Ce) $Ca(Ce,La)_2(CO_3)_3F_2$
Parisite-(Nd) $Ca(Nd,Ce,La)_2(CO_3)_3F_2$
Perrierite $Ce_4Mg(Ti_3Mg)Si_4O_{22}$ synthetic or $(Ca,Ce,Th)_4(Mg,Fe^{2+})(Ti,Mg,Fe)_4Si_4O_{22}$; dimorph with chevkinite
Petersite-(Y) $(Y,Ce,Nd,Ca)Cu_6(PO_4)_3(OH)_6\cdot 3H_2O$; mixite group
Phosinaite $Na_{11}(Na,Ca)_2Ca_2Ce_{0.67}(Si_4O_{12})(PO_4)_4$
Plumbopyrochlore $(Pb,Y,U,Ca)_{2-x}Nb_2O_6(OH)$
Polycrase-(Y) $YTiNbO_6$ (simplified w. $Ti > Nb + Ta$); or $(Y,Ca,Ce,U,Th)(Ti,Nb,Ta)_2O_6$
Polymignite $(Ca,Na,Ce,Th)_2Zr_2(Ti,Nb,Ta)_3([],Fe)_4O_{14}$

Remondite-(Ce) $Na_3(Ce,La,Ca,Na,Sr)_3(CO_3)_5$
Retzian-(Ce) $Mn_2Ce(AsO_4)(OH)_4$
Retzian-(La) $(Mn,Mg)_2(La,Ce)(AsO_4)(OH)_4$
Retzian-(Nd) $Mn_2(Nd,Ce,La)(AsO_4)(OH)_4$
Rhabdophane-(Ce) $(Ce,La)PO_4\cdot nH_2O$; rhapbophane group
Rhabdophane-(La) $(La,Ce)PO_4\cdot nH_2O$
Rhabdophane-(Nd) $(Nd,Ce,La)PO_4\cdot nH_2O$
Rinkite-(Ce) $Na_2Ca_4CeTi(Si_2O_7)_2OF_3$ (Petersen et al., 1989); compare nacareniobsite-(Ce)
Röntgenite-(Ce) $Ca_2(Ce,La)_3(CO_3)_5F_3$
Rowlandite-(Y) $(Y,Ca,Na,Th)_{14}(Fe,Mn,Mg)_2(Si,Al)_{13}(F,OH,Cl)_9O_{45}$?

Sahamalite-(Ce) $(Mg,Fe)Ce_2(CO_3)_4$
Samarskite-(Y) $Fe^{2+}_2YNb_5O_{16}$ (simplified) or $(Y,Ce,U,Fe^{3+},Fe^{2+})_3(Nb,Ta,Ti)_5O_{16}$
Saryarkite-(Y) $(Ca,Y,Th)_2Al_5[(Si,P,S)O_4]_4(OH)_7\cdot 6H_2O$
Sazhinite-(Ce) $Na_2CeSi_6O_{14}(OH)\cdot nH2O$ $(n > 1.5)$
Scheteligite $(Ca,Y,Sb,Mn)_2(Ti,Ta,Nb,W)_2O_6(O,OH)$? (MF)
Schuilingite-(Nd) $PbCu(Nd,Gd,Sm,Y)(CO_3)_3(OH)\cdot 1.5H_2O$
Semenovite-(La) $(Ce,Na,[])_2(Na,[],La,Ce)_2(Fe,Mn,[])([],Fe,Mn)(Ca,Na)_8Si_8(Si,Be)_6(Be,Si)_6O_{40}(OH,O)_2(OH,F)_6$
Shabaite-(Nd) $(Ca,RE)UO_2(CO_3)_4(OH)_2\cdot 6H_2O$ (Deliens & Piret, 1989)
Steenstrupine-(Ce) $Na_{14}Ce_6Mn_2Fe^{3+}{}_2(Zr,Th)(OH)_2(PO_4)_6(Si_6O_{18})_2\cdot 3H_2O$
Stillwellite-(Ce) $(Ce,La)BSiO_5$
Strontiochevkinite $(Sr,La,Ce,Ca)_4(Fe^{2+},Fe^{3+},Mn)(Ti,Zr)_4Si_4O_{22}$ [MF]
Strontio-joaquinite $(Na,Fe^{2+},[])_2Ba_2(Sr,Ce)_2Ti_2Si_8O_{24}(O,OH)_2[]\cdot 0.8H_2O$
Synchysite-(Ce) $Ca(Ce,La)(CO_3)_2F$
Synchysite-(Nd) $Ca(Nd,La)(CO_3)_2F$
Synchysite-(Y) $Ca(Y,Ce)(CO_3)_2F$

Tadzhikite-(Ce) $(Ca,Ce)_4(Ce,Y)_2(Ti,Fe^{3+},Al)B_4Si_4(O,OH)_2O_{22}$
Tantalaeschynite-(Y) $YTaTiO_6$ (simplified w. $Ta + Nb > Ti$) or $(Y,Ce,Ca,Th)(Ta,Nb,Ti)_2O_6$
Tanteuxenite-(Y) $YTaTiO_6$ (simplified w. $Ta + Nb > Ti$) or $(Y,Ce,Ca)(Ta,Nb,Ti)_2(O,OH)_6$
Tengerite-(Y) $Y_2(CO_3)_3\cdot nH_2O$ (n about 2)
Thalenite-(Y) $Y_3Si_3O_{10}(F,OH)$
Thorbastnaesite $(Th,Ca,La)_2(CO_3)_2F_2\cdot nH_2O$ (n about 3)
Thortveitite $(Sc,Y)_2Si_2O_7$; compare keiviite
Titanite, cerian $(Ce,La,Ca)(Ti,Fe^{2+})SiO_5$ (Exley, 1980)
Tombarthite-(Y) $YSiO_4(OH)$ (greatly simplified) or $Y(Si,H_4)O_{3-x}(OH)_{1+2x}$ [M&N]
Törnebohmite-(Ce) $(Ce,La)_2Al(SiO_4)_2(OH)$
Törnebohmite-(La) $(La,Ce)_2Al(SiO_4)_2(OH)$
Tranquillityite $Fe_8(Zr,Y)_2Ti_3Si_3O_{24}$
Tritomite-(Ce) $(Ce,La,Y,Th)_5(Si,B)_3(O,OH,F)_{13}$; compare fluorapatite
Tritomite-(Y) $(Y,Ca,La,Fe^{2+})_5(Si,B,Al)_3(O,OH,F)_{13}$
Tundrite-(Ce) $Na_2(Ce,La)_2TiO_2[SiO_4](CO_3)_2$
Tundrite-(Nd) $Na_2(Nd,La)_2TiO_2[SiO_4](CO_3)_2$
Tveitite-(Y) $Ca_{14}Y_5F_{43}$

Uranmicrolite $(U,Ca,Ce)(Ta,Nb)_2O_6(OH,F)$

Vesuvianite, cerian $(Ca_{15.5}Ce_{3.5})(Al,Fe^{3+})_4Fe^{2+}(Mg_5Ti_3)Si_{17.5}Al_{0.5}O_{71}(OH)_7$ (Fitzgerald et al., 1987; cf. Himmelberg
   and Miller, 1980)
Vigezzite $(Ca,Ce)(Nb,Ta,Ti)_2O_6$; compare aeschynite
Vitusite-(Ce) $Na_3(Ce,La,Nd)(PO_4)_2$
Vyuntspakhkite-(Y) $Y_4Al_2AlSi_5O_{18}(OH)_5$?

Wakefieldite-(Ce) $(Ce,Pb^{2+},Pb^{4+})VO_4$; compare xenotime and chernovite
Wakefieldite-(Y) YVO_4

Xenotime-(Y) YPO_4; zircon str.; compare chernovite and wakefieldite

Yttrialite-(Y) (Y,Th)$_2$Si$_2$O$_7$
Yttrobetafite-(Y) (Y,U,Ce)$_2$(Ti,Nb,Ta)$_2$O$_6$(OH)
Yttrocolumbite-(Y) (Y,U,Fe)(Nb,Ta)$_2$O$_6$ [M&N] or (Y,U,Fe^{2+})(Nb,Ta)O$_4$ [MF] (both BO$_2$, essentially); compare
    yttrotantalite-(Y); formula of fergusonite-(Y)
Yttrocrasite-(Y) YTi$_2$O$_5$(OH) (simplified) or (Y,Th,Ca,U)(Ti,Fe)$_2$(O,OH)$_6$; compare euxenite-(Y)
Yttropyrochlore-(Y) (Y,Na,Ca,U)$_2$(Nb,Ta,Ti)$_2$O$_6$(OH)
Yttrotantalite-(Y) (Y,U,Ca)(Ta,Nb,Fe)$_2$O$_6$ [M&N] or (Y,U,Fe^{2+})(Ta,Nb)O$_4$ [MF] (both BO$_2$, essentially); compare
    yttrocolumbite-(Y); formula of formanite-(Y)
Yttrotungstite-(Y) YW$_2$O$_6$(OH)$_3$

Zhonghuacerite-(Ce) Ba$_2$Ce(CO$_3$)$_3$F
Zirconolite CaZrTi$_2$O$_7$ (simplified) or (Ca,Th,U,Ce)$_2$(Zr,Ti)$_2$(Ti,Nb,Zr,Fe)$_2$([],Fe,Ti)$_2$(Ti,Nb)O$_{14}$; compare formulas of
    calciobetafite (pyrochlore group), zirkelite, polymignite
Zirkelite (Ca,Na,Ce)$_2$Zr$_2$(Ti,Nb,Ta)$_3$(Fe,[])$_2$O$_{14}$

Table 2. Discredited rare-earth-element mineral names (synonyms, varieties, etc.). Mostly from Miyawaki and Nakai (1987; 1988); also Fleischer (1987; 1989); Nickel and Mandarino (1987).

Discredited Name	Accepted Name
absite	= brannerite
abukumalite	= britholite-(Y)
ampangabeite	= samarskite
barsanovite	= eudialyte
beckelite	= britholite
blomstrandine	= aeschynite-(Y)
brocenite	= beta fergusonite-(Ce)
caryocerite	= thorian melanocerite?
cenosite	= kainosite
coutinite	= lanthanite-(Nd)
delorenzite	= tanteuxenite
djalmaite	= uranmicrolite
doverite	= synchysite-(Y)
eucolite	= eudialyte
ferutile	= davidite
huangheite	= huanghoite-(Ce)
johnstrupite	= rinkite-(Ce)
keilhauite	= yttrian titanite
knopite	= cerian perovskite
koppite	= cerian pyrochlore
kozhanovite	= karnasurtite
kularite	= monazite-(Ce)
kusuite	= wakefieldite-(Ce)
lessingite	= britholite-(Ce)
lombaardite	= allanite
marignacite	= ceriopyrochlore
mohsite	= plumboan crichtonite
mosandrite	= altered rinkite
nuevite	= samarskite
obruchevite	= yttropyrochlore
orthite	= allanite-(Ce)
pravdite	= britholite-(Y)
priorite	= aeschynite-(Y)
rinkolite	= rinkite-(Ce)
scheteligite	none (discredited)
spencite	= tritomite-(Y)
sphene	= titanite
taiyite	= aeschynite-(Y)
texasite	none (discredited)
treanorite	= allanite-(Ce)
tscheffkinite	= chevkinite
tysonite	= fluocerite-(Ce)
ufertite	= davidite-(La)
weinschenkite	= churchite-(Y)
xinganite	= hingganite-(Y)
yftisite	none (discredited)
yttroceberysite	= hingganite-(Y)
yttrofluorite	= yttrian fluorite
yttromicrolite	none (discredited)

COUPLED SUBSTITUTIONS

The smaller or Y-group (heavy) REE exhibit irregular coordination numbers with oxygen of 6 to 9, most commonly 8, whereas the larger or Ce-group (light) REE exhibit larger coordination numbers of from 7 to 12, most commonly 9 (e.g., Miyawaki and Nakai, 1987, p. 122). The trivalent Ce-group REE are very similar crystal-chemically to Ca^{2+}, and they commonly substitute for Ca in rock-forming minerals.

Substitution of a trivalent REE cation for divalent Ca requires some sort of a charge-compensating mechanism, i.e., a coupled substitution. Simple or coupled substitutions can be represented conveniently and compactly as "exchange operators," containing negative quantities of the elements replaced (Burt, 1974, based on a concept introduced in class lectures in the late 1960s by J.B. Thompson, Jr.). Simple exchange operations involving REE can be represented by the operators $EuCa_{-1}$ (for Eu^{2+}; "add one Eu, remove one Ca" or "exchange an Eu for a Ca"), YCe_{-1} (discussed separately below), and $CeTh_{-1}$ (for Ce^{4+}).

As an example of more complex exchange operations, consider how to substitute Y^{3+} for Ca^{2+} in grossular garnet, $Ca_3Al_2Si_3O_{12}$. Charge can be compensated in the 8-coordinated Ca-site, the 6-coordinated Al-site, or the 4-coordinated Si site. In the Ca-site, charge compensation could occur by a vacancy [ ], via the exchange operator $([\ ]Y_2)Ca_{-3}$, but in the dense crystal structure of garnet it is more likely to occur by concomitant substitution of a univalent ion such as a Na^+, via the operator $(NaY)Ca_{-2}$. This would lead to the theoretical garnet end-member $(Na_{1.5}Y_{1.5})Al_2Si_3O_{12}$; the other two sites remain unaffected. Charge compensation in the octahedral Al-site could occur via the exchange operator $YMg(CaAl)_{-1}$, leading to the theoretical garnet end-member $(Y_2Ca)Mg_2Si_3O_{12}$. Charge compensation in the tetrahedral Si-site could occur via $YAl(CaSi)_{-1}$, leading to the well-known synthetic oxide garnet end-member $Y_3Al_2Al_3O_{12}$ (YAG). These substitutions are illustrated graphically below; in nature all might occur simultaneously (in solid solution in the same crystal).

The possibilities for such coupled substitutions can generally be estimated from tables of ionic radii. A convenient table is given in somewhat inconvenient (alphabetic) form by Shannon (1976). Shannon's extensive tabulation of effective ionic radii is reproduced in part in Table 3, where cations (including the REE) are sorted both by valence and ionic radius into three sub-tables, respectively for 8-, 6-, and 4-coordination (as in the garnet example above). These coordination groups are here called cation groups A, B, and C. The REE obviously typically belong to large cation group "A," although the Y-group of smaller REE verges on fitting in octahedral group "B" (where the smaller Sc^{3+} fits very well).

Such tables of ionic radii provide only a general guide to substitutional tendencies; they ignore factors such as the extremely irregular coordination polyhedra present in many minerals (particularly for the "A" cations), the tendencies to covalency of elements, and the geochemical segregation of the elements (i.e., the fact that many ions of the same size are extremely unlikely to occur together in the same mineral-growing environment).

Coupled substitutions involving REE for which there is some natural or experimental evidence are listed in Table 4. The basic classification is similar to that employed by Miyawaki and Nakai (1987); the majority of the examples are from Solodov et al. (1987) and especially Töpper (1979), which should be consulted for references. Note that in addition to the types of substitutions mentioned above, Table 4 lists coupled substitutions that involve anions, anion vacancies, or a change in valence.

An interesting example of this last type of substitution is the last listed, $Ce^{4+}O(Ce^{3+}OH)_{-1}$, compositionally equivalent to H^0_{-1} (minus atomic H; the Ce's and O's

Table 3. Sorted effective ionic radii (in Å) of some typical cubic A(VIII), octahedral B(VI), and tetrahedral C(IV) cations potentially involved in coupled substitutions with REE (data from Shannon, 1976, with "HS" designating high spin radius.). REE are in bold type.

A(VIII) Cations

A+	A2+	A3+	A4+	A6+	Radius
			Zr^{4+}		0.84
				U^{6+}	0.86
		Sc^{3+}			0.870
	Mg^{2+}				0.89
	Zn^{2+}				0.90
		Sb^{3+}			0.91
Li^+					0.92
	Fe^{2+}(HS)				0.92
			Pb^{4+}		0.94
	Mn^{2+}				0.96
			Ce^{4+}		**0.97**
		Lu^{3+}			**0.977**
		Yb^{3+}			**0.985**
		Tm^{3+}			**0.994**
			U^{4+}		1.00
		Er^{3+}			**1.004**
		Ho^{3+}			**1.015**
		Y^{3+}			**1.019**
		Dy^{3+}			**1.027**
		Tb^{3+}			**1.040**
			Th^{4+}		1.05
		Gd^{3+}			**1.053**
		Eu^{3+}			**1.066**
		Sm^{3+}			**1.079**
		Pm^{3+}			**1.093**
	Cd^{2+}				1.10
		Nd^{3+}			**1.109**
	Ca^{2+}				1.12
		Pr^{3+}			**1.126**
		Ce^{3+}			**1.143**
		La^{3+}			**1.160**
		Bi^{3+}			1.17
Na^+					1.18
	Eu^{2+}				1.25
	Sr^{2+}				1.26
	Pb^{2+}				1.29
	Ba^{2+}				1.42
K^+					1.51
Rb^+					1.61
Cs^+					1.74

B(VI) Cations

B+	B2+	B3+	B4+	B5+	B6+	Radius
			Si^{4+}			0.400
			Mn^{4+}			0.530
		Al^{3+}				0.535
					Mo^{6+}	0.59
					W^{6+}	0.60
				Sb^{5+}		0.60
			Ti^{4+}			0.605
				Nb^{5+}		0.64
				Ta^{5+}		0.64
		Fe^{3+}(HS)				0.645
		Mn^{3+}(HS)				0.645
			Mo^{4+}			0.65
			W^{4+}			0.66
			Sn^{4+}			0.690
			Zr^{4+}			0.72
	Mg^{2+}					0.72
					U^{6+}	0.73
	Zn^{2+}					0.740
		Sc^{3+}				0.745
		Sb^{3+}				0.76
Li^+						0.76
				Bi^{5+}		0.76
			Pb^{4+}			0.775
	Fe^{2+}(HS)					0.780
	Mn^{2+}(HS)					0.830
		Lu^{3+}				**0.861**
		Yb^{3+}				**0.868**
			Ce^{4+}			**0.867**
		Y^{3+}				**0.87**

C(IV) Cations

C+	C2+	C3+	C4+	C5+	C6+	Radius
		B^{3+}				0.11
					S^{6+}	0.12
			C^{4+}			0.15
				P^{5+}		0.17
			Si^{4+}			0.26
	Be^{2+}					0.27
					Se^{6+}	0.28
				As^{5+}		0.335
				V^{5+}		0.355
			Ge^{4+}			0.390
			Mn^{4+}			0.39
		Al^{3+}				0.39
					Mo^{6+}	0.41
					W^{6+}	0.42
			Ti^{4+}			0.42
		Fe^{3+}(HS)				0.49
			Sn^{4+}			0.55
	Mg^{2+}					0.57
			Zr^{4+}			0.59
Li^+						0.590
	Zn^{2+}					0.60
	Fe^{2+}(HS)					0.63

Table 4. Substitutions involving rare earth elements in minerals, with examples (greatly modified from Topper, 1979; Solodov et al., 1987; Miyawaki and Nakai, 1987).

A. Simple, within-site substitutions

$Eu^{2+}Ca_{-1}$	Plagioclase, many others
$Ce^{4+}Th_{-1}$	Cerianite
$NaCeCa_{-2}$	Apatite, loparite, many others
$CaThCe_{-2}$	Monazite, britholite, many others

B. Substitutions involving cation vacancies

$[]Ce_2Ca_{-3}$	Hellandite, semenovite
$[]Ce(NaCa)_{-1}$	Gagarinite
$Ce_4([]Th_3)_{-1}$	Thorianite

C. Coupled substitutions involving cations

1. Second site large (VIII, IX, or X), but distinct

$CeNa(SrCa)_{-1}$	Carbocernaite, burbankite

2. Second site octahedral (VI)

$YNb(FeW)_{-1}$	Wolframite
$CeNb(UTi)_{-1}$	Brannerite
$YTi(FeNb)_{-1}$	Columbite
$CeTi(CaNb)_{-1}$	Euxenite/aeschynite, pyrochlore
$YFe^{3+}(Fe^{2+}Ti)_{-1}$	Ilmenite
$CeFe^{3+}(CaTi)_{-1}$	Perovskite, titanite
$CeMg(CaAl)_{-1}$	Dollaseite

3. Second site tetrahedral (IV)

$YP(CaS)_{-1}$	Churchite-gypsum
$CeP_2(KS_2)_{-1}$	Florencite-alunite
$YP(ZrSi)_{-1}$	Xenotime-zircon
$CeNb(CaW)_{-1}$	Scheelite
$CeSi(CaP)_{-1}$	Britholite
$CeSi(ThB)_{-1}$	Tritomite
$CeAl(CaSi)_{-1}$	Chukhrovite
$YAl(MnSi)_{-1}$	Spessartine
$CeB(CaSi)_{-1}$	Melanocerite, tritomite
$Ce_2B(Ca_2P)_{-1}$	Melanocerite
$YBe(CaB)_{-1}$	Gadolinite-homilite, hingganite-datolite

4. Second site trigonal (III)

$CeB(CaC)_{-1}$	$CeBO_3$ (synthetic, calcite str.)

D. Coupled substitutions involving anions

$CeO(CaF)_{-1}$	Fersmite, pyrochlore, apatite
$CeO(CaOH)_{-1}$	Hydroxylapatite, etc.
$ThO(CeF)_{-1}$	Fluocerite
$YOH(ZrO)_{-1}$	Zircon
$CeOH(CaH_2O)_{-1}$	Ancylite, hibonite

E. Coupled substitutions involving anion vacancies (interstitials)

$YF(Ca[])_{-1}$	Tveitite, fluorite
$Y_2[](U_2O)_{-1}$	Uraninite
$Y_2S(Pb_2[])_{-1}$	Galena

F. Coupled substitutions involving a change in valence

$CeFe^{2+}(CaFe^{3+})_{-1}$	Allanite, gadolinite
$Ce^{4+}O(Ce^{3+}OH)_{-1}$	Heated metamict phases?

cancel). This type of change probably affects metamict, hydrous REE minerals that are heated in air or vacuum in an attempt to restore their original crystal structure; they are instead oxidized as they lose hydrogen (cf. Burt, 1988).

VECTOR TREATMENT

A typical ionic substitution in REE minerals is YCe_{-1}. Inasmuch as such ionic substitutions have both a direction (the direction opposite to YCe_{-1} is CeY_{-1}) and a magnitude, it is logical to treat exchange operators as vectors. In terms of magnitude, not only can the extent of CeY_{-1} substitution vary, but also one has only to apply the operation once to change bastnaesite-(Ce) into bastnaesite-(Y), or the formula of monazite-(Ce) into that of xenotime-(Y). On the other hand, one has to apply it twice to change, e.g., gadolinite-(Ce) into gadolinite-(Y) (Table 1). I developed this vector treatment in 1981 in China for Li-micas, shortly after J.B. Thompson, Jr. (1981, 1982, following J.V. Smith, 1959) independently applied the concept to amphiboles. I and others had been using the chemical potentials of exchange operators ("exchange potentials") as vectors for years (see also below); this is not the same as using the operators themselves as vectors ("exchange vectors") in crystal chemistry.

Diagrams involving exchange vectors are generally either anion conservative or cation conservative or both (cf. Brady and Stout, 1980). A useful property, copiously illustrated below, is that straight-line contours can be drawn for the formula contents of individual ions. Extensions to the third dimension are also illustrated.

The complex coupled substitutions involving REE mean that the accessible composition space of REE minerals in a plane is not necessarily a triangle or a reciprocal ternary square. Similarly, in three dimensions, it need not be a tetrahedron or a cube. This complexity, as as well the convenience of direct plotting of formula contents, is the main justification for the vector approach.

In what follows, mineral names may be loosely used to refer to end-member compositions. "Derived from monazite-(Ce)," for example, means derived from the composition or the formula of ideal monazite, $CePO_4$. Similarly, inasmuch as the vector approach deals with compositions, not structures, minerals of diverse structeres may be mixed on the same diagram. This does not imply that the end-members are completely miscible, that they are isostructural (e.g., monazite has a different structure from xenotime), or even that the end-member compositions are stable phases. In many cases, of course, not only are the end-members stable, but polymorphs occur (Table 1). The formula type $RE_2Si_2O_7$, for example, has at least four stable polymorphs (e.g., Ito and Johnson, 1968).

A final vector concept, which should be familiar, is that of "condensation" down simple exchange vectors, that is, combining ions of similar size and coordination (A, B, or C as defined above) and valence. To simplify the following discussion, A+ will be generally be considered to be Na, A^{2+} Ca (or rarely Sr or Ba), A^{3+} Y (or Ce), A^{4+} Th (or U), B^{5+} Nb (or Ta), B^{4+} Ti (or Sn), B^{3+} Al (or Fe), B^{2+} Mg (or Fe), B+ Li, C^{6+} S, C^{5+} P (or As), C^{4+} Si (or rarely C, carbon), C^{3+} Al (or Fe^{3+} or rarely B, boron), and C^{2+} Be. Due to lack of abundant natural examples, B^{6+}, W (or Mo), will generally be ignored. Appropriate substitutions in the A, B, and C groups can be surmised from Table 3. (Note, however, that certain substitutions, such as Ti for Ta, only rarely occur in nature.)

In Table 3, A cations smaller than Mn^{2+} are small for 8-coordination, unless they are combined with larger cations (as in couples such as $LiSr(NaCa)_{-1}$ or $MgSrCa_{-2}$); in contrast, K+ and especially Rb+ and Cs+ are rather large. Similarly, B cations smaller than W^{6+} are small for 6-coordination (except possibly in fergusonite and at high pressure), and those larger than Zr^{4+} are rather too large. The radius ratios among the A, B, and C cations also exert a strong influence on the structures Mn^{2+} adopted; therefore a single site cannot be considered alone (e.g., Muller and Roy, 1974).

A problem is that of "intermediate" ions such as $Fe^{2+}Mn^{2+}$, and to a certain extent, Y^{3+}. They are small for A and large for B; their partitioning between A and B sites is therefore difficult to predict and is likely to involve both sites (disordering, as in samarskite-(Y) at high temperatures; Sugitani et al., 1985). Unambiguous assignment of other ions of variable valence is also difficult in phases containing them in minor amounts; examples include Sn^{2+} or Pb^{2+} (A) vs. Sn^{4+} or Pb^{4+} (B), Sb^{3+} (A or B) vs. Sb^{5+} (B), and U^{4+} (A) vs. U^{6+} (A or B; Lumpkin et al., 1986, favor A).

APPLICATION TO SELECTED MINERAL GROUPS

Fluorides

Fluocerite-(Ce), gagarinite-(Y), and tveitite-(Y) (Table 1) are the simple rare earth fluorides, and fluorite, CaF_2, commonly contains REE and Th in solid solution for its 8-coordinated Ca. The simplest way to substitute Ce^{3+} and Th^{4+} in a Ca-phase is by a coupled substitution involving Na^+, as shown by the vector diagram at the top of Figure 1. (The same vector diagram can be applied to any Na,REE,Th-bearing calcic mineral.) The two orthogonal basis vectors are $NaCeCa_{-2}$ (horizontal) and Na_2ThCa_{-3} (vertical); these respectively are unit vectors of constant Th and of constant Ce, as labelled. Twice the first minus the second cancels Na in the vector of constant Na, namely $Ce_2(CaTh)_{-1}$, which has a slope of -1/2. Three times the first minus twice the second cancels Ca in the vector of constant Ca, namely $Ce_3(NaTh_2)_{-1}$, which has a slope of -2/3. Similar relations will be labelled without discussion on other vector diagrams throughout this section.

The two basis vectors are here drawn orthogonal and of unit length for convenience in plotting only; if they were drawn at 60° and with the vertical vector 1.5 times the horizontal, a conventional equilateral triangle would result below. Also, any two vectors could have been chosen as a basis; those chosen describe simple Na-Ce and Na-Th solid solution in fluorite. ($NaCeCa_{-2}$ and Na_2ThCa_{-3} are the electrically neutral substitutions of this type.)

When these vectors are applied to the "additive component" CaF_2 (the fluorite formula, from which all other formulas are derived by substitution; it is multiplied by 6 to avoid fractional coefficients), the triangle at the bottom of the figure results. It has the new "end-member" compositions (vertices) $3NaCeF_4$ (generated by applying the horizontal basis vector three times to Ca_6F_{12}) and $2Na_2ThF_6$ (generated by applying the vertical basis vector twice to Ca_6F_{12}). These "end-members" are not necessarily isostructural with fluorite, inasumuch as this is a composition diagram. The sides of the triangle are lines of Th = 0 (horizontal base), Ce = 0 (vertical left side), and Ca = 0 (slanting top of slope -2/3); dashed lines of constant Th, Ce, and Ca within the diagram are parallel to the respective sides (which in turn are parallel to the corresponding vectors in the diagram above). Lines of constant Na have a slope of -1/2, corresponding to that of the constant-Na vector above, although there is no side corresponding to Na = 0 (only the point $6CaF_2$). The dashed line for Na = 2 passes through the point $(Ca_3Na_2Th)F_{12}$ on the left side and $2NaCaCeF_6$, labelled "G," on the base. This point would correspond to a Ce-analog of gagarinite-(Y) (not yet found in nature). Formula contents of any fluorite solid solution $(Ca,Na,Ce,Th)F_2$ could be plotted directly on this diagram by multiplying the formula by 6.

The crystal structure of fluorite commonly contains vacancies; these offer another way to insert REE, via a coupled substitution such as $[\ ]Y_2Ca_{-3}$, where $[\ ]$ is a cationic vacancy. This substitution is analogous to that relating dioctahedral to trioctahedral micas, $[\ ]Al_2Mg_{-3}$ (e.g., Burt, 1988), and is depicted as the vertical basis vector at the top of Figure 2A. The horizontal vector of constant vacancies is $NaYCa_{-2}$, where Y is used in

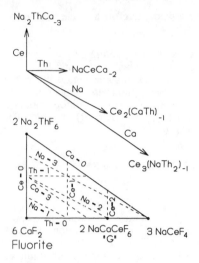

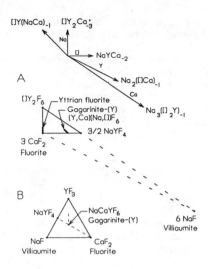

Figure 1. Planar vector diagram showing substitution of Ce and Th in fluorite via coupled substitution with Na. See text for discussion.

Figure 2. Coupled substitutions in fluorite of formula $(Ca,Na,Y,[\ ])F_2$. (A) Planar vector diagram. (B) Barycentric triangle NaF-CaF_2-YF_3. See text for discussion.

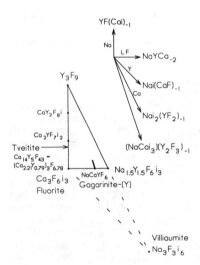

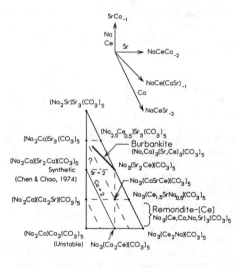

Figure 3. Planar vector diagram showing generation of the tveitite composition by the substitution $YF(Cai)_{-1}$ in fluorite, where i is an interstitial position.

Figure 4. Planar vector diagram showing the composition plane accessible to burbankite and related phases. Dashed lines on these and following figures are element contours; numerical values are obtainable from formula labels and vector slopes.

place of Ce to allow the depiction of gagarinite-(Y). Vectors of constant Y and Ca can be derived as shown. When these vectors are applied to the additive component $3CaF_2$ (I have divided by 2 to get some room), a new triangle is generated, that includes gagarinite-(Y) 2/3 across the base, and a Y compositional analog to fluocerite-(Ce), CeF_3, at the top. The vertical minus the horizontal basis vectors yields a substitution $[\ ]Y(NaCa)_{-1}$ that is observed in gagarinite-(Y), and is shown by making the gagarinite composition a diagonal line rather than a point on this and the following figure (data cited in Miyawaki and Nakai, 1987). Note that extending the two vectors of constant Y and Ca to their intersection yields the composition 6NaF (villiaumite). This triangle, involving dissimilar crystal structures, could just as well be depicted barycentrically, as shown in Figure 2B, the system NaF-CaF_2-YF_3.

Fluorite can have anionic as well as cationic defects; in fact, anionic defects (interstitials, here abbreviated i) are typical (see review by, e.g., Catlow, 1985). Its formula can then be represented as $Ca_{1-x}Y_xF_{2+x}$, where x indicates a variable composition. These can be generated by the vector $YF(Cai)_{-1}$, where i is an interstitial site, as shown on Figure 3. This yields essentially the same composition triangle as Figure 2, but with different proportions. Tveitite, $Ca_{14}Y_5F_{43}$, is an ordered defect structure of fluorite, one of a series of such between CaF_2 and YF_3 (Bevan et al., 1982).

An interesting exercise is to simultaneously apply the vectors $[c]Y_2Ca_{-3}$ and $Ca[a](YF)_{-1}$, where [c] and [a] are respectively cationic and anionic vacancies, to the composition of fluorite, $3CaF_2$. You will generate the two triangular corner compositions $(Y_2[c])F_6$, as expected, and also $3[c][a]2$, or nothing (a vacuum). In other words, a proper composition diagram cannot be constructed because, on the basis of analyses alone, one cannot distinguish between cationic vacancies and anionic interstitials (that is, one knows only the ratio of anions to cations). Fluorite is the only mineral discussed in which interstitials are considered.

In terms of relevant experimental work, Thoma et al. (1963a) studied the system NaF-YF_3 and Thoma et al. (1963b) the system NaF-ThF_4-UF_4. Brown (1968) reviewed the structures of various forms of $(RE)F_3$ and Sobolev and Fedorov (1978) reviewed the complexities of phase diagrams for the various binary systems CaF_2-$(RE)F_3$ (cf. Greis and Haschke, 1982; Posypaiko and Alekseeva, 1987).

Carbonates

Anhydrous $RE_2(CO_3)_3$, known synthetically, has not yet been found in nature; its hydrated analogs are calkinsite-(Ce), with $4H_2O$ and lanthanite-(Ce), -(La), and -(Nd), with 8. Space does not permit a discussion of these and other hydrated carbonates or of sahamalite-(Ce), $Ce_2Mg(CO_3)_4$, equivalent to $Ce_2(CO_3)_3 + MgCO_3$.

What is left might be called the A-B carbonates, where A and B are both large cations. Burbankite, khanneshite (its Ba-analog), and remondite-(Ce) (a more Na,REE-rich analog) share the formula $A_3B_3(CO_3)_5$, where A is 8-coordinated (Na,Ca) and B is 10-coordinated (Sr,Ce,Ba,Ca...) (e.g., Effenberger et al., 1985). Carbocernaite and ewaldite share the formula $BA(CO_3)_2$, but have different structures. In carbocernaite B is 10-coordinated (Sr,Ce,Ba,Ca,...) and A is 7-coordinated (Ca,Na) (Voronkov and Pyatenko, 1967). In ewaldite B is 9-coordinated Ba and A is 6-coordinated (Ca,Y,Na...) (Donnay and Preston, 1971).

The composition plane accessible to burbankite is depicted on a vector diagram in Figure 4. Burbankite of ideal composition $(Na,Ca)_3(Sr,Ce)_3(CO_3)_5$ is free to vary only along the vector $NaCe(CaSr)_{-1}$, between the two end-members $(Na_2Ca)Sr_3(CO_3)_5$ and $Na_3(Sr_2Ce)(CO_3)_5$ (dark line segment on Fig. 4). However, Chen and Chao (1974) synthesized a REE-free burbankite of composition $(Na_2Ca)(Sr_2Ca)(CO_3)_5$, and reported that attempts to synthesize the theoretical end-member $(Na_2Ca)Sr_3(CO_3)_5$ (just above it on

Fig. 4) did not yield a phase of burbankite structure, nor did the composition just below it $(Na_2Ca)(Ca_2Sr)(CO_3)_5$. Clearly, some Ca can fit in the B site, and presumably some smaller Y-group REE could fit in the A site, so that burbankite compositions could cover a large part of the triangle of Figure 4 (for instance, above and to the right of the diagonal line for Ca = 2). The theoretical composition of end-member remondite-(Ce) (a new mineral of burbankite-like structure determined by Ginderow, 1989) is seen to be $Na_3(Ce_2Na)(CO_3)_5$. It is unknown if this Na-Ce end-member composition would have the burbankite structure (the Na-Ca and probably Na-Sr end-member compositions do not). Figures 5A and 5B respectively depict the composition planes theoretically accessible to carbocernaite and ewaldite. The vectors used in 5A are the same used in Figure 4; only the starting composition (formula type) is different. The two planes of Figures 5A and 5B are analogous; Figure 5B is related to 5A by the exchange operator $BaSr_{-1}$. Ewaldite can be derived from the composition $BaCa(CO_3)_2$ (the polymorph alstonite, barytocalcite, or paralstonite) by the substitution vector $NaCeCa_{-2}$, whereas carbocernaite is derived by from the analogous Sr composition $SrCa(CO_3)_2$ by the substitution vector $NaCe(CaSr)_{-1}$ (also used for burbankite). Carbocernaite could therefore coexist with ewaldite. These systems have not been studied experimentally; the REE-free carbonate system $CaCO_3$-$SrCO_3$-$BaCO_3$ has been studied by Chang (1971).

The similarities between the burbankite and carbocernatite composition spaces (Figs. 4 and 5A) suggest some compositional relation between the two phases. Both phases can contain cation vacancies; this fact allows us to relate them via a vector such as $[\ ]_2Ce_3(NaSr_4)_{-1}$, as shown on Figure 6, for which the horizontal basis vector is $CaSr(NaCe)_{-1}$, common to both phases (ideally). The resulting composition space is a quadrilateral, with sides marked by lines of Na, Ca, Sr, and Ce = 0, as shown. The inset (below right) shows that this quadrilateral is a planar section through the tetrahedron $CaCO_3$-$SrCO_3$-Na_2CO_3-$Ce_2(CO_3)_3$, at points marked as 1/2, 1/3, 1/2, and 2/3 along the respective edges. The triangular plane is that of Figure 5A; the two planes intersect along a carbocernaite line. The triangular plane of Figure 4 (not shown) would lie parallel and below and would intersect the quadrilateral plane of Figure 6 along a burbankite line. Both burbankite and carbocernatite could occupy spaces inside the Figure 6 (inset) tetrahedron; given sufficient cation vacancies burbankite compositions could approach those of carbocernatite.

Fluorocarbonates

The REE-fluorocarbonates include bastnaesite, parisite, röntgenite, and synchysite, all discussed below, and the similar Ba-phases huanghoite, $BaCe(CO_3)_2F$, zhonghuacerite, $Ba_2Ce(CO_3)_3F$, cordylite, $BaCe_2(CO_3)_3F_2$, and cebaite, $Ba_3Ce_2(CO_3)_5F_2$ (all at least with -(Ce) suffixes: Table 1). Bastnaesite also has a thorian variety, thorbastnaesite, which is reportedly hydrated; it also has the OH-analogs hydroxylbastnaesite-(Ce), -(La), and -(Na) (Table 1).

Crystallographic relations among bastnaesite, $CeCO_3F$, parisite, $CaCe_2(CO_3)_3F_2$, röntgenite, $Ca_2Ce_3(CO_3)_5F_3$, and synchysite, $CaCe(CO_3)_2F$, were reported by Donnay and Donnay (1953, summarized in Miyawaki and Nakai, 1987). Bastnaesite is basically a sheet structure, with Ce in irregular coordination with eight O and three F. In the other fluorocarbonates, bastnaesite-like sheets are interlayered with sheets of $CaCO_3$, in the pattern BBCBBC... in parisite, BBCBCBBCBC...in röntgenite, and BCBC... in synchysite (where B indicates a bastnaesite-like sheet and C a $CaCO_3$ sheet). Cordylite is chemically analogous to parisite and huanghoite to synchysite, among the Ba-fluorocarbonates. Van Landuyt and Amelinckx (1975) used TEM to show that the ordering pattern in naturally-occurring fluorocarbonate minerals is highly imperfect.

Chemically, bastnaesite-(Ce) is $1/3[Ce_2(CO_3)_3 + CeF_3]$. The other REE-fluorocarbonate compositions are linear combinations of $CaCO_3$ and $CeCO_3F$, in the ratio of 1:1 for synchysite-(Ce), 2:3 for röntgenite-(Ce), and 1:2 for parisite-(Ce). These relations are

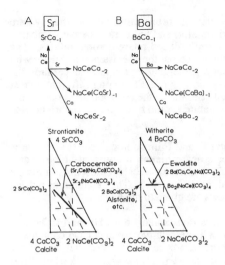

Figure 5 (left). Planar vector diagrams showing the analogous composition planes for carbocernaite (A, with Sr) and ewaldite (B, with Ba). (A) Carbocernaite plane. Slanting dark line satisfies formula. (B) Ewaldite plane. Horizontal dark line satisfies formula.

Figure 6. Planar vector diagram showing how the composition of carbocernaite can be derived from that of burbankite via introduction of cation vacancies. Inset shows how the plane of this figure (quadrilateral) and that of Figure 5A (triangle) intersect along a carbocernaite line in the tetrahedron $CaCO_3$-$SrCO_3$-Na_2CO_3-$Ce_2(CO_3)_3$.

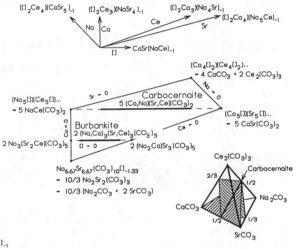

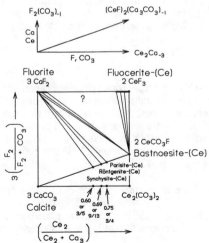

Figure 7. Planar vector diagram showing the reciprocal ternary system /Ca,Ce//F$_2$,CO$_3$/ as a square.

shown on Figure 7, a reciprocal ternary vector representation of the system /Ca,Ce//F$_2$,CO$_3$/, where cation ratios vary horizontally, and anion ratios vary vertically, resulting in a square (or rectangle). Tie lines show deduced phase compatibilities. The stability of the intermediate fluorocarbonates should rule out the coexistence of calcite with bastnaesite. However, bastnaesite is a common mineral in carbonatites; perhaps the intermediate phases are not always stable, or partitioning of REE among different phases allows tie lines in the simple system to be broken. Similarly, the common and presumably stable coexistence of bastnaesite with fluorite in nature should rule out any occurrences of the other REE-fluorocarbonates with fluocerite-(Ce).

In order to show the true proportions among O and F in fluorocarbonates (the same method can be used for other minerals or for silicate melts), a constant-anion vector representation is preferable. Figure 8 gives such an alternate vector representation for the fluorocarbonates and related phases, with O and F totalled to 18. The calculations needed to plot the Ce-content of the fluorocarbonate phases are given to the lower right.

Fluorocarbonates obviously grow from fluids that contain both HF and CO$_2$; less obviously, perhaps, they have the potential to indicate relative HF and CO$_2$ contents in such fluids. This potential is indicated in Figure 9, a schematic vector diagram of the chemical potentials of CO$_2$ and F$_2$O$_{-1}$ (shorthand for μ(F$_2$) - 1/2 μ(O$_2$); think of F$_2$O$_{-1}$ as the "acid anhydride" of HF). The numbered slanting lines of slope +1 each represent an exchange reaction of the type

$$CaCO_3 + 2HF = CaF_2 + CO_2 + H_2O . \tag{1}$$

The reaction above is labelled "1" or Cal/Flu on the diagram. Reactions 2 to 4 represent the successive "burning off" of the CaCO$_3$ component of the calcic REE-fluorcarbonates synchysite, röntgenite, and parisite, most CaCO$_3$-rich first, and reaction 5 represents the much more difficult fluorination of bastnaesite itself

$$CeCO_3F + 2HF = CeF_3 + CO_2 + H_2O . \tag{2}$$

The actual (quantitative) spacing of these five lines is unknown, but is likely to be much less regular than shown, with lines 2, 3, and 4 bunched closely together (because the phases involved have such similar compositions). In any case, fluocerite indicates much more HF-rich, CO$_2$-poor fluids than does fluorite.

Although the above system has apparently not been studied experimentally, bastnaesite analogs have been synthesized (e.g., Haschke, 1975), and the hydrothermal phase relations of hydroxylbastnaesite analogs in the system RE$_2$O$_3$-CO$_2$-H$_2$O have been extensively studied (Chai and Mroczkowski, 1978); hydroxylbastnaesite is stable to above 500°C.

Monazite, xenotime, zircon, and related phases

The tetragonal REE-phosphate xenotime-(Y) and its corresponding arsenate, chernovite-(Y) and vanadate, wakefieldite-(Y) are isostructural with zircon and thorite, and have Y and small REE in 8-fold coordination (see, e.g, Speer, 1982a,b for a discussion of the corresponding silicates). The possible coupled substitutions among these phases are indicated as vectors in Figure 10A. Extensive solid solutions involving REE in thorite-group minerals are documented by, e.g, Lumpkin and Chakoumakos (1988); for zircon data see Speer (1982a, p. 74-76) and Deer et al. (1982a, p. 423).

The larger LREE (Ce-group) phosphates and the arsenate gasparite-(Ce) adopt the monoclinic structure of monazite-(Ce), -(La), and -(Nd), in which the RE^{3+} cation is in irregular 8 to 9-fold coordination with O (see again Speer, 1982b); these are isostructural with huttonite, a dimorph of ThSiO$_4$. The coupled substitution CaThCe$_{-2}$ in monazite-(Ce)

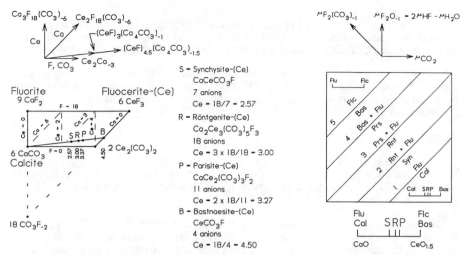

Figure 8 (left). Planar vector diagram showing the reciprocal ternary system /Ca,Ce//F$_2$,CO$_3$/ as a triangle (accessible portion is a quadrilateral) in which the total number of (O + F) is constant at 18. This diagram shows the relative proportions of O and F in the minerals.

Figure 9 (right). Chemical potential vector diagram (μF$_2$O$_{-1}$ vs. μCO$_2$) showing how equilibria involving phases in the reciprocal ternary system /Ca,Ce//F$_2$,CO$_3$/ indicate the relative proportions of HF and CO$_2$ in aqueous fluids (μF$_2$(CO$_3$)$_{-1}$).; calcite and the REE-fluorocarbonates are fluorinated in the sequence of HF-CO$_2$ exchange reactions labelled 1 to 5. Inset at base shows phases along the composition coordinate CaO-CeO$_{1.5}$ (drawn only for the two extreme divariant fields on the diagram, upper left and lower right). Abbreviations: Flu = fluorite, Cal = calcite, S or Syn = synchysite-(Ce), R or Rnt = röntgenite-(Ce), P or Prs = parisite-(Ce), Bas = bastnaesite-(Ce), Flc = fluocerite-(Ce).

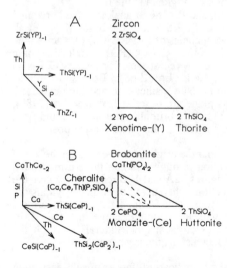

Figure 10. Planar vector diagrams for simple REE-phosphates. (A) Diagram relating xenotime-(Y) to thorite and zircon. (B) Diagram relating monazite-(Ce) to huttonite and brabantite.

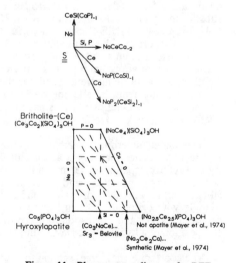

Figure 11. Planar vector diagram for REE substitutions in hydroxylapatite, showing the generation of britholite and of related end-member compositions. This is a constant S (sulfur) section (at S = 0) through a 3-D polyhedron (hydroxylapatite space).

yields the solid solution termed cheralite and the end-member brabantite. The possible coupled substitutions among these phases are indicated as vectors in Figure 10B. The same substitution $CaThCe_{-2}$ relates the hydrous phosphate brockite to rhabdophane-(Ce).

Miyawaki and Nakai (1987) cite Neumann and Nilssen (1968) in proposing that the hydrous monoclinic silicate tombarthite-(Y) might be derivable structurally from monazite by a substitution such as $SiOH(PO)_{-1}$ (my notation), yielding the much simplified composition $YSiO_3(OH)$.

Apatites

Apatite chemistry is wonderfully complex and substitution of Ca by REE is only the least of what can occur. A comprehensive review of the apatites is provided in a book by McConnell (1973); a more concise summary of silicate apatites (including those of REE) is given by Speer and Ribbe (1982, p. 413-417).

For the purposes of this discussion, a general formula for apatite is $A_5C_3O_{12}X$, where A can be (Ca,Na,Ce,Th), C can be (P,Si,S,B; also C, carbon), and X can be (F,OH,Cl,O). Minor vacancies, not considered here, can also occur in the A site, leading to the synthetic oxyapatite end-member $Ce_{9.33}[ ]_{0.67}(SiO_4)_6O_2$, which has the simple oxide composition $7Ce_2O_3·9SiO_2$ (Felsche, 1972). The A or Ca-sites in apatite are not all equivalent; two are 9 and three 7 to 8 coordinated. However, inasmuch as REE appear to be able to occupy both types of sites (e.g., Felsche, 1973; Miyawaki and Nakai, 1987), the two sites won't be distinguished in what follows. Degrees of F, OH, and Cl dominance in the "X" site (generally indicated by prefixes to the mineral name) generally will not be differentiated either.

Trivalent Ce can substitute for divalent Ca via a substitution in the Ca site alone, $NaCeCa_{-2}$, considered above for fluorite, or by a coupled substitution involving the tetrahedral site, such as $CeSi(CaP)_{-1}$, considered above in monazite. These vectors are shown acting on the hydroxylapatite composition on Figure 11; the resulting figure is a quadrilateral bounded by lines of zero Na, Ca, Si, and P. The Sr analog of the horizontal substitution yields belovite, $(Sr_3NaCe)(PO_4)_3(OH)$, 2/5 across the base. Mayer et al. (1974) synthesized REE-fluorapatites with A occupancies of (Na_2RE_2Ca) (4/5 across the base), but report that the end-member composition $(Na_{2.5}RE_{2.5})(PO_4)F$ yielded a phase mixture instead of an apatite. The vertical substitution vector yields britholite-(Ce) as an end-member; analogs have been synthesized by, e.g., Ito (1968). Both substitutions together yield a fourth theoretical end-member, of the Na-Ce silicate composition indicated. Natural apatites from peralkaline rocks on this plane are described by Ronsbo (1989); apatites from silica-deficient rocks have up to 25 mol % britholite end-member and from quartz-bearing rocks up to 25 mol % of the unnamed Na-Ce phosphate end-member.

The double underline beneath the "S" (look under the origin of the vectors) at the top of Figure 11 indicates that this is a constant S section through a larger, three-dimensional figure (hydroxylapatite space, drawn below). Figures 12A and 12B respectively give the corresponding constant Na and constant Ce sections; the substitution $SiSP_{-2}$ yields the silicate-sulfate apatite hydroxylellestadite (cf. Rouse and Dunn, 1982), which forms a triangle with britholite-(Ce) in Figure 12A or with the sulfate cesanite in Figure 12B. Figure 13 gives the corresponding constant Si section, a quadrilateral in which the composition of a hypothetical Na-Ce sulfate end-member apatite is generated.

The constant P (or P-free) section for end-member silicate-sulfate apatites is derived in Figure 14, starting from hydroxylellestadite. (Try deriving these compositions on a triangle!) The plane of accessible compositions is completely non-degenerate; that is, a pentagon, bounded by lines of zero Na, Ca, Ce, Si, and S. Three of its vertices are named and two, for the Na-Ce compositions, await names. This figure truly demonstrates the

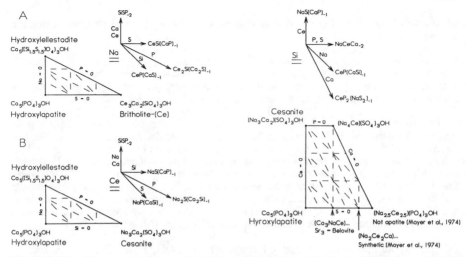

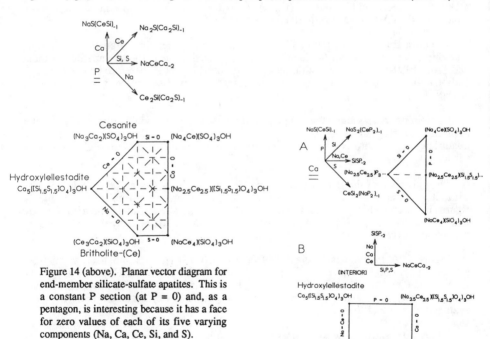

Figure 12 (left). Planar vector diagrams for silicate-sulfate apatites. (A) Constant Na section (at Na = 0) showing the generation of hydroxylellestadite. (B) Constant Ce section (at Ce = 0) showing the generation of cesanite.

Figure 13 (right). Planar vector diagram for sulfate-phosphate apatites: a constant Si section (at Si = 0).

Figure 14 (above). Planar vector diagram for end-member silicate-sulfate apatites. This is a constant P section (at P = 0) and, as a pentagon, is interesting because it has a face for zero values of each of its five varying components (Na, Ca, Ce, Si, and S).

Figure 15 (right). Miscellaneous vector sections through apatite space. (A) Constant Ca section (at Ca = 0). (B) Interior section generated by vectors $NaCaCe_{-2}$ and $SiSP_{-2}$.

advantage of thinking of coupled substitutions as vectors. Several similar figures are presented below.

Two miscellaneous sections through hydroxyl apatite space are given in Figure 15. Figure 15A is the constant Ca (Ca-free) plane, with the Na-Ce end-members as vertices on a triangle. Figure 15B is an interior reciprocal ternary rectangle, generated by independently varying $NaCeCa_{-2}$ in the Ca site and $SiSP_{-2}$ in the tetrahedral site.

Put all of these sections together, and you get Figure 16, a vector depiction of (hydroxyl) apatite space. It is an irregular polyhedron bounded by 6 faces of zero Na, Ca, Ce, Si, P, and S, all depicted in Figures 11-15, and is bisected by the plane of Figure 15B (partly outlined in long dashes). One of the bounding planes is a pentagon (zero P, sloping front, Fig. 14), two are quadrilaterals (zero S, the base, Fig. 11, and zero Si, the back, Fig. 13); the rest are triangular. It has seven vertices, four named and three (the Na-Ce end-members) unnamed. It is unknown if any of the unnamed end-members are stable as apatite structures; the Na-Ce fluorphosphate, as mentioned above, apparently is not, although compositions very close to it (with only 1/5 Ca) are (Mayer et al., 1974).

The apatite compositions derived above all have OH, F, or Cl in the X-site. In oxyapatites this site is instead occupied by O, as by the substitution $CeO(CaF)_{-1}$. This vector is depicted on Figure 17A, a constant (zero) Na section through fluor, oxy- apatite space; from fluorapatite it generates $(Ca_4Ce)(PO_4)_3O$ and from "B" (hypothetical fluorine analog of britholite) it generates $(CaCe_4)(SiO_4)_3O$; both end-member oxyapatites were synthesized by Ito (1968). Figure 17B provides two parallel constant Si,P sections, drawn at zero P (left) and zero Si (right). The Na-Ce end-member oxyapatite $(Na_{0.5}Ce_{4.5})(SiO_4)_3O$ was synthesized by Felsche (1972).

The dashed lines to the left of Figure 17A show the generation of apatites with more than one F per formula unit (a phenomenon recognized in the carbonate apatites: Binder and Troll, 1989). With the substitution CSi_{-1}, Figure 17A could therefore be regarded as a carbonate-apatite plane. In carbonate-apatites, the inferred substitution is $CF(PO)_{-1}$, or $(CO_3F)(PO_4)_{-1}$.

Combining these sections with one already drawn (Fig. 11) yields Figure 18, a vector depiction of fluor, oxy apatite space. The space is bounded by six planes, all quadrilaterals, for zero Na, Ca, Si, P, O, and F (there is no plane of zero Ce) and has 8 vertices.

The existence of the B-rich silicate apatite melanocerite-(Ce) and the Th and B-rich tritomite-(Ce) and tritomite-(Y) gives me an excuse to present a final group of apatite figures involving Th^{4+} in the Ca site and/or B^{3+} in the tetrahedral site. The tritomites are listed by Fleischer (1987) as oxyapatites, so that figures for both hydroxyl- and oxy-apatites are derived.

Figure 19A introduces B via the vector $CeB(CaSi)_{-1}$; this substitution changes britholite-(Ce) into melanocerite-(Ce); its end-member composition is $Ce_5(B_2Si)O_{12}(OH)$. A mixed B-P end-member $Ce_3(B_{2.5}P_{0.5})O_{12}(OH)$ is also derived. The oxyapatite Figure 19B is similar; the Y analog of the end-member borosilicate oxyapatite $Ce_5(BSi_2)O_{13}$ was synthesized by Ito (1968).

Figure 20A shows how melanocerite-(Ce) can be changed into tritomite-(Ce) by the substitution $ThB(CeSi)_{-1}$; the borate end-member is $(Ce_4Th)B_3O_{12}(OH)$ (possibly unstable). Two Ce-free Th end-members, a borate and a silicate, are also derived. The oxyapatite Figure 20B is similar, except that the zero Ca plane of ideal (oxy)tritomite-(Ce) has increased in size. Both figures are pentagons, with a face for zero values of each of the varying components (Ca, Ce, Th, B, and Si), analogous to Figure 14.

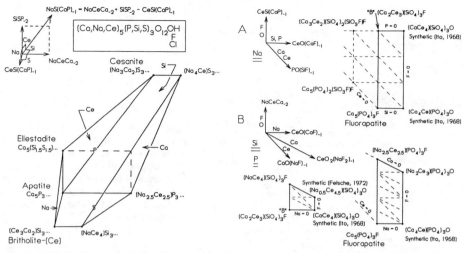

Figure 16 (left). Vector depiction of hydroxylapatite space, as generated by the orthogonal basis vectors NaCeCa$_{-2}$, SiSP$_{-2}$, and CeSi(CaP)$_{-1}$ acting on the composition of hydroxylapatite. See text for discussion.

Figure 17 (right). Planar vector diagrams involving oxyapatites and the vector CeO(CaF)$_{-1}$. "B" = composition of hypothetical F-analog of britholite-(Ce). (A) Constant Na section (at Na = 0). (B) Parallel constant Si and P sections (at P = 0; Si = 3 left and Si = 0; P = 3 right).

Figure 18 (below). Vector depiction of fluor, oxy-apatite space, as generated by the orthogonal basis vectors CeSi(CaP)$_{-1}$, NaCeCa$_{-2}$, and CeO(CaF)$_{-1}$ acting on the composition of fluorapatite. "B" = hypothetical F-analog of britholite-(Ce).

Figure 19 (below). Vector diagrams involving B-substituted silicate apatites (melanocerites) and the vector CeB(CaSi)$_{-1}$. These are planes of constant (zero) Th. (A) Hyroxylapatite plane. (B) Oxyapatite plane.

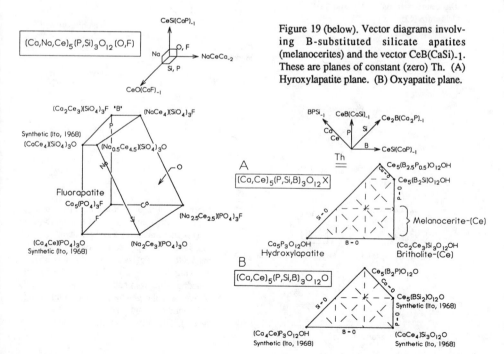

280

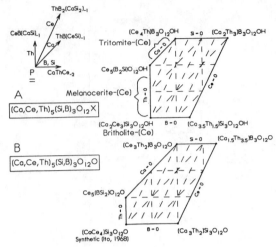

Figure 20. Vector diagrams involving B- and Th-substituted silicate apatites (tritomites) and the vector ThB(CeSi)$_{-1}$. These are planes of constant (zero) P. These pentagons have faces for zero values for each of their five varying components (Ca, Ce, Th, B, and Si). (A) Hydroxylapatite plane. (B) Oxyapatite plane.

Figure 21. Vector depiction of Th,B- substituted hydroxylapatite space, generated by the orthogonal basis vectors CeB(CaSi)$_{-1}$, CeSi(CaP)$_{-1}$, and CaThCe$_{-1}$ acting on the hydroxylapatite composition, to yield limits on the compositions of melanocerite-(Ce) and tritomite-(Ce).

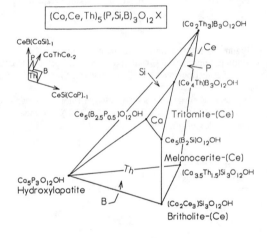

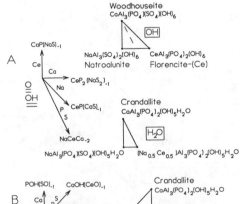

Figure 22. Planar vector diagrams showing the compositional relations of florencite-(Ce) to isostructural sulfates and phosphates. (A) Constant O,OH planes generated by the basis vectors CeP$_2$(NaS$_2$)$_{-1}$ and CaP(NaS)$_{-1}$, showing the generation of florencite-(Ce) and woodhouseite from natroalunite (OH-triangle above) and the generation of two unnamed end-members from crandallite (H$_2$O-triangle below). (B) Constant (zero) Na plane.

Two undepicted vectors, $ThSi(CeP)_{-1}$ (constant Ca) and $ThSi_2(CaP_2)_{-1}$ (constant Ce) allow Th to be introduced into B-free apatites. Similarly, Th and B can be introduced into phosphate apatites (Si-free), via undepicted vectors including $ThB(CaP)_{-1}$ (constant Ce) and $Th_2B(Ce_2P)_{-1}$ (constant Ca).

Combining the hydroxylapatite or top planes of Figures 19 and 20 (and several undepicted planes) yields Figure 21, for (hydroxyl) melanocerite-tritomite space. It has 6 faces, including the pentagon of zero P (Figure 20A). It is uncertain how many (if any) of the five unnamed vertices deserve the name melanocerite-(Ce) or tritomite-(Ce). The analogous oxyapatite space (not depicted to save space) likewise has 6 faces, including two pentagons, for zero P and zero Si; it also has an extra vertex (8 instead of 7). The zero Ca face is considerably larger than on Figure 21 (the oxyapatite-yielding vector $CeO(CaF)_{-1}$ implies that there should be fewer compositions containing Ca).

Florencite and related phases

Florencite-(Ce), -(La), and -(Nd) are REE phosphates isostructural with sulfates of the alunite group. Florencite was found to hydrothermally stable to more than 500°C by McKie (1962); a more recent source of references is Lefebvre and Gasparrini (1980).

As shown in Figure 22A, the florencite-(Ce) formula can be derived from that of natroalunite by the vector $CeP_2(NaS_2)_{-1}$; $CaP(NaS)_{-1}$ yields woodhouseite, a mixed sulfate-phosphate of the beudantite group (Fleischer, 1987). Florencite-(Ce) is placed in the crandallite group; crandallite is a more hydrous phosphate whose formula can be derived from that of florencite-(Ce) by operation of the vector $CaOH(CeO)_{-1}$, as shown in Figure 22B. This vector could also be written as $CaH_2O(CeOH)_{-1}$, and the crandallite formula is generally written as $CaAl_3(PO_4)_2(OH)_5H_2O$, which would (incorrectly) imply water of hydration. A more correct structural formula (Blount, 1974) is $CaAl_3[PO_3O_{0.5}(OH)_{0.5}]_2(OH)_6$; that is, some of the O's in the PO_4 groups are replaced by OH, as they are replaced by F in carbonate apatites (Binder and Troll, 1989). A structural refinement is lacking for florencite-(Ce); in the similar crandallite Ca is coordinated to 6 oxygens and 6 hydroxyls in a distorted polyhedron, but only 4 oxygens and 4 hydroxyls are nearest neighbors (Blount, 1974).

The planes of Figures 22A and 22B are respectively planes of constant O,OH and Na; planes of constant Ca, Ce, and S are similarly depicted on Figures 23A, 23B, and 23C. Combining these planes yields the "florencite space" depicted in Figure 24, which shows the mutual relations among some phases in the alunite, crandallite, and beudantite groups (respectively sulfates, phosphates, and mixed sulfate-phosphates). In addition to the four named end-members, an unnamed Na end-member of the beudantite group, and an unnamed Na-Ce end-member of the crandallite group, are implied.

The Sr-analog of woodhouseite is svanbergite, and of crandallite is goyazite; extensive solid solution among these phases and alunites was suggested by Wise (1975). He drew the square of Figure 23B as a single line, and then drew (p. 544) a triangle with crandallite, goyazite, and alunite at the corners (also drawn by Stoffregen and Alpers, 1987, p. 206). I would instead draw this figure as a triangular prism (planar Fig. 23B with the vertical vector $SrCa_{-1}$ affecting the two calcic phases on the right side). Try this simple drawing exercise yourself.

A-B oxides (niobates, tantalates, titanates, ferrites)

The A-B oxides are Nb-Ta-Ti minerals that consist of combinations of relatively large, low-valence A(VIII) cations, modelled as above by Na^+, Ca^{2+}, Y^{3+} (Ce^{3+}), and U^{4+} (Th^{4+}), with smaller, high-valence B(VI) cations, modelled as Nb^{5+}, Ti^{4+}, and Fe^{3+}. The major mineral groups are ABO_4 (fergusonite), ABO_3 (perovskite), AB_2O_6 (aeschynite or euxenite), and $A_2B_2O_6F$ (pyrochlore). In these formulas, O can be replaced by F or OH or

282

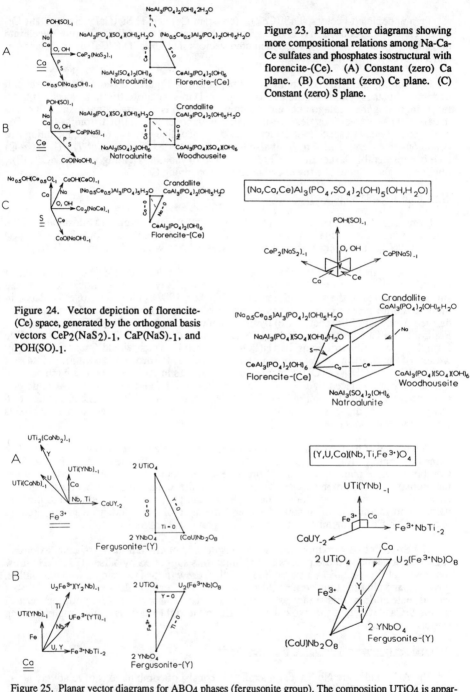

Figure 23. Planar vector diagrams showing more compositional relations among Na-Ca-Ce sulfates and phosphates isostructural with florencite-(Ce). (A) Constant (zero) Ca plane. (B) Constant (zero) Ce plane. (C) Constant (zero) S plane.

Figure 24. Vector depiction of florencite-(Ce) space, generated by the orthogonal basis vectors $CeP_2(NaS_2)_{-1}$, $CaP(NaS)_{-1}$, and $POH(SO)_{-1}$.

Figure 25. Planar vector diagrams for ABO_4 phases (fergusonite group). The composition $UTiO_4$ is apparently unstable. (A) Constant (zero) Fe^{3+} section. (B) Constant (zero) Ca section.

Figure 26. Vector depiction of fergusonite space, generated by the orthogonal basis vectors $FeNbTi_{-2}$, $UTi(YNb)_{-1}$, and $CaUY_{-2}$ operating on the composition of fergusonite-(Y).

a vacancy; vacancies are common in A sites. The potential composition spaces of these formula types have been explored using exchange vectors by Burt (1989b); only a summary regarding REE is given here. Other A-B oxides such as samarskite (e.g., Sugitani et al., 1985), or phases with essential Zr such as zirconolite or zirkelite (e.g., White, 1984) are not considered.

The A-B oxides are infamous for their difficult nomenclature (cf. Ewing, 1976; Hogarth, 1977; Fleischer, 1987, 1989; Cerny and Ercit, 1989) as much as for the complexity of their coupled substitutions, their polymorphism, their defect chemistry, and their tendency to become metamict. The A cations, of radius about 1.0 to 1.3 Å, are generally in irregular 8-fold coordination with oxygen (ranging from 8 up to 12 in "ideal" perovskite); the B cations, of radius about 0.6 to 0.7 Å, are in irregular 6-fold coordination (down to 4 nearest neighbors in fergusonite-(Y)).

In general, the small, highly-charged B cations provide the strong framework that holds the structures together; the larger, weakly-bonded A sites commonly contain vacancies, especially in the pyrochlore group, and the A cations are easily leached or exchanged. Classification, especially in the aeschynite/euxenite and pyrochlore groups, is therefore based primarily on the content of B cations. Anions may include OH^- and F^- replacing the dominant O^{2-} (typical of the pyrochlore group; also of some members of the aeschynite/euxenite group). Some of the anion sites may be vacant, as in the perovskite and pyrochlore groups; this situation is best-documented in synthetic phases.

Some useful reviews of A-B oxides include Ginzburg et al. (1958) on the perovskite and pyrochlore groups, Goodenough and Longo (1970), Nomura (1978), and Hazen (1988) on the perovksite group, Ewing (1976) on the euxenite and aeschynite groups, Hogarth (1977), Subramanian et al. (1983), and Chakoumakos (1984, 1986) on the pyrochlore group, and Muller and Roy (1974), Ewing and Chakoumakos (1982), and Cerny and Ercit (1989) in general.

Simple linear combinations of A-B oxides allow one to predict phase relations, at least tentatively. For example (London and Burt, 1982; Burt, 1989b) the relation

$$BO_2 + ABO_4 = AB_2O_6 \qquad (3)$$

suggests that rutile or cassiterite (BO_2) should be unstable with fergusonite-(Y) (ABO_4), if aeschynite or euxenite (AB_2O_6) is stable. Similarly (Burt, 1989), the relation

$$ABO_3 + ABO_4 = A_2B_2O_7 \qquad (4)$$

suggests that perovskite or lueshite (ABO_3) should be unstable with fergusonite-(Y) (ABO_4), owing to the formation of yttrobetafite-(Y) or yttropyrochlore-(Y) ($A_2B_2O_7$). Another example (London and Burt, 1982)

$$AB_2O_6 + AF = A_2B_2O_6F \qquad (5)$$

suggests that aeschynite or euxenite should be unstable in Na and F-rich environments, owing to the formation of pyrochlore. For many other such relations see Burt (1989b).

Simple consideration of the formulas likewise allows the prediction of the order of selectivity of A-B oxides for high-valence A cations, such as the REE, in a uniform B environment (Burt, 1989b). For example, the valence-satisfying compositions of the Nb end-members are $YNbO_4$, $CaNb_2O_6$, $NaCaNb_2O_6F$, and $NaNbO_3$ for the four oxide groups considered, suggesting that fergusonite-(Y) is likely to be the niobate richest in HREE. In a pure Ti system, the similar sequence of decreasing valence in the A site is $UTiO_4$ (apparently unstable), UTi_2O_6, $(CaY)Ti_2O_6F$, and $CaTiO_3$; therefore one should look for a maximum content of REE in pyrochlore in a pure titanite system. In an analogous way, looking at the valence-satisfying B cations in a pure Ca system, the formulas are $CaWO_4$, $CaNb_2O_6$, $Ca_2(TiNb)O_6F$, and $CaTiO_3$. In a pure Y system, they are $YNbO_4$, $Y(TiNb)O_6$, $Y_2(Fe^{3+}Ti)O_6F$, $YFe^{3+}O_3$. This is the same order of formula types, and presumably implies that high-valence B-cations are most likely in fergusonite, and least likely in perovskite, for the same A environment. Highly-fractionated melts or aqueous fluids derived from them are, in nature, likely to be reservoirs of all of the elements in question, and the partitioning among phases may be complicated by crystal-chemical factors, but the general tendencies suggested above by simple charge balance should still predominate.

Fergusonite/beta fergusonite, ABO_4. Tetragonal fergusonite-(Y) in nature is in $TaNb_{-1}$ solid solution with formanite-(Y), both with wide lanthanide solid solution; it is polymorphically related to the more common beta-fergusonite, which (Table 1) also has -(Ce) and -(Nd) varieties in nature. Coupled substitutions of the types $CaUY_{-2}$ and $UTi(YNb)_{-1}$ plus $Fe^{3+}NbTi_{-2}$, occur, but apparently not to the extent of yielding additional natural end-members. Some substitution of $WTiNb_{-2}$ or $CaW(YNb)_{-1}$ (towards tetragonal scheelite) should also be possible; conversely, scheelite can contain Y (Ce) and Nb (Ta). Note that "yttrocolumbite-(Y)" and "yttrotantalite-(Y)" have similar formulas and appear to be the same phases (Cerny and Ercit, 1989). Polymignite, $(Ca,Fe,Y,Th)(Nb,Ti,Ta,Zr)O_4$, as listed in Fleischer (1987), also would appear to be an ABO_4 mineral; White (1984) considers it a substituted polytype (pseudotype) of zirconolite and zirkelite, $CaZrTi_2O_7$.

Planar sections of the composition space accessible to the ABO_4 formula are depicted on Figure 25. Figure 25A is a section of zero Fe^{3+}; the composition $UTiO_4$ is apparently unstable. Figure 25B is a section of zero Ca; it yields a fourth end-member, $U_2(Fe^{3+}Nb)O_8$. Sections of zero Ti and of zero Y could be drawn, but would yield no additional end-members. The four end-members and four planes, of zero Ca, Fe^{3+}, Y, and Ti, are depicted in Figure 26, the accessible composition space for the formula $(Y,U,Ca)(Nb,Ti,Fe^{3+})O_4$. It is a normal tetrahedron.

Perovskite, ABO_3. A general perovskite formula that covers many natural occurrences is $(Ca,Na,Ce)(Ti,Nb,Fe^{3+})O_3$. Figure 27A derives the composition of loparite-(Ce) and lueshite on a constant Fe^{3+} section; Figure 27B does the same on a Ce-free section for latrappite, which could be defined as the end-member $Ca_2(NbFe)O_6$. Figure 27C derives the unnamed end-member $CeFeO_3$, well-known experimentally. The upper left half of this figure was termed the "forbidden region" for natural perovskite compositions by Nickel and McAdam (1963; cf. Haggerty and Mariano, 1983), although they recognized that trivalent Fe or Al would permit perovskite compositions to fall in this region.

Figure 28 depicts perovskite space; it has six faces, all triangular, for zero contents of Na, Ca, Ce, Nb, Ti, and Fe^{3+}, and five vertices (only one unnamed). Adding YBO_4 to each composition makes it a composition space for yttropyrochlore-(Y) - yttrobetafite-(Y) (equation 4 above); more generally, adding SiO_2 makes it a composition space for F,OH-free titanites of general formula $(Ca,Na,Ce)(Ti,Nb,Fe^{3+})SiO_5$. Other titanite composition spaces are discussed below.

Aeschynite/euxenite, AB_2O_6. The end-members aeschynite-(Y) and euxenite-(Y) are polymorphic forms of $YNbTiO_6$, with euxenite considered the higher temperature form (Seifert and Beck, 1965, cited in Ewing and Chakoumakos, 1982). Similarly, the end-members vigezzite and fersmite apparently are polymorphic forms of $CaNb_2O_6$. Aeschynite in nature tends to be Ce- and Th-selective, euxenite Y- and U-selective (e.g., Ewing, 1976). Figures 29A and 29B give analogous vector representations for the two mineral groups, with the larger Ce and Th depicted for aeschynite, and the smaller Y and U for euxenite. Ignoring different structures, thorutite and brannerite are the Th and U end-members, respectively. Names corresponding to the euxenite group are arbitrarily used for the rest of this discussion.

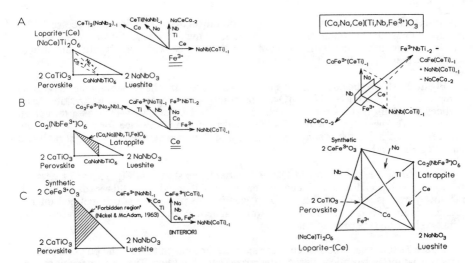

Figure 27 (left). Planar vector diagrams for ABO3 phases (perovskite group). (A) Constant (zero) Fe^{3+} section for lueshite and loparite. (B) Constant (zero) Ce section for latrappite. Shaded area satisfies the latrappite formula as normally given. (C) Interior section including the end-member CeFeO3.

Figure 28 (right). Vector depiction of perovskite space, generated by the orthogonal basis vectors NaNb(CaTi)-1, NaCeCa-2, and CeFe(CaTi)-1 operating on the composition of perovskite; dashed lines in inset show the relation of these vectors to FeNbTi-2. Adding SiO_2 to the formulas makes this a composition space for anhydrous titanite.

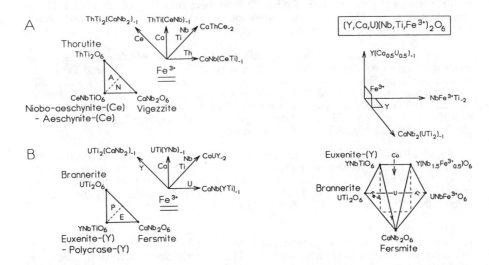

Figure 29 (left). Planar vector diagrams for AB2O6 phases (aeschynite and euxenite groups). (A) Derivation of vigezzite and thorutite compositions from aeschynite. A = aschynite area (Ti > Nb); N = niobo-aeschynite area (Nb > Ti). (B) Derivation of fersmite and brannerite compositions from euxenite. P = polycrase area (Ti > Nb); E = euxenite area (Nb > Ti).

Figure 30 (right). Vector depiction of euxenite (aeschynite) space, generated by the orthogonal basis vectors CaNb2(UTi2)-1, FeNbTi-2, and 1/2 Y2(CaU)-1 operating on the composition of euxenite-(Y).

Strictly speaking, euxenite refers to Nb-dominant solid solutions, polycrase to Ti-dominant ones; the end-member therefore straddles the (arbitrary) division. The same is true of end-member aeschynite and of the division between aeschynite and niobo-aeschynite (aeschynite being Ti-dominant).

Figure 30 shows the composition space accessible to euxenite (or aeschynite) of general composition $(Y,Ca,U)(Nb,Ti,Fe)_2O_6$. The composition space, as for perovskite, has 5 vertices, but only 5 sides for zero values of Y, Ca, U, Fe^{3+}, and Ti (it is a tetragonal pyramid). The back face for zero Ca is a quadrilateral and the face for zero Nb is only intercepted at a point, the brannerite composition. The two vertices to the right represent unnamed Fe^{3+} end-members.

By omitting Fe^{3+} we can add OH as a variable, as shown in Figure 31A, a zero-Ca section in which OH is introduced by the vector $TiOH(NbO)_{-1}$; this leads to the end-member yttrocrasite-(Y). Kobeite-(Y) appears on this plane as a poorly-defined interior phase. Figure 31B is a zero-U section which generates an unnamed calcic end-member with two hydroxyls per formula unit.

Figure 32 (on which the shaded plane separates the euxenite from the polycrase fields) depicts what might be called "yttrocrasite space" and shows that Ti-rich compositions such as polycrase or aeschynite (or yttrocrasite) are more likely to be hydrated than Nb-rich ones (because of the vector $TiOH(NbO)_{-1}$). Figure 32 is, like Figure 30, a tetragonal pyramid (five vertices and five faces). The five faces are zero-U basal quadrilateral, zero-OH left rear, zero-Ca right rear, zero-Y left front, and zero-Nb right front; the zero-Ti plane is only intercepted at the point composition of fersmite.

Pyrochlore, $A_{1-2}B_2O_6(O,F,OH)$. The major division in the pyrochore group is based on the population of the B site (Hogarth, 1977). Any phase with more than 33% Ti (in Nb + Ta + Ti) in the B site is defined as a betafite (few analyses show more than 50% Ti; a possible reason is given below); if Nb + Ta total more than 67% of Nb + Ta + Ti, the phase is defined as either pyrochlore (Nb-dominant) or microlite (Ta-dominant). Ferric iron is not specifically considered in the definition.

The dominant A-site cations are typically Na^+ and Ca^{2+}, which participate in the coupled substitution $CaTi(NaNb)_{-1}$, described above for perovskite. This and $TaNb_{-1}$ are used as vectors on Figure 33 to generate typical pyrochlore (P), microlite (M), and betafite (B) compositions. The Nb and Ta-rich betafites would both be "calciobetafite," although Ta-rich betafites are hardly common. Hogarth (1977) noted that B-site Ti contents greater than 50% are rare; Figure 33 shows that they are impossible for these "simple" pyrochlores (the Ti "end-member" has negative Na in the A site).

Substitution of trivalent or quadrivalent ions in the A site allows the generation of Ti end-member betafites, as shown on Figure 34 A and B for Y and U, respectively. Only 20 % substitution in the A site is needed to define a new (prefixed) mineral name in the pyrochlore group (Hogarth, 1977). The horizontal dotted lines define such divisions; the vertical dotted lines the 33% Ti division between pyrochlore and betafite.

Figure 35 depicts normal pyrochlore space; it is an irregular hexahedron with 8 vertices. The base of the figure corresponds to Figure 34A (zero U), the back to Figure 34B (zero Y). The left front plane is for zero Ti and the right rear for zero Nb. The larger front plane slanting towards the front is for zero Ca; the smaller rear slanting plane is for zero Na. Of the 8 vertices, two correspond to pyrochlore, one to yttropyrochlore, none to uranpyrochlore, one to calciobetafite, and two each to betafite and yttrobetafite. Clearly, mineral names in the pyrochlore group have little to do with logical end-member compositions, and, as for other polycomponent minerals, an analysis or structural formula is essential for knowing the composition.

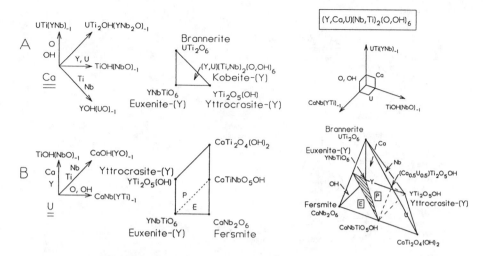

Figure 31 (left). Vector diagrams for hydroxylation of euxenite (aeschynite). (A) Constant (zero) Ca plane with vector TiOH(NbO)-1. (B) Constant (zero) U plane. P = polycrase field; E = euxenite field. This figure is the base of the following figure.

Figure 32 (right). Vector depiction of yttrocrasite space, generated from the euxenite-(Y) composition by the orthogonal basis vectors are CaNb(YTi)-1, TiOH(NbO)-1, and UTi(YNb)-1. Shaded plane separates euxenite ("E") from polycrase ("P") compositions. Dashed triangle markes OH = 1.

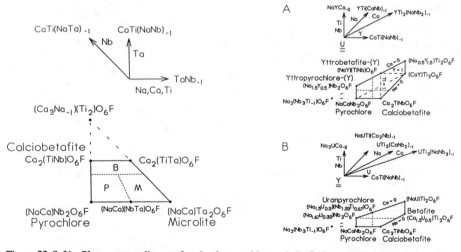

Figure 33 (left). Planar vector diagram for simple pyrochlores, A₂B₂O₆F, generated from endmember pyrochlore via TaNb-1 and CaTi(NaNb)-1. P = pyrochlore area, M = microlite area, and B = betafite area, as defined in Hogarth (1977). Betafite compositions with more than 50% Ti in the B sites are impossible on this plane.

Figure 34 (right). Planar vector diagrams for A-site substitution of Y and U in pyrochlore and calciobetafite. Dotted lines separate various terminological fields as defined by Hogarth (1977). (A) Derivation of yttropyrochlore and yttrobetafite via NaYCa-2. (B) Derivation of uranpyrochlore and betafite compositions via Na₂UCa-3.

288

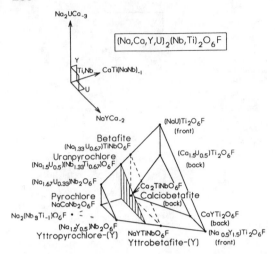

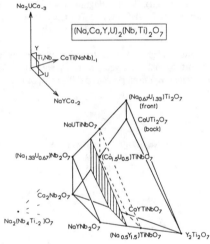

Figure 35 (above). Vector depiction of pyrochlore space for the formula $(Na,Ca,Y,U)_2(Nb,Ti)_2O_6F$. Orthogonal basis vectors are $CaTi(NaNb)_{-1}$, Na_2UCa_{-3}, and $NaYCa_{-2}$. Ruled plane separates pyrochlore from betafite compositions.

Figure 36 (right). Vector depiction of pyrochlore space for the formula $(Na,Ca,Y,U)_2(Nb,Ti)_2O_7$ (F,OH-free pyrochlores). Ruled plane separates pyrochlore from betafite compositions.

Figure 37 (below). Planar vector diagrams for simultaneous substitution of vacancies ([ ]) and Y or U in the A-site of pyrochlore. (A) Constant (zero) U plane. (B) Constant (zero) Y plane.

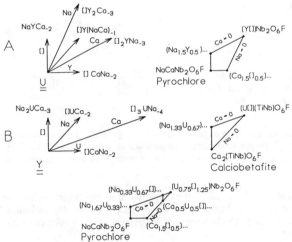

Natural pyrochlore may contain variable O for (F,OH) substitution and vacancies, herein designated [ ], in the A-site, both ignored in the preceding. It also has vacancies in the unique (F,OH)-site, herein designated { }; these can be introduced by, e.g., the vector $O\{ \}F_{-2}$. (The vacancies are likely to be filled with H_2O molecules, herein ignored, or large cations such as Cs in "inverse pyrochlores": Barker et al., 1976; Ercit et al., 1985.) Anionic vacancies will not be treated, although it can be pointed out that coupled substitutions such as $[ ]\{ \}(NaF)_{-1}$ or $[ ]\{ \}(CaO)_{-1}$ change pyrochlore back into an AB_2O_6 compound, compositionally if not structurally, whereas a vacancy in the Y site alone, via exchanges such as $Na\{ \}(CaF)_{-1}$ or $Na\{ \}(YO)_{-1}$, changes it back into the perovskite formula type. The above discussions of ABO_3 and AB_2O_6 compounds therefore also apply to the composition spaces of end-member defect pyrochlores.

A simple half vacancy in the Y site via $1/2\ O\{ \}F_{-2}$ (that is, a Y site half-filled with O instead of completely filled with F) has the same permitted composition range as normal (1-F) pyrochlore. Pyatenko (1961) argues that such 1 F or 1/2-O pyrochlores should be more stable than other compositions.

The composition space accessible to fluorine-free pyrochlores is depicted in Figure 36, to the same scale and orientation as Figure 35. Note that it resembles Figure 35, but the accessible space has been shifted to the right, according to the exchange $NbO(TiF)_{-1}$. That is, the point that was earlier calciobetafite is now $Ca_2Nb_2O_7$ (not a pyrochlore, according to Subramanian et al., 1983, although its Ta^{5+} and Sb^{5+} analogs are, as are its Pb^{2+} and Cd^{2+} analogs). Fluorine apparently stabilizes the pyrochlore structure relative to oxide competitors (such as the weberite, perovskite, or fluorite structures: data summarized in Subramanian et al., 1983).

The effect of A-site vacancies on Y and U-bearing pyrochlores is depicted in Figures 37A and 37B. Other things being equal, Y and U both allow charge balance for a larger content of vacancies in the A-site (which have no charge of their own to contribute). The accessible composition space for pyrochlores of one A-site vacancy is depicted on Figure 38, to the same scale and orientation of Figures 35 and 36. Clearly, A-site vacancies decrease the accessible range of pyrochlore compositions; only uranpyrochlore, yttropyrochlore, and betafite are stable. Increasing vacancies should correspond to increasing contents of U and Y via substitutions such as $[ ]Y(NaCa)_{-1}$ and $[ ]UCa_{-2}$. The Nb content should similarly increase via substitutions such as $[ ]Nb(NaTi)_{-1}$ and $[ ]Nb_2(CaTi_2)_{-1}$.

Combining variable F and variable [ ] could also be done. For F-free pyrochlore with one vacancy, only the composition $[ ]UNb_2O_7$ is possible (i.e., the accessible composition space has contracted to a single point).

Ferric-iron substituted pyrochlores are not depicted; many of them can be generated simply by adding ABO_4 compositions to ABO_3 compositions, or NaF or CaO to the euxenite/aeschynite composition spaces. At constant A-site occupancy, introduction of Fe^{3+} via $FeNbTi_{-2}$ would decrease Ti and increase Nb. In general, Fe^{3+} decreases charge in the B site; this must be compensated via increasing charge in A. The most Fe^{3+}-rich end-members become $YUFe^{3+}_2O_6F$ and $U_2Fe^{3+}_2O_7$. Tungsten as W^{6+} (also molybdenum as Mo^{6+}) could alternatively be added to the B site via substitutions such as $TiWNb_{-2}$. This would have the opposite effect, of driving down charge in the A site, via substitutions such as $NaW(CaNb)_{-1}$ or $CaW(YNb)_{-1}$ (mentioned above under fergusonite).

Allanite

Allanite-(Ce) is a REE-epidote, of idealized composition $(CaCe)(Fe^{2+}Al_2)(SiO_4)_3(OH)$. The irregular Ca site (A1) is approximately 8-coordinated; the distinct Ce (A2) site is 9-11 coordinated, according to different authors. An excellent recent review is by Deer et al.

290

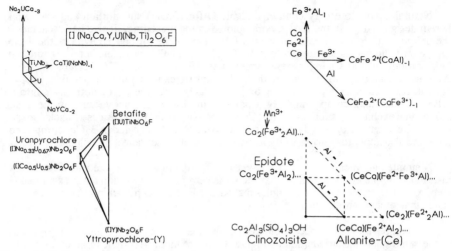

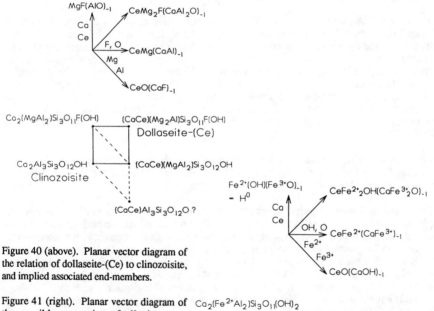

Figure 38 (left). Vector depiction of pyrochlore space for the formula [ ](Na,Ca,Y,U)(Nb,Ti)$_2$O$_6$F (pyrochlores with an A-site vacancy). Ruled plane separates pyrochlore from betafite compositions.

Figure 39 (right). Planar vector diagram of the relation of allanite-(Ce) to epidote and clinozoisite.

Figure 40 (above). Planar vector diagram of the relation of dollaseite-(Ce) to clinozoisite, and implied associated end-members.

Figure 41 (right). Planar vector diagram of the possible generation of allanite-group phases with either two or zero hydroxyls, starting from epidote.

(1986), who for chemistry cite an earlier review by Hasegawa (1960, not consulted). Allanite is related to normal epidote by the substitution $CeFe^{2+}(CaFe^{3+})_{-1}$, and to clinozoisite by the analogous $CeFe^{2+}(CaAl)_{-1}$, as shown on Figure 39. The dashed lines on this figure reflect the fact that, although epidote normally has only 1/3 of clinozoisite's three Al's replaced by Fe^{3+}, the co-occurrence of Fe^{2+} or Mn^{3+} with Fe^{3+} can result in epidote-allanite minerals with Al of less than 2 (Deer et al., 1986). There is therefore a possible additional end-member of composition $(CaCe)(Fe^{2+}Fe^{3+}Al)(SiO_4)_3(OH)$.

That figure might have ended the discussion, except for the recent description of the new species dollaseite-(Ce) by Peacor and Dunn (1988). It is related to clinozoisite by the complicated substitution $CeMg_2F(CaAl_2O)_{-1}$, as shown on Figure 40. This substitution implies two other end-members generated from clinozoisite by $MgF(AlO)_{-1}$ and $CeMg(CaAl)_{-1}$ and possibly an oxy-composition generated by $CeO(CaF)_{-1}$

Figure 41 is the Fe^{2+}, OH analog of Figure 40, starting from epidote instead of clinozoisite. In addition to allanite-(Ce), it implies a ferrous end-member with two hydroxyls per formula unit, and "oxyallanite," a deprotonated allanite reportedly synthesized by Affholter (1987, abstract). She also investigated the hydrothermal stability of allanites, and reported them stable to about 800°C at 1 kb. This high temperature is consistent with the common occurrence of allanite in igneous rocks.

Finally, Figure 42 illustrates the vectors by which Th can be substituted in either epidote or allanite. The vector $CaThCe_{-2}$ obviously only works for allanite-(Ce); the Th end-member has the composition $(Ca_{1.5}Th_{0.5})(Fe^{2+}Al_2)(SiO_4)_3(OH)$.

Titanite

Titanite (sphene), $CaTiSiO_5$, generally contains only small amounts of REE substituting for Ca, perhaps because the Ca is only in 7 coordination with O. On the other hand, because titanite is such an extremely common accessory mineral in a wide variety of igneous and metamorphic rocks, it can be a major carrier of REE. Furthermore, Exley (1980) describes titanites with up to 87% of the Ca substituted by REE, according to the dominant substitution $Ce_2Fe^{2+}(Ca_2Ti)_{-1}$ (my notation); a yttrian variety of sphene was formerly called "keilhauite." Recent reviews of titanite chemistry are provided by Ribbe (1982) and Deer et al. (1982a); many of the end-members derived below were earlier postulated by Zabavnikova (1957).

The major substitutions are for Ca, Ti, and O (not Si), so that a good approximation to some aspects of anhydrous titanite chemistry is given by Figures 27 and 28 above for perovskite (just add SiO_2 to the formulas). These figures did not consider substitution by divalent octahedral cations or by OH (F) for O, however, both of which are important for titanite (not perovskite), so that some additional figures are called for. Figure 43A, a zero-OH plane, depicts the substitution $Ce_2Mg(Ca_2Ti)_{-1}$, with Mg proxying for the Fe^{2+} of Exley (1980), that leads to the theoretical end-member $Ce_2(MgTi)Si_2O_{10}$. Figure 43B, a zero-Mg plane, depicts the substitution $AlOH(TiO)_{-1}$, that leads to the end-member $CaAlSiO_4(OH)$ (in this regard, note that titanite is isostructural with panasqueiraite, $CaMgPO_4(OH)$, isokite, $CaMgPO_4F$, lacroixite, $NaAlPO_4F$, and durangite, $NaAlAsO_4F$: Lahti and Pajunen, 1985). Figure 43C, a zero-Al plane, contains two addition hydrous end-members, $CeMgSiO_4(OH)$, and $Ca_2(MgTi)Si_2O_8(OH)_2$ (or the composition above it, $CaMgSiO_3(OH)_2$: Zabavnikova, 1957). Figure 43D shows two parallel planes, of zero Ca and of zero Ce, but no additional end-members are generated.

Combining these planar faces and others not depicted yields Figure 44, titanite space for the general formula $(Ca,Ce)(Ti,Mg,Al)SiO_4(O,OH)$. (Recall that Figure 28 could be described as a titanite space for the general formula $(Ca,Na,Ce)(Ti,Nb,Fe^{3+})SiO_5$.) This figure has 7 planes and 6 vertices; all of the faces are triangular, except that for zero Al (left front), which is a rectangle.

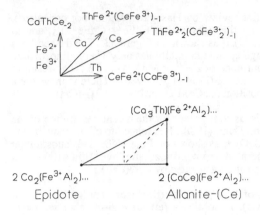

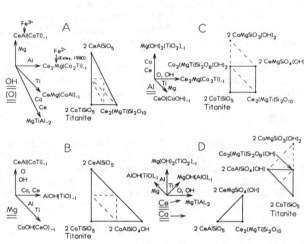

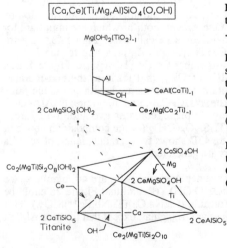

Figure 42 (top). Planar vector diagram of the substitution of Th in epidote and allanite. For all formulas, ...(SiO$_4$)$_3$(OH).

Figure 43 (above). Planar vector diagrams for the substitution of REE, Mg (Fe^{2+}), Al, and OH (F) in titanite. (A) Constant OH,O plane drawn at zero OH. (B) Constant (zero) Mg plane. (C) Constant (zero) Al plane. (D) Constant Ce,Ca planes drawn at zero Ca (below left) and Ce (above right).

Figure 44 (left). Vector depiction of titanite space for the formula (Ca,Ce)(Ti,Mg,Al)SiO$_4$(O,OH). Orthogonal basis vectors are Mg(OH)$_2$(TiO$_2$)$_{-1}$, CeAl(CaTi)$_{-1}$, and Ce$_2$Mg(Ca$_2$Ti)$_{-1}$.

By arbitrarily fixing OH at zero or one, we can include Na in the titanite formula, so as to allow the substitution $NaCeCa_{-2}$. Figure 45A shows zero-Mg planes for OH = 0 (above) and OH = 1 (below); Figure 45B shows analogous zero-Al planes. End-members such as $(NaCe)Ti_2Si_2O_{10}$ (= loparite-(Ce) plus SiO_2), $(NaCe)Al_2Si_2O_8(OH)_2$, and $NaTiSiO_4(OH)$ are generated, as shown.

These end-members are shown as vertices for anyhdrous titanite space in Figure 46A and for hydrous titanite space in Figure 46B. The anhydrous space is a normal tetrahedron; the hydrous space is a more complex hexahedron (topologically similar to Fig. 28 for perovskite), with 5 vertices and 6 triangular faces; the zero-Ce and zero-Ti faces are not present on Figure 46A. Incidentally, by subtraction of SiO_2, Figure 46A is also a composition space for perovskites of general formula $(Ca,Na,Ce)(Ti,Mg,Al)O_3$.

Garnet

When one thinks of REE in garnets, one is more likely to think of synthetic phases such as yttrium-aluminum garnet (YAG) or its ferric analog yttrium-iron garnet (YIG), rather than natural phases. Yet natural garnets, particularly spessartine, may also be yttrian (e.g., Jaffe, 1951; Kasowski and Hogarth, 1968; Wakita et al., 1969), and the first study cited above apparently inspired Yoder and Keith (1951) to synthesize YAG. Recent reviews of the garnet group include Deer et al. (1982) and Meagher (1982) on natural silicates and Geller (1967) on synthetic garnets.

As mentioned regarding coupled substitutions, in the grossular garnet formula, $Ca_3Al_2Si_3O_{12}$, REE^{3+} substitution in the normally divalent 8-coordinated A site can theoretically involve charge compensation in the A site alone, as by $NaYCa_{-2}$, charge compensation in the octahedral B site, as by $YMg(CaAl)_{-1}$, or charge compensation in the tetrahedral C site, as by $YAl(CaSi)_{-1}$. The last two substitutions are shown as basis vectors on the zero-Na plane Figure 47; end-members generated include YAG, $(Y_2Ca)Mg_2Si_3O_{12}$, and $Y_3Mg_2(AlSi_2)O_{12}$. Figure 48A shows the corresponding constant Si plane, in which the end-members $(Na_{1.5}Y_{1.5})Al_2Si_3O_{12}$ and $(Na_{0.5}Y_{2.5})Mg_2Si_3O_{12}$ are generated. The zero-Mg plane of Figure 48B contains no new end-members.

Figure 49 depicts garnet space for the general formula $(Ca,Na,Y)_3(Al,Mg)_2(Si,Al)_3O_{12}$ a pentahedron with 5 vertices. The lack of Na and of octahedral Mg (Fe^{2+}) in most recalculated garnet analyses might imply that $YAl(CaSi)_{-1}$ (or its Mn analog) is the most practical way of substituting REE in garnet. This would result in silicon-deficient formulas.

Several other substitutions also result in Si-deficient garnet formulas. One is the hydrogrossular substitution $[\ ](OH)_4(SiO_4)_{-1}$, depicted on Figure 50. Another is the schorlomite substitution, $TiFe^{3+}(Fe^{3+}Si)_{-1}$, or $TiSi_{-1}$, depicted as acting on andradite in Figure 51 and leading to schorlomite, $Ca_3Ti_2(SiFe_2)O_{12}$. The Zr analog of this substitution leads to kimzeyite, $Ca_3Zr_2(SiFe_2)O_{12}$. The term schorlomite is sometimes applied to the corner composition, $Ca_3Ti_2(TiFe_2)O_{12}$, on Figure 51 (Rickwood, 1968), although attempts to synthesize garnet of this composition (and its Zr analog) by Ito and Frondel (1967a) were unsuccessful. Hydrogarnets containing Ti and Zr were synthesized by Ito and Frondel (1967b), but did not contain REE, so that I shall omit a figure combining the two vertical vectors on Figures 50 and 51 (try drawing one yourself).

Gadolinite

Gadolinite-(Y), $Y_2(Fe^{2+},[\ ])Be_2Si_2O_8(O,OH)_2$, is structurally related to datolite, $CaBSiO_4(OH)$, as can be seen if the latter's formula is written $Ca_2[\ ]B_2Si_2O_8(OH)_2$. Both

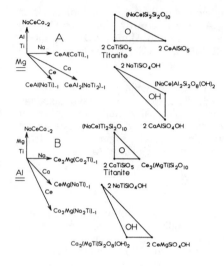

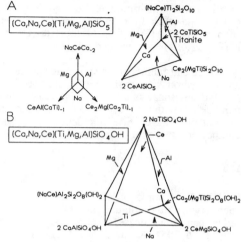

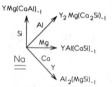

Figure 45 (top left). Planar vector diagrams for Na-substitution in titanites with OH = 0 (labelled "O") and OH = 1 (labelled "OH"). (A) Zero-Mg planes. (B) Zero-Al planes.

Figure 46 (top right). Vector depictions of Na-bearing titanite spaces, generated by applying the orthogonal basis vectors $NaCeCa_{-1}$, $Ce_2Mg(Ca_2Ti)_{-1}$, and $CeAl(CaTi)_{-1}$ to titanite formulas. (A) Titanite formula $(Ca,Na,Ce)(Ti,Mg,Al)SiO_5$. By subtracting SiO_2, this figure is also a space for perovskites of formula $(Ca,Na,Ce)(Ti,Mg,Al)O_3$. (B) Titanite formula $(Ca,Na,Ce)(Ti,Mg,Al)SiO_4(OH)$.

Figure 47 (left). Planar vector diagram (at zero Na) for coupled substitution of Y into garnet.

Figure 48 (below. Planar vector diagrams for coupled substitution of Y into garnet. (A) At constant Si = 3. (B) At zero Mg.

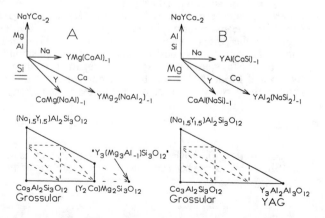

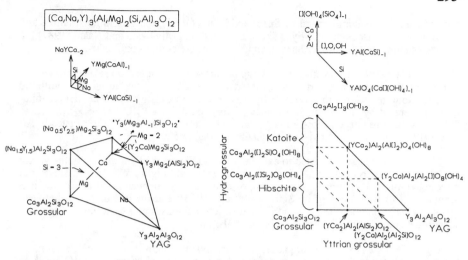

Figure 49 (left). Vector depiction of garnet space for the formula $(Ca,Na,Y)_3(Al,Mg)_2(Si,Al)_3O_{12}$. Orthogonal basis vectors are $NaYCa_{-2}$, $YMg(CaAl)_{-1}$, and $YAl(CaSi)_{-1}$.

Figure 50 (right). Planar vector diagram showing how the vector $YAl(CaSi)_{-1}$ and the hydrogrossular substitution, $[\,](OH)_4(SiO_4)_{-1}$, both lower Si contents of garnet.

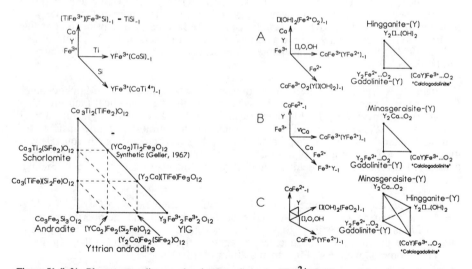

Figure 51 (left). Planar vector diagram showing how the vector $YFe^{3+}(CaSi)_{-1}$ and the schorlomite substitution, $TiFe^{3+}(Fe^{3+}Si)_{-1}$, both lower Si contents of garnet.

Figure 52 (right). Vector diagrams showing gadolinite-(Y) and related phases. (A) Relation of gadolinite-(Y) to hingganite-(Y) and synthetic "calciogadolinite". (B) Relation of gadolinite-(Y) to minasgeraisite-(Y) and synthetic "calciogadolinite". (C) Tetrahedron for the 4 end-members, using $CaFe_{-1}$, $[\,](OH)_2(FeO_2)_{-1}$, and $CaFe^{3+}(YFe^{2+})_{-1}$, as orthogonal basis vectors.

minerals are briefly summarized by Speer and Ribbe (1982); the gadolinite structure and substitutions are discussed by Miyawaki et al. (1984) and Miyawaki and Nakai (1987). Y is 8-coordinated, Fe^{2+} is 6-coordinated, and Be and Si are both 4-coordinated.

Ito and Hafner (1974) synthesized a large number of gadolinites and related phases and suggested the three end-members depicted on Figure 52A, including the synthetic "calciogadolinite" related to gadolinite by the substitution $CaFe^{3+}(YFe^{2+})_{-1}$. Hingganite-(Y) is generated by substituting a vacancy [ ] for the Fe^{2+} via $[ ](OH)_2(FeO_2)_{-1}$. Figure 52B shows that minasgeraisite-(Y) is generated by substituting Ca for Fe^{2+} instead. All four end-members are depicted on a tetrahedron in Figure 52C.

Hingganite-(Y) can be related to datolite, and gadolinite-(Y) to homilite, by the vector $CaB(YBe)_{-1}$, as shown on the reciprocal ternary rectangle Figure 53A (which leaves out "calciogadolinite"). A phase with a composition half-way between that of datolite and hingganite-(Y) was reported by Semenov et al. (1963); I know of none such between gadolinite-(Y) and homilite. A quadrilateral which includes "calciogadolinite" and Ca-B substituted compositions is shown on Figure 53B.

Figure 54 depicts a version of gadolinite space, leaving out minasgeraisite-(Y); it has 5 faces and 6 vertices, four of them named. Applying the substitution $CaB(YBe)_{-1}$ to minasgeraisite-(Y) would yield a hypothetical end-member of composiiton $Ca_2CaB_2(Si_2O_8)O_2$ (a probably unstable calcic analog of homilite).

Chevkinite/perrierite

Chevkinite and perrierite are closely related structurally, share essentially the same formula (are essentially dimorphic, in other words), and may occur together in pegmatites and other igneous rocks. A simplified synthetic composition is $Ce_4Mg(Ti_3Mg)Si_4O_{22}$ (Ito, 1967; Ito and Arem, 1971). Their sheet-like structures, in which Ca and the REE are 9-coordinated, are reviewed by Miyawaki and Nakai (1987), after Calvo and Faggiani (1974). They have two distinct octahedral sites; in the smaller, more distorted site Ti is mixed with Mg. Perrierite over chevkinite seems to be favored by increasing ionic radius of the cations in both the 9-coordinated and octahedral sites (Ito, 1967; Segalstad and Larsen, 1978). A Sr-dominant chevkinite, strontiochevkinite, has been found by Haggerty and Mariano (1983).

Figure 55A depicts Ca and Th substitutions in the ideal synthetic "Ce-Mg-chevkinite" (my term) of Ito and Arem (1971); the substitutions yield hypothetical (and probably unstable) end-members "Th-Mg chevkinite," $Th_4Mg(TiMg_3)Si_4O_{22}$, and "Ca-Ti chevkinite," $Ca_4TiTi_4Si_4O_{22}$. This latter composition is almost certainly unstable (as indicated by the dashed lines on the figure), inasmuch as it is compositionally equivalent to $4CaTiSiO_5 + TiO_2$ (titanite plus rutile). Ferric iron (instead of Th) substitution is considered in Figure 55B; it yields a hypothetical end-member $Ce_4Fe^{3+}(TiFe^{3+}_3)Si_4O_{22}$, possibly also unstable.

Figure 56 depicts combined Th and Fe^{3+} substitution (at zero Ca) in "Ce-Mg-chevkinite"; two new hypothetical (and probably unstable) end-members appear, of compositions $(Ce_3Th)Fe^{3+}Fe^{3+}_4Si_4O_{22}$ and $Th_4Mg(Fe^{3+}_2Mg_2)Si_4O_{22}$. This figure is interesting in that it is a non-degenerate pentagon, with faces of zero Ce, Th, Mg, Fe^{3+}, and Ti (in common with, e.g., Figs. 14 and 20 for apatites).

Figure 57 depicts possible coupled substitutions in hypothetical Ce-free chevkinites (if any exists). A new and probably also unstable end-member is generated: $(Th_{2.5}Ca_{1.5})Fe^{3+}Fe^{3+}_4Si_4O_{22}$.

Combinining the planes of Figures 55-57 (plus others not depicted) yields Figure 58 for chevkinite space. It has the usual 6 faces, with 7 vertices, none named (indicating that

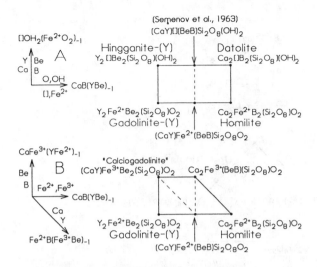

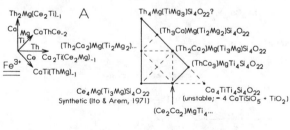

Figure 53. Planar vector diagrams showing Be-B substitutions in gadolinite group minerals. (A) Hingganite-(Y) - datolite and gadolinite-(Y) - homilite relations. (B) Gadolinite-(Y) - homilite - "calciogadolinite" relations.

Figure 54. Vector depiction of gadolinite space. Faces are OH = 0 left front, Fe^{3+} = 0 base, B = 0 left rear, Y = 0 sloping right front, and Fe^{2+} = 0 right rear.

Figure 55. Planar vector diagrams showing potential chevkinite (or perrierite) compositions. (A) At zero Fe^{3+} with Ca and Th substitution. (B) At zero Th with Ca and Fe^{3+} substitution.

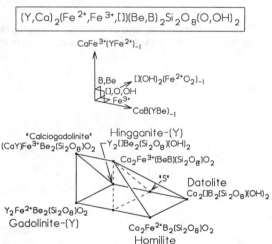

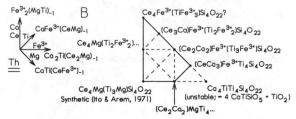

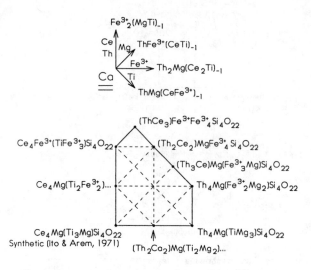

Figure 56. Planar vector diagram (at zero Ca) showing Fe^{3+} and Th substitution in chevkinite (or perrierite). This pentagon has faces of zero values of each of its five varying ions (Ce, Th, Mg, Fe^{3+}, and Ti).

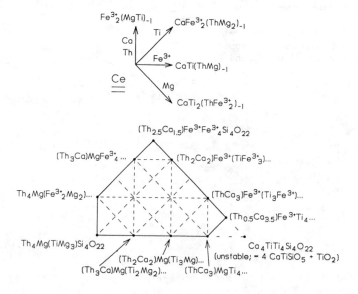

Figure 57. Planar vector diagram showing possible coupled substitutions in hypothetical Ce-free chevkinite (or perrierite).

maybe this would be a good place to stop drawing such diagrams!). Cutting off the "Ca-Ti chevkinite" corner as clearly unstable (along a plane "Mg or Fe^{3+} = 1") adds one face and 4 vertices. Of course, none of the vertices except the starting composition may be stable as chevkinite/perrierite structures; experimental work is lacking since the pioneering studies of Jun Ito (1967) on these and so many other REE minerals.

ELEMENT DISTRIBUTIONS: ACID-BASE RELATIONS

The major type of ionic substitution omitted above is that of one REE for another, modelled by YCe_{-1}. It has long been recognized that Ce-group REE are more "basic" than Y-group REE (e.g., Moeller and Kremers, 1945), and that certain REE-minerals, such as monazite-(Ce), bastnaesite-(Ce), and allanite-(Ce), are relatively Ce-selective and prefer the LREE, whereas others, such as xenotime-(Y) and fergusonite-(Y) are relatively Y-selective and prefer the HREE (e.g., Mineyev, 1965, 1968; Fleischer, 1965; Adams, 1969). Some minerals of intermediate selectivity (e.g., apatite and titanite), have been described as having "complex" behavior, and their REE-distribution largely depends on that of the rock in which they occur. These tendencies are generally ascribed to fitting the variations of ionic radii of the REE (lanthanide contraction with increasing atomic no.) to different irregular site geometries (especially coordination nos.) of REE in different minerals. Some authors (e.g., Mineyev, 1963) have also emphasized the role of variations of acidity and its effect on REE activities and complexing in mineral-forming fluids.

Based on the averages of hundreds of REE analyses of rock-forming minerals, Mineyev (1968) gives the following series, of most Ce-selective mineral to most Y-selective (capitalized phases being the common):

Perovskite > PYROCHLORE >> calcite >> APATITE > titanite = eudialyte > epidote >> andradite >> fluorite > ZIRCON > almandine-spessartine.

For REE minerals proper, he gives the following series:

Cerianite > LOPARITE > rhabdophane > BASTNAESITE > allanite > MONAZITE >> britholite > ceriopyrochlore >> aeschynite >> GADOLINITE > gagarinite >> SAMARSKITE > yttropyrochlore >> euxenite > churchite > XENOTIME.

He also states that all available analyses of coexisting minerals are consistent with these two series. The following similar series (containing REE and other minerals) was independently derived by Khomyakov (1970, 1972):

Lamprophyllite, barian celestine > bastnaesite, monazite, allanite, loparite > apatite, perovskite, pyrochlore > titanite, fluorite, eudialyte > schorlomite, andradite, gagarinite > zircon, xenotime, spessartine.

The most Ce-selective minerals are likely to have LREE in solid solution with Ba and Sr, whereas the most Y-selective are likely to have HREE in solid solution with Mn and Fe or Zr.

These tendencies have been represented graphically by plotting the REE proportions of various minerals on an equilateral triangle (Mineyev, 1965, 1968), with SUM(Ce) at the left corner (= La + Ce + Pr + Nd), $SUM(Y_1)$ at the top (= Sm + Eu + Gd + Tb + Dy + Ho), and $SUM(Y_2)$ at the right (= Er + Tm + Yb + Lu). Sometimes the element Y is combined with $SUM(Y_2)$ at the top corner (Vladykin et al., 1982). Vertical lines on the triangle roughly represent tendencies to be Ce-selective (on the left) vs. Y-selective (on the right).

300

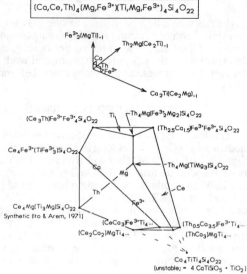

$(Ca,Ce,Th)_4(Mg,Fe^{3+})(Ti,Mg,Fe^{3+})_4Si_4O_{22}$

Figure 58 (right). Vector depiction of chevkinite (perrierite) space, for the general formula $(Ca,Ce,Th)_4(Mg,Fe^{3+})(Ti,Mg,Fe^{3+})_4Si_4O_{22}$. The space is generated by the orthogonal basis vectors $Fe_2(MgTi)_{-1}$, $Th_2Mg(Ce_2Ti)_{-1}$, and $Ca_2Ti(Ce_2Mg)_{-1}$ operating on the Ce-Mg composition of synthetic chevkinite of Ito and Arem (1971).

"Acidic"

(Anion prefers larger cation)

μ MgCa$_{-1}$

MgF$_2$	Sellaite
CaF$_2$	Fluorite
MgCO$_3$	Magnesite
CaCO$_3$	Calcite
MgSiO$_3$	Enstatite
CaSiO$_3$	Wollastonite
Mg$_2$SiO$_4$	Forsterite
Ca$_2$SiO$_4$	Larnite
MgO	Periclase
CaO	Lime

A

μ YCe$_{-1}$

B

YF$_3$	(Synthetic)
CeF$_3$	Fluocerite-(Ce)
YCO$_3$F	Bastnaesite-(Y)
CeCO$_3$F	Bastnaesite-(Ce)
YPO$_4$	Xenotime-(Y)
CePO$_4$	Monazite-(Ce)
YNbO$_4$	Fergusonite-(Y)
CeNbO$_4$	Fergusonite-β-(Ce)
Y$_2$Si$_2$O$_7$	Yttrialite-(Y)
Ce$_2$Si$_2$O$_7$	(Synthetic)

"Basic"

(Anion prefers smaller cation)

Figure 59 (left). Schematic diagrams (not to scale) that define relative anionic basicities in terms of cation exchange operators.
(A) Anionic basicity of some Mg and Ca minerals on a scale of μ(MgCa$_{-1}$), based on data in Robie et al. (1978).
(B) Anionic basicity of some pegmatitic REE minerals on a scale of μ(YCe$_{-1}$), as deduced by Burt and London (1982).

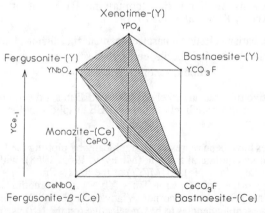

Xenotime-(Y)
YPO$_4$

Fergusonite-(Y)
YNbO$_4$

Bastnaesite-(Y)
YCO$_3$F

YCe$_{-1}$

Monazite-(Ce)
CePO$_4$

CeNbO$_4$
Fergusonite-β-(Ce)

CeCO$_3$F
Bastnaesite-(Ce)

Figure 60. Schematic mineral stability diagram (trigonal prism) showing phases that should occur with coexisting monazite-(Ce) and xenotime-(Y). The stable mineral assemblage is defined by the ruled planes (modified from Burt and London, 1982).

Alternatively, Fleischer (1965; cf. Fleischer and Altschuler, 1986) presented plots of the ratio between two REE, namely La/Nd, vs. the sum of the three lightest REE (La + Ce + Pr). Minerals on such plots tend to occur in a concave-upward trend, with the most Y-selective plotting at the lower left, and the most Ce-selective plotting at the upper right.

Of course, REE data, whether for minerals or rocks, today is almost universally plotted on a chondrite-normalized diagram (Coryell et al., 1963) of REE (in sample)/REE (in meteorite standard) vs. atomic no. of each individual REE. A problem with this method is that the diagrams present so much information at a time that it is sometimes difficult to visually compare one sample with many others. These diagrams are most often used to study geochemical, as opposed to purely crystal-chemical processes (see the majority of other chapters in this book).

Burt and London (1980, 1982) suggested still another method of quantifying tendencies for Y vs. Ce-selectivity, one that will become practical only when thermochemical data for synthetic REE mineral end-members becomes available. This is a plot of $\mu(YCe_{-1})$ for individual minerals (that is, the free energy of the Y end-member minus that of the Ce end-member, each normalized to one REE). These plots indicate acid-base tendencies among mineral groups.

The exhange operator YCe_{-1} is acidic in the electronic or Lewis (1938) sense (Burt, 1974, 1979), because Y, as the smaller cation, "holds on to" its electrons more tightly than Ce; the relative activities of the two cations should therefore change as the acidity of the mineral-growing aqueous fluid or magma changes (Korzhinskii, 1956), and should affect their partitioning between the fluid and the mineral. The exchange operator $MgCa_{-1}$ acts the same way in hydrothermal fluids (that is, dolomitization of calcite could theoretically be promoted by increasing the acidity of the fluid, at a constant X(Ca) and X(Mg), where X is the concentration in the fluid (Burt, 1979).

Figures 59A and 59B give a comparison between the chemical potentials of the two acidic cation exchange operators $MgCa_{-1}$ (on the left) and YCe_{-1} (on the right) used as "probes" as anion basicity. The right-hand figure is modified from Burt and London (1982). (In the same way, the chemical potential between any pair of anions, such as F_2O_{-1}, can be used as a "probe" or standard of cation acidity.) This is a simplified conceptual model; it is not drawn to scale. Ordered intermediate Ca-Mg phases such as monticellite, diopside, and dolomite complicate the picture on the left, and partial to complete solid solutions of REE in most of the minerals complicate it on on the right.

The double line above calcite on the left was drawn because calcite-dolomite (not magnesite, as depicted) equilibria control, or at least indicate, $\mu(MgCa_{-1})$ in many hydrothermal (skarn-forming) environments (that is, $\mu(MgCa_{-1})$) tends to be buffered at the calcite level). Similarly, the common coexistence of monazite-(Ce) and xenotime-(Y) led me to draw a double line between them as a typical control, or at least indicator, of $\mu(YCe_{-1})$ in many igneous environments. In such environments, we would expect bastnaesite-(Ce) or fluocerite-(Ce) to be more common than their Y counterparts, and fergusonite-(Y) and yttrialite-(Y) to be more common than their Ce counterparts. This relationship, as also indicated in the mineral compatibility diagram Figure 60, is certainly valid. In fact, a Y counterpart of fluocerite-(Ce) and a Ce counterpart of yttrialite-(Y) (or else of thorveitite) are not reported in nature. Similarly, the "extremal" phases on Figure 59A, lime and sellaite, are extremely rare in nature.

SUMMARY

We have explored, using exchange vectors to represent complex coupled substitutions and their linear combinations, the composition spaces accessible to a wide variety of REE mineral formulas. Traditional triangles and tetrahedra would have been largely inadequate for this task, inasmuch as the figures derived commonly have 4 or 5 sides on a plane and 6 or more faces in space. Each side or face corresponds, in general, to zero content of a given ion; planar contours of equal content of this ion through the accessible composition space (dashed in the planar figures) are parallel to the limiting face, and increase proportionally away from it. Their orientation (slope) can also be deduced from the vector relation diagram that accompanies each planar figure, and their numerical values from the labelled formulas.

The choice of basis exchange vectors, each of which is electrically neutral, is fairly arbitrary; any small linearly independent set will yield all of the rest. In some cases, restricting at least some vectors to those affecting only the A, B, or C site facilitates plotting mineral compositions. The choice of basis vectors should be made appropriate to the the data available. Making the basis vectors orthogonal and of the same length allows composition planes to be plotted on normal graph paper. Computer plotting routines for x-y or x-y-z diagrams could be substituted for graph paper.

For many mineral groups, REE mineral names may not correspond to logical end-members (vertices) on such figures, and many vertices remain unnamed, either because they have not yet been found in nature or because they are unstable.

Only two mineral stability diagrams involving chemical potentials of exchange operators as vectors were presented. Figure 9 showed that bastnaesite-(Ce) is much more difficult to fluorinate than calcite (or that fluocerite-(Ce) is much easier to carbonate than fluorite), via $HF-CO_2$ exchange. Figure 59 showed qualitatively how $\mu(YCe_{-1})$.can be used to order the relative Lewis acidity (and Y vs. Ce selectivity) of various anion groups in crystals. Much more thermochemical and experimental data on REE phases would be needed to quantify these ideas, or to test several others that are currently sitting around my office on various pieces of scrap paper, and didn't make it into this chapter.

ACKNOWLEDGMENTS

J.B. Thompson, Jr. first suggested to me the concept of exchange operators, and has inspired me to explore their uses. I also am grateful to many previous workers on REE for making this paper an easier one to write, and especially to M. Fleischer, R. Miyawaki and P. Cerny for sending me reprints. Finally, I thank R.L Hervig and the editors for careful and thoughtful reviews of earlier versions of the manuscript, and M. Fleischer for reviewing an early version of Tables 1 and 2. The figures were drafted using FastCAD™ software and printed on a Hewelett-Packard™ Deskjet Plus.

Some of this work (the part dealing with A-B oxides) was performed in 1988 while the author was a Visiting Scientist at the Lunar and Planetary Institute, which is operated by the Universities Space Research Association under Contract No. NASW-4066 with the National Aeronautics and Space Administration. This is Lunar and Planetary Institute Contribution No. 727. Except for the chemical potential diagrams, which date back 10 to 15 years, most of the other research, writing, and drafting was performed over two hectic months in the summer of 1989. This schedule allowed little time for library work, and I therefore apologize in advance to those whose work I neglected to cite or misquoted.

REFERENCES

Adams, J.W. (1969) Distribution of lanthanides in minerals. U.S. Geol. Surv. Prof. Paper 650-C, C38-C44.
Affholter, K.A. (1987) Synthesis and crystal chemistry of lanthanide allanites. Unpub. Ph.D. dissertation, Virginia Polytechnic Inst. and State Univ., Blacksburg, VA, 218 p. (not seen; abstract in Diss. Abstr. Int., B 49, 2098-2099, 1988).

Armbruster, T., Buhler, C., Graeser, S., Stalder, H.A., and Amthauer, G. (1988) Cervandonite-(Ce), $(Ce,Nd,La)(Fe^{3+},Fe^{2+},Ti^{4+},Al)_3SiAs(Si,As)O_{13}$, a new Alpine fissure mineral. Schweiz. Mineral. Petrograph. Mitt. 68, 125-132.

Barker, W.W., White, P.S. and Knop, O. (1976) Pyrochlores X. Madelung energies of pyrochlores and defect fluorites. Canad. J. Chem. 54, 2316-2334.

Bayliss, P. and Levinson, A.A. (1978) A system of nomenclature for rare-earth mineral species: Revision and extension. Am. Mineral. 73, 422-423.

Bevan, D.J.M., Strahle, J., and Greis, O. (1982) The crystal structure of tveitite, an ordered yttrofluorite mineral. J. Solid. State Chem. 44, 75-81.

Binder, G. and Troll, G. (1989) Coupled anion substituion in natural carbon-bearing apatites. Contrib. Mineral. Petrol. 101, 394-401.

Blount, A.M. (1974) The crystal structure of crandallite. Am. Mineral. 59, 41-47.

Brady, J.B. and Stout, J.H. (1980) Normalizations of thermodynamic properties and some implications for graphical and analytical problems in petrology. Am. J. Sci. 280, 173-189.

Brown, D. (1968) Halides of the Lanthanides and Actinides. In: Halides of the Transition Elements, Wiley-Interscience, New York, vol. 1, 280 p.

Burt, D.M. (1974) Concepts of acidity and basicity in petrology - The exchange operator approach (extended abs.). Geol. Soc. Am., Abstracts with Programs 6, 674-676.

Burt, D.M. (1979) Exchange operators, acids, and bases. In: V.A. Zharikov, W.I. Fonarev, and S.P. Korikovskii, Eds., Ocherki Fiziko-Khimicheskoi Petrologii (Problems in Physico-Chemical Petrology) 2, 3-15. Moscow, "Nauka" Press (in Russian).

Burt, D.M. (1988) Vector representation of phyllosilicate compositions. Reviews in Mineralogy, 19, 561-599.

Burt, D.M. (1989a) Vector representation of tourmaline compositions. Am. Mineral. 74, 826-839.

Burt, D.M. (1989b) Vector derivation of composition spaces for some A-B oxide minerals (perovskite, fergusonite, euxenite/aeschynite, and pyrochlore) and for BO_2 and B_2O_3 oxides. Canad. Mineral. (submitted).

Burt, D.M. and London, D. (1980) A model for element partitioning among pegmatitic rare earth minerals (abstr.). EOS,.Am. Geophys. Union, Trans. 61, 404.

Burt, D.M. and London, D. (1982) Subsolidus equilibria [in pegmatites]. Mineral. Assoc. Canada, Short Course Handbook 8, 329-346.

Calvo, C. and Faggiani, R. (1974) A re-investigation of the crystal structures of chevkinite and perrierite. Am. Mineral. 59, 1277-1285.

Catlow, C.R.A. (1985) Defects in solid fluorides. In: Hagenmuller, P., ed., Inorganic Solid Fluorides. Chemistry and Physics. Academic Press, New York, 259-273.

Cerny, P. and Burt, D.M. (1984) Paragenesis, crystallochemical characteristics, and geochemical evolution of micas in granitic pegmatites. Reviews in Mineralogy, v. 13, p. 257-297.

Cerny, P. and Ercit, T.S. (1989) Mineralogy of niobium and tantalum: crystal chemical relationships, paragenetic aspects and their economic implications. In. Muller, P., Cerny, P. and Saupé, F., eds., Lanthanides, Tantalum and Niobium. Springer-Verlag, New York, 26-78.

Chai, B.H.T. and Mroczkowski, S. (1978) Syntheses of rare-earth carbonates under hydrothermal conditions. J. Cryst. Growth 44, 84-97.

Chakoumakos, B.C. (1984) Systematics of the pyrochlore structure, ideal $A_2B_2X_6Y$. J. Solid State Chem., 53, 120-129.

Chakoumakos, B.C. (1986) Pyrochlore. In: Parker, S.P., ed., McGraw-Hill Yearbook of Science and Technology 1987. McGraw-Hill, New York, 393-395.

Chang, L.L.Y. (1971) Subsolidus phase relations in the aragonite-type carbonates: I. The system $CaCO_3$-$SrCO_3$-$BaCO_3$. Am. Mineral. 56, 1660-1673.

Chen, T.T. and Chao, G.Y. (1974) Burbankite from Mont St. Hilaire, Quebec. Canad. Mineral. 12, 342-345.

Clark, A.M. (1984) Mineralogy of the rare earth elements. In: Henderson, P., ed., Rare Earth Element Geochemistry. Elsevier, Amsterdam, 33-61.

Coryell, C.D., Chase, J.W. and Winchester, J.W. (1963) A procedure for geochemical interpretation of terrestrial rare-earth abundance patterns. J. Geophys. Res. 68, 559-566.

Deer, W.A., Howie, R.A. and Zussman, J. (1982a) Sphene (titanite). In: Rock-Forming Minerals, 2nd ed., vol. 1A, Orthosilicates. Longman, London, 443-466.

Deer, W.A., Howie, R.A. and Zussman, J. (1982b) Zircon. In: Rock-Forming Minerals, 2nd ed., vol. 1A, Orthosilicates. Longman, London, 418-442.

Deer, W.A., Howie, R.A. and Zussman, J. (1982c) Garnet group. In: Rock-Forming Minerals, 2nd ed., vol. 1A, Orthosilicates. Longman, London, 467-698.

Deer, W.A., Howie, R.A. and Zussman, J. (1986) Allanite. In: Rock-Forming Minerals, 2nd ed., vol. 1B, Disilicates and Ring Silicates. Wiley, New York, 151-179.

304

Deliens, M. and Piret, P. (1989) La shabaite-(Nd), $Ca(TR)_2UO_2(CO_3)_4(OH)_2 \cdot 6H_2O$, nouvelle espece minérale de Kamoto, Shaba, Zaire. Europ. J. Mineral. 1, 85-88.

Donnay, G. and Donnay, J.D.H. (1953) The crystallography of bastnaesite, parisite, roentgenite, and synchysite. Am. Mineral. 8, 932-963.

Donnay, G. and Preston, H. (1971) Ewaldite, a new barium calcium carbonate II. Its crystal structure. Tschermaks Mineral. Petrogr. Mitteilungen 15, 201-212.

Effenberger, H., Kluger, F., Paulus, H., and Wolfel, E.R. (1985) Crystal structure refinement of burbankite. Neues Jahrb. Mineral., Monatsh., 161-170.

Ercit, T.S., Cerny, P. and Hawthorne, F.C. (1985) Normal and inverse pyrochlore group minerals. Geol. Assn. Canada-Mineral. Assn. Canada, Progr. Abstr. 10, A17.

Ewing, R.C. (1976) A numerical approach toward the classification of complex, orthorhombic, rare-earth, AB_2O_6-type Nb-Ta-Ti oxides. Canad. Mineral. 14, 111-119.

Ewing, R.C. and Chakoumakos, B.C. (1982) Lanthanide, Y, Th, U, Zr, and Hf minerals: selected structure descriptions. Mineral. Assoc. Canada, Short Course Handbook 8, 239-265.

Exley, R.A. (1980) Microprobe studies of REE-rich accessory minerals: implications for Skye granite petrogenesis and REE mobility in hydrothermal systems. Earth Planet. Sci. Lett. 48, 97-110.

Felsche, J. (1972) Rare earth silicates with the apatite structure. J. Solid State Chem. 5, 266-275.

Felsche, J. (1973) Crystal chemistry of the rare-earth silicates. Structure and Bonding, 13 (Rare Earths), 99-197. Springer-Verlag, New York.

Felsche, J. (1978) Crystal chemistry, Part A in chapters 39, 57-71 (Yttrium and Lanthanides) of Handbook of Geochemistry: Springer-Verlag, New York, A1-A42.

Fitzgerald, S., Leavens, P.B., Reingold, A.L. and Nelen, J.A. (1987) Crystal structure of a REE-bearing vesuvianite from San Benito County, California. Am. Mineral. 72, 625-628.

Fleischer, M. (1965) Some aspects of the geochemistry of yttrium and the lanthanides. Geochim. Cosmochim. Acta 29, 755-772.

Fleischer, M. (1987) Glossary of Mineral Species, 5th ed. Mineralogic Record, Inc., Tucson, AZ, 227 p.

Fleischer, M. (1989) Additions and corrections to the Glossary of Mineral Species, 5th Edition (1987). Mineral. Rec. 20, 289-298.

Fleischer, M. and Altschuler, Z.S. (1986) The lanthanides and yttrium in minerals of the apatite group. An analysis of the available data. Neues Jahrb. Mineral., Monatsh., 467-480.

Geller, S. (1967) Crystal chemistry of the garnets. Z. Krist., 125, 1-47.

Ginderow, D. (1989) Structure de $Na_3M_3(CO_3)_5$ (M = Terre Rare,Ca,Na,Sr), rattaché a la burbankite. Acta Crystallogr., Sect. C: Cryst. Struct. Commun. C45, 185-187.

Ginzburg, A.I., Gorzhevskaya, S.A., Erofeeva, E.A. and Sidorenko, G.A. (1958) The chemical composition of isometric titanium-tantalum niobates. Geochimistry, no. 5, 615-636 (transl. from Geokimiya, no. 5, 486-50, 1958).

Goodenough, J.B and Longo, J.M. (1970) Crystallographic and magnetic properties of perovskite and perovskite-related compounds. In: Landold-Bornstein Numerical Data and Functional Relationships in Science in Technology, New Series, Group III, Vol. 4a, p. 126-314. Springer-Verlag, Berlin.

Greis, O and Haschke, J.M. (1982) Rare earth fluorides. In: Gschneider, K.A., Jr. and Eyring, L., eds., Handbook on the Physics and Chemistry of Rare Earths, vol. 5, p. 387-460.

Haggerty, S.E. and Mariano, A.N. (1983) Strontian-loparite and strontio-chevkinite: Two new minerals in rheomorphic fenites from the Parana Basin carbonatites, South America. Contrib. Mineral. Petrol. 84, 365-381.

Harris, C. and Rickard, R.S. (1987) Rare-earth-rich eudialyte and dalyite from a peralkaline granite dike at Straumsvola, Dronning Maud Land, Antarctica. Canad. Mineral. 25, 755-762.

Harris, C., Cressey, G., Bell, J.D., Atkins, F.B. and Beswetherich, S. (1982) An occurrence of rare-earth-rich eudialyte from Ascension Island, South Atlantic. Mineral. Mag. 46, 421-425.

Haschke, J.M. (1975) Lanthanum hydroxide fluoride carbonate system. Preparation of synthetic bastnaesite. J. Solid State Chem. 12, 114-121.

Hasegawa, S. (1960) Chemical composition of allanite. Sci. Rep. Tohoku Univ., 3rd ser. (Min. Petr. Econ. Geol.) 6, 331-334 (not seen; cited in Deer et al., 1986).

Hazen, R.M. (1988) Perovskites. Sci. Am. 258 (6), 74-81.

Himmelberg, G.R. and Miller, T.P. (1980) Uranium- and thorium-rich vesuvianite from the Seward Peninsula, Alaska. Am. Mineral. 65, 1020-1025.

Hogarth, D.D. (1977) Classification and nomenclature of the pyrochlore group. Am. Mineral., 62, 403-410.

Ito, J. (1967) A study of chevkinite and perrierite. Am. Mineral. 52, 1094-1104.

Ito, J. (1968) Silicate apatites and oxyapatites. Am. Mineral. 53, 890-907.

Ito, J. and Arem, J.E. (1971) Chevkinite and perrierite; synthesis, crystal growth, and polymorphism. Am. Mineral. 56, 307-319.

Ito, J. and Frondel, C. (1967a) Synthetic zirconium and titanium garnets. Am. Mineral. 52, 773-781.

Ito, J. and Frondel, C. (1967b) New synthetic hydrogarnets. Am. Mineral. 52, 1105-1109.

Ito, J. and Hafner, S.S. (1974) Synthesis and study of gadolinites. Am. Mineral. 59, 700-708.

Ito, J. and Johnson, H. (1968) Synthesis and study of yttrialite. Am. Mineral. 53, 1940-1952.

Jaffe, H.W. (1951) The role of yttrium and other minor elements in the garnet group. Am. Mineral. 36, 133-155.

Kasowski, M.A. and Hogarth, D.D. (1968) Yttrian andradite from the Gatineau Park, Quebec. Canad. Mineral. 9, 552-558.

Khomyakov, A.P. (1970) Derivation of a series of relative rare-earth affinities of minerals. Dokl. Acad. Sci. USSR, Earth Sci. Sect. 190, 142-145 (from Doklady Akademii Nauk SSSR 190, 940-943, 1970).

Khomyakov, A.P. (1972) Typomorphism of the composition of lanthanides in coexisting minerals. In: Chukhrov, F.V., ed., Tipomorfizm Mineralov i ego Prakticheskoye Znachenie (Typomorphism of Minerals and its Practical Significance), Nedra Press, Moscow, p. 50-51 (in Russian).

Korzhinskii, D.S. (1956) Dependence of the activity of components upon the solution acidity and reaction sequence in postmagmatic processes. Geochemistry (in transl.), p. 643-652.

Kubach, I. and Schubert, P. (1984) Y, La, and the Lanthanoids; Minerals (excluding silicates). Gemelin Handbook of Inorganic Chemistry, 8th ed., System No. 39, vol. A7, 248 p. Springer-Verlag, Berlin.

Kubach, I. and Töpper, W. (1984) Y, La, and the Lanthanoids; Minerals (silicates). Deposits. Mineral Index. Gemelin Handbook of Inorganic Chemistry, 8th ed., System No. 39, vol. A8, 413 p. Springer-Verlag, Berlin.

Lahti, S.I. and Pajunen, A. (1985) New data on lacroixite, NaAlFPO₄. Am. Mineral. 70, 849-855.

Lefebvre, J.-J. and Gasparrini, C. (1980) Florencite, an occurrence in the Zairian copperbelt. Canad. Mineral. 18, 301-311.

Leskela, M. and Niinisto, L. (1986) Inorganic complex compounds I. In: Gschneidner, K.A., Jr. and Eyring, L., eds., Handbook on the Physics and Chemistry of Rare Earths 8, 203-334. North-Holland (Elsevier), Amsterdam.

Levinson, A.A. (1966) A system of nomenclature for rare-earth minerals. Am. Mineral. 51, 152-158.

Lewis, G.N. (1938) Acids and bases. J. Franklin Institute 226, 293-313.

Lumpkin, G.R. and Chakoumakos, B.C. (1988) Chemistry and radiation effects of thorite-group minerals from the harding pegmatite, Taos County, New Mexico. Am. Mineral. 73, 1405-1419.

Lumpkin, G.R., Chakoumakos, B.C. and Ewing, R.C. (1986) Mineralogy and radiation effects of microlite from the Harding pegmatite, Taos County, New Mexico. Am. Mineral. 71, 569-588.

Mayer, I., Roth, R.S., and Brown, W.E. (1974) Rare earth substituted fluoride-phospate apatites. J. Solid State Chem. 11, 33-37.

McConnell, D. (1973) Apatite. Its Crystal Chemistry, Mineralogy, Utilization, and Geologic and Biologic Occurrences. Springer-Verlag, New York, 111 p.

McKie, D. (1962) Goyazite and florencite from two African carbonatites. Mineral. Mag. 33, 280-297.

Meagher, E.P. (1982) Silicate garnets. Reviews in Mineralogy, 5, 2nd ed., 25-66.

Mineyev, D.A. (1963) Geochemical differentiation of the rare earths. Geochemistry, Engl. transl., 1129-1149.

Mineyev, D.A. (1965) A study of the properties and the possibilities of the ternary diagram SUM(Ce)-SUM(Y₁)-SUM(Y₂). Geokhimiya Translations (A Supplement to Geochem. Int'l, vol. 2), part 2, p. 1098-1133 (from Geokhimiya, 1423-1438, 1965).

Mineyev, D.A. (1968) Mean compositions of the lanthanides in minerals. Geochemiya Translations 1968 (A Supplement to Geochem. Internat., vol. 5), p. 354-366 (from Geokhimiya, 825-835, 1968).

Miyawaki, R. and Nakai, I. (1987) Crystal structures of rare-earth minerals. Rare Earths (Kidorui), no. 11, 1-133.

Miyawaki, R. and Nakai, I. (1988) Crystal structures of rare earth minerals, 1st supplement (1988 C.S.R.M.). Rare Earths (Kidorui), no. 13, 1-42.

Miyawaki, R., Nakai, I, and Nagashima, K. (1984) A refinement of the crystal structure of gadolinite. Am. Mineral. 69, 948-953.

Moeller, T. and Kremers, H.E. (1945) Basicity characteristics of scandium, yttrium, and the rare earth elements. Chem. Reviews 37, 97-159.

Muller, O. and Roy, R. (1974) The Major Ternary Structural Families. Springer-Verlag, New York, 487 p.

Neumann, H. and Nilssen, B. (1968) Tombarthite, a new mineral from Hogetveit, Evje, South Norway. Lithos 1, 113-123.

Nickel, E.H. and McAdam, R.C. (1963) Niobian perovskite from Oka, Quebec; a new classification of the perovskite group. Canad. Mineral. 7, 683-697.

Nickel, E.H. and Mandarino, J.A. (1987) Procedures involving the IMA Commission on New Minerals and Mineral Names and guidelines on mineral nomenclature. Am. Mineral. 72, 1031-1042.

306

Niinisto, L and Leskela, M. (1987) Inorganic complex compounds II. In: Gschneidner, K.A. and Eyring, L., eds., Handbook on the Physics and Chemistry of Rare Earths 9, 91-320. North-Holland (Elsevier), Amsterdam.

Nomura, S. (1978) Crystallographic and magnetic properties of perovskite and perovskite-related compounds (supplement). In: Landold-Bornstein Numerical Data and Functional Relationships in Science in Technology, New Series, Group III, Vol. 12a, 368-520. Springer-Verlag, Berlin.

Peacor, D.R. and Dunn, P.J. (1988) Dollaseite-(Ce) (magnesium orthite redefined): structure refinement and implications for F + M^{2+} substitutions in epidote-group minerals. Am. Mineral. 73, 838-842.

Petersen, O.V., Ronsbo, J.G. and Leonardsen, E.S. (1980) Nacareniobsite-(Ce), a new mineral from the Ilimaussaq alkaline compex, South Greenland, and its relation to mossandrite and the rinkite series. N. Jahrb. Mineral., Monatsh., 84-96.

Posypaiko, V.I. and Alekseeva, E.A. (1987) Phase Equilibria in Binary Halides. Plenum Publ., Corp., New York (compiled and edited by H.B. Bell; transl. from 1979 Russ. compiliation by B. Indyk), 470 p.

Pyatenko, Yu.A. (1961) Normal and defect structures of the pyrochlore type. J. Struct. Chem. 2, 545-548 (trans. from Zhur. Strukt. Khimii 2, 591-596, 1961).

Ribbe, P.H. (1982) Titanite (sphene). Reviews in Mineralogy 5, 2nd. ed., 137-154.

Rickwood, P.C. (1968) On recasting analyses of garnet into end-member molecules. Contrib. Mineral. Petrol. 18, 175-198.

Robie, R.A., Hemingway, B.S., and Fisher, J.R. (1978) Thermodynamic proberties of minerals and related substances at 298.15 K and 1 bar (105 pascals) pressure and at higher temperatures. U.S. Geol. Surv. Bull. 1452, 456 p.

Ronsbo, J.G. (1989) Coupled substitutions involving REEs and Na and Si in apatites in alkaline rocks from the Ilimaussaq intrusion, South Greenland, and the petrological implications. Am. Mineral. 74, 896-901.

Rouse, R.C. and Dunn, P.J. (1982) A contribution to the crystal chemistry of ellestadite and the silicate sulfate apatites. Am. Mineral. 67, 90-96.

Segalstad, T.V. and Larsen, A.O. (1978) Chevkinite and perrierite from the Oslo region, Norway. Am. Mineral. 63, 499-505.

Seifert, H. and Beck, B. (1965) Zur Kristallchemie und Geochemie der metmikten Minerale der Euxenitgruppe. Neues Jahrb. Mineral., Abhandlungen 103, 1-20.

Semenov, E.I. (1964, transl. 1966) Minerals of yttrium and of yttrium earths. In: Vlasov, K.A., ed., Geochemistry and Mineralogy of Rare Elements and Genetic Types of Their Deposits, vol. 2, Mineralogy of Rare Elements, 220-248; 819-824. Israel Program for Scientific Translations, Jerusalem.

Semenov, E.I. (1964, transl. 1966) Minerals of cerium earths. In: Vlasov, K.A., ed., Geochemistry and Mineralogy of Rare Elements and Genetic Types of Their Deposits, vol. 2, Mineralogy of Rare Elements, 249-328; 824-836. Israel Program for Scientific Translations, Jerusalem.

Semenov, E.I., Dusmatov, V.D., and Samsonova, N.S. (1963) Yttrium-beryllium minerals of the datolite group. Soviet Physics-Crystallogr. 8, 539-541.

Shannon, R.D. (1976) Revised effective ionic radii and systematic studies of interatomic distances in halides and chalcogenides. Acta Cryst. A32, 751-767.

Smith, J.V. (1959) Graphical representation of amphibole compositions. Am. Mineral. 44, 437-440.

Sobolev, B.P. and Fedorov, P.P. (1978) Phase diagrams fo the CaF_2-$(Y,Ln)F_3$ systems. I. Experimental. Journal of the Less-Common Metals 60, 33-46.

Solodov, N.A., Semenov, E.I., and Burkov, V.V. (1987) Geological Handbook for Heavy Lithophile Rare Metals (in Russian). Nedra Press, Moscow, 438 p.

Speer, J.A. (1982a) Zircon. Reviews in Mineralogy 5, 2nd ed., 67-112.

Speer, J.A. (1982b) The actinide orthosilicates. Reviews in Mineralogy 5, 2nd ed., 113-135.

Speer, J.A. and Ribbe (1982) Miscellaneous orthosilicates. Reviews in Mineralogy 5, 2nd ed., 383-427.

Stoffregen, R.E. and Alpers, C.N. (1987) Woodhouseite and svanbergite in hydrothermal ore deposits: products of apatite destruction during advanced argillic alteration. Canad. Mineral. 25, 201-211.

Subramanian, M.A., Aravamudan, G. and Subba Rao, G.V. (1983) Oxide pyrochlores—a review. Progr. Solid State Chem. 15, 55-143.

Sugitani, Y., Suzuki, Y. and Nagashima, K. (1985) Polymorphism of samarskite and its relationship to other structurally related Nb-Ta oxides with the alpha-PbO_2 structure. Am. Mineral. 70, 856-866.

Thoma, R.E., Hebert, G.M., Insley, H. and Weaver, C.F. (1963a) Phase equilibria in the system sodium fluoride-yttrium fluoride. Inorg. Chem. 2, 1007-1012.

Thoma, R.E., Insley, H., Friedman, H.A., Hebert, G.M. and Weaver, C.F. (1963b) Phase equilibria in the system NaF-ThF_4-UF_4. J. Am. Ceram. Soc. 46, 37-42.

Thompson, J.B., Jr. (1981) An introduction to the mineralogy and petrology of the biopyriboles. Reviews in Mineralogy, 9A, 141-188.

Thompson, J.B., Jr. (1982) Composition space: An algebraic and geometric approach. Reviews in Mineralogy 10, 1-31.

Töpper, W. (1979) Y, La, und Lanthanide; Kristallchemische Grundlagen, in Gmelin Handbuch der Anorganischen Chemie, 8te Auflage, System No. 39, Sc, Y, La-Lu Seltenerdelemente, vol. A4, 242 p. Springer-Verlag, Berlin.

Van Landuyt, J. and Amelinckx, S. (1975) Multiple beam direct lattice imaging of new mixed-layer compounds of the bastnaesite-synchisite series. Am. Mineral. 60, 351-358.

Vladykin, N.V. Smirnova, E.V. and Kovalenko, V.I. (1982) Rare earth elements in minerals of Mongolia. In: Samoilov, V.S., ed., Geokhimiya Redkozemel'nykh Elementov v Endogennykh Protsessakh (Geochemistry of Rare Earth Elements in Endogenic Processes). Nauka Press, Siberian Branch, Novosibirsk, 178-205 (in Russian).

Voronkov, A.A. and Pyatenko, Y.A. (1967) Crystal structure of carbocernaite $(Na,Ca)(TR,Sr,Ca,Ba)(CO_3)_2$. J. Struct. Chem. USSR 8, 835-840 (from Zhurnal Strukturnoi Khimii 8, 935-942, 1967).

Wakita, H., Shibao, K., Nagashima, K. (1969) Yttrian spessartine from Suishoyama, Fukushima Prefecture. Am. Mineral. 54, 1678-1683.

White, T.J. (1984) The microstructure and microchemistry of synthetic zirconolite, zirkelite and related phases. Am. Mineral. 69, 1156-1172.

Wise, W.S. (1975) Solid solution between the alunite, woodhouseite, and crandallite mineral series. N. Jahrb. Mineral., Monatsh., 540-545.

Yoder, H.S. and Keith, M.L. (1951) Complete substitution of aluminum for silicon: the system $3MnO \cdot Al_2O_3 \cdot 3SiO_2 - 3Y_2O_3 \cdot 5Al_2O_3$. Am. Mineral. 36, 519-533.

Zabavnikova, I.I. (1957) Diadochic substitutions in sphene. Geochemistry, 271-278 (transl. from Geokhimiya (3) 226-232, 1957).

ECONOMIC GEOLOGY OF RARE EARTH ELEMENTS

INTRODUCTION

Although the term rare earth elements (REE) is considered to be a misnomer to many, a comparison of their crustal abundance with other better known rare elements shows that at least some are indeed relatively rare. None are as rare as gold, but their extraction as pure elements requires a higher technology than the winning of gold. The combination of rarity and high expense of isolating some of the REE precludes their use in technological areas that require high volumes of these materials.

The elusiveness of the REE is also a consequence of their dispersion as substitutional impurities in many common rock-forming minerals. The mineralogy of the REE is extensive (Clark, 1984), but only a few minerals contain major quantities of REE and are sufficiently concentrated to constitute economic deposits. Some of the major REE and Y-bearing minerals are listed in Table 1. Because of its geochemical properties, Y is included here as an REE. Of the minerals listed in Table 1, bastnaesite, monazite and xenotime account for most of the recorded world production.

As shown in other chapters, the REE and Y are a geochemically coherent group of elements because they are similar in their chemical properties. Despite chemical similarities, the REE may show significant distribution variations in different minerals or in the same mineral crystallized under different environmental conditions. Some of the factors that control the distribution of REE in minerals include: (1) ionic radius differences; (2) crystal structure; (3) basicity differences; (4) oxidation states; (5) stability of complexes; (6) melt structure in magmas; (7) REE content and distribution in source fluids; (8) the role of temperature and pressure; and (9) relative solubility of REE and their migratory capacity in hydrothermal solutions and in the weathering environment.

Most of these topics have been described in various chapters of Henderson (1984) and other chapters in this book.

Structural constraints in minerals have a major control on the REE distribution. As pointed out by Semenov (1958), minerals with high coordination numbers (CN) of the site occupied by REE are LREE selective (i.e., bastnaesite-CN of 11, monazite-9), whereas minerals with low CN are HREE selective (i.e., xenotime-CN of 8, kainosite-8). Minerals that accept REE in several sites with different CN often show complex REE distribution (i.e., britholite-7,9). The mineral bastnaesite with a CN of 11 prefers LREE despite the environment of occurrence that could have a bulk chemistry of LREE or HREE dominance, as shown by Fleischer (1978), whereas REE distributions in monazite with a CN of 9 more closely reflect the geochemistry of their environment (Fleischer, 1965). Therefore, monazite from carbonatites that have a prominent enrichment in LREE show a strong LREE preference, and monazite from granitic pegmatites that are enriched in HREE have a stronger HREE preference.

A notable exception to the extreme LREE preference of the bastnaesite structure is shown in Figure 1 for this mineral from a granite pegmatite in Kazakhstan, U.S.S.R. (Mineev et al., 1970).

Although bastnaesite and the related bastnaesite-type minerals (Table 2) are most often LREE selective, chondrite normalized comparison diagrams for parisite from different environments (Fig. 2) reveal distinct differences in the REE distribution. Thus, the CN of minerals is not always the dominant control. Calcite is rarely included in studies

Table 1. Major REE and Y minerals.*

Mineral	Formula	Theoretical RE_2O_3
Bastnaesite	$(REE)(CO_3)F$	74.81
Monazite	$(REE)PO_4$	69.73
Parisite	$Ca(REE)_2(CO_3)_3F_2$	60.89
Synchysite	$Ca(REE)(CO_3)_2F$	52.64
Ancylite	$Sr(REE)(CO_3)_2(OH)\cdot H_2O$	47.98
Xenotime	YPO_4	61.40[†]
Florencite	$(REE)Al_3(PO_4)_2(OH)_6$	31.99
Rhapdophane	$(REE)PO_4\cdot H_2O$	64.83
Britholite	$(REE,Ca)_5(SiO_4,PO_4)_3(OH,F)$	~60
Kainosite	$Ca_2(Y,REE)_2(Si_4O_{12})CO_3\cdot H_2O$	~38
Allanite	$(REE,Ca,Y)_2(Al,Fe^{3+})_3(SiO_4)_4(OH)$	variable (as high as ~17)

* Minerals where REE are major cations and that are known to, or have the potential of occurring in ore quantities.
† Y_2O_3

Table 2. Bastnaesite and chemically analogous minerals: composition and crystal system.

Mineral	Composition	System
bastnaesite	$(REE)(CO_2)F$	hex.
parisite	$(REE)_2Ca(CO_3)_3F_2$	trig.
synchysite	$(REE)Ca(CO_3)_2F$	orth.
röntgenite	$(REE)_3Ca_2(CO_3)_5F_3$	trig.
cordylite	$(REE)_2Ba(CO_3)_3F_2$	hex.
huanghoite	$(REE)Ba(CO_3)_2F$	trig.
zhonghuacerite	$(REE)Ba_2(CO_3)_3F$	trig.

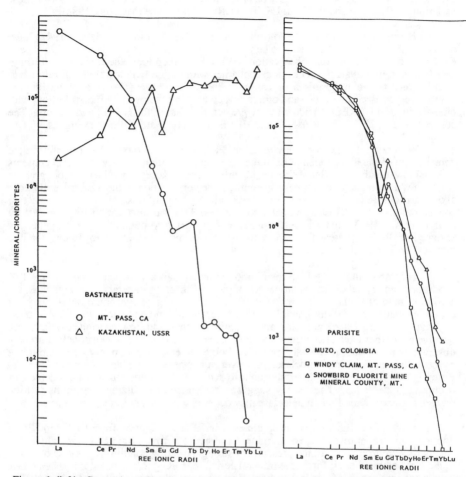

Figure 1 (left). Comparison diagram for bastnaesite from Mt. Pass, CA, and Kazakhstan, U.S.S.R. Bastnaesite from most environments usually shows an REE distribution similar to Mt. Pass with strong LREE selectivity despite the geochemistry of the source fluids. The Kazakhstan bastnaesite (Mineev et al., 1970) is a notable exception.

Figure 2 (right). Comparison diagram for parisite from varied geologic environments. The Mt. Pass parisite is from the Windy Claim, a carbonatite dike south of the main ore body (Olsen et al., 1954). Independent REE minerals from Mt. Pass all show strong LREE selectivity. the Muzo occurrence is the type locality for parisite where it is cocrystallized with emeralds in the Cretaceous Villeta shale. Fluorite from this same occurrence shows HREE selectivity in chemical analyses and in cathodoluminescence emission spectra (Mariano, 1988). The Snowbird parasite occurrence also is characterized by HREE selectivity in gangue calcite and fluorite (Mariano, 1988), and xenotime is also reported in this environment (Metz et al., 1985). Parisite with a coordination number of 11 in similarity to bastnaesite, is strongly LREE selective despite the bulk geochemistry of the environment. Nevertheless, chondrite-normalized REE plots for parisite and bastnaesite usually show small differences in distribution that reflect the bulk geochemistry.

312

concerning REE-bearing minerals, but in early carbonatite rock units it can be the major carrier of REE. Despite the low CN of 6, calcite from different environments can show variations in REE distribution from LREE to HREE dominance (Mariano, 1988).

With respect to basicity, Goldschmidt et al. (1933) and Haberlandt (1947) have noted that the more basic REE tend to be predominant in alkalic rocks. Murata (1957) gave evidence to substantiate this observation. With few exceptions, the minerals with high LREE were found in alkalic rocks or carbonatites while those with lower LREE relative to HREE were found in granites or rocks of lower basicity. The role of basicity is emphasized by Borodin (1960) and others as a controlling factor in REE distribution, with the basicity of the LREE considered to be greater than that of the HREE, and yttrium. The concentration of LREE in bastnaesite and HREE in xenotime are given as examples.

With respect to geologic environment, Fleischer (1965) reports a strong shift in the composition of the independent REE minerals (minerals with major and essential contents of REE) towards the HREE in monazite, allanite and other accessory minerals from granitic pegmatites as compared to granites. In contrast, the composition of the REE minerals from alkaline rocks (e.g., bastnaesite, monazite, allanite, apatite) show a general shift in concentration to the LREE as compared to granites. This was also demonstrated in data by Murata et al. (1953) that showed monazites from granitic pegmatites to be generally enriched in HREE, whereas those from alkaline rocks and carbonatites (excluding alkaline granites) are enriched in the LREE.

The variability of the REE distribution in minerals is also dependent on both P-T-X conditions and crystal-chemical properties of the REE-bearing minerals involved. thus, the oxidation state of Eu is a primary factor that takes precedence over ionic radius, and, for example, apatite may be dominant in Eu^{2+} or Eu^{3+} (Mariano and Ring, 1975). Although negative Eu anomalies are predominant in apatite (Puchelt and Emmermann, 1976), strong positive anomalies have also been reported (Roeder et al., 1987). The negative Eu anomaly observed in many bulk rock analyses and for some minerals is commonly attributed to the concentration of Eu^{2+} in plagioclase. However, it has been shown recently by D'Arco and Piriou (1989) that synthetic anorthite can also accept Eu^{3+} trapped in structural defects or occupying Ca^{2+} sites. Their study suggests that under crystallizing conditions of high oxygen fugacity, some Eu in plagioclase may occur in the trivalent state.

Some of the most pronounced changes in REE distribution from LREE to HREE dominance is sometimes observed in minerals of hydrothermal origin and especially in the weathering environment. In the presence of strong oxidizing conditions, Ce^{3+} is oxidized to the Ce^{4+} state where it forms the mineral cerianite (CeO_2) that is not stable when redox conditions are such that Ce^{3+} is dominant. In such cases, those minerals that accept REE^{3+} over REE^{4+} show a strong Ce-depletion. The term *Ce-depletion* is used in this chapter to represent a negative anomaly for Ce that is demonstrated in the absence of the $CeL\alpha_1$ emission in energy dispersive x-ray spectra.

Geochemical differentiation based on the relative solubilities and migratory capacity of the REE and Y is especially emphasized in the weathering zone of some carbonatites where supergene enrichment of REE minerals derived from the chemical dissolution of calcite, dolomite and apatite can cause radical changes in REE distribution and mineralogy. Examples of this condition will be given in a later section of this chapter.

RARE EARTH MINERALS OF ECONOMIC IMPORTANCE

Bastnaesite

Bastnaesite [(REE)(CO₃)F] is the most important REE-bearing mineral because of its occurrence as a high-grade accessory mineral of igneous or hydrothermal origin from Mt. Pass, California; Bayan Obo, China; Wigu Hill, Tanzania; and Karonge, Burundi. At

Mt. Pass, bastnaesite is the major REE mineral, and this deposit contains 31 million tons of 8.86 wt % rare earth oxides (REO) (J.O. Landret, pers. comm., Molycorp, Inc., 1989). In the Bayan Obo deposit, bastnaesite is also the major REE ore mineral with proven reserves of 37 million tons of ore (*Mining Annual Review*, 1986). Grade and tonnage data are not available for Wigu Hill and Karonge, but at Wigu bastnaesite and related REE minerals are visible in outcrops over a large area, and at Karonge rich bastnaesite veins have been mined intermittently. The 1978 production of bastnaesite concentrate at Karonge was 404 tons (J. Brinckmann, pers. comm., 1983).

Several other minerals are chemically analogous to bastnaesite (Table 2), and collectively they were listed as minerals of the bastnaesite group (Fleischer, 1978). Some of these minerals have been assigned to different crystal systems and space groups (Donnay and Donnay, 1953). Because of the recognized structural differences, the previously designated bastnaesite group is no longer referred to as a single structural group (Fleischer, 1987). In its pure form, bastnaesite is a simple member of the carbonates. Despite reported space group differences, closely related structures are indicated by the syntaxial intergrowth relationship that is sometimes observed in these minerals (Donnay and Donnay, 1953).

Considerable variables are reported in the composition of bastnaesite including members that have Ce, La, Nd and Y as the dominant elements in the REE crystallographic sites, and others that have (OH) > F (Fleischer, 1978). However, a strong LREE enrichment is encountered for this mineral in all deposits where bastnaesite occurs in established or near ore quantities. Figure 3 shows composition plots for bastnaesite from Mt. Pass; Lincoln Co., New Mexico (Glass and Smalley, 1945); Karonge (Wambeke, 1977); Wigu Hill (James, 1966); and Bayan Obo (Wang, 1981). The yttrian bastnaesite from Kazakhstan reported by Mineev et al. (1970) is indeed rare, appearing as an alteration product of gagarinite [(NaCaY(F,CL)]. Yttrobastnaesite was also reported in the Ukrainian shield by Rozanov et al. (1983).

For most bastnaesites, including those shown in Figure 3, the ThO_2 content rarely exceeds 0.1 wt %.

The theoretical rare earth oxide (REO) content for bastnaesite [(Ce,La)(CO$_3$)F with Ce:La = 1:1] is 74.8 wt %. A concentrate of hand-picked bastnaesite hexagonal prisms from the ore body at Mt. Pass gave a REO content of 74.6 wt %, attesting to the near exclusivity of REE cations in the structure.

<u>Syntaxic growth in bastnaesite-type minerals</u>. The term *syntaxy* as used by Donnay and Donnay (1954) refers to an oriented intergrowth between two chemically analogous minerals that are also related structurally. Because of the structural similarity, the intergrowths will show crystallographic continuity. Reports on syntaxial intergrowth for different pairs of bastnaesite-related minerals are abundant in the literature (Ungemach, 1935; Donnay and Donnay, 1954; Landuyt and Amelinckx, 1975; Gamaccioli, 1977; McLaughlin and Mitchell, 1989). Syntaxial intergrowths can sometimes be seen with the naked eye as demonstrated by Donnay and Donnay (1954, Fig. 6) for intergrowths of röntgenite-synchysite from Narsarsuk, Greenland. Fine-scale syntaxy of bastnaesite-type mineral pairs have been studied using multiple beam direct lattice imaging by Landuyt and Amelinckx (1975). However, a most rapid and efficient method for revealing syntaxy in minerals is through the observation of Z affects in secondary electron or back scattered electron images on a scanning electron microscope and with the combined use of energy dispersive x-ray detection. Figure 4 shows an example of syntaxy between parisite and bastnaesite in rødbergite from the Fen, Norway.

From various account in the literature, the impression is given that syntaxy in bastnaesite-type minerals is common (Landuyt and Amelinckx, 1975). Although most of the crystals studied by the Donnays showed syntaxial growth, the phenomenon is actually rare. When REE mineralization is being considered on an economic level, the presence of

314

Figure 3. Plots of La/Nd versus Σ(La+Ce+Pr) for bastnaesites from environments with substantial mineralization. In all cases, the LREE-selectivity is pronounced. This plotting method is taken from Fleisher (1965): (1) Mt. Pass, California; (2) Lincoln County, New Mexico; (3) Wigu Hill, Tanzania; (4) Karonge, Burundi; (5) Bayan Obo, China.

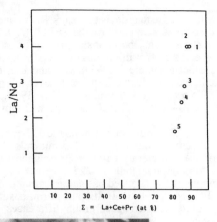

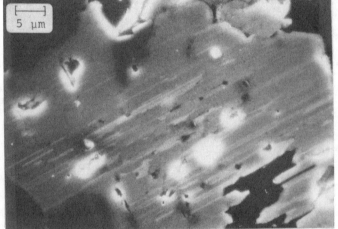

Figure 4. Syntaxial intergrowth of bastnaesite and parisite in røbergite from Fen, Norway. The light-colored bands in the secondary electron image are bastnaesite layers in parisite.

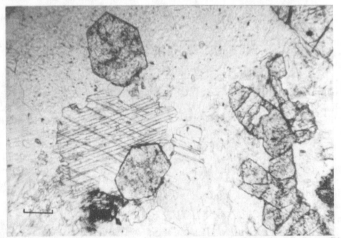

Figure 5. Plane-polarized light micrograph of bastnaesite in søvite from the Mt. Pass, CA, carbonatite body. The hexagonal prisms with high relief are bastnaesite. The surrounding matrix is calcite. Textures of this type suggest that bastnaesite is a primary mineral cocrystallized with calcite. Scale bar is 0.14 mm.

syntaxy is significant because the minerals that comprise the syntaxial pair have different REO values that obviously affect the REO content of a mineral concentrate. As mentioned previously, concentrates of hand-picked bastnaesite grains from the Mt. Pass ore body gave REO values for pure bastnaesite with no Ca. In general, bastnaesite from Mt. Pass, Gallinas Mt. and Karonge show no syntaxy. parisite from the Snowbird Fluorite Mine, Montana, Mt. Pass, and Muzo, Colombia, also show virtually no syntaxy.

Occurrence. The most common occurrence for bastnaesite and other independent rare earth minerals is in hydrothermal systems, although bastnaesite from Mt. Pass is an exception as primary igneous. As indicated by textural evidence (Fig. 5), it has cocrystallized with calcite, barite, etc. A recent experiment by Jones and Wyllie (1986), designed to synthesize a fluid similar to the Mt. Pass primary carbonatite magma, demonstrates that bastnaesite and calcite can precipitate together. Primary igneous parisite is also encountered in substantial quantities in the Mt. Pass carbonatite. The more common hydrothermal origin of bastnaesite-type minerals observed in most environments other than Mt. Pass is demonstrated by their habit as fine-grained fibrous or platey masses in vugs, microfractures, and veinlets, where they are associated with quartz, fluorite, strontianite, barite and hematite. Examples include the Wigu Hill carbonatite, Tanzania (Deans, 1966) and the Bayan Obo iron ore deposit with the world's largest known reserves of REE (Chao et al., 1989).

The formation of bastnaesite-type minerals in the weathering environment is less frequently encountered than monazite, rhabdophane and the crandallite group minerals. this is a result of the chemical instability of the carbonate minerals in the weathering zone and the higher affinity that REE have for available PO_4 during initial crystallization. A rare example of supergene bastnaesite with strong Ce-depletion from the carbonatite-derived laterite of Cerro Impacto, Estado Bolivar, Venezuela is shown in Figure 6.

Monazite

Monazite [(REE)PO_4] is the most common and abundant independent REE mineral, occurring as an accessory mineral in granitic rocks and in many metamorphic rocks. Monazite is chemically and mechanically stable, and in the weathering cycle it becomes concentrated in beach sands and river placers throughout the world. In some beach sands and placer deposits, monazite is a by-product of Ti or Sn mining.

The theoretical REO content for a monazite with Ce:La = 1:1 is 69.73 wt %, but in most monazites, Ca and Th substitute for REE that lower the REO content, e.g., monazite from beach sands of Western Australia has an average REO of about 55 wt %. In contrast, monazite crystallized as a primary mineral in carbonatites is low in Ca and Th, and the REO content more closely approaches the theoretical value (e.g., Mt. Pass. 66.2 wt %; Kangankunde, Malawi. 68.6 wt %).

Monazite almost invariable shows LREE selectivity. The REE distribution for monazite with a coordination number of 9 (Ueda, 1967) supports the structural control on REE distribution proposed by Semenov (1958). The use of a sigma (Σ) factor was introduced by Murata et al. (1953) to designate the sum of the *atomic* percentages (*not* weight) of La, Ce and Pr in minerals, and it serves as an index of the REE distribution in minerals. In a later study, Murata et al. (1957) discussed the limits of Σ in monazite including the constraints of ionic radii on the formation of monazite. They pointed out that for 26 samples analyzed, monazite never showed a Σ value lower than 58 atom % and that a lower value was not encountered in 23 samples previously analyzed by Vainshtein et al. (1955). Subsequently, Rosenblum and Mosier (1983) included Eu-rich dark monazite from several countries plotted on a compositional diagram showing La/Nd versus Σ(La+Ce+Pr) as proposed by Fleischer (1965). In all of the plotted analyses, the value never appears lower than 58 atom %, and for most of them the value exceeds 70 atom %, indicating that monazite is almost invariably LREE dominant.

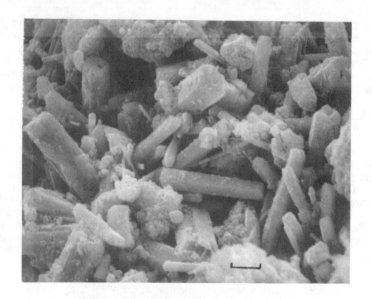

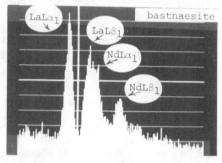

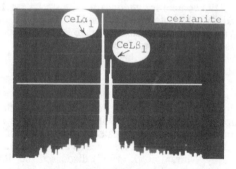

Figure 6. Secondary electron micrograph and accompanying EDS for euhedral hexagonal bastnaesite prisms and clustered aggregates of cerianite from laterite overlying carbonatite; Cerro Impacto, Estado Bolivar, Venezuela. The bastnaesite EDS shows an absence of the CeL series emission energies. The vertical line between the $LaL\alpha_1$ and $LaL\beta_1$ emissions in the bastnaesite spectrum is the marker designating the absence of the $CeL\alpha_1$ emission. The cerianite spectrum is almost exclusively CeL series emissions. This type of REE separation is commonly encountered in the zone of weathering for most carbonatites and is the result of Ce^{3+} oxidizing to Ce^{4+} and consequently forming a separate mineral. Scale bar is 2 μm.

Occurrence. Historically, the major source for monazite worldwide has been from beach sands and in all cases the REE distribution is always LREE dominant.

A rare occurrence of mid-atomic number REE dominance was reported by Graeser and Schwander (1987) for monazite-(Nd). The appended Nd in parentheses is the accepted means of designating a mineral species where the element in parentheses is greater in atomic percent than any other element within a single set of crystal structure sites (Levinson, 1966). A chondrite-normalized REE plot of this monazite is compared with beach sand monazite from Western Australia that constitutes the largest economic source of this mineral (Fig. 7). The monazite-(Nd) has a Σ(La+Ce+Pr) of 42.9 atom %, significantly below the lowest value reported by Murata et al. (1957) and Rosenblum (1983). This represents a rare case where the REE distribution in monazite shows anomalous values for Nd and Sm relative to the lighter REE. Both Nd and Sm are in great demand for use in the manufacture of high strength permanent magnets.

The lower solubility of the LREE relative to the HREE in mineralizing fluids also plays an important role in the formation of monazite with LREE selectivity in varied geologic environments. The bulk REE distribution for the peralkaline granite and quartz syenite complexes of Strange Lake, Quebec-Labrador (Currie, 1985) and Pajarito Mt.,

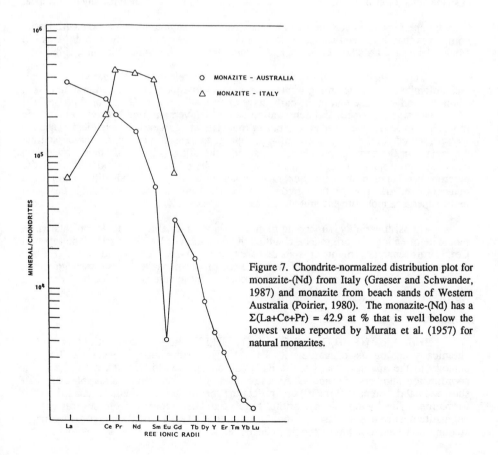

Figure 7. Chondrite-normalized distribution plot for monazite-(Nd) from Italy (Graeser and Schwander, 1987) and monazite from beach sands of Western Australia (Poirier, 1980). The monazite-(Nd) has a Σ(La+Ce+Pr) = 42.9 at % that is well below the lowest value reported by Murata et al. (1957) for natural monazites.

New Mexico (Moore et al., 1988) is HREE dominant. Both of these complexes are impoverished in phosphate, but in the early stages of crystallization monazite and apatite precipitate with strong LREE preference (Roeder et al., 1987; Mariano and Brikett, 1988). The strong affinity of REE for PO_4^{3-} is especially emphasized in the paragenesis of REE minerals in carbonatite laterites and hydrothermal fluids where the dominant REE minerals are usually monazite, rhapdophane and REE species of the crandallite group. The bastnaesite-type minerals are typically found in hydrothermal systems that are relatively impoverished in PO_4^{3-}.

An unusual type of monazite referred to as *dark monazite* is known to occur in beach sands and river placers in various parts of the world including Taiwan, France, Zaire, Peru, Pakistan and the U.S.A. (Rosenblum and Mosier, 1983). This monazite has a pellet-like form and is dark gray to black in color. Compositionally, they contain more SiO_2 and Eu_2O_3 than the yellow monazites and have considerably less Th. According to Rosenblum and Mosier (1983), the dark monazite is of low-grade metamorphic origin where the source rock is a weakly metamorphosed carbonaceous shale or phyllite.

The average Eu_2O_3 content for carbonatite-derived and placer monazites is about 0.10 and 0.05 wt %, respectively, whereas for dark monazites the worldwide average is 0.36 wt % (Rosenblum and Mosier, 1983). Since europium is one of the most valued REE, dark monazites offer an intriguing target for the exploration geologist, but at this time sustained quantities of economic grade have not been firmly established for any occurrence.

Supergene and hydrothermal monazite constitute large tonnage in some carbonatite complexes that have been subjected to deep lateritic weathering. They include Araxá in Minas Gerais and Catalão I in Goiás, Brazil, and Mt. Weld, Australia (Mariano, 1989).

In these deposits monazite is produced from the release of REE derived from calcite, dolomite and apatite during weathering and its subsequent reconstitution with PO_4^{3-}. There is good evidence that in the early stage of crystallization the hexagonal phase REE $PO_4 \cdot nH_2O$ was produced, and dehydration caused conversion to anhydrous monoclinic monazite. In some cases, pseudomorphs of monazite after an earlier hexagonal phase are well preserved (Fig. 8). Most commonly, supergene monazite occurs as polycrystalline aggregates in the form of platelets, crusts and spherulites (Fig. 9) in laterite overlying carbonatite in tropical environments, or on the weathered surface of exposed carbonatite outcrops in other environments. Supergene monazite coatings on carbonatite outcrop from magnet Cove, Arkansas, first reported by Rose et al. (1958) is almost identical in character to the supergene monazite encountered in tropical regions.

Almost invariably, supergene monazite shows a notable Ce-depletion relative to monazite of endogenic origin as a result of the oxidation of Ce^{3+} to Ce^{4+}. The released Ce^{4+} forms separate cerianite crystals that are typically associated with monazite (Fig. 9). A detailed account of the geology of vein-type and placer monazite deposits throughout the world is given by Overstreet (1967), and a more recent update is provided by Neary and Highley (1984).

Britholite

Britholite, $[(REE,Ca)_5(SiO_4,PO_4)_3(OH,F)]$, has the same structure and is chemically analogous to apatite with $REE^{3+} + Si^{4+}$ substituting for $Ca^{2+} + P^{5+}$. By analogy to the apatite structure, the REE can occupy two different cation sites with coordination numbers of 7 and 9. As a result, the REE distribution is complex and can show radical differences (Fig. 10) that reflect the specific physicochemical conditions for formation. The Y-rich variety, britholite-(Y) has been reported from several granite pegmatite occurrences (Vlasov, 1966; Griddin et al., 1979). Britholite contains major amounts of Th and is always strongly radioactive.

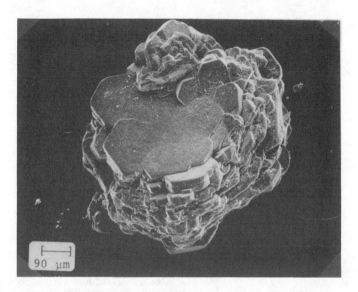

Figure 8. Secondary electron micrograph of supergene monazite from Mt. Weld laterite, Western Australia. The material was originally crystallized as REE(PO$_4$)·nH$_2$O with a hexagonal structure. Dehydration caused a conversion to anhydrous monoclinic monazite.

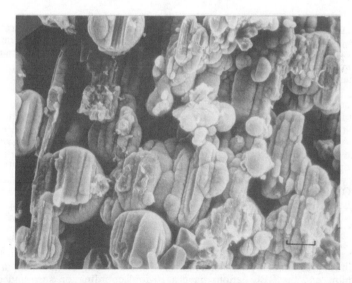

Figure 9. Secondary electron image of supergene monazite globular aggregates from a depth of 190 m in laterite overlying carbonatite Araxá, Brazil. The small spherical forms on the larger aggregates are cerianite, and the euhedral prisms in the lower center are strontianite. Scale bar is 20 µm.

320

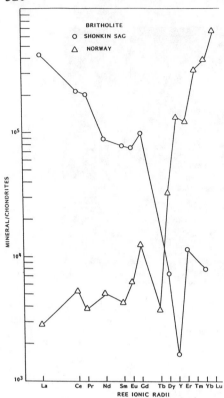

Figure 10. Chondrite-normalized REE distribution plot for britholite from the Shonkin Sag lacolith (Nash, 1972) and from a granite pegmatite, Norway (Griffin et al., 1979). The extreme contrast in REE distribution shows the large variability possible for REE composition in britholite. In these cases the REE distribution reflects the bulk chemistry of their environment.

Occurrence. Britholite occurs in many different environments as a trace mineral. Deposits approaching economic quantities are confined to alkaline rocks (both miaskitic and agpaitic) and in one carbonatite at Oka, Quebec, where the mineral is a thorian member of the britholite-apatite series (Hughson and Sen Gupta, 1964). It occurs in a late-stage dike injected during the last stages of the carbonatite intrusion.

Concentrations of britholite are known in the miaskitic nepheline syenite complexes of the Urals, Siberia, and Soviet Central Asia (Vlasov, 1966). In agpaitic nepheline syenites, it occurs as Ilimaussaq, South Greenland, Kipawa, Quebec and the Pilanesberg of South Africa. In the Pilanesberg complex, britholite occupies a series of near vertical fissures in tinguaite and foyaite. Associated minerals in the fissures include magnetite allanite, apatite, calcite, strontianite, fluorite, and aegirine. The britholite is LREE dominant with 56.36% REO and 1.56% ThO_2. Estimated reserves are 13.5 million tons at 0.7% REO + ThO_2, 1.2 million tons at 6.54% REO + ThO_2 and 24,000 tons at 10% REO + ThO_2 (Lurie, 1985).

Ancylite

From its composition [$Sr(REE)(CO_3)_2(OH) \cdot H_2O$] and paragenesis, it is apparent that ancylite is predominantly a mineral of hydrothermal origin. The structure of ancylite is orthorhombic and the REE cations have a 10-fold coordination surrounded by eight oxygens and two hydroxyls or water molecules (Dal Negro and Rossi, 1975). Complete chemical analyses for ancylite are sparse in the literature, and REE distributions have not been reported. Incomplete REE analyses listed by Vlasov (1966), Keidel et al. (1971) and Kapustin (1980) suggest LREE selectivity. Sarp and Bertrand (1985) have described an isostructural and chemically analogous mineral gysinite [$Pb(Nd,La)CO_3)_2(OH) \cdot H_2O$] from

Zaire with a variable range for Nd_2O_3 and La_2O_3 of 35.0-43.0 and 4.0-7.5 wt %, respectively.

Ancylite commonly exhibits euhedral morphology and typically shows pseudo-octahedral forms (Fig. 11). According to Dal negro and Rossi (1975), there is a close structural relationship between ancylite and strontianite. At Ravalli, Montana, Gem Park, Colorado, and Bear Lodge, Wyoming, ancylite is encountered intergrowth with strontianite. The intergrowth pair exhibits a quasi-graphic texture as pseudomorphs after an earlier hexagonal mineral. The texture suggests that the pseudomorphs were originally crystallized as a REE-Sr carbonate of trigonal or hexagonal morphology and eventually through cooling or the introduction of water they exsolved, forming a symplectic intergrowth.

Occurrence. Ancylite occurs in alkalic complexes containing nepheline syenites and related alkaline rocks. It commonly appears in cavities associated with aegirine, albite, fluorite, natrolite, strontianite, quartz and epididymite. These occurrences include Narssarssuk, Greenland (type locality), Mount St. Hilaire, Quebec, and Lovozero in the Kola Peninsula. It has also been reported in cavities in granite at Baveno, Italy (Palache et al., 1951) and cavities in trap rock from Cornog, Pennsylvania (Keidel et al., 1971). In all the above occurrences, ancylite is a rare mineral of academic interest. In carbonatites, however, ancylite may occur in ore quantities.

In the Brazil carbonatites of Araxá and Catalão I, ancylite has been observed in carbonatite drill core well below the weathering zone (Mariano, unpublished data). In these occurrences ancylite appears in microfractures and cavities associated with fluorite, quartz and an ubiquitous hematite dusting. The presence of euhedral ancylite with quartz in cavities establishes the hydrothermal origin for the mineralization.

In the Gem Park carbonatite of Colorado, rare earth veins have been reported by Parker and Sharp (1970) where the dominant REE mineral is ancylite. Other carbonatites

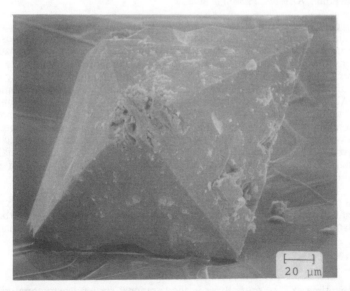

Figure 11. Euhedral ancylite crystal from the Bear Lodge carbonatite, Wyoming. This pseudo-octahedral morphology is frequently observed for ancylite from carbonatites and nepheline syenites. It appears to be the general form {hkl}.

and related occurrences where ancylite appears at least locally in substantial quantities include Bear Lodge, Wyoming, Rocky Boy, Montana, and Ravalli, Montana.

Allanite

The composition for allanite given by Fleischer (1987) is $(Ce,Ca,Y)_2(Al,Fe^{3+})_3(SiO_4)_3(OH)$. The mineral is monoclinic, belonging to the epidote group, but unlike epidote and zoicite it can accept variable amounts of REE ranging between 10 and 12% REO (Vlasov, 1966). An allanite occurrence in hydrothermal veins adjacent to the Mt. Pass carbonatite gave an REO content of 27 wt % with a strong LREE selectivity. Exley (1980) reported a maximum REO for Skye Granite allanites of 26.07 wt %. In a structural refinement study on allanite, Dollase (1971) describes two Ca sites, A(1) with CN = 9 and A(2) with CN = 11. Site A(2) was found to predominantly contain REE. In a crystal/liquid partition study including allanite, apatite and zircon, Brooks et al. (1981) found allanite to show the highest partition coefficients for LREE. They explained this behavior based on the structural control of the A(2) sites of allanite accepting the large ionic radius LREE. A strong LREE selectivity for allanite was also reported by Fleischer (1965) in allanite from varied geologic environments that all showed a Σ(La+Ce+Pr) 62 atom % relative to total REE. Contrary to these data are some Y-rich allanites reported from North Karelia, U.S.S.R., that also show high contents of Nd, Sm, Gd, and one analysis where Dy is the major REE (Vlasov, 1966).

The allanite used in the structure refinement by Dollase (1971) was LREE selective. The behavior of REE site preference in allanite with HREE dominance may be complex and may not necessarily be confined to the A(2) sites. The allanite analyses listed in Vlasov (1966) were performed on mineral concentrates, and it is well known that allanite, particularly in granite pegmatite environments, can have numerous mineral inclusions with HREE that can influence the analyses. However, more recent electron microprobe analyses reported by Exley (1980) show hydrothermal and metamorphic allanites from the Skye area with REE fractionation varying on chondrite-normalized plots from near flat slopes to strongly enriched HREE patterns. Clearly, more research is needed to understand the factors governing REE distribution in allanite.

Occurrence. As an accessory mineral allanite has a widespread occurrence in igneous and metamorphic rocks and in hydrothermal environments, but in most cases the mineralization is spotty or localized and lacks volume. In some granite pegmatites, alkali granite environments and skarns, allanite is a major REE-bearing phase in grades approaching economic potential. In addition to several granite pegmatite occurrences in Canada, allanite mineralization also appears in Bokan Mountain and Mount Prindle, Alaska. Allanite concentrates with strong LREE selectivity were produced from the Mary Kathleen uranium-bearing skarns in Queensland (Hawkins, 1975). Another possible economic allanite source in Australia was recently discovered in a prospect 100 km northeast of Alice Springs in the Northern Territory. Allanite was reported to occur in a coarse crystalline pegmatite at 1 million tons with a grade of 4% allanite containing 20% REE and 1.5% Th. Preliminary beneficiation tests including leaching of REE gave encouraging results (O'Driscoll, 1988).

In carbonatites, allanite is always of hydrothermal origin, usually as vein fillings associated with quartz, fluorite, apatite and monazite. Within the Mt. Pass carbonatite ore body, allanite appears sporadically as a hydrothermal overprint on the primary igneous minerals. Allanite has not been reported in economic quantities in carbonatites.

Xenotime

Xenotime, YPO_4, crystallizes in the tetragonal system isostructural with zircon. Synthetic $REE(PO_4)$ compounds that are HREE dominant, i.e., Tb through Lu, are tetragonal, whereas LREE dominant compounds La through Gd are monoclinic and have the monazite structure. The monoclinic structure of monazite with CN = 9 will accept larger REE ions that occupy large sites of occupancy, while the higher atomic number REE including Tb through Lu that have a smaller ionic radius occupy the smaller sites of tetragonal xenotime with CN = 8. Yttrium with an ionic radius similar to Ho also assumes the smaller site occupancy.

The highest Y_2O_3 content for pure YPO_4 is 61.4 wt %. In nature, xenotime always contains REE substituting for Y with HREE selectivity. The most common dominant REE is Dy or Yb. Values for Y_2O_3 + REO in xenotime reach as high as 68 wt % (Palache et al., 1951). The higher atomic weight of the HREE as compared to Y is responsible for the higher Y_2O_3 + REO value. Mineral/matrix (fluid) partition coefficients for REE in xenotime have not been measured, but considering the similarity of ionic radii and the same metal coordination number of zircon and xenotime, it is logical to assume similar partition coefficients.

Other elements that can proxy for Y in xenotime include Th and U (Palache et al., 1951). Their presence in xenotime obviously dilutes the Y and REE content.

Occurrence. Xenotime can be found in small quantities in diverse geologic environments including acidic and alkaline igneous rocks, pegmatites, gneisses and schists.

The major economic source of xenotime has been from beach sands and placer deposits in Malaysia, Thailand and Australia where it is a by-product of titanium and tin. The availability of by-product xenotime from placer deposits is largely governed by the demand for the major commodity in the ore. In the Younger granites of northern Nigeria, xenotime has been a by-product of cassiterite and columbite from the biotite granites of Ropp, Jos-Bukuru and the Afu Hills at Odegi (Williams et al., 1956).

Epigenetic mineralization related to Proterozoic arkosic-conglomeritic sandstones offers some potential for economic xenotime deposits. Xenotime associated with uranium mineralization precipitated from hydrothermal solutions in sedimentary environments has produced xenotime in near economic accumulations in the Athabasca sediments near Wheeler River in Alberta. In Western Australia a xenotime occurrence as veinlets or pods with quartz occupies fractures in quartzite of the Carpentarien Red Rock Beds in the Halls Creek Province (Hill, 1976). In both of these occurrences, the mobilization of Y relative to the other REE, especially the LREE, is indicated by the association of xenotime with uranium mineralization. In a study of REE mobility associated with uranium, McLennan and Taylor (1979) point out that U and REE were transported together in a hydrothermal fluid as soluble carbonate complexes. They noted that the HREE tend to be more soluble than LREE, and they show REE patterns for rocks with high U content normalized to average Australian sedimentary rocks that exhibit a strong HREE selectivity. The higher migratory capacity of Y and HREE complexes relative to the LREE has been noted by Mineyev (1963) and Balashov et al. (1964).

Higher solubility and migratory capacity of Y and the HREE in certain environments of supergene weathering, or the activity of hydrothermal solutions has caused the isolation of these elements from LREE-dominant sources and their subsequent precipitation as xenotime in grades approaching economic values. Deposits of this type that contain xenotime and churchite ($YPO_4 \cdot 2H_2O$) mineralization include the Mt. Weld carbonatite laterite, Western Australia, and Hicks Dome, Illinois (Mariano, unpublished data). Churchite mineralization is also reported fro the Sandkopsdrif carbonatite in South Africa (Verwoerd, 1986). More recently, Adrian et al. (1989) report a strong enrichment of Y_2O_3 in weathered carbonatized and silicified tuffs in the Goudini carbonatite complex of the Transvaal, South Africa.

The examples given above demonstrate that in complexes with a bulk geochemistry of LREE dominance, hydrothermal and weathering processes can cause a separation of the REE and a redistribution that can developed areas of HREE enrichment. The major Y mineral in these environments is usually xenotime or churchite.

Comparisons for xenotime and monazite concentrates (Rhône-Poulenc, 1987) from placer sources of Malaysia and Australia are given in Figure 12. Figure 13 shows similar REE distribution trends for xenotime from one of the Younger granites of Nigeria and xenotime from quartz veins of hydrothermal origin from Western Australia.

Crandallite group minerals

The REE-bearing members of the trigonal crandallite group are as follows:

florencite	- REE	$Al_3(PO_4)_2(OH)_6$
gorceixite	- (Ba,REE)	$Al_3(PO_4)_2(OH_5·H_2O)$
goyazite	- (Sr,REE)	$Al_3(PO_4)_2(OH_5·H_2O)$
crandallite	- (Ca,REE)	$Al_3(PO_4)_2(OH_5·H_2O)$

Cation interchange of REE for Ca, Ba and Sr between these species is common, although nearly pure florencite with total REO approaching 30 wt % can occur (LeFebvre and Gasparrini, 1980).

Occurrence. The crandallite minerals are confined to the zone of weathering and have been found as small scale mineralization in sedimentary beds in many parts of the world. Gorceixite and goyazite were first described from placers derived from the weathered portions of kimberlites in Brazil. In central West Africa they were used as indicators for kimberlite prospecting by United Nations exploration teams.

Although the crandallite group minerals are currently not the source of any economic REE ores, they occur in the weathered portions of some carbonatites in grade and tonnage that approach economic levels. In the weathered laterite residuum of Mrima, Kenya, gorceixite, goyazite and monazite are the major REE-bearing minerals. According to Deans (1966), reserves grading about 5% REO comprise at least 6 million tons with isolated areas containing from 10 to 25% REO. Large accumulations of gorceixite and goyazite are also reported in the northern edge of the Araxá complex, Brazil, in a location known as *Area Zero*. This area has been studied as a potential source of U, Nb and REE. Reports on the geology and mineralogy are given by DeSouza and Castro (1968) and Mendes et al. (1968). They report average grades for 800,000 metric tons o measured resources as 13.5% REO, 2% Nb_2O_5 and 0.05% U_3O_8. In Area Zero, gorceixite occurs mostly as pisolites with an average diameter between 0.5-1.0 mm. At high magnifications, the interior of broken pisolites are found to consist of aggregates of rhombohedra less than 1 μm in diameter (Fig. 14).

Florencite has been observed in many carbonatite laterites in substantial quantities including Cerro Impacto, Venezuela, Mt. Weld, Australia, Morro do Seis Lagos, Amazonas, Brazil and Twareietau, Guyana (Mariano, 1989). Extreme lateritic weathering in these complexes has destroyed magnetite and pyrochlore, and some of the liberated Nb is incorporated in supergene florencite and goyazite. At Mt. Weld these minerals are found as pseudomorphs after pyrochlore in well-preserved octahedra.

The strong oxidation in the weathering zone causes Ce^{4+} to be separated from the other trivalent REE. Cerianite is usually formed intimately crystallized with the crandallite group minerals, and in such cases the crandallite group exhibits Ce-depletion (negative Ce anomaly).

Several other minerals have been mined for REE on a small scale. They include gadolinite, $(REE,Y)_2Fe^{2+}Be_2Si_2O_{10}$, fergusonite, $YNbO_4$, euxenite, $(Y,Ca,REE,U,Th)(Nb,Ta,Ti)_2O_6$, and yttrotantalite, $(Y,U,Fe^{2+})(Ta,Nb)O_4$. These minerals occur as local concentrations in some granites, in granitic pegmatites and placer deposits. They are commonly enriched in HREE and Y. The REE distribution is economically attractive, but these minerals are not known to occur in large tonnage.

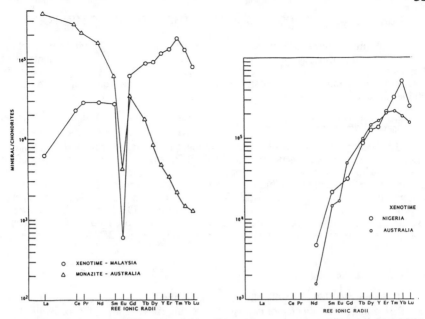

Figure 12 (left). Chondrite-normalized comparison diagram for concentrates of xenotime from Malaysia and monazite from Australia (Rhône-Poulenc, 1987). These minerals are some of the main ores processed by Rhône-Poulenc. The monazite shows the steep slope of LREE enrichment typical of beach sand sources on a world level. Xenotime exhibits the inverse trend of HREE enrichment. Both minerals have a strong negative europium anomaly.

Figure 13 (right). Chondrite-normalized comparison diagram for xenotime from a biotite granite of northern Nigeria (Jefford, 1962) and from hydrothermal quartz veins in Western Australia (Hill, 1976). Despite petrogenetic differences, the xenotimes have similar REE trends with steep slopes of HREE enrichment. Unlike xenotime from Malaysia (Fig. 12), these show no Eu anomaly.

Figure 14. Secondary electron image of the exposed center of a gorceixite pisolite from Araxá, Brazil, laterite. The pisolites are composed euhedral gorceixite rhombohedra whose average dimension is <0.5 μm. The clustered rods or nodes on the gorceixite rhombohedra are goethite. Scare bar is 1 μm.

REE AS A MINING BY-PRODUCT OF OTHER ECONOMIC MINERALS

By-product mining already accounts for a major portion of the world production of REE and Y from placer deposits. Yttrium is also a by-product of uranium mining in the Elliot lake deposit of Ontario. New trends in mining, particularly of phosphate deposits in several continents and anatase deposits in Brazil, can supply vast quantities of by-product REE into the market place in the future.

By-product yttrium from Elliot Lake

The Proterozoic quartz-pebble conglomerates of Blind River in the Elliot Lake area of Ontario contains brannerite $(U,Ca,Y,REE)(Ti,Fe)_2O_6$, uraninite and uraniferous monazite. The deposit has been mined for uranium since 1955, and a by-product concentrate of acid-leached residue rich in yttrium has been recovered intermittently as a major world source of Y. A composition for the REE residue given by Rose (1979) shows 14.3% LREE, 85.7% HREE, half of which (51.4%) is Y_2O_3. Production temporarily ceased in 1978, but it has recently been resumed. The current production of Y_2O_3 from the Elliot Lake occurrence is 300,000 pounds per year, representing 25% of the world market for Y.

By-product REE from loparite

Loparite, $(REE,Na,Ca)_2(Ti,Nb)_2O_6$, is a member of the perovskite group $(CaTiO_3)$ where REE and Nb occur as major cations substituting for Ca and Ti, respectively. Values of REO for loparite from the Kola Peninsula and Siberia are in excess of 30 wt % (Vlasov, 1966). The REE distribution of loparites are strongly LREE selective (Semenov and Barinskii, 1958). In agpaitic nepheline syenite complexes of the Kola Peninsula, loparite is mined for Nb and REE (Vlasov, 1968; O'Driscoll, 1988).

By-product REE from apatite

Apatite from carbonatites and alkaline igneous rocks can contain major amounts of REE. The highest recorded REO (19.2 wt %) values for apatite is from the Proterozoic peralkaline granite complex of Pajarito, New Mexico (Roeder et al., 1987). Apatite with major amounts of REE have also been detected in the Strange Lake, Quebec-Labrador, peralkaline granite complex and various agpaitic nepheline-syenite complexes in North America (Mariano, unpublished data). Recently, apatite with >16.0 wt % REO has been reported from syenites of Ilimaussaq, Greenland (Rønsbo, 1989).

The major source of phosphate has been from marine phosphorite where the average REE content ranges between 0.01 and 0.1 wt %. Altschuler et al. (1967) recognized the REE potential from marine phosphorites, but since the content is low and other REE sources were less costly, no attempts have been made for commercial recovery. However, apatite in carbonatite deposits, alkaline igneous rocks, Kiruna-type ores and nelsonites have REO values ranging between 0.4 wt % and slightly greater than 5 wt %. A carbonatite apatite from the Bond Zone of Oka, Quebec, was found to average 8.6 wt % REO (Mariano, 1985), while apatite from the Mineville, New York, magnetite deposit contains about 4 wt % REO (Roeder et al., 1987). Chondrite-normalized REE distribution plots for Oka and Mineville apatite are shown in Figure 15. Girault (1966) analyzed six apatites from søvite at Oka and reported REE concentrations ranging from 2.5 to 8.25 wt %. Eby (1971) examined nine apatite samples from Oka søvite and reported a ΣREE range of 1.3-4 with an average of 2.4 wt %. In the Palaborwa carbonatite of South Africa, apatite concentrates from phoscorite and pyroxenite contain 0.6 and 0.85 wt % REO, respectively (Russell, 1977). At Mineville, New York, apatite tailings from magnetite mining in metagabbros and syenites intermixed with Grenville sedimentary rocks have been studied by various companies interested in the economic recovery of REE and Y. In Brazil, apatite is mined from carbonatite complexes in Araxá and Tapira, Minas Gerais, and also from Catalão I, Goiás and Jacupiranga, São Paulo. Large resources of apatite also

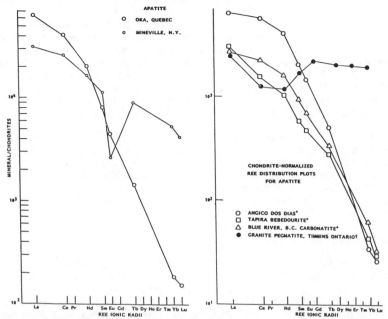

Figure 15 (left). Chondrite-normalized comparison diagram for apatite from Oka, Quebec søvite (Mariano, 1985) and Mineville, NY, magnetite deposit (Roeder et al., 1987). Notice the absence of an Eu anomaly in the Oka apatite. This apatite shows a steep slope of LREE enrichment with ΣREE at 7 wt %. In contrast, the Mineville apatite shows the more usual negative Eu anomaly and a noticeable enrichment of the mid-atomic number REE. The apatites were analyzed using neutron activation by Nelson Eby.

Figure 16 (right). Comparison diagram of chondrite-normalized REE distribution plots for apatites from carbonatite complexes and a granite pegmatite occurrence. The REO for these apatites ranges between 0.4 (Tapira) and >1 wt % (Angico dos Dias, Bahia, Brazil). The carbonatite apatites show the typical LREE dominance, while the granite pegmatite apatite has an HREE enrichment. None of these apatites show a europium anomaly. † microprobe analysis (Roeder et al., 1987). * NA analysis (N. Eby).

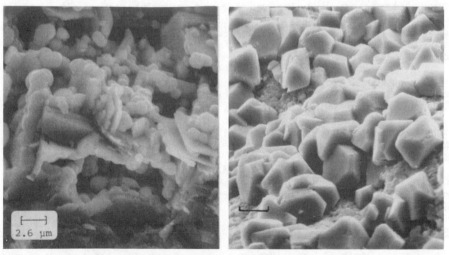

Figure 17. Secondary electron micrographs of REE minerals on supergene anatase from Brazil laterites.
(a) Tetragonal platelets of anatase with rounded disk-like growths of cerianite. Serra Negra, Minas Gerais.
(b) Euhedral florencite on anatase. Galeria 7, Catalão I, Goiás. Scale bar for (b) is 7 μm.

occur in Salitre and Serra Negra, Minas Gerais; Anitapolis, Santa Catarina; Angico dos Dias, Bahia; and Maicuru, Pará (Mariano, 1989). Many apatite deposits also occur in Africa and are being mined (Deans, 1978; Mariano, 1989). In North America, large reserves of eluvial apatite in carbonatite are known in Cargill and Martison Lake, Ontario (Mariano, 1989). In all of these occurrences, the REO content of apatite ranges between 0.4 and 1 wt %. The REE distribution in carbonatite apatites is always LREE dominant with a radically different slope compared to apatite from granite pegmatites (Fig. 16). This feature is useful for identification of carbonatite and mantle-derived apatite.

By the next decade the increasing world demand for phosphates will not be met by production of U.S. marine phosphate reserves (currently the world's largest producer) because of land use restrictions, restrictions for phosphate mining and processing, increasing environmental regulations, and high production costs. Therefore, the outlook is for a world increase in phosphate mining of carbonatite apatite. Despite the relatively high REE content for apatite from carbonatites, at present REE recovery from apatite mining is confined to the Kola Peninsula alkaline complexes.

By-product REE from anatase

Some of the largest resources of Ti in the world occur in various parts of Brazil in carbonatite complexes where perovskite, a major mineral in pyroxenites, has been decalcified through lateritic weathering, leaving a supergene residuum of high grade anatase (TiO_2) (Mariano, 1989). Perovskite below the zone of weathering has about 1.5 wt % REE. In the decalcification process, REE are liberated and combine with phosphate from accessory apatite to form supergene REE-bearing crandallite group minerals or monazite. These minerals are associated with the anatase (Fig. 17) and commonly show a strong Ce-depletion. Cerianite is almost always found intimately associated with the anatase and Ce-depleted REE minerals. Cerianite is almost exclusively a mineral of supergene origin crystallizing only under conditions of strong oxidation. It is not encountered below the weathering zone in the anatase-bearing carbonatites of Brazil. The presence of cerianite together with anatase indicates a supergene rather than hydrothermal origin for the anatase. The alteration of perovskite to anatase as a product of weathering is demonstrated by examining drill core through the laterite down to depths below the zone of weathering, typically 20 to 120 m below the surface. In the weathering zone only anatase is encountered. At some depth, residual central cores of unweathered perovskite are found surrounded by anatase. Below the weathering zone, anatase is not present. Pseudomorphs of anatase octahedra after perovskite are abundant in the weathered pyroxenite of the Fazenda Boa Vista area in Tapira, Brazil (Cassedanne and Cassedanne, 1973). REE abundance plots for perovskite and anatase from Salitre II are almost identical (Fig. 18), demonstrating that the REE were immobile during weathering.

Considerable progress has already been made on the recovery of a TiO_2 product from the anatase deposits in Brazil that include the following carbonatite complexes: Tapira, Serra Negra, Salitre I and II, Minas Gerais; Catalão I, Goiás and Maicurú and Maraconai, Pará. The anatase mineralization in all of the complexes is amenable to open pit mining. Grade and tonnage data (Mariano, 1989) demonstrate that they are major world resources for Ti. Pilot plant operations have been established for several of the complexes, and concentrates ranging between 84-94 wt % TiO_2 have been stockpiled. The concentrates contain in excess of 3 wt % REE in addition to other deleterious components (for TiO_2 production) including Th. Commercial utilization of the anatase will require removal of the REE. If this is successful, large quantities of by-product LREE will enter the market place.

By-product REE and Y from eudialyte

Eudialyte is a trigonal silicate with a general composition of $Na_4(Ca,Ce)_2(Fe,Mn,Y)ZrSi_8O_{22}(OH,CL)_2$ (Fleischer, 1987). Substitution of REE in the structure takes place in moderate quantities with values reaching as high as >10% REO +

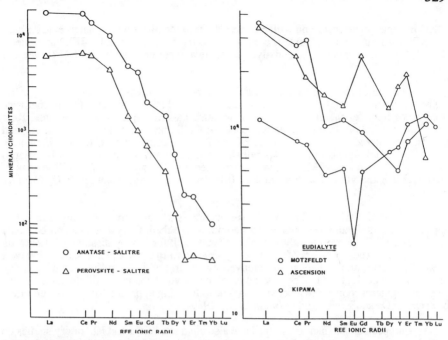

Figure 18 (left). Chondrite-normalized REE distribution for anatase and perovskite from the Salitre carbonatite complex, Minas Gerais, Brazil. The slope labeled anatase is actually from submicroscopic independent REE minerals that are inextricably crystallized with the anatase. The close similarity in the slope of these distributions is geochemical proof that anatase is an alteration product of perovskite. The REE content is upgraded in anatase as a result of the decalcification of perovskite. Although the Ca is totally removed in the groundwater, the REE are essentially immobile.

Figure 19 (right). Chondrite-normalized REE distribution comparisons for eudialyte. The Motzfeld and Kipawa occurrences are in agpaitic nepheline syenite complexes. The Motzfeld plot (Jones and Larsen, 1985) shows LREE enrichment with REO + Y_2O_3 = 3.75 wt %. Associated zircon crystallization in the Ascension eudialyte may account for partitioning of HREE in zircon and the LREE enrichment of the eudialyte.

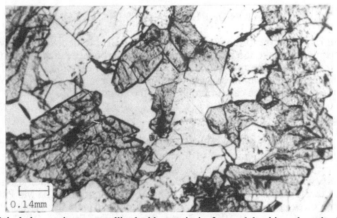

Figure 20. Euhedral monazite co-crystallized with strontianite from a dolomitic carbonatite, Kangankunde Hill, Malawi. High relief grains are monazite; others are dolomite and strontianite. Plane polarized light.

Y2O3 in some occurrences and with reported Y_2O_3 exceeding 4 wt % (Harris et al., 1982). The REE distribution in eudialyte is complex, showing both LREE and HREE enrichments (Fig. 19). Almost invariably, eudialyte contains anomalous quantities of Y (Fryer and Edgar, 1977).

The major occurrences of eudialyte appear in silica-undersaturated agpaitic nepheline syenite complexes and silica-saturated peralkaline granites. The former type occurrence includes the eudialyte magmatic cumulates of the Lovozero complex, Kola Peninsula; the kakortokite-lujavrite sequences of the Ilimaussaq complex, Greenland, and the metasomatized nepheline syenites of the Kipawa complex in Quebec. Economic quantities of eudialyte occur in the peralkaline granite-quartz syenite complex of Pajarito Mt., Otero County, New Mexico, where reported estimates include 24 million tons of recoverable eudialyte ore with a grade of 0.18 wt % Y_2O_3 and 1.2 wt % ZrO_2 (*Mining Engineering*, 1989).

Almost invariably where eudialyte occurs as a major rock-forming mineral, it appears as coarse-grained ovoids or, less commonly, euhedral crystals that can be easily concentrated. The name *eudialyte* is taken from Greek for "easily dissolved." Leaching experiments on rock powders of aegirine lujavrite from Ilimaussaq have shown that over 50% of the ZrO_2 and Y_2O_3 can be removed by stirring with 2% sulfuric acid for a few hours (Bailey et al., 1981). The ease at which eudialyte dissolves in dilute acids has led mineral engineers to propose a conventional heap leaching approach for the concentration of ZrO_2, Y_2O_3 and REO (*Mining Engineering*, 1989).

European and Canadian groups have also conducted field and laboratory studies on eudialyte mineralization from Ilimaussaq. Reported values for some of the eudialyte ore includes 4% Y and other REE. One ore body is reported to contain consistent grades of 1.2% ZrO_2, 0.12 Y_2O_3 and 0.7% REO (O'Driscoll, 1988).

According to Sinclair et al. (1989), the demand for Zr is projected to increase by more than 0.2 million tons (mt) from 0.17 mt in 1983 to more than 0.4 mt for the year 2000. The utilization of eudialyte will depend largely on the demand for ZrO_2 as opposed to zircon and increasing demands for Y. Markets for ZrO_2 include abrasives, ceramics and various applications in the glass industry. The metal is used principally in nuclear power reactors and superalloys. The mining of eudialyte from peralkaline granites and nepheline syenite complexes has the potential of providing the world demands for Y and some REE that are less abundant or not found in the currently utilized REE sources.

CLASSIFICATION OF REE DEPOSITS IN CARBONATITES

Based on our current knowledge of existing REE ore deposits on a world level, carbonatites appear to have the best resources. From carbonatite environments REE can appear as primary, hydrothermal or weathering-type mineralization. Hydrothermal and weathering mineralization can form within the complex, or in cases where mobilization of REE occur, the mineralization can form outside of the complex.

Primary mineralization

Mt. Pass, California. The sill at Mt. Pass, California, is the only carbonatite occurrence where REE minerals are of primary igneous crystallization, as indicated by textural features (Fig. 5) that demonstrate that bastnaesite and parisite are cocrystallized predominantly with calcite, barite, and dolomite. Fine-grained interstitial minerals including quartz and talc are a late overprint of hydrothermal origin. The Mt. Pass complex is a composite of separate bodies within a small area that includes shonkinite, syenite, granite and fenites (Olson et al., 1954). Recent dating by DeWitt et al. (1987) gave an age of 1400 Ma for the alkalic complex. The REE minerals are confined to the carbonatite sill and dikes. The major ore mineral is bastnaesite. Current reported reserves are 31 mt of 8.86% REO with a 5% REO grade cutoff for mining.

Kangankunde Hill, Malawi. A dolomitic carbonatite rock unit at Kangankunde Hill contains an average of 7% monazite as green euhedral crystals. The texture of the rock including dolomite, monazite, strontianite and barite suggests an assemblage where monazite may be of primary crystallization from a carbonatite melt (Fig. 20). Florencite and bastnaesite also occur in other rocks at Kangankunde, but they are mostly fine-grained drusy mineralization associated with quartz and are of hydrothermal origin (Deans, 1966).

Hydrothermal mineralization

Wigu Hill, Tanzania. Wigu Hill consists of massively fenitized granitic gneisses (country rock) that have been invaded by a number of dike carbonatite intrusions that show extensive replacement by hydrothermal solutions rich in REE, Sr, F, Ba and silica (James, 1966). Large pseudomorphs of bastnaesite, quartz, fluorite, strontianite and dolomite after an earlier hexagonal mineral are conspicuous in outcrops that assay as high as 15 wt % REO. In addition to bastnaesite, other REE minerals include synchysite, parisite and monazite.

Itapirapua, São Paulo, Brazil. At Itapirapua, carbonatite consisting of both calcite and dolomite is flooded with late hydrated ferric iron oxides, silica and polycrystalline clusters of synchysite, parisite and bastnaesite (Loureiro and Tavares, 1983). The REE minerals are intimately crystallized in clusters with quartz, barite, fluorite and strontianite. A hydrothermal genesis is indicated by the concentration of the REE minerals in late veinlets and void fillings and by crystallization with quartz.

Karonge, Burundi. At Karonge, REE veins composed almost exclusively of massive bastnaesite and lesser amounts of monazite occur about 35 km southeast of Bujambura in four major zones coincident with Lake Tanganyika rift fracturing. The mineralization appears as a stockwork in the south and up to a meter thick by tens of meters long bastnaesite veins in the north. The veins are discordant to the country rock fabric that includes Ruzizian biotite and muscovite gneisses and schists and minor pegmatite bands. Deans (1966) and Wambecke (1977) first postulated an origin from hydrothermal alteration of fluids derived from a carbonatite. No relationships are found between the pegmatite and the REE mineralization. Associated minerals in the veins include quartz, goethite and barite. More recently, Fe^{3+}-activated orthoclase has been revealed by cathodoluminescence examination (Mariano, 1989) that is a good indication of conditions influenced by strong alkalinity (Mariano, 1988). Although no carbonatite is known to occur in the vicinity of the Karonge mineralization, a carbonatite-alkaline igneous complex does exist at Matongo-Bandaga, approximately 50 km to the north. Chondrite-normalized REE distributions for the Karonge bastnaesite show an LREE dominance, while the country rock and pegmatite bands show an HREE dominance typical of granitic pegmatites and sialic rocks of crustal origin. Lateritic weathering of the veins has produced some rhapdophane-(La) and cerianite (Wambeke, 1977).

The following sequence is postulated for the origin of the Karonge bastnaesite occurrence: (1) Large quantities of a buried carbonatite that is near the surface, has been subjected to karst weathering and as a result, REE are released and mobilized from the dissolved carbonates. They are transported in groundwater and fed into a hydrothermal system that is channeled along a series of fault-controlled fractures discordant to the country rock foliation. The REE are deposited as bastnaesite and silica within the fault fractures. (2) At a later period, an apatite residuum from the same original carbonatite source is also decomposed by weathering, and phosphate solutions follow the same channels that the REE assumed. When the phosphate solutions encounter the bastnaesite, they react with the REE to form monazite after bastnaesite. This is only partially developed. (3) The last stage is lateritic weathering of the bastnaesite veins that produces rhapdophanite-(La), cerianite, and other supergene minerals. The Karonge veins are mined on a small scale by hand labor. They yield small tonnage of high-grade bastnaesite ore for European markets. Irrespective of the origin of the mineralization, Karonge is an example of REE mobilization and redeposition to form small but rich ore veins.

Bayan Obo, China. The Fe-Nb-REE ore body of Inner Mongolia, China, is reported to be the largest known REE deposit in the world, with estimated reserves given as 37 mt of REO. The iron ore formations are described as being an east-west trending folded sequence of interbedded mica schists, quartzites, slates and dolomites occurring in a doubly plunging.synclinal structure. They belong to the middle Proterozoic Bayan Obo group of sedimentary rocks dated between 1350 and 1650 Ma. The Basement to this sequence is an amphibolitic gneiss dated at 2000 Ma. The REE minerals occur in lenticular bodies of magnetite-hematite ore and consist predominantly of granular bastnaesite and monazite intimately crystallized with dolomite, fluorite, quartz, calcite and apatite. Hydrothermal activity is generally accepted for the origin of the REE mineralization, but the source of the REE is most likely mantle derived. Zhou et al. (1980) proposed a carbonatite origin. Philpotts et al. (1989) suggested a mantle origin for the REE based on extreme LREE enrichment of the ore minerals and low $^{87}Sr/^{86}Sr$ ratios Other characteristics that suggest an origin from carbonatite include the presence of Nb mineralization with high Nb/Ta ratios and the widespread occurrence of alkali amphiboles and pyroxenes. Although the deposit has undergone several stages of metamorphism, tectonic disruptions and hydrothermal episodes, the geochemistry and mineralogy supports a carbonatite, mantle-derived origin for the REE and Nb.

Supergene REE mineralization in carbonatites

In most carbonatites, the major source of REE is in calcite, dolomite and apatite. during lateritic weathering, these minerals are the most susceptible to dissolution. The major cations Ca and Mg are totally removed from the system in the groundwater, while the less mobile elements including REE, Sr, Ba and Th are incorporated in supergene phosphates, carbonates and sulfates.

The best conditions for the development of large quantities of supergene REE mineralization in carbonatites exist in humid tropical climates with moderate to high rainfall conditions and in complexes where interior drainage and a basin-type topography allows the entrapment of newly-formed residual minerals. The extent of development of supergene mineralization can vary from thin surface crustifications in cool temperate climates to depths exceeding 300 m in tropical regions.

Common residual REE minerals that form and remain stable in conditions of lateritic weathering include monazite, rhapdophanite, florencite, gorceixite, goyazite, and cerianite. Bastnaesite-type minerals are also formed but are most susceptible to dissolution. If a karst system is developed, the liberated REE can be flushed out of the system along with the Ca and Mg. Depending on the degree of REE mobility, the newly formed supergene minerals can show an REE distribution similar to the primary minerals (e.g., Fig. 17), or there may be extensive separation of REE with the formation of xenotime and churchite with HREE enrichment.

The supergene minerals are typically fine grained, with individual crystals in sub-micron dimensions that occur in polycrystalline aggregates that range between 0.1-5 mm in diameter. Although they constitute large tonnage and show economically attractive grades in several deposits (Mariano, 1989), none are currently being mined for REE, although feasibility studies are being conducted for the beneficiation of REE from the Mt. Weld carbonatite of Western Australia where exploration has established resources of 15.4 mt of 11.2% REO + Y_2O_3, 7.43 mt of 15.7% REO + Y_2O_3 and 3.47 mt of 19.6% REO + Y_2O_3 (Carr Boyd Minerals Ltd., *Quarterly Report*, Oct.-Dec. 1988).

A few occurrences of supergene REE mineralization in carbonatites is briefly described below.

Araxá. This carbonatite in Minas Gerais, Brazil, is a circular structure of approximately 5 km in diameter. It consists almost totally of calcitic and dolomitic carbonatite with areas that grade into pyroxenite and glimmerite. The central core with a diameter of about 1 km contains the largest accumulation of Nb in the mineral bariopyrochlore. The depth of lateritic weathering averages 150 m, but along a major fault the weathering extends to 300 m. The laterite averages 7 wt % monazite and 5 wt % gorceixite;goyazite (Jaffe and Selchow, 1960). Below the zone of weathering, only REE minerals of hydrothermal origin are encountered in diminished quantities. The Area Zero gorceixite/goyazite body, mentioned earlier in the crandallite section of this chapter, is another supergene REE resource in the Araxá complex.

Evidence of some REE mobilization in the Araxá complex is indicated by the presence of polycrystalline aggregates of yellow-green monazite in the silicified peripheral country rock ridges of the Precambrian Araxá Group.

Catalão I. This circular carbonatite complex in Goiás, Brazil, has some similarity to Araxá. Both rim the periphery of the Paraná Basin and are Cretaceous in age (Ulbrich and Gomes, 1981). Unlike Araxá, Catalão I contains an intermediate annulus of pyroxenite with major amounts of perovskite and magnetite (Mariano, 1988), but the central core consists of carbonatite that has been subjected to deep lateritic weathering and is mined for bariopyrochlore. Lateritized carbonatite in Catalão I contains supergene monazite, rhapdophane, florencite and gorceixite/goyazite. Tonnage and grade estimations are not available.

Carbonatites of the Amazon. Several carbonatites are known in the Amazon of South America. In Brazil they include Morro do Seis Lagos in the Amazonas, and Maicuru and Maraconai in Pará (Ulbrich and Gomes, 1981). Cerro Impacto in Estado Bolivar, Venezuela, has been described by Aarden et al. (1973), and for Twareitau on the border of Guyana and Pará, Brazil, REE mineralization was described from a suspected carbonatite in the Muri Mountains (Mariano, 1981). In all of these occurrences, intense lateritic weathering has produced a deep regolith of insoluble iron and aluminum oxides accompanied by supergene REE minerals, mostly of the crandallite group and including cerianite, monazite, rhapdophane, and some bastnaesite-type minerals. Because of the remoteness of these occurrences, they have been explored only on a cursory level, and logistic constraints preclude their consideration as viable economic REE deposits.

Mt. Weld. Tonnage and grade estimations of REO reserves have already been given for the Mt. Weld carbonatite of Western Australia. The complex is a circular structure with a diameter of about 4 km. It has no topographic expression and was found as a result of an airborne magnetic survey. Subsequent drilling established a concealed carbonatite-derived laterite overlying fresh carbonatite. The complex has been dated at 2064 Ma (Goode, 1981). The area has been a stable carton well into the Precambrian, and based on current knowledge it has the oldest surviving laterite over any known carbonatite. The Deposit is extraordinary in the concentrated development of supergene REE minerals in laterite, consisting mostly of monazite, but REE-bearing crandallite-group minerals and cerianite are also abundant. Some of the crandallite minerals also have Nb impurities derived from the chemical breakdown of pyrochlore in the upper zones of the laterite blanket. They occur as sub-micron crystallites that form polycrystalline aggregates with spongy texture and as pseudomorphs after pyrochlore.

In some isolated areas, supergene calcite and dolomite occur in the laterite with sufficient amounts of LREE for activation by cathodoluminescence (Mariano, 1988). This is rarely encountered in carbonatite-derived laterites and may imply greater mobility of REE in the vadose zone of the Mt. Weld laterite. Supergene xenotime and churchite are also concentrated in some areas of the laterite. The presence of secondary Y and HREE mineralization in a carbonatite environment can be explained based on their higher migratory capacity relative to the LREE.

334

CONCLUSIONS

Although REE are found in many minerals in major quantities, historical production has been confined to beach sand monazite and xenotime and bastnaesite from carbonatites. In this chapter, the distribution of REE in these minerals and other potential mineral sources and geologic occurrences have been discussed, but they are by no means inclusive. At this time, new types of occurrences are being evaluated as economic sources, including peralkaline granite and quartz syenite environments such as Pajarito Mt., New Mexico, Strange Lake, Quebec-Labrador, and Thor Lake, Northwest Territories, Canada. In addition, the enormous agpaitic nepheline syenite complexes of Greenland are also being investigated as possible sources for REE, Y and other high technology metals, similar to these deposits from the Kola Peninsula, U.S.S.R.

The demand for REE and Y is projected to show a steady increase, but the relative demand for individual REE is dynamic and subject to change. The major sources for the REE from monazite and bastnaesite from beach sands and carbonatites are strongly LREE dominant, with La, Ce and Pr constituting between 75 and 88 wt % of the total REO. Currently, the REE in greatest demand include Nd, sm, Eu, Tb and Dy. It is conceivable that low-grade REE sources with a more favorable REE distribution may constitute the best future economic deposits.

ACKNOWLEDGMENTS

I wish to thank B.R. Lipin and P.H. Ribbe for their help in editing and patience with my efforts to complete this chapter.

REFERENCES

Aarden, H.M., Arozena, J.M., Moticska, P., Navarro, J., Pascauli, J. and Sifontes, R.S. (1973) El complejo geológico del area de El Impacto, Distrito Cedeño, Estado Bolivar, Venezuela. Informe interno. Direccion de Geologia, Ministerio de Energia y Minas, Caracas, 53 p.

Adrian, J., Winfield, G.M., Boshoff, O., Bristow, J.W. and Grutter, H.S. (1989) Geochemical and mineralogical features of a RE-enriched zone with the Goudini carbonatite complex, Transvaal, South Africa. Int'l Geochem. Explor. Symp. XIII, Brazilian Geochem. Congress (abstr.), 61-62.

Alschuler, Z.S., Berman, S. and Cuttitta, F. (1967) Rare Earths in phosphorite -- geochemistry and potential recovery. U.S. Geol. Surv. Prof. Paper 575-B, B1-B9.

von Backstrom, J.W. (1976) Thorium. In: C.B. Coetzee, ed., Mineral Resources of the Republic of South Africa. Dept. of Mines, Geol. Surv., Pretoria 5th ed., 209-212

Bailey, J.C., Bohsc, H. and Demina, A. (1981) Extension of Zr-REE-Nb resources at Kangerdluarssuk, Ilimaussaq intrusion. Rapp. Grønlands geol. Unders. 103, 63-67.

Balashov, Yu. A., Ronov, A.B., Migdisov, A.A. and Turanskaya, N.V. (1964) The effect of climate and facies environment on the fractionation of the rare earths during sedimentation. Geochem. Int'l 10, 951-969.

Borodin, L.S. (1960) Correlations among rare earth elements and some characteristics of their fractionation under endogenic conditions. Geochemistry (transl.) 604-616.

Brooks, C.K., Henderson, P. and Rønsbo, J.G. (1981). Rare-earth partitioning between allanite and glass in the obsidian of Sandy Braes, Northern Ireland. Mineral. Mag. 44, 157-160.

Cassedanne, J.P. and Cassadanne, O. (1973) Note sur l'anatase de Tapira (Minas Gerais, Brésil) Bull. Soc. franc. Mineral. Crist. 96, 316-318.

Chao, E.C.T., Minkin, J.A., Back, J.M., Okita, P.M., McKee, E.H., Tosdal, R.M., Tatsumoto, M., Junwen, W., Edwards, C.A., Yingzhen, R. and Weijun, S. (1989) The H8 dolomite host rock of the Bayan Obo iron-niobium-rare-earth-element ore deposit of Inner Mongolia, China-origin, episodic mineralization, and implications. U.S. Geol. Surv. Circular 1035, B-10.

Clark, A.M. (1984) Mineralogy of the rare earth elements. In: P. Henderson, ed., Rare Earth Element Geochemistry, Elsevier, New York, 33-61.

Currie, K.L. (1985) An unusual peralkaline granite near Lac Brisson, Quebec-Labrador. Geol. Surv. Canada, Paper 85-1A, 73-80.

D'Arco, P. and Piriou, B. (1989) Fluorescence spectra of Eu^{3+} in synthetic polycrystalline anorthite: Distribution of Eu^{3+} in the structure. Amer. Mineral., 74, 191-199.

Dal Negro, A., Rossi, G. and Tazzoli, V. (1975) The crystal structure of ancylite $(RE)_x(Ca,Sr)_{2-x}(CO_3)_2(OH)_x(2-x)H_2O$. Amer. Mineral., 60, 280-284.

Deans, T. (1966) Economic mineralogy of African carbonatites. In: O.F. Tuttle and J. Gittins, eds., Carbonatites. Wiley Interscience, New York.

Deans, T. (1978) Mineral production from carbonatite complexes: A world review. In: J.R. 'de Andrade Ramos, ed., Proceedings of the First Int'l Symposium on Carbonatites, p. 123-133.

DeWitt, E., Kwak, L.M. and Zartman, R.E. (1987) U-Th-Pb and $^{40}Ar/^{39}Ar$ dating of the Mountain Pass carbonatite and alkalic igneous rocks, S.E. California. Geol. Soc. Am., Abstracts with Program, 19, 642.

Dollase, W.A. (1971) Refinement of the crystal structures of epidote, allanite and hancockite. Amer. Mineral., 56, 447-464.

Donnay, G. and Donnay, J.D.H. (1953) The crystallography of bastnaesite, parisite, roentgenite and synchisite. Amer. Mineral., 38, 932-963.

Eby, N.G. (1971) Rare-earth, yttrium and scandium geochemistry of the Oka carbonatite complex, Oka, Quebec. Unpublished Ph.D. thesis, Boston University.

Exley, R.A. (1980) Microprobe studies of REE-rich accessory minerals: Implications for Skye Granite petrogenesis and REE mobility in hydrothermal systems. Earth Planet. Sci. Lett., 48, 97-110.

Fleischer, M. (1965) Some aspects of the geochemistry of yttrium and the lanthanides. Geochim. Cosmochim. Acta, 29, 755-772.

Fleischer, M. (1978) Relative proportions of the lanthanides in minerals of the bastnaesite group. Can. Mineral., 16, 361-363.

Fleischer, M. (1987) Glossary of Mineral Species, 5th edition. Mineralogical Record Inc., Tucson.

Fryer, B.J. and Edgar, A.D. (1977) Significance of rare earth distributions in coexisting minerals of peralkaline undersaturated rocks. Contrib. Mineral. Petrol., 61, 35-48.

Girault, J. (1966) Genese et geochimie de l'apatite et de la calcite dans les roches liees au complexe carbonatitique et hyperalcalin d'Oka Canada. Bull. Soc. franc. Mineral. Cristallogr., LXXXIX, 496-513.

Glass, J.J. and Smalley, R.G. (1945) Bastnaesite. Amer. Mineral., 30, 601-615.

Goldschmidt, V.M., Hauptmann, H. and Peters, C. (1933) Ueber die bereucksichtigung seltener elemente bei gesteins-analysen. Naturwissenschaften, 21, 362-365.

Goode, A.D.T. (1981) Proterozoic geology of Western Australia. In: D.R. Hunter, ed., Precambrian of the Southern Hemisphere. Elsevier, Amsterdam, p. 105-203.

Graeser, S. and Schwander, H. (1987) Gasparite-(Ce) and monazite-(Nd): Two new minerals to the monazite group for the Alps. Schweiz. Mineral. Petrogr. Mitt., 67, 103-113.

Gramaccioli, C.M. (1977) Rare earth minerals in the alpine and subalpine region. Mineral. Record, 8, 287-293.

Griffin, W.L., Nilssen, B. and Jensen, B.B. (1979) Britholite(-Y) and its alteration: Reiarsdal Vest-Agder, south Norway. Contrib. Mineral. Norway No. 64, Norsk Geol. Tidsskrift, 3, 265-271.

Haberlandt, H. (1947) Die bedeutung der spurenelements in der geochemischen forschung. Akad. Wiss. Wien, Math. naturwiss. K. Sitzungsber. Abt. IIb, Bd. 156, 293-323.

Harris, C., Cressey, G., Bell, J.D., Atkins, F.B. and Beswetherick, S. (1982) An occurrence of rare-earth-rich eudialyte from Ascension Island, South Atlantic. Mineral. Mag., 46, 421-425.

Hawkins, B.W. (1975) Mary Kathleen uranium deposit. In: C.L. Knight, ed., Economic Geology of Australia and Papua, New Guinea. 1. Metals. Australasian Inst. Mining and Metal., Parkville, Vict., p. 398-402.

Henderson, P., editor (1984) Rare Earth Element Geochemistry. Elsevier Science Publications, New York.

Hill, W.B. (1976) Rare earths Western Australia. In: C.L. Knight, ed., Economic Geology of Australia and Papua, New Guinea. 4. Industrial Minerals and Rocks. Australasian Inst. Mining and Metal., Parkville, Vict., p. 331.

Hughson, M.R. and Sen Gupta, J.G. (1964) A thorian intermediate member of the britholite-apatite series. Amer. Mineral., 49, 937-951.

Jaffe, H.W. and Selchow, D.H. (1960) Mineralogy of the Araxá columbium deposit. Union Carbide Ore Co. Research Report 4, Tuxedo, NY.

James, T.C. (1966) The carbonatites of Tanganika: A phase of continental-type volcanism. Unpublished Ph.D. thesis, Imperial College, London.

Jefford, G. (1962) Xenotime from Rayfield, northern Nigeria. Amer. Mineral., 47, 1467-1473.

Jones, A.P. and Larsen, L.M. (1985) Geochemistry and REE minerals of nepheline syenites from the Motzfeldt Centre, South Greenland. Amer. Mineral., 70, 1087-1100.

Jones, A.P. and Wyllie, P.J. (1986) Solubility of rare earth elements in carbonatite magmas, indicated by the liquidus surface in $CaCO_3-Ca(OH)_2-La(OH)_3$ at 1 kbar pressure. Applied Geochem., 1, 95-102.

336

Kapustin, Yu.L. (1980) Mineralogy of Carbonatites translated from Russian. Smithsonian Institution and National Science Foundation, Washington, DC, by Amerind Publishing Co. Pvt. Ltd., New Delhi, India.

Keidel, F.A., Montgomery, A., Wolfe, C.W. and Christian, R.P. (1971) Calcian ancylite from Pennsylvania: New data. Mineral. Record, 2, 18-25.

Kudrina, M.A., Kudrin, V.S. and Sidorenko, G.A. (1961) Britholite and alumobritholite from alkalic pegmatites of Siberia (in Russian). Geol. Mestorozhdenii Redkikh Elementov, 9, 108-120.

Landuyt, J. Van and Amelinckx, S. (1975) Multiple beam direct lattice imaging of new mixed-layer compounds of the bastnaesite-synchisite series. Amer. Mineral., 60, 351-358.

LeFabvre, J.J. and Gasparrini, C. (1980) Florencite, an occurrence in the Zairian copperbelt. Can. Mineral., 18, 301-311.

Levinson, A.A. (1966) A system of nomenclature for rare-earth minerals. Amer. Mineral., 51, 152-158.

Loureiro, F.L. and Tavares, J.R. (1983) Duas novas ocorrências de carbonatitos: Mato Preto e Barra do Itapirapuã. Revista Brasileira de Geosciencias, 13, 7-11.

Lurie, J. (1985) Mineralization of the Pilanesberg alkaline complex. In: C.P. Anhaeusser, C.P. and S. Maske, eds., Mineral Deposits of Southern Africa. Geol. Soc. S. Africa Johannesburg.

Mariano, A.N. and Ring, P.J. (1975) Europium-activated cathodoluminescence in minerals. Geochim. Cosmochim. Acta, 39, 649-660.

Mariano, A.N. (1981) Carbonatite exploration report on the Muri Mountain Region, Guyana. Report to the United Nations Revolving Fund for Natural Resources Exploration.

Mariano, A.N. (1988) Some further geologic applications of cathodoluminescence. In: D.J. Marshall, ed., An Introduction to Geological Applications of Cathodoluminescence, Chapter 8. Unwin Hyman, London.

Mariano, A.N. (1989) Nature of economic mineralization in carbonatites and related rocks. In: K. Bell, ed., Carbonatites—Genesis and Evolution, Unwin Hyman, London.

Mariano, A.N. and Birkett, T.C. (1988) Cathodoluminescence in minerals of the Strange Lake alkalic complex. Geol. Assoc. Canada-Mineral. Assoc. Canada Ann. Mtg., Program with Abstracts, 13, A79.

Mariano, C. (1985) The paragenesis of the Bond Zone of the Oka carbonatite complex, Oka, Quebec. Unpublished B.Sc. thesis, Dept. of Geology, Bucknell University, Pennsylvania.

McLaughlin, R.M. and Mitchell, R.H. (1989) Rare metal mineralization in the Coldwell Alkaline complex, northwestern Ontario. Geol. Assoc. Canada-Minral. Assoc. Canada Ann. Mtg., Program with Abstracts, A1.

McLennan, S.M. and Taylor, S.R. (1979) Rare earth element mobility associated with uranium mineralization. Nature, 282, 247-250.

Mendes, M.J.C., Murta, C.C. and Castro, L.O. (1968) Mineralogic do Depósito de terras raras, nióbio e urânio da Area Zero, Araxá, MG. Anais do XXII Contr. Brasiliero de Geologia, Belo Horizonite.

Metz, M.C., Brookins, D.G., Rosenberg, P.E. and Zartman, R.E. (1985) Geology and geochemistry of the Snowbird Deposit, Mineral County, Montana. Econ. Geol., 80, 394-409.

Mineev, D.A., Larrishscheva, T.L. and Bykova, A.V. (1979) Yttrium bastnaesite—a product of gagarinite alteration. Zap. Vses. Mineral. O-va, 99, 328-332.

Mineyev, D.A. (1963) Geochemical differentiation of rare-earth elements. Geokhimiya, 12, 1082-1100.

Mining Annual Review (1986) Mining Journal of London.

Mining Engineering (1989) Molycorp, Apache Indians to develop yttrium/zirconium deposit. July, p. 515.

Moore, S.L., Foord, E.E. and Meyer, G.A. (1988) Geologic and aeromagnetic map of a part of the Mescalero Apache Indian Reservation, Otero County, New Mexico. U.S. Geol. Surv. Misc. Invest. Series Map 1-1775.

Murata, K.J., Rose, H.J., Jr., Carron, M.K. and Glass, J.J. (1957) Systematic variation of rare earth elements in cerium-earth minerals. Geochim. Cosmochim. Acta, 11, 141-161.

Nash, W.P. (1972) Apatite chemistry and phosphorus fugacity in a differentiated igneous intrusion. Amer. Mineral., 57, 877-886.

Neary, C.R. and Highley, D.E. (1984) The economic importance of the rare earth elements. In: P. Henderson, ed., Rare Earth Element Geochemistry, Elsevier, New York, p. 423-466.

O'Driscoll, M. (1988) Rare earths. Enter the dragon. Indus. Mineral., 254, 21-55.

Olson, J.C., Shawe, D.R., Pray, L.C. and Sharp, W.N. (1954) Rare-earth mineral deposits of the Mountain Pass District, San Bernardino County, California. U.S. Geol. Surv. Prof. Paper 261.

Overstreet, W.C. (1967) The geologic occurrence of monazite. U.S. Geol. Surv. Prof. Paper 530, 327.

Palache, C., Berman, H. and Frondel, C. (1951) The System of Mineralogy, Vol. II, 7th edition. John Wiley & Sons, New York.

Parker, R.L. and Sharp, W.N. (1970) Mafic-ultramafic igneous rocks and associated carbonatites of the Gem Park complex, Custer and Fremont Counties, Colorado. U.S. Geol. Surv. Prof. Paper 649.

Philpotts, J.A., Tatsumoto, M., Wang, K. and Fan, P.F. (1989) Petrography, chemistry, age and origin of the rare-earth iron deposit at Bayan Obo, China, and implications of Proterozoic iron ores in earth evolution. U.S. Geol. Surv. Circ. 1035, 53-55.

337

Poirier, P. (1980) New developments in rare earth markets. In: Proc. 4th Industrial Minerals Int'l Congr., Atlanta, Georgia, Metal. Bull. PLC (London), p. 205-209.

Puchelt, H. and Emmermann, R. (1976) Bearing of rare earth patterns of apatites from igneous and metamorphic rocks. Earth Planet. Sci. Lett., 31, 279-286.

Rhône-Poulenc (1987) Rare earths reminder. Rhône-Poulenc Chimie, Div. Minerale Fine, Dept. Terres Rares, Les Miroirs-La Défenses 92400 Courbevoie, France.

Roeder, P.K., MacArthur, D., Ma, X.P., Palmer, G.L. and Mariano, A.N. (1987) Cathodoluminescence and microprobe study of rare earth elements in apatite. Amer. Mineral., 72, 801-811.

Rønsbo, J.G. (1989) Coupled substitutions involving REEs and Na and Si in apatites in alkaline rocks from the Ilimaussaq intrusion, South Greenland, and the petrological implications. Amer. Mineral., 74, 896-901.

Rose, E.R. (1979) Rare-earth prospects in Canada. Can. Inst. Min. Metall. Bull., 72, 110-116.

Rose, H.J., Jr., Blade, L.V. and Ross, M. (1958) Earthy monazite at Magnet Cove, Arkansas. Amer. Mineral., 43, 995-997.

Rosenblum, S. and Mosier, E.L. (1983) Mineralogy and occurrence of europium-rich dark monazite. U.S. Geol. Surv. Prof. Paper 1181.

Rozanov, K.I., Flerova, L.B., Khomich, P.Z. and Zingerman, A.Y. (1983) Fractionation and concentration of lanthanides and yttrium in Precambrian complexes of the western part of the Russian Platform. Doklady, Acad. Sci., USSR Earth Sci. Sec., 258, 177-181.

Russell, B.G. (1977) The possible recovery, during the manufacture of phosphoric acid, of rare earths from Foskor concentrate. Bull. Mineral. Bur., Dept. Mines S. Africa, 1, 8.

Sarp, H. and Bertrand, J. (1985) Gysinite, $Pb(Nd,La)(CO_3)_2(OH)\cdot H_2O$, a new lead, rare-earth carbonate from Shinkolobwe, Shaba, Zaire and its relationship to ancylite. Amer. Mineral., 70, 1314-1317.

Semenov, E.I. (1958) Relationship between composition of rare earths and structure of minerals. Geokhimiya, 5, 574-586.

Semenov, E.I. and Barinskii, R.L. (1958) Characteristics of the composition of rare earths in minerals. Geochim., 4, 398-419.

Sinclair, W.D., Richardson, Z.D.G., Birkett, T.C. and Truman, D.L. (1989) High-technology metals from a Canadian perspective: definition, uses and aspects of future demand, supply and marketing. Geol. Assoc. Canada-Minral. Assoc. Canada Ann. Mtg., Program with Abstracts, A1.

DeSouza, J.M. de and Castro, L.O. (1968) Geologia do deposito de terras raras, niobio e uranio da Area Zero, Araxá (MG). Anais do XXII Congr. Brasiliero de Geologia, Belo Horizonte.

Ueda, T. (1967) Reexamination of the crystal structure of a monazite. J. Japan Assoc. Mineral. Petrol. Econ. Geol., 58, 170-179.

Ulbrich, H.H.G.J. and Gomes, C.B. (1981) Alkaline rocks from continental Brasil: a review. Earth Sci. Rev., 17, 135-154.

Ungemach, H. (1935) Sur certains mineraux sulfate's du Chili. Bull. Soc. fr. Min., 58, 97-221.

Vainshtein, E.E., Tugarinov, A.I. and Turanskaya, N.V. (1955) The distribution of the rare earths in monazite. Doklady Akad. Nauk. SSSR, 104, 268-271.

Vlasov, K.A., editor (1966) Geochemistry and Mineralogy of Rare Elements and Genetic Types of Their Deposits. Vol. II, Mineralogy of Rare Elements. Israel program for Scientific Translations, Jerusalem.

Vlasov, K.A. (1968) Geochemistry and Mineralogy of Rare Elements and Genetic Types of Their Deposits. Vol. III, Genetic Types of Rare-element Deposits. Israel Program for Scientific Translations, Jerusalem, 916 p.

Van Wambeke, L. (1977) The Karonge rare earth deposits, Republic of Burundi; new mineralogical-geochemical data and origin of the mineralization. Mineral. Deposit., 12, 373-380.

Wang, K. (1981) Distribution characteristics of the rare earth elements in Bayan Obo iron ore deposit (Chinese with English abstract). Scientia Geologica Sinica, p. 360-367.

Williams, F.A., Meehan, J.A., Paulo, K.L., John, T.U. and Rushton, H.G. (1956) Economic geology of the decomposed columbite-bearing granites, Jos Plateau, Nigeria. Econ. Geol., 51, 303-332.

Zhou, Z., Gongyuan, L., Tongyun, S. and Yuguan, L. (1980) On the geological characteristics and the genesis of the dolomitic carbonatites at Bayan Obo, Inner Mongolia. Geol. Rev., 26, 35-42.

Appendix

CATHODOLUMINESCENCE EMISSION SPECTRA OF
RARE EARTH ELEMENT ACTIVATORS IN MINERALS

A. N. Mariano

INTRODUCTION

Cathodoluminescence (CL) is becoming a more widespread tool in the study of minerals and is beginning to have a larger application in various fields of geologic research, as evidenced by increasing publications on the subject, including the book *Cathodoluminescence of Geological Materials* by D.J. Marshall (1988).

A major cause of visible CL in minerals is from REE activators, including Sm^{3+}, Eu^{2+}, Eu^{3+}, Th^{3+}, and Dy^{3+}. Because the valence electrons of the trivalent REE are inner orbital electrons, their wavelength of emission is only slightly affected by the crystal field of the host, and therefore they always show multiple line emissions that allow them to be identified with little ambiguity. Divalent europium is an exception, although in most cases it gives a broad band of emission in the blue part of the visible spectrum. In the following pages, CL emission spectra in the visible spectrum are shown for a number of minerals that exhibit REE activation. Verification for identifying some of the REE activators was made by correlation with synthetic phases that contain known impurities.

For basic information on CL physics and a comprehensive reference list, see Marshall (1988).

The instrumentation used to generate the spectra includes an MAAS, Inc. (Nuclide) Luminoscope with a cold cathode electron gun, and all of the spectra were obtained using a Spex Industries, Inc., 1681B Spectrophotometer.

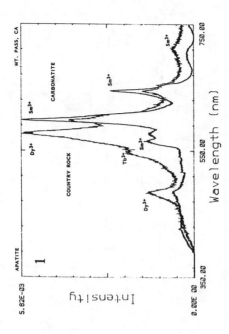

Figure 1. Two superimposed spectra including country rock apatite, $Ca_5(PO_4)_3F$, adjacent to the Mt. Pass carbonatite body and apatite from the carbonatite are presented. The dominant activator in the country rock apatite is Dy^{3+}, whereas in the carbonatite apatite it is Sm^{3+}. A comparison of the major intensity emissions of Dy^{3+} and Sm^{3+} can be used as a guide for differentiating between LREE and HREE dominance in apatite (Roeder et al., 1987). CL conditions: country rock, 4.5 kV, 0.60 mA; carbonatite, 10 kV, 0.50 mA.

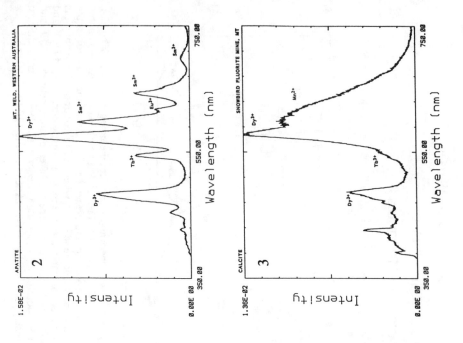

Figure 2. Supergene apatite, $Ca_5(PO_4,CO_3)_3F$, occurs as crusts on botryoidal laterite material derived from carbonatite at Mt. Weld, Western Australia. Under strong oxidizing conditions, Mn occurs in a higher valence state that Mn^{2+} and therefore does not occur as an activator in apatite. This accounts for the excellent resolution of the REE activators line emissions. Also, Eu occurs as Eu^{3+}, and Eu^{2+} is absent. The dominance of HREE is indicated by the relative intensities of the strongest lines of Sm^{3+} and Dy^{3+}. In this case, $Dy^{3+} > Sm^{3+}$. Notice also the well-resolved strongest line emission for Tb^{3+} at 544 nm. This apatite occurs in an area of the laterite that is enriched in Y and HREE. CL conditions: 5 kV, 0.40 mA.

Figure 3. The Snowbird Fluorite mine in Mineral County, Montana, is a hydrothermal vein deposit with abundant gangue calcite (Metz et al., 1985). The CL emission spectrum reveals an HREE enrichment with Dy^{3+} as the major REE activator superimposed on the Mn^{2+} broad band in calcite. Activation from Sm^{3+} is virtually absent (Mariano, 1988). CL conditions: 6.5 kV, 0.60 mA.

Figure 4. Some of the optical quality calcite from Chihuahua, Mexico, shows well-resolved REE activators and an absence of the typical Mn^{2+} activation. Chemical analyses show Dy and Sm at slightly less than 1 ppm, while Tb gave 0.1 ppm (R.A. Mason, pers. comm.). CL conditions: 8.5 kV, 0.30 mA.

Figure 5. Supergene dolomite, $CaMg(CO_3)_2$, from laterite overlying carbonatite at Mt. Weld, Australia, shows only the line emissions of Sm^{3+} activation. This area of the laterite is enriched in LREE. CL conditions: 7.5 kV, 0.45 mA.

Figure 6. Strontianite, $SrCO_3$, from the Bear Lodge carbonatite in southeastern Wyoming is intergrown with ancylite. It shows only Sm^{3+} activation. This spectrum is identical to strontianite spectra from the Jacupiranga carbonatite of São Paulo State, Brazil. CL conditions: 15 kV, 0.20 mA.

341

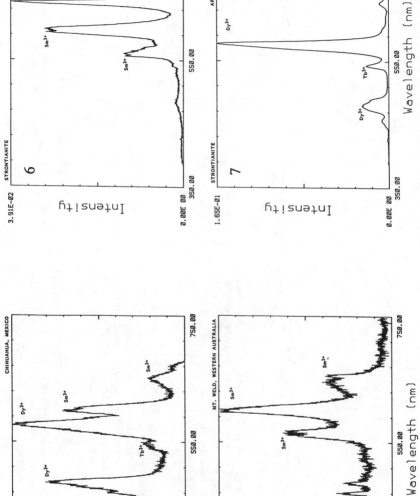

Figure 7. Strontianite from the classic locality in Argyllshire, Scotland (U.S.N.M. #48,904) shows only Dy3+ activation and the strongest emission of Tb3+. CL conditions: 6 kV, 0.80 mA.

342

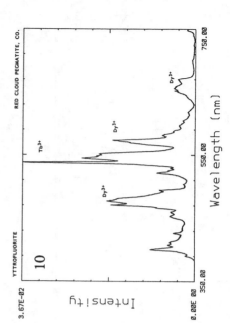

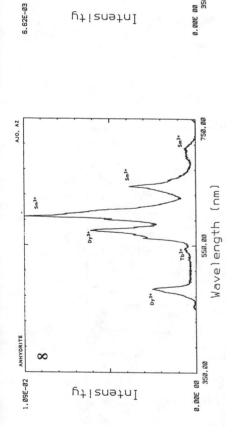

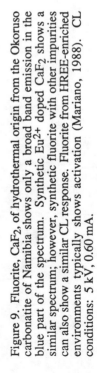

Figure 8. Anhydrite, CaSO₄, of hydrothermal origin associated with porphyry copper deposits in various parts of the world typically shows REE activation with variations in REE distribution. In some cases, sharp CL zonations are observed. Activator concentrations are only a few ppm (Mariano, 1988). CL conditions: 12 kV, 0.90 mA.

Figure 9. Fluorite, CaF₂, of hydrothermal origin from the Okoruso carbonatite of Namibia shows only a broad band emission in the blue part of the spectrum. Synthetic Eu²⁺ doped CaF₂ shows a similar spectrum; however, synthetic fluorite with other impurities can also show a similar CL response. Fluorite from HREE-enriched environments typically shows activation (Mariano, 1988). CL conditions: 5 kV, 0.60 mA.

Figure 10. Yttrofluorite typically shows HREE activation with either Tb³⁺ or Dy³⁺ as the dominant activator. In this spectrum, notice the resolution and fine structure of the line emissions (sample from D.M. Wayne). CL conditions: 5.5 kV, 0.80 mA.

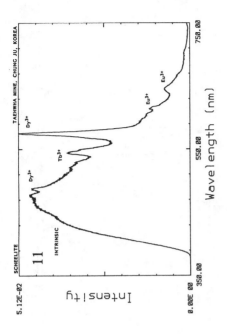

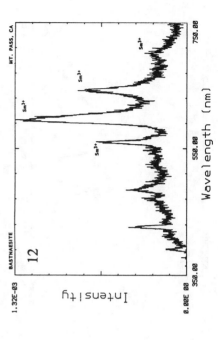

Figure 11. The dominant cause of luminescence in scheelite, $CaWO_4$, is intrinsic due to deviations from stoichiometry or lattice defects from nonactivator impurities such as Mo. This accounts for the broad band observed in CL and UV emission spectra in scheelite. However, the spectra commonly also show superimposed line emissions from the trivalent REE activators. Their wavelengths of emission and relative intensities correlate well with synthetic $CaWO_4$ with controlled impurities. In this spectrum, the major REE activators are Dy^{3+} and Tb^{3+}, with only weak emissions from Eu^{3+}. Scheelite typically shows strong Eu^{3+} emissions (Mariano and Ring, 1975). CL conditions: 5 kV, 0.25 mA.

Figure 12. Minerals that contain major amounts of REE rarely show noticeable CL as a result of *concentration quenching* (see Marshall, 1988). However, under strong excitation conditions and maximum focusing of the electron beam, a dull or weak luminescence is discernable. This spectrum for bastnaesite, $REECO_3F$, from the Birthday Claim dike of the Mt. Pass carbonatite shows only Sm^{3+} activation. This is generally representative for bastnaesite from any locality, reflecting the strong crystal chemical control on LREE preference. CL conditions: 13.5 kV, 0.45 mA.

Figure 13. Similar to bastnaesite, CL spectra in parisite, $Ca(REE)_2(CO_3)_3F_2$, is very weak and only observed under strong excitation conditions. The spectra are always LREE dominant even in environments that show a bulk HREE dominance such as the Snowbird Fluorite mine where gangue calcite and fluorite are HREE enriched (Mariano, 1988). CL conditions: 11 kV, 0.60 mA.

Figure 14. Mosandrite, $(Na,Ca,REE)_3Ti(SiO_4)_2F$, from the Kipawa agpaitic syenite complex shows only dull CL under conditions of strong excitation. The REE distribution is variable but most often shows a mid-atomic number enrichment. The major activators are Dy^{3+} and Sm^{3+}, but the well-resolved emission of Tb^{3+} in the CL spectrum reflects the mid-atomic number enrichment. CL conditions: 7.5 kV, 0.55 mA.

344

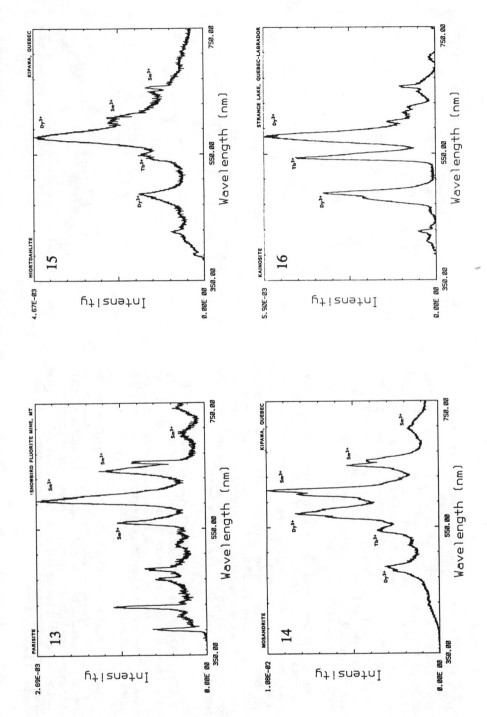

345

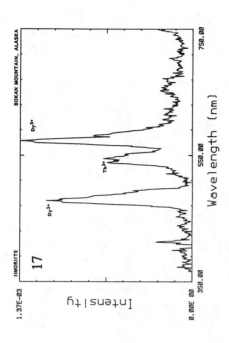

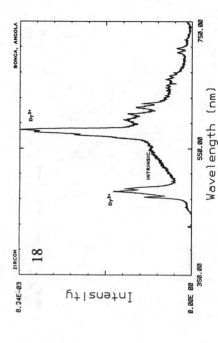

Figure 15. Hiortdahlite, $(Ca,Na)_3(Zr,Ti)Si_2O_7(O,OH,F)_2$, from the Kipawa complex and other agpaitic nepheline syenite complexes has a diagnostic CL spectrum that serves to differentiate it from other REE-bearing minerals. The CL spectrum consistently shows Dy^{3+} as the major emission. CL conditions: 8.5 kV, 0.90 mA.

Figure 16. Kainosite, $Ca_2(Y,REE)_2Si_4O_{12}(CO_3)\cdot H_2O$, is a secondary mineral of hydrothermal origin in HREE-enriched environments. Its identification at Strange Lake, Quebec-Laborador border, Pajarito Mt., New Mexico, and Bokan Mt., Alaska, was made by CL examination (Mariano and Birkett, 1988). The CL is a green color because of the strong Tb^{3+} line emissions that sometimes are greater than the Dy^{3+} emissions. CL conditions: 4.5 kV, 0.95 mA.

Figure 17. This spectrum was generated from mottled green and dull brown CL on iimoriite, $Y_2(SiO_4)(CO_3)$, grains from the Bokan Mt. peralkaline granite complex. Sample from E.E. Foord. CL conditions: 11.5 kV, 0.05 mA.

Figure 18. The cause of both UV and CL excited luminescence in zircon, $ZrSiO_4$, is intrinsic, but the line emissions from Dy^{3+} commonly superimpose the intrinsic broad band emission. The crystal chemical preference of zircon for HREE restricts the presence of appreciable Sm^{3+} activation irrespective of the environment of origin. Therefore, the CL spectra of zircon cannot be used as a guide for REE distribution. This zircon is from a calcitic of Bonga Mt. in Angola whose bulk chemistry is LREE enriched. CL conditions: 6.5 kV, 0.60 mA.

Figure 19. At the Thor Lake peralkaline granite-syenite complex, both zircon and xenotime, YPO_4, are accessories. Their morphology and optical properties are often the same, but their CL spectra are radically different. Xenotime shows only Dy^{3+}, Tb^{3+} and Eu^{3+} emissions, but the intrinsic band that is always present in zircon is absent. Trivalent Eu emissions in xenotime are typically observed in xenotime from other localities. CL conditions: 11 kV, 0.80 mA.

346

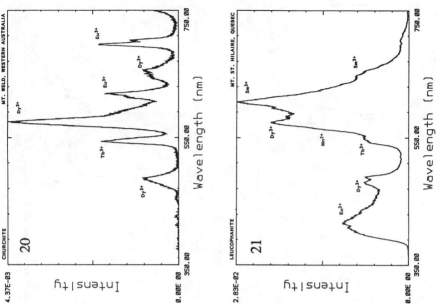

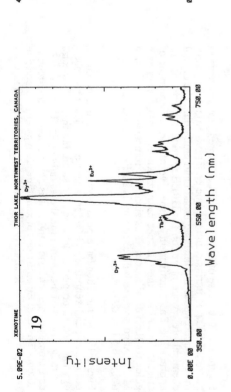

Figure 20. The CL emission spectra of churchite, $YPO_4 \cdot 2H_2O_2$, is similar to xenotime. Although their crystal morphology is distinctly different, churchite is typically replaced pseudomorphically by xenotime. Identification requires x-ray diffraction. The Mt. Weld churchite comes from the laterite that is HREE enriched. CL conditions: 5 kV, 0.35 mA.

Figure 21. Leucophanite, $(Na,Ca)_2BeSi_2(O,OH,F)_7$, from the Mount Saint Hilaire alkaline gabbro-peralkaline syenite complex occurs as euhedral crystals of late hydrothermal origin. The bulk REE chemistry shows mostly LREE enrichment. This spectrum is LREE dominant, but Dy^{3+} and Tb^{3+} are also present. The broad band at approximately 600 nm is assumed to be from Mn^{2+} activation. In some crystals, Eu^{2+} is the major activator. Sample from A.T. Grant. CL conditions: 10 kV, 0.15 mA.

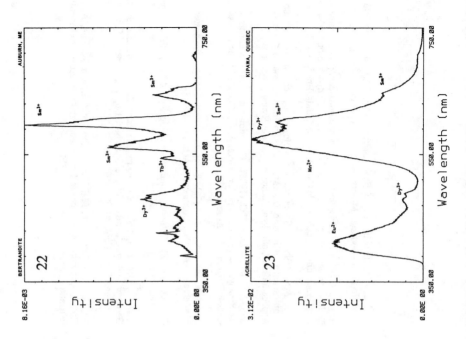

Figure 22. Bertrandite, $Be_4Si_2O_7(OH)_2$, is difficult to identify by petrographic methods, but it commonly exhibits CL that aids in its recognition. From the pegmatite in Auburn, Maine, euhedral bertrandite gives well-resolved REE emissions with LREE dominance. Sample from R.M. Beger. CL conditions: 7 kV, 0.70 mA.

Figure 23. Agrellite, $NaCa_2Si_4O_{10}F$, from the Kipawa agpaitic alkaline complex has been known to show vivid UV luminescence. In petrographic examination and using CL, it can be found in diverse areas of the complex. The CL spectrum shows REE activation superimposing a strong Mn^{2+} broad band. Sample from J. Gittins. CL conditions: 5 kV, 0.60 mA.

Figure 24. Isokite, $CaMg(PO_4)F$, is a rare mineral that was first described as hydrothermal crystallization from the Nkombwa carbonatite of Zambia (Deans and McConnell, 1955). It has a diagnostic CL spectrum from both Eu^{2+} and Eu^{3+} that allowed it to be recognized in drill core from the carbonatite of Araxá, Brazil (Mariano, 1978). Nkombwa and Araxá isokite give the same CL spectra. In contrast, isokite from a pegmatite in Portugal gave activation only from the Mn^{2+} band (see Fig. 25). CL conditions: 10 kV, 0.70 mA.

Figure 25. Isokite from the Mangualde Pegmatite of Portugal gives only Mn^{2+} activation, in sharp contrast to the Eu-activated isokite from carbonatite environments. Sample from T. Deans. CL conditions: 5 kV, 0.70 mA.

348

REFERENCES

Deans, T. and McConnell, J.D.C. (1955). Isokite, CaMg(PO$_4$)F, a new mineral from Northern Rhodesia. Mineral. Mag. 30, 681-690.

Mariano, A.N. (1978). The application of cathodoluminescence for carbonatite exploration and characterization. In: International Symposium on Carbonatites, Proceedings 1st, Pocos de Caldas, Minas Gerais, Brasil, 39-57.

Mariano, A.N. (1988). Some further geologic applications of cathodoluminescence. Ch. 8 in D.J. Marshall, An Introduction to Geological Applications of Cathodoluminescence. Unwin Hyman, London.

Mariano, A.N. and Birkett, T.C. (1988). Cathodoluminescence in minerals of the Strange Lake alkalic complex. Geol. Assoc. Canada / Mineral. Assoc. Canada Program with Abstracts 13, A79.

Mariano, A.N. and Ring, P.J. (1975). Europium-activated cathodoluminescence in minerals. Geochim. Cosmochim. Acta 39, 649-660.

Marshall, D.J. (1988). Cathodoluminescence of Geological Materials. Unwin Hyman, London

Metz, M.C., Brookins, D.G., Rosenberg, P.E. and Zartman, R.E. (1985). Geology and geochemistry of the Snowbird Deposit, Mineral County, Montana. Econ. Geol. 80, 394-409.

Roeder, P.L., MacArthur, D., Ma, Xin-Pei, Palmer, G.L. and Mariano, A.N. (1987). Cathodoluminescence and microprobe study of rare earth elements in apatite. Amer. Mineral. 72, 801-811.

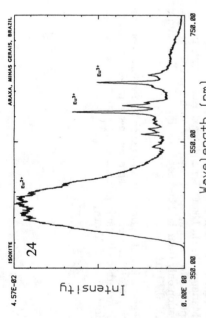

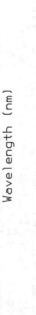

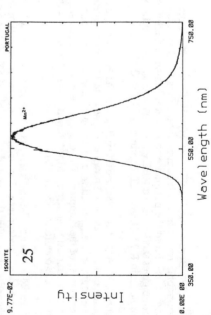